Springer Series in Computational Neuroscience

Volume 11

For further volumes:
http://www.springer.com/series/8164

Hermann Cuntz • Michiel W.H. Remme
Benjamin Torben-Nielsen
Editors

The Computing Dendrite

From Structure to Function

Springer

Editors
Hermann Cuntz
Ernst Strüngmann Institute (ESI)
for Neuroscience in Cooperation
with Max Planck Society
Frankfurt/Main, Germany

Institute of Clinical Neuroanatomy
Goethe University
Frankfurt/Main, Germany

Michiel W.H. Remme
Institute for Theoretical Biology
Humboldt-Universität zu Berlin
Berlin, Germany

Benjamin Torben-Nielsen
Department of Neurobiology
Hebrew University of Jerusalem
Jerusalem, Israel

Blue Brain Project
Swiss Federal Institute of Technology
in Lausanne
Lausanne, Switzerland

ISSN 2197-1900 ISSN 2197-1919 (electronic)
ISBN 978-1-4939-5353-0 ISBN 978-1-4614-8094-5 (eBook)
DOI 10.1007/978-1-4614-8094-5
Springer New York Heidelberg Dordrecht London

Printed on acid-free paper

Springer is part of Springer Science+Business Media (www.springer.com)

Foreword

Dendrites have always been there—for hundreds of millions of years—but we became closely familiar with their anatomical esthetics only at the end of the nineteenth century, following the intellectual fascination and artistic capabilities of several great anatomists, in particular Santiago Ramón y Cajal and Camillo Golgi. Like the naming of the trees in the forest based on the shape of their crown, many of the names given to neurons rely on the particular shape of the dendritic tree—retinal amacrine cells, cortical spiny stellate cells, and cerebellar Purkinje cells. Naming the trees makes the forest more familiar to us; classifying neurons into anatomical subclasses is a natural path towards "knowing your brain".

Next came the understanding that dendrites serve as the major region of neurons to "accept" their (pre) synaptic partners and that neurons are polygamist, each interacting with thousands of other neurons—forming large interactive networks. In 1959, Wilfrid Rall realized that dendrites are electrically distributed devices (rather than an isopotential "point"); this ignited the need to understand how the synaptic current spreads from the synaptic input site to other dendritic regions, in particular to the soma-axon where the output is generated. Rall's passive cable theory for dendrites has provided the theoretical foundation for this biophysical understanding. The experimental finding that the dendritic membrane is endowed with a rich repertoire of nonlinear voltage- and ligand-gated ion channels, and that synaptic inputs (inhibitory and excitatory) from different input sources target specific dendritic subdomains, suggested that dendrites (and their synapses) may empower neurons with enhanced computational capabilities.

With this experimental and theoretical foundation, a new perspective has emerged regarding the possibility that dendrites might implement computational functions. McCulloch and Pitts (1943) inspired this conceptual jump, demonstrating that logical operation could be performed by neurons (with their "all or none" "0" and "1" spikes, and their excitatory and inhibitory synapses). In 1964 Rall showed in a computational study that the soma voltage is sensitive to the temporal sequence of synaptic activation "swiping" over the dendritic tree, and that this property could be used to perform a fundamental computation—sensitivity to the direction of motion.

This was the first convincing demonstration that dendrites and their synapses could perform elementary computations; it marked the beginning of the 50-year search for "dendritic computation".

Remarkable theoretical ideas were published regarding the role of dendrites in performing specific computations, including (1) the detection of motion direction; (2) their key role in collision avoidance; (3) storage and classification of multiple input features; (4) calculation of position variables; (5) recovering input signals in the presence of strong noise; (6) enhancing temporal resolution for coincidence detection. But do dendrites really perform these computations in the behaving brain or are these ungrounded (but nevertheless intellectually valuable) theories? To answer this we need to connect the (e.g., sensory) information impinging on the dendritic tree via its synaptic inputs to the behaviorally relevant information represented by the neuron's output. If supplemented with knowledge about synaptic location on dendrites, then we could fully link dendritic structure, synaptic activity/connectivity to the neuron's function.

Finally with the development of new technological (e.g., new refined anatomy available in this age of "connectomics") and with the inspiration of theories that suggested specific mechanisms that might underlie dendritic computation, we are now in a position to directly examine whether dendrites compute. Such direct demonstration was first provided by the in vivo study of Single and Borst (1998) and, very recently, additional concrete examples were offered as summarized in the present book.

This book is therefore a highly timely reflection of "dendrites coming of age"; from an anatomical appreciation of dendritic structure (which started 130 years ago and are ongoing) via theoretical ideas about dendritic function(s) (which started 50 years ago and are ongoing), to state-of-the art experimental approaches, enabling for the first time to directly link dendritic electro-anatomical structure, synaptic loci and activity, to the computational functions of dendrites.

In his small insightful book ("An Advice to a Young Scientist") Ramón y Cajal wrote very beautifully the following: "*Basically, the theorist is a lazy person masquerading as a diligent one. He unconsciously obeys the law of minimum effort because it is easier to fashion a theory than to discover a phenomenon*". I am confident that after reading the present book Cajal would be both totally amazed and inspired by what we have learned about the anatomical, physiological, and molecular secrets that were hidden within the intricacies of *his dendrites*. He might also admit that he himself was a theoretician. By proposing the "theory of dynamic polarization" (the direction of signal flow from axon-to-dendrites-to-soma-to-axon), he laid the foundations for present-day (and future) theories about information flow in dendrites and about how dendrites process this information—about dendritic computation.

On a personal note—it is enchanting for a dendritic person like myself to have such a fresh view on dendrites, as this book offers—a result of a collaborative efforts among an enthusiastic group of young "dendritic lovers" (perhaps inspired by

Cajal's advices to the young). To learn about new developmental principles underlying the unique structure of dendrites, to see experimental validation of earlier theoretical ideas, and to realize as emphasized in this book what is still missing, in both theoretical and experimental realms, until dendrites (of both the researcher and those of the experimental system studied) are "forced" to fully expose their secrets.

Jerusalem, Israel Idan Segev

Preface

In the spring of 2011, two workshop proposals landed on Idan Segev's desk: both were virtually identical, suggesting a workshop for the computational neuroscience conference (CNS) in Stockholm later that year. They both proposed a workshop on computational approaches to study dendrites and both asked him, Idan, to give a keynote presentation. Idan brought us together in what was to become the beginning of an ongoing, fruitful collaboration between the three of us, which has presently resulted in the book that lies before you.

Our goal with this book was to provide a résumé of the state of the art in experimental and theoretical investigations into the functions of dendrites in a variety of neural systems. We are happy that so many authors enthusiastically contributed to this endeavor, resulting in a total of 29 chapters. The book is divided into two parts, each starting with an introductory chapter.

The first part of this book focuses on the morphological properties of dendrites and summarizes the approaches to measure dendrite morphology quantitatively and to actually generate synthetic dendrite morphologies in computer models. This morphological characterization ranges from studying fractal principles to describe dendrite topologies, to the consequences of optimization principles for dendrite shape. We collected individual approaches to study which aspects of dendrite shape can be related directly to underlying circuit constraints and computation.

The second part of this book focuses on how dendrites contribute to the computations that neurons perform. What role do dendritic morphology and the distributions of synapses and membrane properties over the dendritic tree have in determining the output of a neuron in response to its input? We brought together representative studies, with topics ranging from general to system-specific phenomena, some having a strong experimental component, and others being completely theoretical. The studies come from many different neural systems and animal species ranging from invertebrates to mammals. With this broad focus we aim to give an overview of the diversity of mechanisms that dendrites can employ to shape neural computations.

We would like to emphasize that in all our projects together, our contributions were exactly equal, in fact, we drew straws to decide on the editor order on the cover of this book (which turned out to be alphabetical). We would like to thank all the authors and anonymous reviewers who helped to make this book the fantastic work that it is. We would also like to give special thanks to Idan Segev for writing the foreword and Alain Destexhe for supporting this volume. Finally, we are particularly grateful to Ann Avouris from Springer who invited us to put this book together and who has continuously helped us throughout its preparation.

Dear dendritic lovers, enjoy!!

Frankfurt, Germany — Hermann Cuntz
Berlin, Germany — Michiel W.H. Remme
Lausanne, Switzerland — Benjamin Torben-Nielsen

Contents

Part II Dendritic Computation

Contributors

Rod Adams Biocomputation Group, Science and Technology Research Institute, University of Hertfordshire, Hatfield, Herts, UK

Marlene Bartos Institute for Physiology I, Albert-Ludwigs-Universität Freiburg, Freiburg, Germany

Alexander Borst Department of Systems and Computational Neurobiology, Max Planck Institute of Neurobiology, Martinsried, Germany

Tiago Branco MRC Laboratory of Molecular Biology, Francis Crick Avenue, Cambridge Biomedical Campus, Cambridge, UK

Romain D. Cazé INSERM U960, École Normale Supérieure, Paris, France

Paris Diderot 7, 5 rue Thomas Mann, Paris, France

Cesar H. Comin Institute of Physics at Sao Carlos, Sao Carlos, Brazil

Stephen Coombes School of Mathematical Sciences, University of Nottingham, Nottingham, UK

Luciano da F. Costa Institute of Physics at Sao Carlos, Sao Carlos, Brazil

Steven J. Cox Rice University, Houston, TX, USA

Hermann Cuntz Ernst Strüngmann Institute (ESI) for Neuroscience in Cooperation with Max Planck Society, Frankfurt/Main, Germany

Institute of Clinical Neuroanatomy, Goethe University Theodor-Stern-Kai, Frankfurt/Main, Germany

Neil Davey Biocomputation Group, Science and Technology Research Institute, University of Hertfordshire, Hatfield, Herts, UK

Vincent J. Dercksen Department of Visualization and Data Analysis, Zuse Institute Berlin, Berlin, Germany

Alain Destexhe Unit for Neuroscience, Information and Complexity (UNIC), UPR-3293, CNRS, Gif-sur-Yvette, France

Jeffrey S. Diamond Synaptic Physiology Section, NINDS Intramural Research Program, 35 Convent Drive, Bethesda, MD, USA

Shaul Druckmann Janelia Farm Research Campus, HHMI, Ashburn, USA

Robert Egger Computational Neuroanatomy Group, Max Planck Institute for Biological Cybernetics, Tuebingen, Germany

Graduate School of Neural Information Processing, University of Tuebingen, Tuebingen, Germany

Ronald A. J. van Elburg Institute of Artificial Intelligence and Cognitive Engineering (ALICE), University of Groningen, Groningen, The Netherlands

Albert Gidon Department of Neurobiology, Alexander Silberman Institute of Life Sciences, The Hebrew University of Jerusalem, Jerusalem, Israel

William N. Grimes Department of Physiology and Biophysics, Howard Hughes Medical Institute, University of Washington, Seattle, WA, USA

Boris S. Gutkin Departement d'Etudes Cognitives, Group for Neural Theory, Ecole Normale Superieure, Paris, France

Juergen Haag Department of Systems and Computational Neurobiology, Max Planck Institute of Neurobiology, Martinsried, Germany

Kathryn R. Hedrick Johns Hopkins University, 720 Rutland Avenue, Baltimore, MD, USA

Mark D. Humphries Faculty of Life Sciences, University of Manchester, Manchester, UK

George Kastellakis Institute of Molecular Biology and Biotechnology, Foundation for Research and Technology, Heraklion, Crete, Greece

Department of Biology, University of Crete, Heraklion, Crete, Greece

Christiaan P. J. de Kock Center for Neurogenomics and Cognitive Research, Neuroscience Campus Amsterdam, VU University Amsterdam, Amsterdam, The Netherlands

Máté Lengyel Department of Engineering, Computational and Biological Learning Lab, University of Cambridge, Cambridge, UK

Artur Luczak Department of Neuroscience, Canadian Centre for Behavioural Neuroscience, University of Lethbridge, Lethbridge, AB, Canada

Reinoud Maex Group for Neural Theory, Department of Cognitive Sciences, Ecole Normale Superieure, Paris, France

Biocomputation Group, Science and Technology Research Institute, University of Hertfordshire, Hatfield, Herts, UK

Anja Matthiä Cluster of Excellence NeuroCure, Institute of Integrative Neuroanatomy, Charité Berlin, Charitéplatz 1, Berlin, Germany

Matthew J. McGinley Neurobiology Department, School of Medicine, Yale University, New Haven, CT, USA

Matthew F. Nolan Centre for Integrative Physiology, University of Edinburgh, Hugh Robson Building, George Square, Edinburgh, UK

Cian O'Donnell Centre for Integrative Physiology, University of Edinburgh, Hugh Robson Building, George Square, Edinburgh, UK

Marcel Oberlaender Computational Neuroanatomy Group, Max Planck Institute for Biological Cybernetics, Tuebingen, Germany

Bernstein Center for Computational Neuroscience, Tuebingen, Germany

Digital Neuroanatomy, Max Planck Florida Institute for Neuroscience, Jupiter, FL, USA

Arjen van Ooyen Computational Neuroscience Group, Department of Integrative Neurophysiology, CNCR, VU University, Amsterdam, The Netherlands

Jaap van Pelt Computational Neuroscience Group, Department of Integrative Neurophysiology, CNCR, VU University, Amsterdam, The Netherlands

Simon P. Peron Janelia Farm Research Campus, Ashburn, VA, USA

Panayiota Poirazi Institute of Molecular Biology and Biotechnology, Foundation for Research and Technology, Heraklion, Crete, Greece

Michiel W.H. Remme Institute for Theoretical Biology, Humboldt-Universität zu Berlin, Invalidenstrasse 43, Berlin, Germany

Antonio C. Roque Department of Physics, School of Philosophy, Science and Letters of Ribeirão Preto, University of São Paulo, Ribeirão Preto, Brazil

Michelle Rudolph-Lilith Unit for Neuroscience, Information and Complexity (UNIC), UPR-3293, CNRS, Gif-sur-Yvette, France

Susanne Schreiber Institute for Theoretical Biology, Humboldt-Universität zu Berlin, Invalidenstrasse 43, 10115 Berlin, Germany

Robert G. Smith Department of Neuroscience, University of Pennsylvania, Philadelphia, PA, USA

Volker Steuber Biocomputation Group, Science and Technology Research Institute, University of Hertfordshire, Hatfield, Herts, UK

Giseli de Sousa Connectionism and Cognitive Science Laboratory, Department of Informatics and Statistics, Federal University of Santa Catarina, Florianopolis SC, Brazil

Biocomputation Group, Science and Technology Research Institute, University of Hertfordshire, Hatfield, Herts, UK

Klaus M. Stiefel The MARCS Institute, University of Western Sydney, Sydney, Australia

Gaia Tavosanis Deutsches Zentrum für Neurodegenerative Erkrankungen e.V., Center for Neurodegenerative Diseases (DZNE), Bonn, Germany

W. Rowland Taylor Department of Ophthalmology, School of Medicine, Portland, OR, USA

Julian Tejada Department of Physics, School of Philosophy, Science and Letters of Ribeirão Preto, University of São Paulo, Ribeirão Preto, Brazil

Yulia Timofeeva Department of Computer Science and Centre for Complexity Science, University of Warwick, Coventry, UK

Benjamin Torben-Nielsen Department of Neurobiology, Hebrew University of Jerusalem, Edmond J. Safra Campus, Givat Ram, Jerusalem, Israel

Blue Brain Project, Swiss Federal Institute of Technology in Lausanne, Quartier d'Innovation, Batiment J, Lausanne, Switzerland

Harry B. M. Uylings Department of Anatomy and Neuroscience, VU University Medical Center, Amsterdam, The Netherlands

Matheus P. Viana Institute of Physics at Sao Carlos, Sao Carlos, Brazil

Imre Vida Cluster of Excellence NeuroCure, Institute of Integrative Neuroanatomy, Charité Berlin, Charitéplatz 1, Berlin, Germany

Quan Wen Department of Physics and Center for Brain Science, Harvard University, Cambridge, MA, USA

Ekaterina Zhuchkova Institute for Theoretical Biology, Humboldt-Universität zu Berlin, Invalidenstrasse 43, 10115 Berlin, Germany

Part I
Dendritic Morphology

Chapter 1
Introduction to Dendritic Morphology

Benjamin Torben-Nielsen and Hermann Cuntz

Abstract Dendrites play an important role in neuronal function and connectivity. This chapter introduces the first section of the book focusing on the morphological features of dendritic tree structures and the role of dendritic trees in the circuit. We provide an overview of quantitative procedures for data collection, analysis, and modeling of dendrite shape. Our main focus lies on the description of morphological complexity and how one can use this description to unravel neuronal function in dendritic trees and neural circuits.

1.1 Introduction

Probably the most striking feature of a neuron is its characteristic morphology: dendritic and axonal processes sprout as intricate tree structures to enable connections with other neurons. Through their dendrites, neurons receive signals from

---

B. Torben-Nielsen (✉)
Department of Neurobiology, Hebrew University of Jerusalem,
Edmond J. Safra Campus, Givat Ram, 91904 Jerusalem, Israel

Blue Brain Project, Swiss Federal Institute of Technology in Lausanne,
Quartier d'Innovation, Batiment J, 1015 Lausanne, Switzerland
e-mail: btorbennielsen@gmail.com

H. Cuntz
Ernst Strüngmann Institute (ESI) for Neuroscience in Cooperation with Max
Planck Society, Frankfurt/Main, Germany

Institute of Clinical Neuroanatomy, Goethe University, Theodor-Stern-Kai 7,
Frankfurt/Main, D-60590, Germany
e-mail: hermann.neuro@gmail.com

H. Cuntz et al. (eds.), *The Computing Dendrite: From Structure to Function*,
Springer Series in Computational Neuroscience 11, DOI 10.1007/978-1-4614-8094-5_1,

other neurons, and via their axons they transmit signals to other neurons. Historically, research on neuronal morphologies has focused more strongly on dendrites because the larger diameters of their branches make them more amenable experimentally and dendrites cover a more restricted space compared to axons. Only recently, full reconstructions of long-ranging axons have become available (Oberlaender et al. 2011; Ropireddy et al. 2011). The increase in the quantity and quality of neuronal staining and microscopy methods sparked a revived interest in morphological analysis and anatomical circuits such as in the projects of the Blue Brain Project (Markram 2006), the cortical column in silico project by Bert Sakmann (Oberlaender et al. 2012), and the connectomics approach (Helmstaedter et al. 2011). In this chapter we summarize the methods of morphological analysis of dendritic tree structures and argue that knowledge obtained through these methods will be invaluable for resolving the circuitry and function of the nervous system.

Dendritic trees come in all shapes and sizes (Fig. 1.1). They range from a total length of a few tens of micrometers to a few millimeters. Some neurons have only one main dendritic branch, while others possess up to 15–20. Some branches meander strongly, while others are approximately straight. Dendritic morphologies vary significantly even within one neuronal class (Ramaswamy et al. 2012; Soltesz 2005). In addition to this morphological diversity, the molecular composition of ion channels in the membrane strongly differs along the stretch of one dendrite (Migliore and Shepherd 2002), and more pronounced differences even exist between neurons of different types. Why such a diversity?

To answer this question it is necessary to consider the functions pertaining to the dendritic tree. Dendrites clearly serve two pivotal roles in the process of signal integration. First, neuronal morphology defines and is defined by the circuitry. The major element of neuronal connectivity is the synaptic contact between the output axon of one neuron and the input dendrite of another. As such, a precise morphology is crucial to establish the connectivity required for the nervous system to operate normally. Secondly, the precise morphology of a dendrite and its membrane's ion channel composition set the computation that a neuron performs on its inputs, i.e., the propagation and integration of synaptic input signals along the dendritic membrane up to the axon initial segment, the location where the neuronal output is typically generated (Van Elburg and Van Ooyen 2010; Mainen and Sejnowski 1996; Silver 2010; Torben-Nielsen and Stiefel 2010).

Studying dendritic trees thus reveals mechanisms of function in a neuron in terms of its connectivity and computation. Neurons of different types serving different functions should therefore noticeably differ in the morphology and/or physiology of their dendrites. Indeed, as Ramón y Cajal already illustrated more than 100 years ago, dendritic morphology is a defining feature of neuronal classes upon which neurons can be categorized. Up to this day, dendrite morphology represents one of the main criteria for classification of neurons into individual types (Cannon et al. 1999; Migliore and Shepherd 2005). At the same time, due to its wide implication in neuronal functioning, dendritic morphology plays a role in many pathological cases. Neurodegenerative diseases, autism, epilepsy, Parkinson, Alzheimer, and many others have been linked to changes in dendritic and axonal morphology

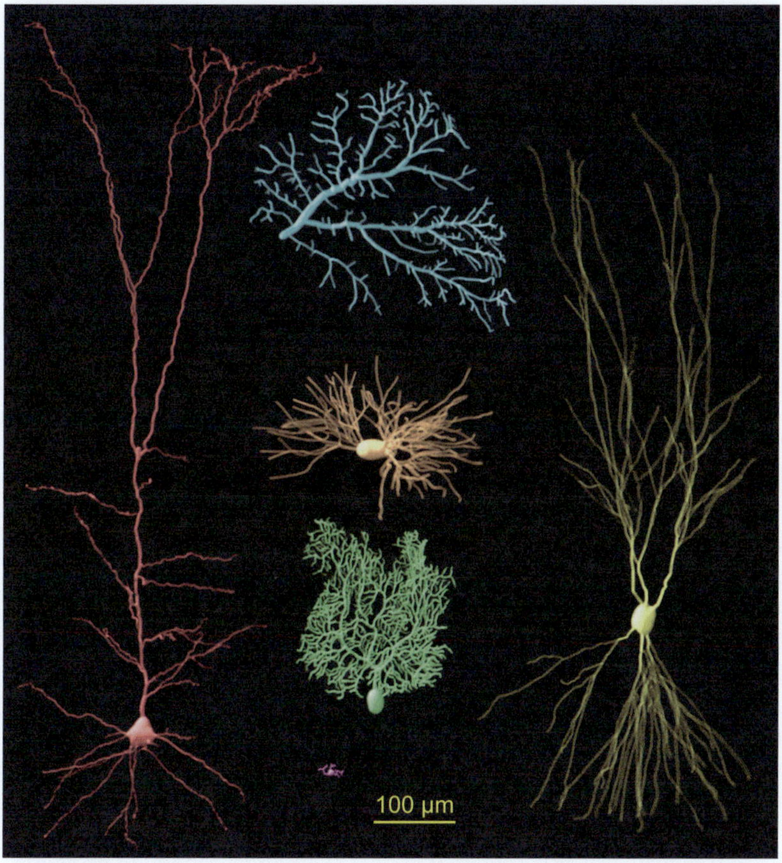

Fig. 1.1 Diversity of dendrite morphology. Different dendritic morphologies illustrating their wide diversity in neural systems. Dendrites are laid out on the same scale: (*red*) rat cortical pyramidal cell (Wang et al. 2002); (*cyan*) fly lobula plate HSN cell (Cuntz et al., 2008); (*orange*) rat thalamic relay neuron (Destexhe et al. 1998); (*yellow*) rat hippocampal pyramidal cell (Ishizuka et al. 1995); (*green*) rat cerebellar Purkinje cell (Vetter et al. 2001); (*pink*) rat neocortical neurogliaform cell (Furtak et al. 2007). Note the differences in size, overall shape, and diameters. Data downloaded from Neuromorpho.org (Ascoli et al. 2007) with reference to the works in which they originally appeared

(Kaufmann and Moser 2000; Moolman et al. 2004; Srivastava et al. 2012). Also, many genes and proteins involved in dendrite formation have recently been identified enabling the study of dendritic dysfunction in a more systematic manner (Jan and Jan 2010; Nagel et al. 2012). Different facets of neural function can therefore be studied directly taking advantage of knowledge of dendrite morphology: the role of different cell types, malfunctions in nervous tissue, development of neural function, and emergence of function in the single cell and in the circuitry. For all these reasons, neuronal morphology lies at the core of many studies in neuroscience. But are there really objective measures to quantify neuronal morphology per se?

1.2 Dendrite Reconstruction, the Data Collection Process

Ultimately, all quantitative measures of neurite morphology are extracted from microscopy data. After an initial stage in which neuronal tissue is prepared and neurons are stained or labeled, a neuron's most prominent features are accessible by visually inspecting it under the microscope. Some general features such as overall size, spatial embedding, and branching complexity can already be resolved at this stage.

However, for a thorough quantification of the dendrite structure, a reconstruction, i.e., a digital representation, of the morphology is required. Reconstructions can then be used in detailed multi-compartmental simulations to calculate the current flow within the tree structure (see chapter "Dendritic Computation") or, as discussed in this chapter, be used for detailed morphological analysis. Because the choice of method used for digitizing a neuronal morphology has consequences for the further analysis, we briefly describe the most common procedures. The very first digital reconstructions were obtained by controlling the microscope's focus with the computer using an electrical stepper (Capowski 1989; Glaser and Glaser 1990). Nowadays, three-dimensional image stacks obtained from confocal and multiphoton microscopes are standard to resolve the z-dimension in the tissue. Advances in staining methods, in particular using genetic tools, allow for staining distinct neurons or neural populations in different fluorescent colors (Lee and Luo 1999; Livet et al. 2007). In addition, novel algorithms and software packages have recently been developed to facilitate the reconstruction process of morphologies. These include the most popular commercial one called Neurolucida (Microbrightfield) and many freely available tools such as VAA3D (Peng et al. 2010), the Trees Toolbox (Cuntz et al. 2010), Neuromantic (Myatt et al. 2012), and the FIJI neurite tracer (Longair et al. 2011) amongst others. In principle, automatic reconstructions of morphologies from neural tissue preparations could provide objective criteria and relieve the human labor associated with manual reconstruction. However, none of the software packages available at present provide tools to flawlessly reconstruct the entire cell, and manual intervention is still required in most cases.

A recent technical development has led the connectome (i.e., the complete reconstruction of all neurons and their connections within a small chunk of neural tissue) to become experimentally accessible at histological scales. This is being made possible for example by combining the resolution of electron microscopy with ion beam or microtome sectioning (Denk and Horstmann 2004; Knott et al. 2008). It is important to note the different resolution of the current reconstruction techniques: while confocal and light microscopes achieve spatial resolutions of a fraction of a micrometer, electron microscopes reach a level of detail in the nanometer range. Dendritic spines or synaptic puncta can therefore hardly be resolved with a light microscope. Electron microscopes have a much higher spatial resolution, but the resulting amounts of data are huge. Obviously, the more detailed the analysis, the finer grained the reconstructions need to be.

1.3 Digital Storage of Neuronal Morphologies

The de facto standard to describe neuronal morphologies is the SWC format (Cannon et al. 1998), where a neuronal morphology (be it dendrite or axon) is a set of connected nodes directed away from a root node. Since each node is attributed one diameter value, the segments in the graph each describes a frustum, i.e., a truncated cone, where the starting diameter of one frustum is the ending diameter of the parent frustum. The morphologies are encoded as plain ASCII text files that contain seven values to describe each node: (1) the node index starting at the value "1"; (2) a region code (or "type") describing whether a node belongs to the soma, the dendrite, or any other region of the neuron; (3–5) x-, y-, and z-coordinates; (6) the diameter at the node location; and (7) the index of the parent node where the root parent index is "−1." In principle, most neuronal structures (dendrites and axons) can be represented in sufficient detail although an accurate description of the soma and spines is hard to represent by connected frusta. Apart from the SWC format, digital reconstructions obtained by (recent) commercial software programs such as NeuroLucida (Microbrightfield) are provided in the software's internal proprietary file types, and they contain additional options to describe a morphology including metadata. For instance, the network context, e.g., laminar structures and tissue boundaries, can be annotated separately.

An entirely different way to represent neurons is to describe their shape as a mesh. A mesh is a detailed description of the surface of a body by means of a list of vertices that are connected to polygons (usually triangles). The level of detail can be adjusted by the number of vertices used to cover the surface and is generally much higher compared to the SWC description mentioned above. Consequentially, mesh representations of neurites are often used in simulations that require high spatial accuracy such as reaction–diffusion of molecules along and across membranes (Hepburn et al. 2012; Kerr et al. 2008).

In an effort to standardize neuroscientific models that span single neurons to entire networks the NeuroML initiative in the field of neuroinformatics has introduced its own formalism describing neuronal morphology, MorphML (Gleeson et al. 2010). The latest version, NeuroML 2.0, includes the possibility to describe functional characteristics of a model (e.g., the Hodgkin–Huxley equation), thereby making it possible to link morphological attributes with ion channel features and dynamics in one unified format.

Another development in the field of neuroinformatics is the advent of databases containing neuronal data and models that are open to the public. Thanks to this trend, reconstructed morphologies from different labs are widely shared in the scientific community. Several labs host their own databases that can be accessed through the Internet. The most complete database, NeuroMorpho.org (Ascoli et al. 2007), shares morphology files from a large number of different labs, standardizes them, and makes them available freely in the public domain. At the time of writing, the database contains almost 9,000 reconstructed morphologies in a standardized

SWC format. Among the available databases some also combine morphologies of single neurons with contextual circuit information, e.g., the standard brain databases for a large variety of insects (http://www.neurobiologie.fu-berlin.de/beebrain/Links.html). Some morphological reconstructions are also made available in combination with their electrical models, e.g., ModelDB (Migliore et al. 2003), allowing the morphology/function relations to be studied in detail.

1.4 Single-Neuron Morphometry and Quantification

Once digital reconstructions are obtained and stored, they can be used for further analysis and quantification. Quantitative analysis of neuronal morphologies can be used to address distinct research questions. One such question addresses the categorization of neurons into types. For this purpose, one discriminative measure can be sufficient. Another analysis strategy, however, is needed when investigating the differences in neuronal phenotype after genetic manipulation of a neuron. The latter might require a far more sensitive analysis likely involving multiple morphometric measures. Due to this diversity of scientific questions relying on morphometric analysis, not a single standard approach has yet emerged to quantify neuronal morphologies. In this section we discuss the possible methods to quantify morphologies.

However, before describing quantitative morphometric analyses in detail, we first introduce some terminology pertaining to morphometrics. The Petilla convention (Ascoli et al. 2008), a nomenclature specifically designed to describe the features of interneurons, sets a standard for some morphological features. However, a complete convention is still missing to our knowledge. As mentioned above, a neuron's morphology is described as a mathematical tree that is generally rooted at the cell body or the soma (Fig. 1.2). A tree is formally defined as a noncyclic graph. The dendritic stem segments (also know as "trunks" or "initial segments") sprout from this root, and each give rise to a branch. Branches in turn are composed of dendritic segments and branch points. Branch points are physical bifurcations at which a parent segment divides into two daughter segments. A branch has at least one termination point, the point at which the branch ends. In the terminology from graph theory, the root, branch, and termination points constitute the set of nodes in the tree, while the dendritic segments are the edges.

Morphometric measures can be divided into two main categories: topological and geometric morphometrics (Uylings et al. 1986; Verwer and van Pelt 1983). We further discuss functional measures that connect morphology directly to its function and measures to quantify similarity between neurons. The topological morphometrics deal with the branching structure of the tree independently of metric units (e.g., number of branch points and branch order). For geometric analyses, the topology is embedded in real space, thereby giving a shape to the topology. Associated measures consequently have metric or angular units. The functional measures refer to a morphology's function in the circuit and its ability to connect to other neurons and

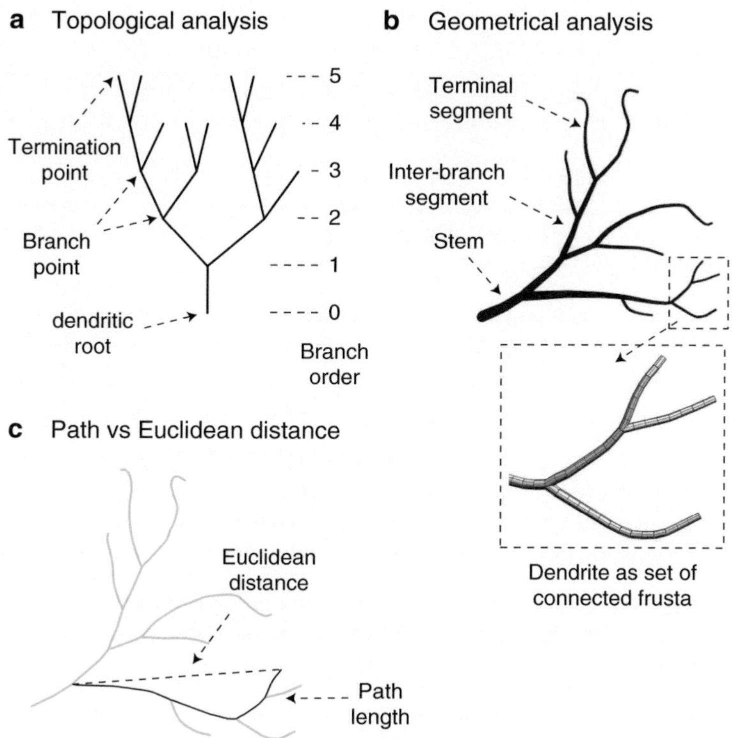

Fig. 1.2 Morphometric analysis of dendrite morphology. (**a**) Topological analyses disregard the metric features and describe the connections in the graph underlying the dendritic tree structure. Terminology shown for an idealized representation of a neuronal tree. (**b**) For a geometrical analysis, the tree is embedded in space and length values, as well as angles play a role. *Inset* shows the frustum-based representation of the dendritic structure in space. (**c**) An important geometric morphometric is the path length in a tree. Path lengths are always longer than the Euclidean distances, and many other morphometrics are calculated as a function of one of these two measures

to integrate inputs. Many other distinctions between types of morphometrics exist, such as the distinction between global/scalar and local/vector morphometric features. An example for the latter is the distinction between total length of a neuron and the individual lengths of all segments in a neuron. The total length is a global, scalar value while the individual lengths are local values. Table 1.1 summarizes the geometric and topological measures presented below.

1.4.1 Topological Measures

Intuitive topological measures are the *number of stems* defined as the number of edges leaving the dendritic root, the *number of branch points*, and the *number of*

Table 1.1 List of frequently used morphometric measures to quantify neuronal morphologies

Property	Brief description
Number of stems[a]	Total number of segments leaving from the dendritic root
Number of branch points[a]	Total number of branch points in the tree
Branch order	Topological distance from the dendritic root. The root has order 0, and the order of a segment in the tree equals the number of branch points along the path to the root + 1
Maximum branch order[a]	Maximum branch order in a neuron
Degree	Number of termination points downstream of the node under investigation
Maximum degree[a]	Maximum degree in a tree. By definition the degree at the dendritic root
Total length[a]	Summed segment lengths of all segments in a tree (see next)
Segment length	Path length of the incoming segment toward a node
Stem length	Path length between a branch point with order = 1 and the dendritic root
Interbranch length	Path length between branch points
Terminal segment length	Path length between the termination point and the last branch point
Euclidean distance	Can be applied in a similar fashion as the path length. Often used to measure the distance between the soma and the termination points
Dimension[a]	Width, height, and depth of the bounding box
Taper rate[a]	The uniform decrease in diameter across a dendritic branch
Somatofugal tropism[a]	Quantification of the preference of a neurite to grow away from the soma. Defined as the ratio of a segment's path length and the Euclidean distance between its starting and end point
Fractal dimension[a]	Fractal dimension used as a measure of space-filling
Contraction	Quick proxy of the fractal dimension: the Euclidean length of a branch divided by the path length
Partition asymmetry[a]	Topological complexity of a tree. A completely asymmetric tree has PA = 1, symmetric has PA = 0
Lacunarity[a]	A measure of "holes" in a volume spanned by a tree. See Sect. 1.4.2
Horton–Strahler index	Measure of topological complexity of a tree relating the order and asymmetry in that tree. Computed for each branch point. See Sect. 1.4.1
Strahler number[a]	The Horton–Strahler index associated with the root of the tree

Light shading—topological measures
Medium shading—geometrical measures
Dark shading—compound measures
[a]Global measure as opposed to distribution of local measures. However, often derived features are used as global feature. For instance, to describe the branch order in a tree, a distribution of all orders can be given, or the distribution can be characterized by considering the maximum branch order, the average branch order, etc. This holds for all local measures

termination points. For the sake of correctness, we use the term "dendritic root" rather than "cell body" or "soma" because in certain neuronal types (for instance many insect neurons) the dendrites do not sprout from the soma. However, in most cases, the soma is the root of the dendritic tree. The number of stems is sometimes used to classify cell types (e.g., bipolar cells in the retina), while the number of branch and termination points represent the extent of branching in a tree. Since branch points typically are bifurcations in all neurons, the number of termination points is the number of branch points plus one. Furthermore, the *branch order* (or "centrifugal order") for each node is computed as follows: the dendritic root has by convention an order of zero so that the order of a node becomes one plus the number of branch points encountered on the path between the inspected node and the dendritic root (Fig. 1.2a). The reciprocal property is the *degree*, which is defined as the number of termination points in the sub-tree rooted at the node under investigation. The distribution (count) of the order and degree of branch points in a tree can be used for classification and description of morphologies (Verwer and van Pelt 1983). Order and degree are often used as auxiliary properties in combination with other morphometric features. For instance, any local morphometric measure can be plotted against order or degree to create a conditional distribution (see Sect. 1.4.4). The order and degree are also used in composite morphometrics. One such composite, topological measure is the *partition asymmetry* that assesses the topological complexity of a neuronal tree based on the normalized difference between the degree of two daughter branches at a branch point. The partition asymmetry index ranges from 0 (completely symmetric) to 1 (completely asymmetric) (van Pelt et al. 1992; van Pelt and Schierwagen 2004). Another composite morphometric based on order and degree is the *Horton–Strahler (HS) index* that relates the asymmetry in a tree with the depth of the tree (Binzegger et al. 2004; Toroczkai 2001). The HS index is computed at each branch point and equals $k+1$ when both daughter branch points have equal HS index of k or as $\max(k_1,k_2)$ when the HS indexes of its daughters k_1 and k_2, respectively, are not equal. The *Strahler number* is defined as the Horton–Strahler index associated with the root of the tree.

1.4.2 Geometric Measures

In contrast to topological properties that have no metric interpretation, geometric properties consider the spatial embedding of a tree. The segment length values and diameters are among the main properties in this category and give rise to a multitude of related morphometric properties. The most basic one is the *total dendritic length*. Distinct parts of the tree can be described in terms of their length as well, e.g., *stem length, interbranch point length,* and *terminal segment length* (Fig. 1.2b). Also, relations between any location in the tree can be described by a length metric in terms of the Euclidean distance or the path length between those locations (Fig. 1.2c).

The (somatofugal) tropism factor relates the segment length to the Euclidean distance from the dendritic root (Marks and Burke 2007; Samsonovich and Ascoli 2003). The ratio between length and distance is 1 for a segment that grows radially away from the dendritic root and 0 for a segment growing concentrically in relation to the dendritic root. Spatial extent and its associated spatial embedding can be quantified in a number of ways. Most straightforward is the *dimension*: the raw bounding box in three dimensions. The *fractal dimension* is a measure of self-similarity and is often used as a measurement of space-filling (Smith et al. 1996). By definition, a straight line has a dimension of 1, a square has 2, and a cube has 3, but dendrites can be associated with fractional dimensions since the space that they cover is not fully filled. However, the interpretation of the fractal dimension is arbitrary and strongly depends on the method used to calculate it. In the "calliper method" (Fernández and Jelinek 2001), the fractal dimension represents the level of meandering of a dendrite, where a straight line has a dimension of 1 and more meandering dendrites receive slightly higher values. Since the validity of the fractal dimension is disputed in the analysis of neuronal morphologies (Cannon et al. 1999; Jelinek and Fernández 1998), *contraction* of a dendrite might be used as a proxy of the fractal dimension (when the fractal dimension is computed using the calliper method). Contraction is defined for a stretch of dendrite between two points as the ratio of the Euclidean distance and the associated path length between those points. A straight line has a contraction of 1, while a meandering dendrite has a slightly lower value. Intuitively, the relation between contraction and the fractal dimension can thus be approximated as fractal dimension $\approx$ (2 − contraction) for planar dendrites. Both contraction and fractal dimension quantify space-filling. The reciprocal morphometric is a measure of "holes" in a morphology and is defined by the *lacunarity* (Smith et al. 1996). Apart from the overall dimension, locally the spatial embedding can be assessed by the angles in three dimensions between parent and daughter branches. Different variants are in use: the amplitude of the angle between the daughters can be measured as well as the angle between the parent segment and the daughters (Scorcioni et al. 2008). Recently, the perceived planarity of dendritic branch points (Kim et al. 2012; Uylings and Smit 1975) has received renewed attention as it was linked to optimal wiring principles and led to the development of detailed morphometrics quantifying the angles of branch points (van Pelt and Uylings 2011).

The diameter can be specified in relation to its change along the neuronal processes, i.e., tapering. Typically diameters reduce along a dendritic cable and can thus be approximated by a *tapering rate:* the linear or the nonlinear rate at which the diameter decreases per unit of length. Discontinuities in the tapering rate occur at branch points and can be referred to by the *child–parent ratio*, the ratio between the diameters of the parent and the child segments.

1.4.3 Functional Measures

Most morphometric properties inherently have some influence on the function and electrotonic structure of that neuron: with longer and/or thinner dendritic segments, input signals are more attenuated than with shorter and/or thicker segments. Hence, distal inputs, all things being equal, contribute less to the electrical activity at the site of action potential initiation compared to proximal inputs. On the other hand, some morphological features do not impact on certain aspects of neural computation: angles in the dendrite do not matter for a current traveling through a dendrite. In some cases, morphometric measures are directly inspired by the function of neurons. An exemplary measure for this is *Rall's power*, a measure relating the diameters of a parent branch and its daughters at the branch point. Wilfrid Rall studied the power relation between parent (D) and daughter branch diameters (d_1 and d_2), $D^r = d_1^r + d_2^r$, with "*r*" now coined Rall's power (Scorcioni et al. 2008). Rall calculated that for the specific power $r = 3/2$, there is continuity in the impedance sensed by centrifugal signal flow at branch points such that a dendritic tree can be collapsed to a non-tapering, non-branching cable piece useful for applying analytical solutions of the cable equation to more complex dendrites (Rall 1959).

The other main function of neurons is to connect to other neurons in order to receive input. Specific measures directly relate a morphology to connectivity. The *critical percolation density* relates the dendritic morphology to the ability to connect to other neurons in the network (Costa and Manoel 2003). To calculate this property one populates a volume with instances of one particular morphology. Then, the percolation density is the average density at which a path suddenly emerges that leads from one side of the volume to the other along dendritic branches. The *excluded volume* (Costa et al. 2005; Wassle et al. 1981) is a related measure that quantifies the part of a dendritic tree's spanning volume that is not easily reachable for axons to make contacts with. The excluded volume is likely at the center of a dendritic structure when surrounded by dense branching.

An overarching metric and first approximation of connectivity can be derived from Peter's rule, which states that the number of synapses is proportional to the overlap between an axon and a dendrite (Binzegger et al. 2004; Peters and Payne 1993). Therefore, the allowed distance between dendrite and axon at which structural appositions are thought to occur can be used as a first approximation of connectivity. Note that Peter's rule associates two morphologies, a dendritic and an axonal one, while the percolation density and the excluded volume are single-neuron metrics.

1.4.4 Similarity

The aforementioned morphometric measures can be used to categorize morphologies and rank them on their similarity. Morphometrics can thus be used as a metric to quantify similarity between morphologies. In a straightforward fashion, the Euclidean distance between scalar morphometrics provides a measure of (dis)similarity. Vector morphometrics comprise a one-dimensional distribution, and hence the similarity between two such distributions can be quantified using hypothesis tests, of which the Student's test (*t*-test) is the most popular parametric test and the Kolmogorov–Smirnov test and ranksum tests are well-known nonparametric alternatives.

An issue is that morphometric measures are often not independent from each other. Thus, morphometric measures can be seen independent from each other (=univariate) or conditioned on other measures (=multivariate). For instance, the segment length might be independent of any other feature of the neurite tree (univariate description), or, might be dependent on the topological order of the segment (conditional, bivariate description), and so forth. Sholl was the first to perform such a two-dimensional analysis as he counted the number of intersections between the dendritic segments and imaginary concentric circles at increasing Euclidean distance from the cell body (Sholl 1953). As briefly mentioned before, any bivariate morphometric can be plotted against another, and this is generally referred to as a Sholl-like analysis. As an example, one can argue that the segment length changes depending on the branch order (Burke et al. 1992; Nowakowski et al. 1992). Such higher order relations surely exist in neurons, but the application of multivariate measurements remains restricted due to limited amount of data (a large sample size is needed to uncover higher order relations).

In addition to comparisons of morphometrics there are dedicated measurements to quantify similarity. A first similarity measure is the *tree edit distance* and formalizes how many operations ("tree edits") have to be made to morph one tree into another (Heumann and Wittum 2009). Another measure of similarity between neurite morphologies is the *shape diffusion index* (Luczak 2010). While technically a bit more complex, intuitively it can be seen as a measure of how easily a morphology can be synthesized using the diffusion-limited aggregation approach (see next section and Chap. 5). Luczak found that some classes of neurons were "easy" to approximate, while others were "harder." This difference can then be used to express similarity.

1.4.5 Further Quantification

The review we presented here is by no means exhaustive; it rather reflects a choice of morphometric measures most commonly used to quantify morphology and link morphology to connectivity. Graph theory itself has many ways to describe tree structures and network topologies. These descriptions, however, often have little or no biological interpretation. A good introduction into more exotic measures to quantify neuronal morphologies is provided by Rocchi et al. (2007). Also, a lot of ad hoc measures are in use. These measures are often proposed to study a particular trait in a particular neuronal type. For instance, somatofugal tropism was first used to quantify the preference of a dendritic tree to direct away from its soma in pyramidal cells. Ad hoc measures have the intrinsic drawback that there is no standard definition of how to compute them, which makes their evaluation and their comparison to well-defined morphometrics harder. Nevertheless, when picked up by others, as in the case of the measure for tropism, these ad hoc measures can make it into the standard repertoire of measures and extend the standard battery of useful measures.

1.5 Algorithms to Synthesize Dendrites

Since morphometric measures are so numerous, variable, and dependent on many factors like context and age, it is difficult to extract principles of dendritic structure directly from these measures. In the last two decades, generative approaches were developed to study dendritic morphologies by synthesizing (parts of) dendritic morphologies in the computer. The rationale behind this approach is that by generating dendrites according to a particular principle, resulting synthetic morphologies may share morphological traits with real dendrites. This finding then can corroborate the initial hypothesis about the underlying principle. In a sense, it is the computational neuroanatomy interpretation of Lord Kelvin's maxim: "I am never content until I have constructed a mechanical model of what I am studying. If I succeed in making one, I understand; otherwise I do not."

One of the ideas that kick-started the generative approach stems from the seminal work of Hillman who described a set of *fundamental parameters* sufficient to generate dendritic morphology (Ascoli and Krichmar 2000; Hillman 1979). Many generative algorithms are based on this idea by iteratively constructing a dendritic tree while sampling from statistical descriptors (Burke et al. 1992; Tamori 1993). One of the first publicly available tools is L-Neuron (Ascoli and Krichmar 2000), which relies on an L-system in combination with stochastic sampling to generate synthetic dendrites. An L-system or a Lindenmayer system (Prusinkiewicz and Lindenmayer 1990) is a formalism that can recursively generate branched structures, such as plants and neurons, from a parsimonious representation. The geometry, however, is sampled from statistical descriptors based on morphometric measures. The differences between existing sampling algorithms mostly lie in the statistical descriptors used: the number and selection of parameters as well as their statistical descriptions as a parametric model (Ascoli and Krichmar 2000; Eberhard et al. 2006) or a non-parametric model (Lindsay et al. 2007; Torben-Nielsen et al. 2008).

A different approach, inspired by principles of neuronal development, has been developed over the years by van Pelt and colleagues (van Pelt et al. 1992; van Pelt and Schierwagen 2004). They followed a generative approach that could be used to synthesize dendritic topology constrained by the branch order and the number of simultaneously developing segments. By adjusting the rate of growth to the number of developing growth cones, a basic interpretation of competition over resources is included. Their work led to the development of a tool, NETMORPH (Koene et al. 2009), that can generate large networks of interconnected synthetic morphologies by addition of a geometric component to the developing branched structures (see Chap. 4).

Note that the aforementioned tools synthesize morphologies without (or to a limited extent) considering the interactions with other parts of the tree, with other neurons, or with physical boundaries within the circuit. Because neurons do not grow in isolation, it can be assumed that not all natural factors are captured by these algorithms. A number of recently developed approaches overcome these limitations in various ways.

One approach that explicitly takes other neurons into account during the generation process is proposed by Luczak (2006) (see Chap. 5) and is based on diffusion-limited aggregation (Hentschel and Fine 1996; Witten and Sander 1981): "neurotrophic particles" are distributed into a bounded space where they randomly move. Then, upon colliding with an n-particle aggregate, the moving particle will stick and form an $n+1$-particle aggregate. Some extra rules are needed to constrain the aggregates to biologically relevant structures, but such an aggregation process yields realistic synthetic dendritic structures. By using several seeds, i.e., initial 1-particle aggregates, at the same time within the same bounded space, virtual morphologies under construction effectively compete over resources and hence a biologically plausible form of interaction is captured in the algorithm.

Recently another approach was proposed by Memelli et al. (2013) where a virtual morphology is grown under the presence of environmental influences. A branching rule determines when a dendritic tree branches, but the direction of growth is solely determined by environmental cues. Cues act as biases on the direction of growth. Additionally, the virtual morphologies are constrained to a bounded space. As a proof of principle, highly realistic morphologies were synthesized by solely taking self-referential (i.e., stiffness, soma tropism, and self-avoidance) cues into account. So far the environmental interaction is thus limited to the neuron's own guidance cues. However, a cue coming from another neuron is conceptually identical to a cue coming from the neuron itself: it is a bias on the direction of growth. Therefore, this approach can be directly used to investigate hypotheses about various environmental cues shaping the final dendrite morphology.

A different approach to generate dendritic morphologies is based on the functional implication of the structure of neural trees, the influence of diameter values on voltage propagation, and circuit connectivity (Cuntz et al. 2007, 2008, 2010; see Chap. 6). Target points are distributed in a volume of interest and connected iteratively to a growing tree in a competitive manner to minimize total wiring cost and conduction times. In a post-processing step, diameters are assigned according to the propagation dynamics, and a synthetic morphology is created. One advantage of using this approach is that it is far less data intensive compared to standard sampling methods: the only statistical descriptor in use is a two- or a three-dimensional density function of the target points. Indeed, no detailed statistics about branching angles, segment length, etc. are used, and hence the algorithm is less prone to researcher-induced biases. It also permits to draw direct conclusions about the connectivity principles in the circuit. Optimization principles of wiring constraints such as the ones used for these models have traditionally been successful at directly predicting scaling relationships in branching statistics (Chklovskii and Koulakov 2004; Klyachko and Stevens 2003; Wen et al. 2009; Wen and Chklovskii 2008). This is described in great detail in Chap. 7.

Beyond the constraint of connectivity, the hypothetical computational function of a neuron can also constrain the morphology. This idea is explored by Torben-Nielsen and Stiefel (2010) and is coined the "inverse approach" (see Chap. 9). In the inverse approach, morphologies are synthesized with a straightforward sampling algorithm. However, the parameters defining the statistical distributions from which they are sampled are not based on biological data. Rather, the parametric distributions are optimized using evolutionary algorithms so that a model neuron endowed with the optimized morphology successfully performs a predefined neuronal computation. The inverse approach can be used in a twofold manner: as a hypothesis tester and a hypothesis generator. In case it is hypothesized that a neuron performs a computation, a synthetic morphology can be optimized to perform the same function. When the resulting morphology resembles the real counterpart, this result can corroborate the hypothesis. On the other hand, morphologies can be optimized to perform a computation of interest. If neurons in the brain then resemble the synthesized morphology, a hypothesis about the function of these neurons can be proposed.

The last approach to be mentioned here is the most elaborate one and is proposed by Zubler and colleagues (Zubler et al. 2011; Zubler and Douglas 2009). They developed a phenomenological growth simulator in which growing neurons can migrate and interact with structural boundaries as well as other cells within the simulated environment. Each growth cone contains the growth rules for this neuron encoded in a gene-like format (Zubler et al. 2011). The growth rules can use environmental cues such as secreted substances or laminar information. In addition a growing neurite can also secrete substances and switch on or off growth rules depending on its environment. With this setup, many of the environmental complexities of cortical circuitry can be modeled (at least in a phenomenological way).

1.6 Future Perspectives

A major problem with studying dendritic morphology is that the shape of a neuron is not a perfect instance of a general blueprint. In the nervous system, individual cells have to compete with others as well as with different structures such as glia cells and blood vessels. Reconstructed morphologies are always mere snapshots of a particular neuron in time and contain the combined effects of development and learning. As a result, large variation exists in their morphology. Variation is commonly seen as the difference in morphology between members of the same type, while diversity is the distinction between different classes (Soltesz 2005). In general, the description of diversity is easier than the description of variation. For instance, it is straightforward to differentiate a motor neuron from a pyramidal neuron. Sometimes, however, the variation within a class is larger than the diversity between classes. Purely based on the morphology it is fairly hard to tell apart a layer five (L5) pyramidal neuron from a layer six (L6) one. It is even harder to tell the difference between L5 pyramidal neurons at different ages. In general, the more detailed the research question becomes, the more sophisticated the quantitative analysis needs to be. The presented morphometric quantification helps considerably: it provides hard numbers to describe neuronal morphologies. But, how to cope with the variance?

Here, we sketch three future strategies that will be useful to quantify and understand variation. Let us for example consider the rat sensorimotor cortex that is roughly 2,000 μm thick and layer 5 (in 2-week-old animals) is 650–700 μm thick. L5 pyramidal cells have an apical dendrite growing to the pia where it branches extensively. A large variation in total length occurs when a neuron's cell body is located deeper in layer 5 and has to extend further to reach the pia (Oberlaender et al. 2012; Ramaswamy et al. 2012). Because the current standard to represent morphological reconstructions (e.g., the aforementioned SWC format) only includes the geometry of a single morphology, there is no direct way of taking the location of the cell body into account. Hence, no statistical model will capture the fact that L5 pyramidal cells closer to the pia have a shorter apical dendrite. This example illustrates a pitfall in the quantification of morphologies: as long as they are considered in isolation, simple biophysical explanations for morphometric variation will be overlooked.

The first strategy to overcome the outlined issue is to include metadata such as laminar position and 3D spatial context in the description of a neuron. This strategy is adopted in the cortical column model by the group of Oberlaender and colleagues. This is summarized in Chap. 8, where the 3D context is stressed in the reconstruction of the rat vibrissal cortex.

A second strategy is a pragmatic approach to investigate the natural variation in neuronal types and relies on the fact that a good statistical model should be able to generalize and infer morphometrics as long as the sample size is large enough. By using larger samples, the statistical models can better generalize. This strategy is adopted by Costa and his colleagues. They analyze morphologies from the complete NeuroMorpho.org database to investigate morphological outliers and archetypes (Costa et al. 2010; see Chap. 3).

A final strategy we discuss here to alleviate the issue of variance and the lack of metadata is to actually study the biophysical mechanisms that underlie neurite morphologies. Our understanding of cell biological mechanisms resulting from genetics is increasing rapidly. In Chap. 2, Tavosanis concisely summarizes some of the prominent molecular mechanisms of morphological differentiation that underlie not only basic properties as branching and elongation but also more complex behaviors requiring interaction between developing dendrites as required for tiling and self-avoidance. If the underlying mechanisms can be expressed as phenomenological rules, morphometrics do not need to be used to deduce morphological traits, because the traits and associated hypothesized rules can be studied using the generative approach. First attempts to generate morphologies according to developmental rules include reports by Memelli et al. (2013) and Zubler and Douglas (2009). In the future, generative approaches in general coupled with better molecular understanding of morphological differentiation will allow us to explain differences in sizes as in the example of the L5 pyramidal cell. Combined with rigorous morphometric analysis, it would then be possible to attribute highly varying morphometric features not to diversity or variation but to, for instance, avoidance of capillaries or the repulsion away from another neuron of a specific class.

In conclusion, morphology of neurons plays a fundamental role in brain functioning. Using morphometric quantification we are currently able to classify some types of neurons. Moreover, rigorous quantification allows us to study morphological traits and their connection to the brain circuit at large. In the future, we expect a synergy between the big data projects generating detailed statistics of neurite morphology and their connections (e.g., connectomics), genetic studies revealing local interaction rules governing dendritic growth, and generative approaches to link large-scale data and local interactions. Please enjoy the following chapters covering the state of the art in the quantitative studies of dendritic morphology.

References

Ascoli GA, Alonso-Nanclares L, Anderson SA et al (2008) Petilla terminology: nomenclature of features of GABAergic interneurons of the cerebral cortex. Nat Rev Neurosci 9:557–568

Ascoli GA, Donohue DE, Halavi M (2007) NeuroMorpho.Org: a central resource for neuronal morphologies. J Neurosci 27:9247–9251

Ascoli GA, Krichmar JL (2000) L-neuron: a modeling tool for the efficient generation and parsimonious description of dendritic morphology. Neurocomputing 33:1003–1011

Binzegger T, Douglas RJ, Martin KAC (2004) A quantitative map of the circuit of cat primary visual cortex. J Neurosci 24:8441–8453

Burke RE, Marks WB, Ulfhake B (1992) A parsimonious description of motoneuron dendritic morphology using computer simulation. J Neurosci 12:2403–2416

Cannon R, Turner D, Pyapali G, Wheal H (1998) An on-line archive of reconstructed hippocampal neurons. J Neurosci Methods 84:49–54

Cannon R, Wheal H, Turner D (1999) Dendrites of classes of hippocampal neurons differ in structural complexity. J Comp Neurol 633:619–633

Capowski JJ (1989) Computer techniques in neuroanatomy. Plenum Press, New York

Chklovskii DB, Koulakov AA (2004) Maps in the brain: what can we learn from them? Annu Rev Neurosci 27:369–392

Costa LDF, Barbosa MS, Coupez V (2005) On the potential of the excluded volume and autocorrelation as neuromorphometric descriptors. Phys Stat Mech Appl 348:317–326

Costa LDF, Manoel ETM (2003) A percolation approach to neural morphometry and connectivity. Neuroinformatics 1:65–80

Costa LDF, Zawadzki K, Miazaki M et al (2010) Unveiling the neuromorphological space. Front Comput Neurosci 4:150

Cuntz H, Borst A, Segev I (2007) Optimization principles of dendritic structure. Theor Biol Med Model 4:21

Cuntz H, Forstner F, Borst A, Häusser M (2010) One rule to grow them all: a general theory of neuronal branching and its practical application. PLoS Comput Biol 6:e1000877

Cuntz H, Forstner F, Haag J, Borst A (2008) The morphological identity of insect dendrites. PLoS Comput Biol 4:e1000251

Denk W, Horstmann H (2004) Serial block-face scanning electron microscopy to reconstruct three-dimensional tissue nanostructure. PLoS Biol 2:e329

Destexhe A, Neubig M, Ulrich D, Huguenard J (1998) Dendritic low-threshold calcium currents in thalamic relay cells. J Neurosci 18:3574–3588

Eberhard JP, Wanner A, Wittum G (2006) NeuGen: a tool for the generation of realistic morphology of cortical neurons and neural networks in 3D. Neurocomputing 70:327–342

Fernández E, Jelinek H (2001) Use of fractal theory in neuroscience: methods, advantages, and potential problems. Methods 24:309–321

Furtak SC, Moyer JR, Brown TH (2007) Morphology and ontogeny of rat perirhinal cortical neurons. J Comp Neurol 505:493–510

Glaser JR, Glaser EM (1990) Neuron imaging with neurolucida—a PC-based system for image combining microscopy. Comput Med Imaging Graph 14:307–317

Gleeson P, Crook S, Cannon RC et al (2010) NeuroML: a language for describing data driven models of neurons and networks with a high degree of biological detail. PLoS Comput Biol 6:e1000815

Helmstaedter M, Briggman KL, Denk W (2011) High-accuracy neurite reconstruction for high-throughput neuroanatomy. Nat Neurosci 14:1081–1088

Hentschel HG, Fine A (1996) Diffusion-regulated control of cellular dendritic morphogenesis. Proc Biol Sci 263:1–8

Hepburn I, Chen W, Wils S, De Schutter E (2012) STEPS: efficient simulation of stochastic reaction—diffusion models in realistic morphologies. BMC Syst Biol 6:36

Heumann H, Wittum G (2009) The tree-edit-distance, a measure for quantifying neuronal morphology. Neuroinformatics 7:179–190

Hillman D (1979) Neuronal shape parameters and substructures as a basis of neuronal form. The Neurosciences, Fourth Study Program. MIT Press, Cambridge, MA, pp 477–498

Ishizuka N, Cowan WM, Amaral DG (1995) A quantitative analysis of the dendritic organization of pyramidal cells in the rat hippocampus. J Comp Neurol 362:17–45

Jan Y-N, Jan LY (2010) Branching out: mechanisms of dendritic arborization. Nat Rev Neurosci 11:316–328

Jelinek HF, Fernández E (1998) Neurons and fractals: how reliable and useful are calculations of fractal dimensions? J Neurosci Methods 81:9–18

Kaufmann WE, Moser HW (2000) Dendritic anomalies in disorders associated with mental retardation. Cereb Cortex 10:981–991

Kerr RA, Bartol TM, Kaminsky B et al (2008) Fast Monte Carlo simulation methods for biological reaction-diffusion systems in solution and on surfaces. SIAM J Sci Comput 30:3126

Kim Y, Sinclair R, Chindapol N et al (2012) Geometric theory predicts bifurcations in minimal wiring cost trees in biology are flat. PLoS Comput Biol 8:e1002474

Klyachko VA, Stevens CF (2003) Connectivity optimization and the positioning of cortical areas. Proc Natl Acad Sci USA 100:7937–7941

Knott G, Marchman H, Wall D, Lich B (2008) Serial section scanning electron microscopy of adult brain tissue using focused ion beam milling. J Neurosci 28:2959–2964

Koene R, Tijms B, Van Hees P et al (2009) NETMORPH: a framework for the stochastic generation of large scale neuronal networks with realistic neuron morphologies. Neuroinformatics 7:195–210

Lee T, Luo L (1999) Mosaic analysis with a repressible cell marker for studies of gene function in neuronal morphogenesis. Neuron 22:451–461

Lindsay KA, Maxwell DJ, Rosenberg JR, Tucker G (2007) A new approach to reconstruction models of dendritic branching patterns. Math Biosci 205:271–296

Livet J, Weissman TA, Kang H et al (2007) Transgenic strategies for combinatorial expression of fluorescent proteins in the nervous system. Nature 450:56–62

Longair MH, Baker DA, Armstrong JD (2011) Simple Neurite Tracer: open source software for reconstruction, visualization and analysis of neuronal processes. Bioinformatics 27:2453–2454

Luczak A (2006) Spatial embedding of neuronal trees modeled by diffusive growth. J Neurosci Methods 157:132–141

Luczak A (2010) Measuring neuronal branching patterns using model-based approach. Front Comput Neurosci 4:10

Mainen ZF, Sejnowski TJ (1996) Influence of dendritic structure on firing pattern in model neocortical neurons. Nature 382:363–366

Markram H (2006) The blue brain project. Nat Rev Neurosci 7:153–160

Marks WB, Burke RE (2007) Simulation of motoneuron morphology in three dimensions. II. Building complete neurons. J Comp Neurol 716:701–716

Memelli H, Torben-Nielsen B, Kozloski J (2013) Self-referential forces are sufficient to explain different dendritic morphologies. Front Neuroinform 7:1–12

Migliore M, Shepherd GM (2002) Emerging rules for the distributions of active dendritic conductances. Nat Rev Neurosci 3:362–370

Migliore M, Shepherd GM (2005) An integrated approach to classifying neuronal phenotypes. Nat Rev Neurosci 6:810–818

Migliore M, Morse TM, Davison AP et al (2003) ModelDB. Neuroinformatics 1:135–139

Moolman DL, Vitolo OV, Vonsattel J-PG, Shelanski ML (2004) Dendrite and dendritic spine alterations in Alzheimer models. J Neurocytol 33:377–387

Myatt DR, Hadlington T, Ascoli GA, Nasuto SJ (2012) Neuromantic—from semi-manual to semi-automatic reconstruction of neuron morphology. Front Neuroinform 6:4

Nagel J, Delandre C, Zhang Y et al (2012) Fascin controls neuronal class-specific dendrite arbor morphology. Development 139:2999–3009

Nowakowski RS, Hayes NL, Egger MD (1992) Competitive interactions during dendritic growth: a simple stochastic growth algorithm. Brain Res 576:152–156

Oberlaender M, Boudewijns ZSRM, Kleele T et al (2011) Three-dimensional axon morphologies of individual layer 5 neurons indicate cell type-specific intracortical pathways for whisker motion and touch. Proc Natl Acad Sci USA 108:4188–4193

Oberlaender M, De Kock CPJ, Bruno RM et al (2012) Cell type-specific three-dimensional structure of thalamocortical circuits in a column of rat vibrissal cortex. Cereb Cortex 22:2375–2391

Peng H, Ruan Z, Long F et al (2010) V3D enables real-time 3D visualization and quantitative analysis of large-scale biological image data sets. Nat Biotechnol 28:348–353

Peters A, Payne B (1993) Numerical relationships between geniculocortical afferents and pyramidal cell modules in cat primary visual cortex. Cereb Cortex 3:69–78

Prusinkiewicz P, Lindenmayer A (1990) The algorithmic beauty of plants. Springer, New York

Rall W (1959) Branching dendritic trees and motorneuron membrane resistivity. Exp Neurol 1:491–527

Ramaswamy S, Hill SL, King JG et al (2012) Intrinsic morphological diversity of thick-tufted layer 5 pyramidal neurons ensures robust and invariant properties of in silico synaptic connections. J Physiol 590:737–752

Rocchi MBL, Sisti D, Albertini MC, Teodori L (2007) Current trends in shape and texture analysis in neurology: aspects of the morphological substrate of volume and wiring transmission. Brain Res Rev 55:97–107

Ropireddy D, Scorcioni R, Lasher B et al (2011) Axonal morphometry of hippocampal pyramidal neurons semi-automatically reconstructed after in vivo labeling in different CA3 locations. Brain Struct Funct 216:1–15

Samsonovich AV, Ascoli GA (2003) Statistical morphological analysis of hippocampal principal neurons indicates cell-specific repulsion of dendrites from their own cell. J Neurosci Res 71:173–187

Scorcioni R, Polavaram S, Ascoli GA (2008) L-Measure: a web-accessible tool for the analysis, comparison and search of digital reconstructions of neuronal morphologies. Nat Protoc 3:866–876

Sholl DA (1953) Dendritic organization in the neurons of the visual and motor cortices of the cat. J Anat 87:387–406

Silver RA (2010) Neuronal arithmetic. Nat Rev Neurosci 11:474–489

Smith TG, Lange GD, Marks WB (1996) Fractal methods and results in cellular morphology–dimensions, lacunarity and multifractals. J Neurosci Methods 69:123–136

Soltesz I (2005) Diversity in the neuronal machine: order and variability in interneuronal microcircuits. Oxford University Press, New York, USA

Srivastava DP, Woolfrey KM, Jones KA et al (2012) An autism-associated variant of Epac2 reveals a role for Ras/Epac2 signaling in controlling basal dendrite maintenance in mice. PLoS Biol 10:e1001350

Tamori Y (1993) Theory of dendritic morphology. Phys Rev E 48:3124–3129

Torben-Nielsen B, Stiefel KM (2010) An inverse approach for elucidating dendritic function. Front Comput Neurosci 4:128

Torben-Nielsen B, Vanderlooy S, Postma E (2008) Non-parametric algorithmic generation of neuronal morphologies. Neuroinformatics 6:257–277

Toroczkai Z (2001) Topological classification of binary trees using the Horton-Strahler index. Phys Rev E 65:1–10

Uylings HB, Ruiz-Marcos A, van Pelt J (1986) The metric analysis of three-dimensional dendritic tree patterns: a methodological review. J Neurosci Methods 18:127–151

Uylings HBM, Smit GJ (1975) The three-dimensional branching structure of cortical pyramidal cells. Brain Res 87:55–60

Van Elburg RAJ, Van Ooyen A (2010) Impact of dendritic size and dendritic topology on burst firing in pyramidal cells. PLoS Comput Biol 6:e1000781

van Pelt J, Schierwagen A (2004) Morphological analysis and modeling of neuronal dendrites. Math Biosci 188:147–155

van Pelt J, Uylings HBM (2011) The flatness of bifurcations in 3D dendritic trees: an optimal design. Front Comput Neurosci 5:54

van Pelt J, Uylings HB, Verwer RW et al (1992) Tree asymmetry—a sensitive and practical measure for binary topological trees. Bull Math Biol 54:759–784

Verwer RW, van Pelt J (1983) A new method for the topological analysis of neuronal tree structures. J Neurosci Methods 8:335–351

Vetter P, Roth A, Häusser M (2001) Propagation of action potentials in dendrites depends on dendritic morphology. J Neurophysiol 85:926–937

Wang Y, Gupta A, Toledo-Rodriguez M et al (2002) Anatomical, physiological, molecular and circuit properties of nest basket cells in the developing somatosensory cortex. Cereb Cortex 12:395–410

Wassle H, Peichl L, Boycott BB (1981) Dendritic territories of cat retinal ganglion cells. Nature 292:344–345

Wen Q, Chklovskii DB (2008) A cost-benefit analysis of neuronal morphology. J Neurophysiol 99:2320–2328

Wen Q, Stepanyants A, Elston GN et al (2009) Maximization of the connectivity repertoire as a statistical principle governing the shapes of dendritic arbors. Proc Natl Acad Sci USA 106:12536–12541

Witten TA, Sander LM (1981) Diffusion-limited aggregation, a kinetic critical phenomenon. Phys Rev Lett 47:1400–1403

Zubler F, Douglas R (2009) A framework for modeling the growth and development of neurons and networks. Front Comput Neurosci 3:25

Zubler F, Hauri A, Whatley AM, Douglas R (2011) An instruction language for self-construction in the context of neural networks. Front Comput Neurosci 5:57

Chapter 2
The Cell Biology of Dendrite Differentiation

Gaia Tavosanis

Abstract The morphology of neuronal dendrites defines the position and extent of input connections that a neuron receives and influences computational aspects of input processing. Establishing appropriate dendrite morphology thus underscores proper neuronal function. Indeed, inappropriate patterning of dendrites is a common feature of conditions that lead to mental retardation. Here, we explore the basic mechanisms that lead to the formation of branched dendrites and the cell biological aspects that underlie this complex process. We summarize some of the major steps that from developmental transcriptional regulation and environmental information modulate the neuron's cytoskeleton to obtain the arborized structures that have fascinated neuroscientists for more than a century.

2.1 Why Is Morphology of Dendrites Relevant?

Neurons of different types display dendrite morphologies that can be strikingly distinct. Some dendrites can be represented by a single unbranched process, while others can reach extreme complexities, as those of Purkinje cells in the cerebellum of a rodent. The resulting wide range of morphological possibilities is tightly related to the pattern of connections that the neuron forms. Some neurons, like *Drosophila* Kenyon cells, make selective contact with a limited number of presynaptic partners, while others tend to cover to a maximal extent the area corresponding to their receptive field. The latter type, or space-filling neurons, can have highly branched dendrites offering surface for synaptic contact to hundreds of thousands of synapses. So the number and distribution of presynaptic partners appears a major determinant

G. Tavosanis (✉)
Deutsches Zentrum für Neurodegenerative Erkrankungen e.V., Center for Neurodegenerative Diseases (DZNE), Carl-Troll-Strasse 31, 53115 Bonn, Germany
e-mail: gaia.tavosanis@dzne.de

H. Cuntz et al. (eds.), *The Computing Dendrite: From Structure to Function*, Springer Series in Computational Neuroscience 11, DOI 10.1007/978-1-4614-8094-5_2,

of the morphology of dendrite trees, suggesting that formation of appropriate contacts underlies correct dendrite differentiation. In addition, basic morphological characteristics of dendrites, including the number of branching points and the thickness and length of dendrites, influence the processing of incoming inputs (see Chap. 10). Thus, the branching pattern of dendrites is tightly linked to connections and processing performed by a neuron of a certain neuronal type.

Given this connection between structure and function, it comes to no surprise that human mental retardation is often accompanied by inappropriately formed neuronal dendrites (Kaufmann and Moser 2000). It is not clear whether the altered dendrite organization is a primary cause of mental retardation or a secondary defect. Genetic analysis of mental retardation syndromes could shed light on this question. This analysis, though, is complicated by the fact that even in cases in which mental retardation is caused by genetic factors, these are often multiple, as in the case of Down syndrome. Fragile X syndrome is a monogenic form of syndromic mental retardation due to lesions in the Fragile X mental retardation protein (FMRP) gene, characterized by immature dendritic spines for instance in layer V pyramidal neurons of the visual cortex (Irwin et al. 2001). Mouse models carrying a deletion of the Fmr1 gene also display immature spines at higher density [reviewed in (De Rubeis et al. 2012)]. Interestingly, *Drosophila FMR1* homologue mutants display dendrites with defective morphology (Lee et al. 2003). It will be interesting to see whether the fly phenotype could help work out some of the basic mechanisms at the origin of the disease. Another form of intellectual disability, Rett syndrome, is linked to mutations in the transcriptional suppressor Mecp2. Mice mutant for Mecp2 display reduced dendrite complexity in layer 2/3 pyramidal cells, a defect that is largely cell autonomous (Amir et al. 1999). Thus, in these two examples, the genes affected encode for proteins that modulate dendrite morphology.

In this chapter, we summarize the basic principles underlying normal dendrite differentiation and the factors that modulate it (Fig. 2.1). Furthermore, we address how variations in dendrite morphology are achieved.

2.2 The Basics of Dendrite Differentiation: Elongation and Branching

Rodent hippocampal neurons in culture differentiate by forming multiple neurites. Among the initially indistinguishable neurites one starts displaying higher actin dynamics at the growth cone and more stabilized microtubules. This process grows faster than the others and will develop into an axon. The axonal fate is suppressed in the other processes, and they will differentiate into dendrites. A similar course is likely to happen also in vivo, as suggested by the analysis in genetically modified mice [reviewed in (Neukirchen and Bradke 2011)].

The initial dendrites extend in a dynamic process that has been observed by time-lapse analysis in various systems, including *Drosophila* larva sensory neurons and tectal neurons of Xenopus tadpole and zebrafish embryo (Hume and Purves 1981;

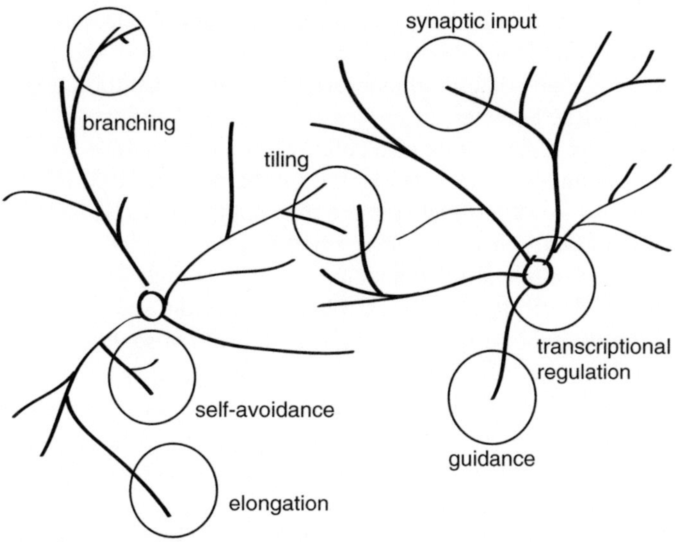

Fig. 2.1 Scheme of the dendrites of two neighboring neurons. Highlighted are the main processes involved in dendrite branching and discussed in this chapter

Kaethner and Stuermer 1997; Wu et al. 1999; Sugimura et al. 2003; Niell et al. 2004; Williams and Truman 2004; Dimitrova et al. 2008). Similar dynamics are shown by the dendrites of newborn neurons that become integrated into the mouse adult olfactory bulb (Mizrahi 2007). Although the systems are different, in all cases it appears that filopodia-like processes dynamically extend and retract. Some of these processes become stabilized and extend into branches (Heiman and Shaham 2010). In zebrafish larva optic tectum neurons, branch stabilization appears to happen after the selective maintenance of an initial synaptic contact (Niell et al. 2004). A recent elegant study combining in vivo time-lapse microscopy and electron microscopy of Xenopus tadpole optic tectum neurons shows a correlation between maturation of synaptic contacts and stabilization of dendrite branches (Li et al. 2011). These observations support a model in which the formation of appropriate synaptic contacts promotes the selective stabilization of subsets of branches (Cline and Haas 2008) (see below).

Forming branches that are contacted by appropriate inputs can thus elongate and become stabilized to extend into dendrites. It is conceivable that a shift from actin-based dynamics to microtubule ingression could accompany this change.

A major determinant of the final morphology of dendrites is their pattern of branching. There are two basic ways of branching that are utilized to different extent by different types of neurons. First, branches can form by the splitting of the growth cone at the tip of an extending dendrite branch. The growth cone of an extending process is decorated by filopodia-like processes in vitro and in vivo. A subset of these filopodia, interestingly usually two, extends into branch-like structures that can acquire in turn a growth cone at their tip and extend further.

Alternatively, lateral branches can be formed along the length of an existing dendrite by interstitial branching. Locally, the organization of the dendrite branch becomes modified, and a lateral filopodium-like process can form and then extend further into a dendrite branch.

In axons, the site of formation of a collateral branch, a process that requires the actin nucleator complex Arp2/3, is marked by the presence of a local actin patch (Spillane et al. 2011). Actin filaments are then suggested to bundle into an initial filopodium. Importantly, it is not clear whether actin patches are also present in dendrites at the site of interstitial branching. Local accumulation of actin predicting the site of lateral branch formation was observed by confocal microscopy in class III sensory neurons of the peripheral nervous system of *Drosophila* larvae (Andersen et al. 2005). Initial observations of actin organization in related class IV neurons of the same system show a very different organization of actin, with no focal accumulation (Lee et al. 2011; Nagel et al. 2012). Further studies combining live imaging and high-resolution microscopy will be necessary to understand the local changes underlying the formation of dendrite branches.

Different neurons utilize these two branching methods to different extents. The combination of initial primary branches formed at the soma, dynamics of extension and retraction of dendrite branches, stabilization of subsets of dendrites, and their branching by growth cone splitting or interstitial branching define the final morphology of the dendrite tree.

These characteristics are in part defined by intrinsic programs established during development, in part result from the interaction of the developing dendrites with the substrate they grow on, the neighboring neurons of same or different type, the guidance signals they encounter, the synaptic contacts they form, and the activity at those contacts.

Even after the completion of differentiation, dendrites can remain dynamic and modify some of their properties to react to injury, to adapt to a modified environment, and to respond to physiological signals including hormones and during learning (Tavosanis 2012).

2.3 The Transcriptional Regulation of Dendrite Differentiation

Some types of neurons in culture seem to preserve part of their morphological characteristics, suggesting an intrinsic determination of dendrite morphology (Montague and Friedlander 1989, 1991). Indeed, neurons are born expressing developmentally defined combinations of transcription factors that will specify their identity, including their functional properties and aspects of their morphology [reviewed in (de la Torre-Ubieta and Bonni 2011)]. In particular, in recent years, it has emerged that transcription factors involved in fate specification display an independent later role in neuronal differentiation, including the establishment of dendrite morphology. The proneural bHLH transcription factor Neurogenin 2, in addition to its role in the specification of deep-layer cortical neurons, controls the

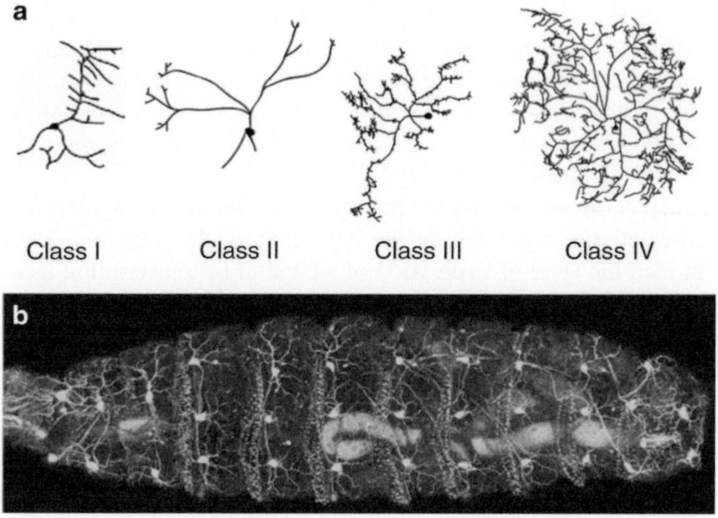

Fig. 2.2 The da neurons of *Drosophila* larvae. (**a**) Tracings of representative da neurons of *Drosophila* third-instar larvae. The four classes of da neurons are shown in order of increasing complexity. (**b**) Live *Drosophila* larva expressing GFP in class IV da neurons, imaged with a confocal microscope

expression of appropriate unipolar dendrite morphology of rodent cortical neurons (Hand et al. 2005), though potentially in a DNA binding-independent manner. In addition, it contributes to the complexity and extension of dendrites of cerebellar Purkinje cells, in part by promoting the expression of genes involved in dendritogenesis (Florio et al. 2012). Similarly, the proneural bHLH transcription factor NeuroD promotes appropriate dendritogenesis of cerebellar granule neurons. These neurons normally display just a few dendrites terminating with typical claws, a morphology that is disrupted in cerebellar slices after NeuroD knockdown (Gaudilliere et al. 2004). The maturation of cerebellar granule cell dendrites is promoted by membrane depolarization in cerebellar slice cultures and could thus depend on activity in vivo. The two transcription factors NeuroD and Sp4, a zinc-finger protein, appear to mediate this maturation downstream of depolarization (Gaudilliere et al. 2004; Ramos et al. 2007).

A comprehensive view of the impact of transcription factor activity on dendrite morphology is afforded by studies of *Drosophila* larva sensory neurons. The da neurons are multimodal sensory neurons that develop dendrite trees in a two-dimensional fashion under the transparent cuticle of the larva. The fact that these neurons can be straightforwardly imaged in live animals and their accessibility to genetic manipulation have made them a most powerful system to understand fundamental aspects of dendrite differentiation (Gao et al. 1999; Jan and Jan 2010). da neurons comprise four morphological and functional categories (Grueber et al. 2002) (Fig. 2.2). In a genome-wide effort, all transcription factors identified in the fly gene genome were knocked down in simple, propioceptive class I neurons (Parrish et al. 2006). 76 of the 730 tested transcription factors produced aberrant dendrite phenotypes, including increased or reduced branching, unbalance between branching and extension, and defective routing (Parrish et al. 2006). While most of the identified factors await detailed analysis, this result certainly supports a major contribution of transcriptional regulation in the establishment of dendrite morphology.

Importantly, comparative analysis between the different da classes revealed a key role of transcriptional regulation in the definition of class-specific dendrite morphology. In fact, the level of expression of a handful of transcription factors determined during earlier development defines, by and large, the distinctions between the four classes of da neurons. The homeodomain-containing transcription factor Cut (a member of the CUX family of transcription factors) is highly expressed in class III da neurons under the control of proneural gene cascades (Blochlinger et al. 1988; Brewster and Bodmer 1995; Brewster et al. 2001; Grueber et al. 2003a). The dendrites of these neurons are decorated by short, straight, and highly dynamic terminal branchlets (Grueber et al. 2002; Nagel et al. 2012). When Cut is genetically removed from these neurons, their typical dendrite characteristics are lost. Conversely, if Cut is expressed at high levels in da neurons that normally do not express it, as class I neurons, their dendrites become more complex and they are now decorated by the short and straight actin-rich branches (Grueber et al. 2003a). Similarly, class I neurons display high levels of expression of the BTB/POZ-zinc-finger protein Abrupt. Loss of Abrupt leads to more complex processes, while Abrupt overexpression in neurons with more complex dendrite organization simplifies their trees (Li et al. 2004). Class IV neurons express the COE family transcription factor Knot/Collier that together with Cut defines their complex morphology (Hattori et al. 2007; Jinushi-Nakao et al. 2007; Crozatier and Vincent 2008). Thus, a combinatorial code of developmentally defined levels of expression of transcription factors is a major determinant of individual class morphology of da neuron dendrites (Parrish et al. 2007b). To understand the mechanism of action of these transcription codes, efforts are being undertaken to identify their downstream effectors. Turtle is an immunoglobulin superfamily protein involved in the regulation of dendrite complexity in class II, III, and IV da neurons of *Drosophila* larvae (Long et al. 2009; Sulkowski et al. 2011). Biochemical and genetic evidence reveals that Turtle expression level is positively regulated by Cut (Sulkowski et al. 2011). Fascin is another downstream effector of Cut. Fascin is a conserved actin-bundling protein, typically associated with filopodia particularly in motile cells, including fly blood cells and metastasizing cancer cells (Zanet et al. 2009; Machesky and Li 2010). In da neurons, it is a specific marker of the short straight terminal branchlets of class III neurons and is required for their formation (Nagel et al. 2012). Fascin is not detected in the dendrites of low-Cut class I neurons. Nonetheless, when Cut is overexpressed in class I neurons, their dendrites become decorated with class III neuron-type short straight branchlets that are now marked by the presence of fascin. In the absence of fascin, Cut cannot operate this morphological transformation in class I neurons, indicating that fascin acts downstream of Cut (Nagel et al. 2012). Nonetheless, whether fascin is directly or indirectly regulated by Cut remains to be demonstrated. Also Knot/

Collier regulates the expression level of a cytoskeletal regulatory protein, the microtubule-severing factor spastin that modulates the complexity of class IV dendrites (Jinushi-Nakao et al. 2007; Ye et al. 2011; Stewart et al. 2012). Spastin expression can also be modulated by Dar1, a transcription regulator of the Krüppel-like factor family (Ye et al. 2011).

In addition to the role of individual transcription factors, chromatin-remodeling complexes modulate dendrite patterning. For instance, Polycomb group genes are transcriptional repressors required to maintain the dendrite arborization of class IV neurons potentially by regulating Bithorax Complex Hox gene expression (Parrish et al. 2007a). The Bap55 chromatin-remodeling factor regulates appropriate dendrite targeting of a subset of olfactory projection neurons during *Drosophila* development (Tea and Luo 2011). Interestingly, related neuron-specific nBAF chromatin-remodeling complexes interact with the calcium-responding transactivator CREST to promote activity-dependent dendrite growth of mouse hippocampal neurons in culture (Wu et al. 2007). CREST supports the extension of the dendrites of cortical layer5 and of hippocampal CA3 neurons in vivo and, with its biochemical interactor CREB-binding protein (CBP), promotes dendrite complexity of cortical neurons in culture after stimulation (Redmond et al. 2002; Aizawa et al. 2004).

Taken together, developmental transcriptional patterning defines major properties of the differentiating neurons, including important aspects of their morphology. However, in an added layer of complexity, dendrites differentiate within a complicated environment, in which they encounter attractive or repulsive signals and interact with substrates or with the branches of same-type or different neurons. The code of intrinsic information that they carry will define how they interact with their environment. An early example of this dialogue between neuron identity and the environment is provided by the response of cortical neurons in the ferret visual cortex to neurotrophin stimulation. In fact, the same neurotrophin alters the dendrite branching pattern in a fashion that is distinctive for pyramidal neurons of different layers (McAllister et al. 1995). Thus, the response of distinct cell types to the same external factor appears to depend on the intrinsic factors unique to each cell type.

2.4 Basic Principles in Dendrite Patterning: Self-Avoidance

While growing and retracting, branches of the same neuron can come in close contact with each other. If the consequence of this contact were adhesion, the whole dendrite tree would collapse. In contrast, branches of the same neuron have a strong tendency to repel each other, allowing dendrites to spread over their appropriate receptive field without crossing each other in a process known as self-avoidance [reviewed in (Grueber and Sagasti 2010)].

All *Drosophila* da neuron classes display self-avoidance, suggesting that this could be a mechanism common to many different neuronal types. *Drosophila* Down syndrome cell-adhesion molecule (Dscam) is a central player promoting self-avoidance in da neurons. The *Dscam* gene, which encodes for an Ig and Fn domain-containing transmembrane protein, is highly complex allowing for the expression of tens of thousands of different isoforms (Schmucker et al. 2000). Dscam is capable of high-affinity homophilic adhesive binding between isoforms containing the same ectodomain variants. Thus, the unique complement of Dscam isoforms expressed by a single neuron represents a sort of individual barcode (Meijers et al. 2007; Wojtowicz et al. 2007; Sawaya et al. 2008; Hattori et al. 2009; Wu et al. 2012). Only branches of the same neuron presenting an overlapping set of Dscam isoforms elicit a repulsive signal. In the absence of Dscam, branches of the same neuron are not repelled and collapse over each other (Hughes et al. 2007; Matthews et al. 2007; Soba et al. 2007). A strikingly similar role to fly Dscam, though with a different molecular makeup, was recently shown for rodent protocadherins. The protocadherin locus encodes 58 transmembrane cadherin-like proteins that are expressed stochastically and display homophilic adhesion. Loss of protocadherins in starbust amacrine cells in the mouse retina leads to collapse of their dendrite tree, while re-expression of a single protocadherin rescues dendrite spreading (Lefebvre et al. 2012). It will now be interesting to identify the cellular pathways that turn the initial homophilic binding of Dscam isoforms or of protocadherins into a repulsive signal.

Vertebrate Dscam and Dscam-like do not display the isoform complexity of insect Dscam. Nevertheless, they are involved in self-avoidance in the mouse retina. In fact, cells that normally express Dscam, including dopaminergic amacrine cells and retinal ganglion cells, have bundled dendrites in the Dscam mutant retina. In addition, although dendrites of same-type cells can overlap quite extensively in the retina, the processes of same-type cells bundle with each other in the mutant. Thus Dscam appears to represent a repulsive signal among same-cell dendrites but could counteract an adhesive signal between same-type cell dendrites (Fuerst et al. 2008, 2009).

Taken together, the presence of a repulsive signal allowing same-cell dendrites to spread extensively appears a common trait of the disparate neuronal types studied so far. To allow for unhindered overlap with dendrites of same-type neurons (or of different types of neurons) while maintaining repulsion of same-cell branches, flies and rodents have evolved systems that guarantee stochastic expression of isoforms capable of homophilic binding. Since only the same barcode guarantees repulsion and there is low chance of finding the same barcode on neighboring neurons, these two opposite requirements can thus be met.

Differentiating dendrites interact extensively with the substrate they grow on. Class IV da neuron dendrites grow on the epidermal layer, and they are enveloped by epidermal cells at discrete positions. Integrins allow dendrites to adhere to the extracellular matrix and thus regulate how much they become enveloped by epidermal cells. Cell-autonomous loss of integrins leads to a broader inclusion of dendrites within the epidermis. Thus, same-neuron dendrites do not come in direct contact anymore and can cross more freely each other on different planes. Therefore, localization to a single plane is a requirement for appropriate coverage of the sensory field in these neurons (Han et al. 2012; Kim et al. 2012).

2.5 The Interaction with Neighboring Neurons: Tiling

Sensory systems might require a complete and unequivocal coverage of a receptive field. Thus, a conserved appearance of these systems is a mosaic tile-like organization of the involved neurons, nicely exemplified by several neuronal types in the mammalian retina, including cholinergic cells (Wässle and Riemann 1978; Galli-Resta 2002). The dendrites of mosaic neurons can overlap to various extents in different systems. In subsets of RGCs and of amacrine cells in the rabbit retina and in class IV da neurons of the *Drosophila* larva, dendritic territories are largely exclusive (Devries and Baylor 1997; MacNeil et al. 1999; Grueber et al. 2002) [extensively reviewed in (Grueber and Sagasti 2010)].

Dendritic tiling among class IV da neurons is likely achieved through local repulsive signals, as the loss of one neuron leaves space for the dendrites of other surrounding same-type neurons to invade the empty area (Grueber et al. 2003a; Sugimura et al. 2003). Several molecules were involved in class IV da neuron tiling in the past years. These include Flamingo, a seven-pass transmembrane cadherin: in *flamingo* mutants the dendrites of class IV neurons overextend and overlap at boundaries (Gao et al. 2000; Kimura et al. 2006). In addition, the *Drosophila* Ser/Thr kinase Tricornered and Furry, two proteins involved in cell morphology control, and their *C. elegans* homologues Sax1 and Sax2 also suppress crosses among same-type neuron dendrite branches (Adler 2002; Emoto et al. 2004; Gallegos and Bargmann 2004). The phosphorylation of Trc in tiling is regulated by the Hippo kinase and by the TORC2 complex (Emoto et al. 2006; Koike-Kumagai et al. 2009). As mentioned above, cell-autonomous loss of integrins makes the dendrites of class IV neurons become largely enveloped by the epidermal cells they grow on. In this situation, no direct contact can be made, and thus neighboring same-type neurons can extend their dendrites over each other on different planes (Han et al. 2012; Kim et al. 2012). Tricornered and TORC2 appear to regulate the level of epidermal envelopment, thus indirectly controlling appropriate tiling (Han et al. 2012).

2.6 Guidance Signals

While extending and retracting, dendrites experience a number of environmental signals that regulate their guidance or their extension and branching.

Dendrites grow to cover their receptive fields and to meet their appropriate partners under the guidance of signals present in their environment.

Semaphorins represent a major family of guidance molecules involved in the directional growth of dendrites. In the rodent cortex the soluble Semaphorin3a produced from the marginal zone covers the dual role of repelling the axon of pyramidal neurons from the pia while attracting their dendrites, thus enforcing the neuron's orientation. The intracellular level of cyclic GMP can shift the guidance response from attractive to repulsive, and so the relative enrichment of soluble guanylate cyclase in the dendrites is at the base of this dual response (Polleux et al. 2000; Komiyama et al. 2007). During the early development of the olfactory system in *Drosophila*, Semaphorins provide important cues for the correct patterning of the dendrites of second-order projection neurons. The level of expression of the transmembrane Semaphorin1a in projection neurons defines the position of dendrites along a dorsolateral-ventromedial axis of the developing antennal lobe (Komiyama et al. 2007). Secreted Semaphorin2a and Semaphorin2b function redundantly as ligands of Semaphorin1a to repel the projection neuron dendrites from the ventromedial region of the antennal lobe (Sweeney et al. 2011).

The conserved soluble Slit ligands and their Robo receptors shape dendrites in vertebrates and flies by promoting dendrite branching or influencing dendrite guidance (Whitford et al. 2002; Gallegos and Bargmann 2004; Furrer et al. 2007; Dimitrova et al. 2008; Brierley et al. 2009; Mauss et al. 2009). *Drosophila* class IV da neurons are space filling: their dendrites branch extensively to achieve high coverage of their receptive field. This coverage is achieved during differentiation via coordination of branching and extension. In *robo* or *slit* mutants this equilibrium is broken: the dendrites extend too fast and branch less and thus do not achieve appropriate coverage of the receptive field (Dimitrova et al. 2008).

Recently, a more subtle form of guidance emerged, in which only subsets of branches of a dendrite tree are positioned at a preferred location but the other branches are not. Branches of class III da neurons project close to the lateral chordotonal organ that expresses attractive Netrin B. The Netrin receptor Frazzeled/DCC in class III neurons promotes dendrite targeting towards the chordotonal organ, but dendrites are kept spread by the action of Dscam, thus avoiding an accumulation of dendrite branches at a single site (Matthews and Grueber 2011).

2.7 Activity Modulates Dendrite Organization During Differentiation

A major piece of information that dendrites receive while growing is whether they are making appropriate connections that lead to functional activity. Activity modulates the morphology of dendrites also in the adult animal (Tavosanis 2012). But during differentiation it might contribute to selecting which branches are stabilized and which are pruned, sculpting the morphology of dendrites. According to the long-standing synaptotrophic hypothesis, dynamic branches become stabilized if they form a functional synaptic connection (Vaughn 1989). Experimental support to this theory was provided by time-lapse experiments in which nascent synapses could be identified using fluorescently tagged synaptic elements or with electron microscopy to reveal sites of synaptic contact [see above; (Li et al. 2011)]. Indeed, dendrites of many different neuronal types appear to grow preferentially towards a source of afferent input (Greenough and Chang 1988; Katz and Constantine-Paton 1988; Kossel et al. 1995; Malun and Brunjes 1996; Inglis et al. 2000).

Supporting the role of activity in shaping dendrite arbors, blocking NMDA receptors genetically strongly affects dendrite remodeling in rodents, yielding longer dendrites of barrelette neurons of the trigeminal principal nucleus and supernumerary primary dendrites of dentate gyrus granule cells (dGCs) (Lee et al. 2005; Espinosa et al. 2009). Contact with input neurons is important for the refinement of a particular serotoninergic neuron in *Drosophila* (Singh et al. 2010). Input activity refines the dendrite tree of the aCC motor neuron in fly larvae and the input circuit of the adult fly mushroom body input region (Kremer et al. 2010). Visual stimulation elicits increased dendrite arbor growth rate of Xenopus optical tectal neurons, but activation of photoreceptors suppresses branching of their postsynaptic partners in the visual system of the fly larva (Sin et al. 2002; Yuan et al. 2011). Thus, although the effect of input activation might vary with the system and with the neurotransmitter involved, input activity seems to be a major factor modeling dendrite guidance, growth, and branching.

Downstream of input activation, calcium concentration within dendrites increases due to direct influx through calcium-permeable receptors, voltage-gated calcium channels, or release from intracellular calcium stores (Lohmann and Wong 2005). The resulting local increase in calcium concentration modulates dendrite dynamics, for instance by stabilizing filopodia of retinal ganglion cells and hippocampal neurons (Lohmann et al. 2002; Lohmann and Wong 2005; Lohmann and Bonhoeffer 2008). Importantly, several transcriptional regulators involved in shaping dendrites are sensitive to variations of calcium levels (see above), showing that in addition to a local transient effect on branch dynamics, calcium increase can induce a more general and lasting control of dendrite morphology.

2.8 The Cytoskeleton: A Dynamic Backbone for Branching

A trait that defines each individual neuron type is the spacing between branching points. This affects the coverage and the overall morphology of dendrites. So how are branching points defined? As mentioned above, work on axonal branching suggests the presence of Arp2/3-dependent actin patches that predict the site at which a lateral branch will be formed (Korobova and Svitkina 2008; Gallo 2011; Spillane et al. 2011). Assuming that a similar mechanism was involved in specifying the position of new branches also in dendrites, then Arp2/3-activating WASP protein recruitment and activation could be a major predictor of branch positioning (Campellone and Welch 2010). The recruitment and activation of WASP family proteins could be a major way that Rac1 Rho family GTPases promote formation of branches, for instance in da neurons (Lee et al. 2003; Andersen et al. 2005; Nagel et al. 2012). Small GTPases of the Rho family orchestrate the branching of dendrites in many different systems, either promoting (Rac1 and Cdc42) or restraining branching (Rho) (Govek et al. 2005). As an alternative to the active recruitment of a complex positively promoting the formation of a branch, it is possible that the capacity to form filopodia is a general property of dendrites that must be actively suppressed, except at specific sites.

In addition to the reorganization of the actin cytoskeleton, microtubules are also involved in the formation of a branch. Microtubules are hollow tubes of 25 nm diameter that extend in a polarized fashion. They are nucleated at their minus end from a template structure, the γ-tubulin ring complex, and then extend by addition of α/β tubulin dimers at their plus end. During collateral branch formation in the axons of hippocampal neurons in culture, microtubules splay at potential sites of lateral branch formation (Gallo 2011). It is possible that this locally destabilizes the microtubules that will then be needed to penetrate into the forming lateral branch. Whether similar phenomena happen in dendrite branching remains to be analyzed. Nonetheless, spastin and katanin-p60-like1, microtubule-severing factors, promote branch formation in class IV da neurons of *Drosophila* larva, suggesting that microtubule destabilization allows for branch formation (Jinushi-Nakao et al. 2007; Stewart et al. 2012). Potentially, severed microtubules can offer templates for de novo microtubule nucleation or elongation. In fact, loss of γ-tubulin in fly class IV da neurons leads to reduced branching (Ori-McKenney et al. 2012). Nonetheless, how and when microtubules become relevant during the process of branch formation remains an open question.

2.9 Microtubules Display Mixed Polarity Within Dendrites

The organization of microtubules within dendrites has been for a long time puzzling. Early electron microscopy data showed that microtubules within the central stretch of a dendrite branch have a mixed orientation in hippocampal neurons in culture (Baas et al. 1988). More recently, this observation has been extended also to fly neurons (Stone et al. 2008). In mitotic cells the γ-tubulin ring complex that nucleates microtubules is attached to the centrosome, formed by a matrix surrounding the centriolar pair, found in a juxta-nuclear position (Gonzalez et al. 1998; Kuijpers and Hoogenraad 2011). Only recently it has been shown that in post-mitotic differentiating hippocampal neurons in culture, the centrosome is not required for microtubule nucleation. Neurons can still differentiate and extend their processes after the centrosome has been ablated. Indeed, in these neurons following chemical depolymerization of existing microtubules, newly forming microtubules can be seen polymerizing in a distributed fashion in the cytoplasm (Stiess et al. 2010). The non-centrosomal nucleation of microtubules during neuronal differentiation is conserved. Flies that lack bona fide centrosomes due to a mutation in the *sas-4* gene display normal neuronal differentiation (Basto et al. 2006; Nguyen et al. 2011).

The fact that microtubules are nucleated in a distributed fashion does not address yet the mechanism leading to their mixed polarity in dendrites. This long-standing question was investigated in fly class IV neurons. In this system, microtubules can be nucleated at Golgi outposts. These are membranous compartments with Golgi characteristics that are localized within dendrites. As dendrites grow, they require membrane and membrane-associated proteins that derive from endoplasmic reticulum and Golgi. The presence of small Golgi elements within dendrites was initially shown by electron microscopy and live cell imaging (Gardiol et al. 1999; Horton and Ehlers 2003). These Golgi outposts are present at dendrite branching points in rodent and fly neurons, and they are required for dendrite polarization and dendrite branch extension (Horton et al. 2005; Ye et al. 2007).

Microtubules nucleated at Golgi outposts in class IV fly neurons can grow with their plus end extending towards the soma, giving rise to mixed microtubule polarity (Ori-McKenney et al. 2012). Whether such a mechanism is used in other neuronal types and in rodent neurons remains to be addressed. In rat hippocampal neurons in culture the Golgi outposts are sparser than in fly neurons and thus might not be enough to represent a widespread platform for microtubule nucleation (Horton and Ehlers 2003; Horton et al. 2005).

The mixed orientation of microtubules in dendrites is fundamental for the distinction between the axon and the dendrites. Molecular motors that translocate along microtubule tracks do so in a polarized fashion: some of them (most kinesins) move towards the plus end of microtubules, while dyneins move towards the minus end of microtubules. Therefore, cargos moving on kinesins or dyneins will be differentially distributed to axon or dendrites in rodent and fly neurons (Zheng et al. 2008; Kapitein et al. 2010). This differential trafficking is important not only for the polarized organization of neurons but also for the appropriate differentiation of dendrites. Indeed, loss of the kinesin MKLP1 disrupts dendrite differentiation in rat hippocampal neurons in culture, and loss of components of the dynein complex strikingly reduces the extension of the dendrite tree in da neurons (Yu et al. 2000; Kuijpers and Hoogenraad 2011).

2.10 Conclusion and Future Perspectives

Research on the mechanisms of dendrite branching has revealed the basic logic of a number of fundamental steps. We now know that neurons are born with a defined transcriptional identity that specifies important aspects of dendrite arborization. Further, transcriptional regulation can be modulated by activity. It will be essential in the next years to identify the cascades that implement the transcriptional codes and how they affect morphology. We have learned the basic principles of how same-neuron dendrite branches avoid overlapping in different systems. Nonetheless, we still need to understand how specific self-recognition is turned into repulsion. Finally, although a number of molecules modulate one aspect or another of branching we still have not addressed the basic mechanisms of formation of a new branch or its stabilization. For instance, how is the cytoskeleton reorganized at a branching point? And are such mechanisms common to all different types of dendrite branches, or are they specific to certain neuronal types? Importantly, high resolution in vivo imaging will help addressing several of these issues.

Molecular understanding of dendrite branching promises to shed light on defects leading to mental function impairment and could suggest ways to promote repair after injury in the nervous system.

Acknowledgements I am grateful to my lab for discussions, to Dr. Linne for Fig. 2.1, and Dr. Nagel for the tracings in Fig. 2.2. I am thankful to the reviewers for their insight and to Dr. Cuntz for his patience. I apologize to the colleagues whose work could not be included due to space reasons. Work in my lab is funded by the DZNE and by the DFG (SPP 1464).

References

Adler PN (2002) Planar signaling and morphogenesis in Drosophila. Dev Cell 2(5):525–535

Aizawa H, Hu SC et al (2004) Dendrite development regulated by CREST, a calcium-regulated transcriptional activator. Science 303(5655):197–202

Amir RE, Van den Veyver IB et al (1999) Rett syndrome is caused by mutations in X-linked MECP2, encoding methyl-CpG-binding protein 2. Nat Genet 23(2):185–188

Andersen R, Li Y et al (2005) Calcium/calmodulin-dependent protein kinase II alters structural plasticity and cytoskeletal dynamics in Drosophila. J Neurosci 25(39):8878–8888

Baas PW, Deitch JS et al (1988) Polarity orientation of microtubules in hippocampal neurons: uniformity in the axon and nonuniformity in the dendrite. Proc Natl Acad Sci USA 85(21):8335–8339

Basto R, Lau J et al (2006) Flies without centrioles. Cell 125(7):1375–1386

Blochlinger K, Bodmer R et al (1988) Primary structure and expression of a product from cut, a locus involved in specifying sensory organ identity in Drosophila. Nature 333(6174):629–635

Brewster R, Bodmer R (1995) Origin and specification of type II sensory neurons in Drosophila. Development 121(9):2923–2936

Brewster R, Hardiman K et al (2001) The selector gene cut represses a neural cell fate that is specified independently of the Achaete-Scute-Complex and atonal. Mech Dev 105(1–2):57–68

Brierley DJ, Blanc E et al (2009) Dendritic targeting in the leg neuropil of Drosophila: the role of midline signalling molecules in generating a myotopic map. PLoS Biol 7(9):e1000199

Campellone KG, Welch MD (2010) A nucleator arms race: cellular control of actin assembly. Nat Rev Mol Cell Biol 11(4):237–251

Cline H, Haas K (2008) The regulation of dendritic arbor development and plasticity by glutamatergic synaptic input: a review of the synaptotrophic hypothesis. J Physiol 586(6):1509–1517

Crozatier M, Vincent A (2008) Control of multidendritic neuron differentiation in Drosophila: the role of Collier. Dev Biol 315(1):232–242

de la Torre-Ubieta L, Bonni A (2011) Transcriptional regulation of neuronal polarity and morphogenesis in the mammalian brain. Neuron 72(1):22–40

De Rubeis S, Fernandez E et al (2012) Molecular and cellular aspects of mental retardation in the Fragile X syndrome: from gene mutation/s to spine dysmorphogenesis. Adv Exp Med Biol 970:517–551

Devries SH, Baylor DA (1997) Mosaic arrangement of ganglion cell receptive fields in rabbit retina. J Neurophysiol 78(4):2048–2060

Dimitrova S, Reissaus A et al (2008) Slit and Robo regulate dendrite branching and elongation of space-filling neurons in Drosophila. Dev Biol 324(1):18–30

Emoto K, He Y et al (2004) Control of dendritic branching and tiling by the Tricornered-kinase/Furry signaling pathway in Drosophila sensory neurons. Cell 119(2):245–256

Emoto K, Parrish JZ et al (2006) The tumour suppressor Hippo acts with the NDR kinases in dendritic tiling and maintenance. Nature 443(7108):210–213

Espinosa JS, Wheeler DG et al (2009) Uncoupling dendrite growth and patterning: single-cell knockout analysis of NMDA receptor 2B. Neuron 62(2):205–217

Florio M, Leto K et al (2012) Neurogenin 2 regulates progenitor cell-cycle progression and Purkinje cell dendritogenesis in cerebellar development. Development 139(13):2308–2320

Fuerst PG, Bruce F et al (2009) DSCAM and DSCAML1 function in self-avoidance in multiple cell types in the developing mouse retina. Neuron 64(4):484–497

Fuerst PG, Koizumi A et al (2008) Neurite arborization and mosaic spacing in the mouse retina require DSCAM. Nature 451(7177):470–474

Furrer MP, Vasenkova I et al (2007) Slit and Robo control the development of dendrites in Drosophila CNS. Development 134(21):3795–3804

Gallegos ME, Bargmann CI (2004) Mechanosensory neurite termination and tiling depend on SAX-2 and the SAX-1 kinase. Neuron 44(2):239–249

Galli-Resta L (2002) Putting neurons in the right places: local interactions in the genesis of retinal architecture. Trends Neurosci 25(12):638–643

Gallo G (2011) The cytoskeletal and signaling mechanisms of axon collateral branching. Dev Neurobiol 71(3):201–220

Gao FB, Brenman JE et al (1999) Genes regulating dendritic outgrowth, branching, and routing in Drosophila. Genes Dev 13(19):2549–2561

Gao FB, Kohwi M et al (2000) Control of dendritic field formation in Drosophila: the roles of flamingo and competition between homologous neurons. Neuron 28(1):91–101

Gardiol A, Racca C et al (1999) Dendritic and postsynaptic protein synthetic machinery. J Neurosci 19(1):168–179

Gaudilliere B, Konishi Y et al (2004) A CaMKII-NeuroD signaling pathway specifies dendritic morphogenesis. Neuron 41(2):229–241

Gonzalez C, Tavosanis G et al (1998) Centrosomes and microtubule organisation during Drosophila development. J Cell Sci 111(Pt 18):2697–2706

Govek EE, Newey SE et al (2005) The role of the Rho GTPases in neuronal development. Genes Dev 19(1):1–49

Greenough WT, Chang FL (1988) Dendritic pattern formation involves both oriented regression and oriented growth in the barrels of mouse somatosensory cortex. Brain Res 471(1):148–152

Grueber WB, Sagasti A (2010) Self-avoidance and tiling: mechanisms of dendrite and axon spacing. Cold Spring Harb Perspect Biol 2(9):a001750

Grueber WB, Jan LY et al (2002) Tiling of the Drosophila epidermis by multidendritic sensory neurons. Development 129(12):2867–2878

Grueber WB, Jan LY et al (2003a) Different levels of the homeodomain protein cut regulate distinct dendrite branching patterns of Drosophila multidendritic neurons. Cell 112(6):805–818

Han C, Wang D et al (2012) Integrins regulate repulsion-mediated dendritic patterning of drosophila sensory neurons by restricting dendrites in a 2D space. Neuron 73(1):64–78

Hand R, Bortone D et al (2005) Phosphorylation of Neurogenin2 specifies the migration properties and the dendritic morphology of pyramidal neurons in the neocortex. Neuron 48(1):45–62

Hattori Y, Sugimura K et al (2007) Selective expression of Knot/Collier, a transcriptional regulator of the EBF/Olf-1 family, endows the Drosophila sensory system with neuronal class-specific elaborated dendritic patterns. Genes Cells 12(9):1011–1022

Hattori D, Chen Y et al (2009) Robust discrimination between self and non-self neurites requires thousands of Dscam1 isoforms. Nature 461(7264):644–648

Heiman MG, Shaham S (2010) Twigs into branches: how a filopodium becomes a dendrite. Curr Opin Neurobiol 20(1):86–91

Horton AC, Ehlers MD (2003) Dual modes of endoplasmic reticulum-to-Golgi transport in dendrites revealed by live-cell imaging. J Neurosci 23(15):6188–6199

Horton AC, Racz B et al (2005) Polarized secretory trafficking directs cargo for asymmetric dendrite growth and morphogenesis. Neuron 48(5):757–771

Hughes ME, Bortnick R et al (2007) Homophilic Dscam interactions control complex dendrite morphogenesis. Neuron 54(3):417–427
Hume RI, Purves D (1981) Geometry of neonatal neurones and the regulation of synapse elimination. Nature 293(5832):469–471
Inglis FM, Zuckerman KE et al (2000) Experience-dependent development of spinal motor neurons. Neuron 26(2):299–305
Irwin SA, Patel B et al (2001) Abnormal dendritic spine characteristics in the temporal and visual cortices of patients with fragile-X syndrome: a quantitative examination. Am J Med Genet 98(2):161–167
Jan YN, Jan LY (2010) Branching out: mechanisms of dendritic arborization. Nat Rev Neurosci 11(5):316–328
Jinushi-Nakao S, Arvind R et al (2007) Knot/Collier and cut control different aspects of dendrite cytoskeleton and synergize to define final arbor shape. Neuron 56(6):963–978
Kaethner RJ, Stuermer CA (1997) Dynamics of process formation during differentiation of tectal neurons in embryonic zebrafish. J Neurobiol 32(6):627–639
Kapitein LC, Schlager MA et al (2010) Mixed microtubules steer dynein-driven cargo transport into dendrites. Curr Biol 20(4):290–299
Katz LC, Constantine-Paton M (1988) Relationships between segregated afferents and postsynaptic neurones in the optic tectum of three-eyed frogs. J Neurosci 8(9):3160–3180
Kaufmann WE, Moser HW (2000) Dendritic anomalies in disorders associated with mental retardation. Cereb Cortex 10(10):981–991
Kim ME, Shrestha BR et al (2012) Integrins establish dendrite-substrate relationships that promote dendritic self-avoidance and patterning in drosophila sensory neurons. Neuron 73(1):79–91
Kimura H, Usui T et al (2006) Potential dual molecular interaction of the Drosophila 7-pass transmembrane cadherin Flamingo in dendritic morphogenesis. J Cell Sci 119(Pt 6):1118–1129
Koike-Kumagai M, Yasunaga K et al (2009) The target of rapamycin complex 2 controls dendritic tiling of Drosophila sensory neurons through the Tricornered kinase signalling pathway. EMBO J 28(24):3879–3892
Komiyama T, Sweeney LB et al (2007) Graded expression of semaphorin-1a cell-autonomously directs dendritic targeting of olfactory projection neurons. Cell 128(2):399–410
Korobova F, Svitkina T (2008) Arp2/3 complex is important for filopodia formation, growth cone motility, and neuritogenesis in neuronal cells. Mol Biol Cell 19(4):1561–1574
Kossel A, Lowel S et al (1995) Relationships between dendritic fields and functional architecture in striate cortex of normal and visually deprived cats. J Neurosci 15(5 Pt 2):3913–3926
Kremer MC, Christiansen F et al (2010) Structural long-term changes at mushroom body input synapses. Curr Biol 20(21):1938–1944
Kuijpers M, Hoogenraad CC (2011) Centrosomes, microtubules and neuronal development. Mol Cell Neurosci 48(4):349–358
Lee A, Li W et al (2003) Control of dendritic development by the Drosophila fragile X-related gene involves the small GTPase Rac1. Development 130(22):5543–5552
Lee LJ, Lo FS et al (2005) NMDA receptor-dependent regulation of axonal and dendritic branching. J Neurosci 25(9):2304–2311
Lee SB, Bagley JA et al (2011) Pathogenic polyglutamine proteins cause dendrite defects associated with specific actin cytoskeletal alterations in Drosophila. Proc Natl Acad Sci USA 108(40):16795–16800
Lefebvre JL, Kostadinov D et al (2012) Protocadherins mediate dendritic self-avoidance in the mammalian nervous system. Nature 488(7412):517–521
Li W, Wang F et al (2004) BTB/POZ-zinc finger protein abrupt suppresses dendritic branching in a neuronal subtype-specific and dosage-dependent manner. Neuron 43(6):823–834
Li J, Erisir A et al (2011) In vivo time-lapse imaging and serial section electron microscopy reveal developmental synaptic rearrangements. Neuron 69(2):273–286
Lohmann C, Bonhoeffer T (2008) A role for local calcium signaling in rapid synaptic partner selection by dendritic filopodia. Neuron 59(2):253–260
Lohmann C, Wong RO (2005) Regulation of dendritic growth and plasticity by local and global calcium dynamics. Cell Calcium 37(5):403–409

Lohmann C, Myhr KL et al (2002) Transmitter-evoked local calcium release stabilizes developing dendrites. Nature 418(6894):177–181

Long H, Ou Y et al (2009) Dendrite branching and self-avoidance are controlled by Turtle, a conserved IgSF protein in Drosophila. Development 136(20):3475–3484

Machesky LM, Li A (2010) Fascin: Invasive filopodia promoting metastasis. Commun Integr Biol 3(3):263–270

MacNeil MA, Heussy JK et al (1999) The shapes and numbers of amacrine cells: matching of photofilled with Golgi-stained cells in the rabbit retina and comparison with other mammalian species. J Comp Neurol 413(2):305–326

Malun D, Brunjes PC (1996) Development of olfactory glomeruli: temporal and spatial interactions between olfactory receptor axons and mitral cells in opossums and rats. J Comp Neurol 368(1):1–16

Matthews BJ, Grueber WB (2011) Dscam1-mediated self-avoidance counters netrin-dependent targeting of dendrites in Drosophila. Curr Biol 21(17):1480–1487

Matthews BJ, Kim ME et al (2007) Dendrite self-avoidance is controlled by Dscam. Cell 129(3):593–604

Mauss A, Tripodi M et al (2009) Midline signalling systems direct the formation of a neural map by dendritic targeting in the Drosophila motor system. PLoS Biol 7(9):e1000200

McAllister AK, Lo DC et al (1995) Neurotrophins regulate dendritic growth in developing visual cortex. Neuron 15(4):791–803

Meijers R, Puettmann-Holgado R et al (2007) Structural basis of Dscam isoform specificity. Nature 449(7161):487–491

Mizrahi A (2007) Dendritic development and plasticity of adult-born neurons in the mouse olfactory bulb. Nat Neurosci 10(4):444–452

Montague PR, Friedlander MJ (1989) Expression of an intrinsic growth strategy by mammalian retinal neurons. Proc Natl Acad Sci USA 86(18):7223–7227

Montague PR, Friedlander MJ (1991) Morphogenesis and territorial coverage by isolated mammalian retinal ganglion cells. J Neurosci 11(5):1440–1457

Nagel J, Delandre C et al (2012) Fascin controls neuronal class-specific dendrite arbor morphology. Development 139(16):2999–3009

Neukirchen D, Bradke F (2011) Neuronal polarization and the cytoskeleton. Semin Cell Dev Biol 22(8):825–833

Nguyen MM, Stone MC et al (2011) Microtubules are organized independently of the centrosome in Drosophila neurons. Neural Dev 6:38

Niell CM, Meyer MP et al (2004) In vivo imaging of synapse formation on a growing dendritic arbor. Nat Neurosci 7(3):254–260

Ori-McKenney KM, Jan LY et al (2012) Golgi outposts shape dendrite morphology by functioning as sites of acentrosomal microtubule nucleation in neurons. Neuron 76(5):921–930

Parrish JZ, Kim MD et al (2006) Genome-wide analyses identify transcription factors required for proper morphogenesis of Drosophila sensory neuron dendrites. Genes Dev 20(7):820–835

Parrish JZ, Emoto K et al (2007a) Polycomb genes interact with the tumor suppressor genes hippo and warts in the maintenance of Drosophila sensory neuron dendrites. Genes Dev 21(8):956–972

Parrish JZ, Emoto K et al (2007b) Mechanisms that regulate establishment, maintenance, and remodeling of dendritic fields. Annu Rev Neurosci 30:399–423

Polleux F, Morrow T et al (2000) Semaphorin 3A is a chemoattractant for cortical apical dendrites. Nature 404(6778):567–573

Ramos B, Gaudilliere B et al (2007) Transcription factor Sp4 regulates dendritic patterning during cerebellar maturation. Proc Natl Acad Sci USA 104(23):9882–9887

Redmond L, Kashani AH et al (2002) Calcium regulation of dendritic growth via CaM kinase IV and CREB-mediated transcription. Neuron 34(6):999–1010

Sawaya MR, Wojtowicz WM et al (2008) A double S shape provides the structural basis for the extraordinary binding specificity of Dscam isoforms. Cell 134(6):1007–1018

Schmucker D, Clemens JC et al (2000) Drosophila Dscam is an axon guidance receptor exhibiting extraordinary molecular diversity. Cell 101(6):671–684

Sin WC, Haas K et al (2002) Dendrite growth increased by visual activity requires NMDA receptor and Rho GTPases. Nature 419(6906):475–480
Singh AP, VijayRaghavan K et al (2010) Dendritic refinement of an identified neuron in the Drosophila CNS is regulated by neuronal activity and Wnt signaling. Development 137(8):1351–1360
Soba P, Zhu S et al (2007) Drosophila sensory neurons require Dscam for dendritic self-avoidance and proper dendritic field organization. Neuron 54(3):403–416
Spillane M, Ketschek A et al (2011) The actin nucleating Arp2/3 complex contributes to the formation of axonal filopodia and branches through the regulation of actin patch precursors to filopodia. Dev Neurobiol 71(9):747–758
Stewart A, Tsubouchi A et al (2012) Katanin p60-like1 promotes microtubule growth and terminal dendrite stability in the larval class IV sensory neurons of Drosophila. J Neurosci 32(34):11631–11642
Stiess M, Maghelli N et al (2010) Axon extension occurs independently of centrosomal microtubule nucleation. Science 327(5966):704–707
Stone MC, Roegiers F et al (2008) Microtubules have opposite orientation in axons and dendrites of Drosophila neurons. Mol Biol Cell 19(10):4122–4129
Sugimura K, Yamamoto M et al (2003) Distinct developmental modes and lesion-induced reactions of dendrites of two classes of Drosophila sensory neurons. J Neurosci 23(9):3752–3760
Sulkowski MJ, Iyer SC et al (2011) Turtle functions downstream of Cut in differentially regulating class specific dendrite morphogenesis in drosophila. PLoS One 6(7):e22611
Sweeney LB, Chou YH et al (2011) Secreted semaphorins from degenerating larval ORN axons direct adult projection neuron dendrite targeting. Neuron 72(5):734–747
Tavosanis G (2012) Dendritic structural plasticity. Dev Neurobiol 72(1):73–86
Tea JS, Luo L (2011) The chromatin remodeling factor Bap55 functions through the TIP60 complex to regulate olfactory projection neuron dendrite targeting. Neural Dev 6:5
Vaughn JE (1989) Fine structure of synaptogenesis in the vertebrate central nervous system. Synapse 3(3):255–285
Wässle H, Riemann HJ (1978) The mosaic of nerve cells in the mammalian retina. Proc R Soc Lond B Biol Sci 200(1141):441–461
Whitford KL, Marillat V et al (2002) Regulation of cortical dendrite development by Slit-Robo interactions. Neuron 33(1):47–61
Williams DW, Truman JW (2004) Mechanisms of dendritic elaboration of sensory neurons in Drosophila: insights from in vivo time lapse. J Neurosci 24(7):1541–1550
Wojtowicz WM, Wu W et al (2007) A vast repertoire of Dscam binding specificities arises from modular interactions of variable Ig domains. Cell 130(6):1134–1145
Wu GY, Zou DJ et al (1999) Dendritic dynamics in vivo change during neuronal maturation. J Neurosci 19(11):4472–4483
Wu JI, Lessard J et al (2007) Regulation of dendritic development by neuron-specific chromatin remodeling complexes. Neuron 56(1):94–108
Wu W, Ahlsen G et al (2012) Complementary chimeric isoforms reveal Dscam1 binding specificity in vivo. Neuron 74(2):261–268
Ye B, Zhang Y et al (2007) Growing dendrites and axons differ in their reliance on the secretory pathway. Cell 130(4):717–729
Ye B, Kim JH et al (2011) Differential regulation of dendritic and axonal development by the novel Kruppel-like factor Dar1. J Neurosci 31(9):3309–3319
Yu W, Cook C et al (2000) Depletion of a microtubule-associated motor protein induces the loss of dendritic identity. J Neurosci 20(15):5782–5791
Yuan Q, Xiang Y et al (2011) Light-induced structural and functional plasticity in Drosophila larval visual system. Science 333(6048):1458–1462
Zanet J, Stramer B et al (2009) Fascin is required for blood cell migration during Drosophila embryogenesis. Development 136(15):2557–2565
Zheng Y, Wildonger J et al (2008) Dynein is required for polarized dendritic transport and uniform microtubule orientation in axons. Nat Cell Biol 10(10):1172–1180

Chapter 3
Archetypes and Outliers in the Neuromorphological Space

Cesar H. Comin, Julian Tejada, Matheus P. Viana, Antonio C. Roque, and Luciano da F. Costa

Abstract Neuromorphology has a long history of meticulous analysis and fundamental studies about the intricacies of neuronal shape. These studies converged to a plethora of information describing in detail many neuronal characteristics, as well as comprehensive data about cell localization, animal type, age, among others. Much of this information has notably been compiled through efforts of the Computational Neuroanatomy Group at the Krasnow Institute for Advanced Study, George Mason University, thus originating the NeuroMorpho.org repository, a resource that incorporates a large set of data and related tools. In the current work we present a methodology that can be used to search for novel relationships in cell morphology contained in databases such as the NeuroMorpho.org. More specifically, we try to understand which morphological characteristics can be considered universal for a given cell type, or to what extent we can represent an entire cell class through an archetypal shape. This analysis is done by taking a large number of characteristics from cells into account, and then applying multivariate techniques to analyze the data. The neurons are then classified as archetypes or outliers according to how close they are to the typical shape of the class. We find that granule and medium spiny neurons can be associated with a typical shape, and that different animals and brain regions show distinct distributions of shapes.

C.H. Comin • M.P. Viana • L.d.F. Costa (✉)
Institute of Physics at Sao Carlos, Sao Carlos, Brazil
e-mail: ldfcosta@gmail.com

J. Tejada • A.C. Roque
Department of Physics, School of Philosophy, Science and Letters of Ribeirão Preto, University of São Paulo, Ribeirão Preto, Brazil

H. Cuntz et al. (eds.), *The Computing Dendrite: From Structure to Function*, Springer Series in Computational Neuroscience 11, DOI 10.1007/978-1-4614-8094-5_3,

3.1 Introduction

There are many studies in the literature showing the influence of neuronal morphology on the respective electrical properties (Carnevale et al., 1997) and firing rates (Krichmar et al., 2002; van Elburg and van Ooyen, 2010; Tejada et al., 2012), regulation of network connectivity (Scorcioni et al., 2004; Chklovskii, 2004) and synaptic integration (Poirazi et al., 2003; Zhou et al., 2013) of neurons. These works were sparked by the seminal work of Ramón y Cajal, who described with great detail the structure of a vast number of neuronal cells. Subsequent studies aimed at tracing the shapes of neuronal cells provided surplus information describing a large number of neuronal characteristics. Some interesting approaches in this area include Sholl analysis (Sholl, 1953; Ristanović et al., 2006; Milosević and Ristanović, 2007), the characterization by fractal dimension (Montague and Friedlander, 1991), the influence area analysis (Toris et al., 1995), and the dendrogram representation (Poznanski, 1992). There are also more recent works concerning the development of computational neuromorphometry (Costa et al., 2002; Rodrigues et al., 2005). Such information made possible the categorization of neurons according to their overall shape, localization, development stage, animal species, and others.

However, the proper study of neuronal morphology requires a large amount of information in order to correctly characterize a group of cells. The NeuroMorpho.org repository (Ascoli, 2002; Ascoli et al., 2007) plays an important role on this matter, as its mission is to build an online repository with well-organized neuromorphological information of as many cells as possible. It offers three-dimensional reconstructions of many types of neurons from different brain regions and animal species. The repository is maintained by the Computational Neuroanatomy Group (Krasnow Institute for Advanced Study, George Mason University), which analyzes each new submitted model before its publication, standardizing its format. The NeuroMorpho.org database offers interesting information that can be used to explore the space of parameters that characterizes neurons. This large database was used in a previous work to study the distribution of neuronal geometry (Costa et al., 2010), which proposed a multidimensional projection technique to map each of the NeuroMorpho.org cells into a space of all possible shapes that a neuron could present, showing that real neurons occupy only a small fraction of this space. Nevertheless, that work considered all neurons available at NeuroMorpho.org, which prevented a detailed investigation of the relationship between subgroups of cells inside the same cell class. The main motivation for doing such an investigation is to approach questions such as if two neurons can be considered to be of the same type despite being traced by different laboratories, or how similar are interneurons of rats and mice when compared to interneurons of, for example, a blowfly.

Here we use a multivariate technique of visualization and classification, known as Principal Component Analysis (PCA) (Hardle and Simar, 2012), to study the morphological characteristics of groups of neuronal cells. Our main point of interest is in searching for typical shapes that can represent an entire class of neurons, which we call archetypes, as well as morphological features of cells that are far from being archetypal and can be understood as outlier cells. We seek to discover similarities and differences inside cell classes when considering traditional metrics present in the literature, as explained in Sect. 3.2.

As an initial step towards the planned investigation of the entire NeuroMorpho.org repository, we decided to study three large classes of neurons, namely interneurons, granule, and medium spiny cells. These classes represent 23 % of the models available in NeuroMorpho.org and are known for having distinct morphological characteristics. Through these three classes we seek to answer the next two questions: (1) Are there morphologies that can be considered as archetypes? (2) Is it possible to associate common characteristics to the cell models considered to be outliers?

In order to present our methodology we first define basic concepts related to morphometry characterization and multivariate analysis. These concepts will be presented in the next section.

3.2 Materials and Methods

3.2.1 Data Acquisition: The NeuroMorpho.org Database

The first version of the NeuroMorpho.org repository was released on 1st August 2006 (Ascoli et al., 2007) with the main goal of making publicly available digital reconstructions of neuronal cells. The repository is very complete and well organized, in the sense that it provides many useful information, such as the cell type, brain region, and animal species for each cell. In addition, 20 morphological measurements are provided for each neuron. All of these measurements were extracted by using the software *L–measure*, following the methods described in Halavi et al. (2008) and Scorcioni et al. (2008a). Both metadata and morphological measurements can be seen in HTML format at the web page of each cell. For instance, in order to access the information of the neuron *n10fts*, one can go to the address http://neuromorpho.org/neuroMorpho/neuron_info.jsp?neuron_name=n10fts, where all information will be shown. Thanks to this pattern used for the web page address of all cells in the database, it is easy to automatically collect the desired information of all cells. In order to obtain the name of the cells, it is possible to use the search engine available in the NeuroMorpho.org website.

3.2.2 Morphological Measurements

In Table 3.1, we list all the 20 measurements collected from the NeuroMorpho.org repository. Since these measurements are well established in the field, we will not describe them here, but their detailed information and significance can be found in Ascoli (2002), Scorcioni et al. (2004, 2008b), and Costa et al. (2010).

Table 3.1 Morphological measurements applied

Measurements	
Soma surface	Total volume
Number of stems	Max Euclidean distance
Number of bifurcations	Max path distance
Number of branches	Max branch order
Overall width	Average contraction
Overall height	Total fragmentation
Overall depth	Partition asymmetry
Average diameter	Average Rall's ratio
Total length	Average bif. angle local
Total surface	Average bif. angle remote

3.2.3 Data Processing

Considering the set of morphological measurements defined above, we have a 20-dimensional space. This high dimensionality makes the visualization and analysis of the data a hard task. In order to decrease the total dimension of our problem, we employed the PCA technique (Hardle and Simar, 2012) that reduces the dimensionality from m to n with minimal loss of information by exploring the variance of the data. In this paper, we assume $n = 2$ and then the data is always projected into a two-dimensional plane whose axes are defined by the two new variables obtained through PCA, which we call PCA1 and PCA2. We chose to project into two dimensions because besides being easier to visualize the results, we found that in all cases the first two axis explained at least 50 % of the variation of the data, while the third axis only added 10 % of explanation.

Upon having the projected data, it is possible to estimate its probability density function by using a technique called kernel density estimation (Duda et al., 2000). Basically, we consider the projected points as Kronecker deltas and convolute these deltas with a Gaussian kernel with given standard deviation σ. The result is a smooth continuous function that tells us not only the probability density of the measured points but also the estimate probabilities over entire space. Note that this is significantly distinct from simply taking the histogram of the data, as besides using the frequency of occurrence we are also taking into account the typical position of the points. If there is a high concentration of points in some region *and* they are close together, this region holds a higher probability of representing the neurons than would be otherwise suggested by the histogram.

3.2.4 Morphology Projection

Our entire database consists of 1,260 interneurons, 426 medium spiny, and 341 ganglion cells, but since some reconstructions in NeuroMorpho.org show additional information about the neurons besides the cell compartments (e.g., spatial localization guides), 72 models had to be discarded (1 medium spiny cell and 71 interneurons).

Therefore, the analyzed data consists of 1,189 interneurons, 425 medium spiny, and 341 ganglion cells. In addition, considering that not all reconstructions had the axonal morphology described, only the soma and the basal and apical dendritic trees were used to obtain the morphological characteristics.

After applying the morphological metrics described in Table 3.1 we created high dimensional spaces that were projected into two dimensions by using PCA. We ended up with one 2D space for each type of cell. Then we estimated the probability density functions through kernel density estimation for these 2D spaces, which allowed us to seek for relationships between high or low density morphological regions.

For each cell on the 2D space we observe what is the probability density of the region where the cell was projected into, that is, we look at the value of the probability density function at the point (PCA1(i),PCA2(i)), where PCA1(i) is the value of the first PCA component and PCA2(i) is the value of the second PCA component for cell i. Through this procedure, we obtain the morphological probability density for each cell, which can be used to classify the cell as an archetype or outlier. If the cell is in a region of high probability, it means that its morphology can be used to characterize the overall shape of the class it belongs to, allowing to confirm if the typical picture one has about the studied class is in fact correct. Additionally, since we project the data in 2D, it is possible to color the points representing the cells according to some other characteristic (i.e., animal species, brain region, or laboratory) and visualize their influence on the projected position of the cell.

In order to compare cells reconstructed from different laboratories, it is important to notice that size-related measurements may represent a source of bias because they are strongly affected by distinct experimental conditions (Scorcioni et al., 2004). Therefore, it is advisable that an initial analysis is done in order to verify for trivial morphological differences. In the case of the three cell classes studied here, we found that there was not a trivial feature responsible for significant differences between laboratories. Nevertheless, in case a size-related measurement, e.g., the average diameter, is found to strongly influence the result, a good approach would be to separate the cells according to size (e.g., small, medium, and large) and project each group separately.

3.3 Results

We first consider the class of granule cells, of which the PCA projections and classifications are shown in Fig. 3.1. In Fig. 3.1a, cells colored with dark gray are considered to be outliers, while archetype cells are represented in black. The projection shows high dispersion with three nearly distinct groups, although only one of them have archetypal cells. An example of a typical archetype granule cell is shown in Fig. 3.1b. In order to identify the main features of each group in the projection, we used the meta information provided by NeuroMorpho. By separating the cells according to species (Fig. 3.1c), one of the groups is mainly composed by neurons of mice, while rat neurons are scattered among the three groups. At first sight this can mean that granule cells show more variation in rats than in mice, but we also

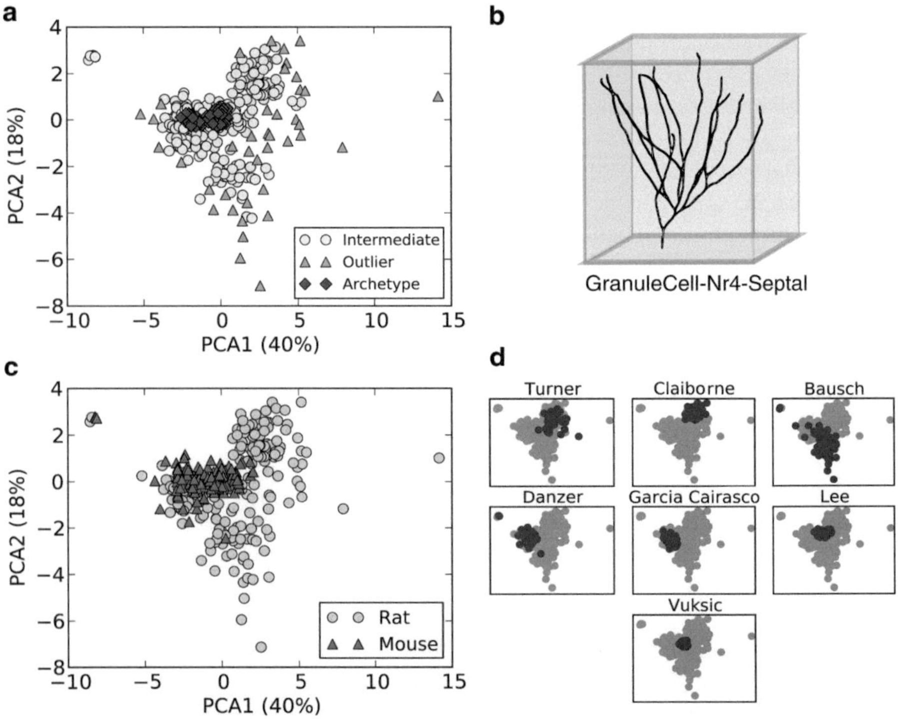

Fig. 3.1 Processing steps for the analysis of the granule cells. (**a**) The 2D PCA projection, showing the identified archetypes and outliers in different *colors* and *marks*. Also shown is the PCA significance for each axis. (**b**) Example of archetype neuron. (**c**) The same projection as in (**a**), but with points colored according to the animal species where the neurons were taken from. (**d**) Again, the same projection as in (**a**), but showing in *dark gray* the different laboratories where neurons were traced (see footnote 1)

have to consider that the experimental procedures used to acquire these neurons are not the same for all cells.

In Fig. 3.1d we color the points according to the laboratory where the data were produced (Rihn and Claiborne, 1990; Cannon et al., 1998; Bausch et al., 2006; Arisi and Garcia-Cairasco, 2007; Vuksic et al., 2008; Carim-Todd et al., 2009; Pierce et al., 2011; Murphy et al., 2011; Gupta et al., 2012). As expected, cells obtained with the same experimental conditions are close to one another in the projection. What is interesting to study here is the relationship between the laboratories. Most cells of the work of Bausch et al. (2006) seem to form a group on their own, and this happens because they are hippocampal culture cells treated chronically with a sodium channel blocker, which rendered some cells far from what would be expected to be a typical granule cell. Another group is formed by cells from Turner and Claiborne,[1] which were all acquired from rats, and may represent the true archetypal cells of this species. The third group is formed by mouse cells from Danzer, Lee, and Vuksic and rat cells from Garcia-Cairasco. Interestingly, the rat cells from

[1] Here we use the lab names that appear on NeuroMorpho.org.

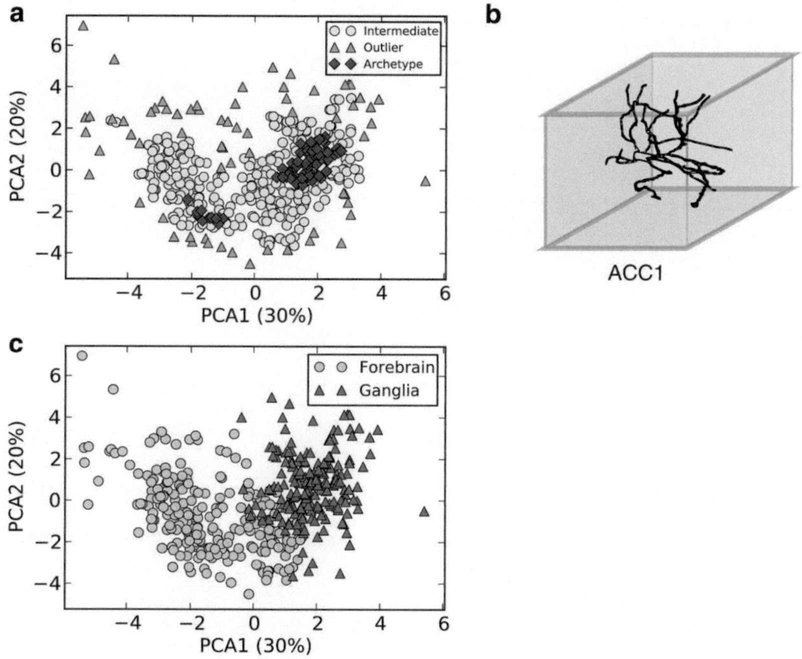

Fig. 3.2 Processing steps for the analysis of the medium spiny cells. (**a**) The 2D PCA projection, showing the identified archetypes and outliers in different *colors* and *marks*. Also shown is the PCA significance for each axis. (**b**) Example of archetype neuron. (**c**) The same projection as in (**a**), but with points colored according to the brain region (basal ganglia and basal forebrain) where the neurons were acquired from

Garcia-Cairasco's group have altered morphology, very likely a consequence of *status epilepticus* induced by pilocarpine (Arisi and Garcia-Cairasco, 2007), which caused them to be more similar to mouse cells.

In Fig. 3.2 we show the projections for the medium spiny neurons. We see that they form two nearly distinct groups. In the same way of the granule cells, the two groups are related to the type of cells and the laboratories from which the cells were reconstructed. One group is composed of neurons from the basal forebrain, reconstructed by Smith, and the other has mostly basal ganglia neurons from Kellendonk. It is also possible that the cells considered as archetypes correspond to cells from control animals, while the outliers are cells that showed morphological alterations, as reported in the original papers (McDonald et al., 2005; Cazorla et al., 2012).

In contrast with the previous results, the interneurons yielded a large group (Fig. 3.3), with few separated cells. This is an interesting result, since interneurons are the cells we could expect to be the most diverse. If we color the points according to the two main brain regions the neurons were acquired from (Fig. 3.3c), we see that there is some discrimination, but it is not strong. By separating the cells according to subclasses described in the NeuroMorpho repository, we see that there are still no noticeable differences between the projections. The only exception is the tangential

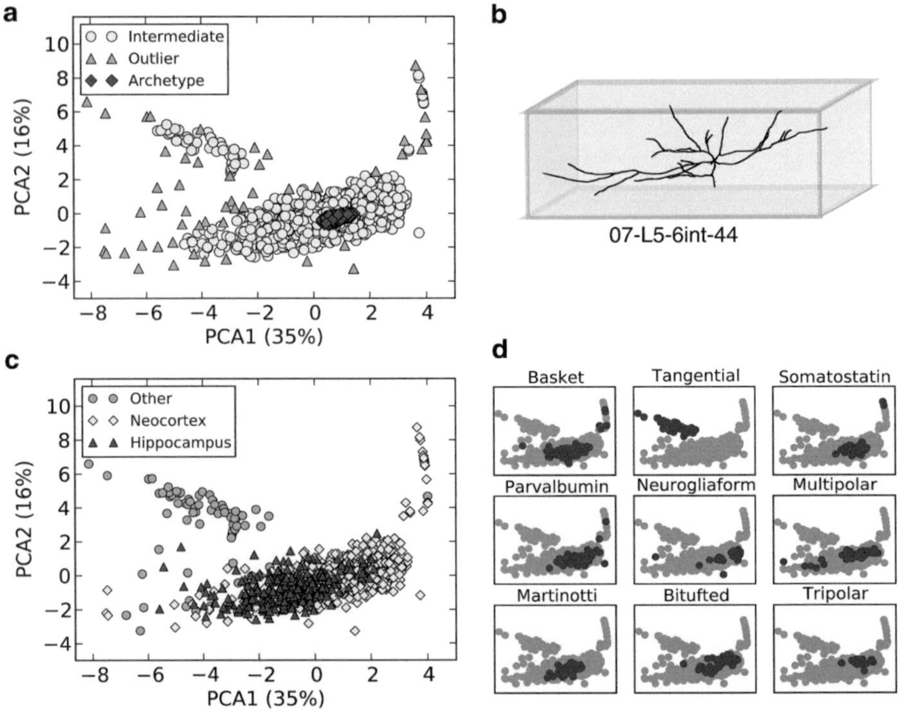

Fig. 3.3 Processing steps for the analysis of the interneuron cells. (**a**) The 2D PCA projection, showing the identified archetypes and outliers in different *colors* and *marks*. Also shown is the PCA significance for each axis. (**b**) Example of archetype neuron. (**c**) The same projection as in (**a**), but with points colored according to the brain region where the neurons were taken from. (**d**) Again, the same projection as in (**a**), but showing in *dark gray* the different subclasses of interneurons according to NeuroMorpho

cells, but they are the only cells obtained from the blowfly, while most of the others are from rats or mice. The reason for not observing clear separations between interneurons probably comes from the fact that they are so diverse that it is difficult to define a characteristic shape for them, even if we consider specific subtypes, i.e., neurons of the same subtype are so diverse that even they are not able to form distinct groups in the projection. This means that it is unlikely that the archetype shown in Fig. 3.3b correctly represents the overall shape of an interneuron.

Concerning only the archetypes and outliers found in the three classes, the importance of the groups found is clear: the archetypes of the granule and medium spiny cells have the typical shape of their cell class and can be used (and usually are) as examples of what characteristics are expected to be found in the respective cell class. The outliers are usually related to morphological alterations provoked by disease or trace of the cell. They may show slight deviations from the cell class, usually in the form of specific features that are not visually apparent but completely change some other feature (e.g., a neuron having slightly larger mean compartment radius and

Table 3.2 Morphological characterizations of the archetypes of the three types of cell analyzed

Measurements	Interneurons	Medium spiny	Granule
Soma surface (μm^2)	721.46(280.15)	638.70(256.25)	471.11(205.77)
Number of stems	4.92(1.79)	5.99(2.31)	1.25(0.52)
Number of bifurcations	23.38(11.90)	17.13(6.58)	10.53(2.75)
Number of branches	51.62(24.48)	39.76(13.53)	22.30(5.23)
Overall width (μm)	204.08(76.47)	166.41(39.83)	156.77(48.41)
Overall height (μm)	300.14(115.94)	160.46(32.23)	198.96(33.85)
Overall depth (μm)	73.24(74.00)	76.71(27.75)	38.81(23.21)
Average diameter (μm)	0.64(0.37)	0.40(0.33)	0.93(0.64)
Total length (μm)	2287.85(772.39)	1778.82(519.74)	1562.05(601.52)
Total surface (μm^2)	4843.67(2265.38)	2132.52(976.02)	5597.07(4344.29)
Total volume (μm^3)	3788.25(1988.49)	2310.17(1175.50)	3359.65(2452.29)
Max Euclidean distance (μm)	229.93(50.55)	143.03(21.81)	235.83(26.22)
Max path distance (μm)	115.78(25.28)	81.49(10.59)	140.41(22.49)
Max branch order	5.87(2.42)	4.30(1.35)	4.56(0.96)
Average contraction	0.84(0.05)	0.84(0.03)	0.91(0.03)
Total fragmentation	1948.26(2250.36)	841.88(354.94)	1002.78(1036.20)
Partition asymmetry	0.50(0.08)	0.45(0.09)	0.43(0.14)
Average Rall	1.47(0.71)	0.36(0.71)	1.94(1.42)
Avg bif. angle local	78.27°(10.10)	77.95°(11.08)	66.12°(9.13)
Avg bif. angle remote	65.39°(11.43)	70.23°(11.45)	33.79°(5.64)

The first number is the mean value and the value within parenthesis corresponds to the standard deviation

length can have a much larger volume), or have a completely different morphology. Nevertheless, it is worth noting that the classification as an outlier can be an effect of a small sample size of the respective cell class in the NeuroMorpho repository, so care must be taken to not classify a shape as atypical simply because it is not usually submitted to the repository (the interneurons of the blowfly in Fig. 3.3 are an example of this). Finally, an important information that can be obtained for the outliers is their distance to the archetype of the associated control group, which represents a measure of the impact of the changes on the cell morphology.

In Table 3.2 we show the mean and standard deviation for measurements taken on the archetypes of the three types of cells analyzed (examples of these cells are shown in panel b of Figs. 3.1–3.3). As is expected for granule cells, they have the smallest number of stems and bifurcations when compared with interneurons or medium spiny cells. At the same time, looking for the overall width, height, and depth of the interneurons, we observe that they are usually bigger and more planar, as can be seen in the example of the morphological archetype shown in Fig. 3.3b.

Considering a visual inspection of the cells classified as morphological archetypes, the groups obtained with the PCA were consistent with the main morphological characteristics of the three cell types. The archetype of the granule cells have the typical symmetric cone-shaped dendritic tree (Cannon et al., 1999; Andersen et al., 2006), the medium spiny cells of the archetype group have the characteristic spherical dendritic branch (Preston et al., 1980; Bishop et al., 1982), and the

Table 3.3 Morphological characterizations of the outliers of the three types of cell analyzed

Measurements	Interneurons	Medium spiny	Granule
Soma surface (μm²)	2930.41(3114.72)	860.35(517.94)	876.02(882.46)
Number of stems	4.74(5.83)	5.56(2.69)	2.38(1.39)
Number of bifurcations	190.00(585.39)	18.58(13.16)	16.88(8.30)
Number of branches	384.70(1169.99)	42.48(27.92)	35.63(16.65)
Overall width (μm)	551.30(387.28)	214.00(102.30)	291.27(176.70)
Overall height (μm)	619.36(418.41)	202.65(99.87)	286.38(177.35)
Overall depth (μm)	115.98(146.55)	80.25(41.59)	95.27(95.26)
Average diameter (μm)	2.31(2.90)	0.64(0.49)	1.19(0.59)
Total length (μm)	8708.91(8277.31)	1948.45(942.43)	2915.05(1948.36)
Total surface (μm²)	44166.33(44134.98)	4075.09(3232.89)	11287.89(7904.89)
Total volume (μm³)	59150.61(73045.68)	4897.61(4799.38)	9019.18(10253.03)
Max Euclidean distance (μm)	530.06(325.97)	189.56(85.40)	310.27(153.05)
Max path distance (μm)	346.82(257.60)	98.35(32.86)	191.80(94.56)
Max branch order	15.64(19.57)	4.45(1.89)	5.93(2.87)
Average contraction	0.86(0.13)	0.85(0.04)	0.83(0.15)
Total fragmentation	6590.36(12772.60)	1276.95(986.31)	1940.60(1449.17)
Partition asymmetry	0.60(0.18)	0.47(0.11)	0.50(0.14)
Average Rall	1.42(0.83)	0.84(1.04)	1.64(0.83)
Avg Bif. angle local	68.24°(27.34)	77.87°(11.99)	75.57°(20.90)
Avg Bif. angle remote	58.30°(25.11)	66.42°(14.63)	44.55°(16.16)

The first number is the mean value and the value within parenthesis corresponds to the standard deviation

interneurons classified as archetype present a spread out dendritic tree, a common characteristic of interneurons connected with pyramidal cells (Andersen et al., 2006). The names of the archetypes and outliers found for each cell class are included in Appendix of this text.

In contrast, Table 3.3 shows the mean and standard deviation for measurements taken on the outliers found for each one of the cell types. It is clear that the values are very different from those found for the archetypes, specially regarding measurements related to size. That is, outlier cells seem to be much larger than the typical size of the class. As expected, the standard deviation of the values is also large, meaning that outliers have many different shapes and sizes and cannot be represented by characteristic structures in the same way that archetypes can.

3.4 Conclusion

We have reported a technique that can be used to analyze a large number of neuronal data. We believe that each procedure presented, namely PCA, kernel density estimation and NeuroMorpho data acquisition can be used to great extent, both individually and in combined form, to study the morphology of neural cells. Here we presented one of many ways in which these tools can be integrated, and exemplified the methodology for three classes of neurons: granule, medium spiny, and interneuron.

We found that for granule cells it is possible to define an archetypal shape of the class, mainly related to a cone-shaped structure. These cells also have profound differences depending if they were obtained from rats or mice, with the interesting property that rat cells altered by *status epilepticus* become more similar to mouse cells. This means that the methodology can be used to quantify to which extent genetic or environmental alterations can modify the overall morphology of the cell when compared to the archetypal shape of the class to which it belongs.

Clearly, it is also possible for a given cell type to exhibit more than one characteristic morphology. In such cases the methodology presented here can help ascertaining if the traditional classification used for the studied cells is in fact differentiating between different morphologies or, more importantly, may even help in the creation of new categories of neurons that are too different from the original class. Here we showed that this may be the case for medium spiny cells occurring in different brain regions.

Also, it can happen that no trivial classification scheme can be defined solely based on the morphology of the cell. This happens for cases where the topology is so diverse that there is actually no motivation in separating cells by small details that can easily change between species or brain regions. This property is clear for the interneurons, which present a large diversity. But even if a given cell class has a typical shape, it is still difficult to define a single structure that can completely represent the class across a variety of animal species or brain areas. Nevertheless, it is still possible to find combinations of characteristics that are similar for the entire class. The PCA method was presented here as the heart of this idea.

Of particular interest is the effect of experimental conditions on the groups observed for each class. In Figs. 3.1 and 3.3 it is clear that the laboratories had an influence on the morphology of the reconstructed cells. This is an interesting result, since, as indicated in Sect. 3.2.4, no trivial feature strongly discriminates between laboratories. Another means to search for trivial effects is to look at the contribution of specific features on the PCA axes and observe if some feature has a prominent influence on the result. In our case, no feature had a significantly larger influence on the projection, meaning that the separation observed between laboratories is caused by a combination of distinct morphological characteristics.

We also note that the method can be applied in a multiscale fashion. That is, after doing the projection for the entire class, it is possible to define a subgroup of cells through some criteria (e.g., brain region, cells near a group of archetypes) and define a new axis of projection based on this group. This new projection can bring novel insights about the data, like the presence of groups of cells from the same brain region. This process can be repeated until there are too few cells to allow for a good projection. But now, since there are just a few cells to characterize, it is possible to study them individually and look at specific measured characteristics.

The minimum amount of cells necessary for a proper definition of the new axis depends on the distribution of the data. If there are seven or eight points scattered in a 2D space, it is difficult to conclude whether they are forming subclasses or are just a single class that has a large variance. On the other hand, if we have two compact groups with four points each, and a large distance between the groups, it is more plausible to conclude that the data has two subclasses. This means that if the amount of data is small, the projection has to be carefully analyzed with respect to the variance of the observed groups. Still, since the NeuroMorpho repository has a sufficiently large database, it is possible to apply the presented methodology to multiple scales of the data.

Acknowledgements CHC is grateful to FAPESP (2011/22639-8 and 2011/50761-2) for sponsorship. JT is grateful to CNPq (560353/2010-3) and FAPESP (2012/17057-2) for a postdoctoral scholarship. MPV thanks to FAPESP for financial support (2010/16310-0). ACR is grateful to CNPq (306040/2010-7) for financial support. LdFC is grateful to FAPESP (05/00587- 5 and 2011/50761-2) and CNPq (301303/06-1 and 573583/2008-0) for the financial support.

Appendix: Tables with the Names of the Archetypes and Outliers Found

Table 3.4 Names of the *archetypes* identified in each cell class

Interneurons		Medium spiny		Granule
020315-2-ST	C060400B2	050803b_finaltrace	WT-P90-01	040726TTX1
020801-3-ST	C060400C1	050803c_finaltrace	WT-P90-03	DGC-610
030223-4-PV	C070600A4	31-9b5AKE	WT-P90-08-	DGC-614
030224-4-PV	C070600B2	35-8a5AKE	WT-P90-11	DGC179
030225-5-PV	C150897B-I1	37-9a2AKE	WT-P90-13	DGC380
030430-5	C170998D-I3	41-8a3AKE	WT-P90-14	DGC385
030501-1	C181000A-I3	45-6a3AKE	WT-P90-16	DGC386
040408C1	C181200A2	45-6b5AKE	WT-P90-19	DGC452
042602-LM-LoDe	C190898A-I1	A6b3AKE	WT-P90-20	DGC507
06-L5int-55	C230300D1	ACC1	WT-P90-23	DGC515
07-L5-6int-44	C240298B-I2	D2OE-AAV-GFP-02	WT-P90-24	DGC607
104-SR-HiDe	C240300C1	D2OE-AAV-GFP-03	WT-P90-25	DGC609
2001-05-29-B-L23-dendax	C250298A-I4	D2OE-AAV-GFP-06	WT-P90-26	DGC648
2001-05-30-B-L23-dendax	C260199A-I3	D2OE-AAV-GFP-09	WT-P90-29	DGC654
2001-06-26-D-L23-dendax	C260897C-I1	D2OE-AAV-GFP-11	WT-P90-30	DGC655
2001-07-05-C-L23dend	C281197A-I4	D2OE-AAV-GFP-14	WT-P90-33	DGC664
2001-09-26-A-L23-dendax	C281199A-IN	D2OE-AAV-GFP-15		DGC665
2001-09-27-A-L23dend	C300898C-I3	D2OE-AAV-GFP-19		DGC671
2001-10-04-A-L23-dendax	DAG052C-1	D2OE-AAV-GFP-29		DGC672
2001-10-08-A-L23-dendax	DAG061C	D2OE-P270-02		DGC673
2001-10-09-A-L23-dendax	DD030225-5-PV	D2OE-P270-04		DGC677
2001-11-09-B-L23-dendax	DEV127T1	D2OE-P270-05		DGC678
2002-03-25-B-L23-dendax	DS081004	D2OE-P270-06		DGC682
2002-04-03-A-L23dend	EF092408n4-5cell2	D2OE-P270-08		DGC685
2002-11-25-A-L23-dendax	EF160608-1	D2OE-P270-13		DGC687
2002-11-28-A-L23-dendax	GA01611c3-a	D2OE-P270-16		DGC696trace

(continued)

Table 3.4 (continued)

Interneurons		Medium spiny	Granule
2002-12-09-A-L23-dendax	GA01913c1	D2OE-P270-17	DGC699trace
2003-02-21-A-L23-dendax	GA02402c1n2-2	D2OE-P270-18	GranuleCell-Nr1-Septal
2003-02-24-A1-L23-dendax	GA05505-1	D2OE-P30-05	GranuleCell-Nr10-Septal
2003-02-24-A2-L23-dendax	Golgi-cell-040408-C1	D2OE-P30-07	GranuleCell-Nr11-Septal
2003-12-02-A-L23-dendax	Golgi-cell-051108-C0-cell1	D2OE-P30-09	GranuleCell-Nr12-Septal
2003-12-02-C-L23dend	JM072203-10	D2OE-P30-10	GranuleCell-Nr2
2003-12-03-B1-L23-dendax	JM080803-20	D2OE-P30-12	GranuleCell-Nr2-Septal
2003-12-03-B2-L23-dendax	JM082703-20-3	D2OE-P90-01	GranuleCell-Nr3-Septal
2003-12-03-C-L23-dendax	MD070412-5	D2OE-P90-04	GranuleCell-Nr4-Septal
2003-12-03-D1-L23dend	MD070419-2	D2OE-P90-05	GranuleCell-Nr5
2003-12-03-D2-L23-dendax	MD070501-4-2	D2OE-P90-06	GranuleCell-Nr5-Septal
2004-02-12-C-L23-dendax	MD070517-1	D2OE-P90-07	GranuleCell-Nr6-Septal
2004-02-14-B-L23-dendax	MD070604-4	D2OE-P90-08	GranuleCell-Nr7
210710C1	MJ062904-2	D2OE-P90-09	GranuleCell-Nr8
24-L5int-120	MJ122304-1-2	D2OE-P90-20	GranuleCell-Nr9-Septal
240710C0	n256	D2OE-P90-22	n04-cont02-sl4
960427c	P18-DEV226	DATKOK_finaltrace	N10ttwt
AK163lay	P36-DEV13	G8s1LS	n11TTKO
AM101-3-1	P9-DEV306	H5b1AKE	n12TTKO
AM50-2-3	PE04B	H6a3AKE	N12ttwt
AM81-2-2	qqscell001	H6a4AKE	n13TTKO
AM85-2-1	qqscell003	I51bAKE	N13ttwt
AP030507-2-2	RosaA251A-incomplete	J62aAKE	n14TTKO
AP030707-4	s030510s1	N7a2AKE	n15TTKO
AP042010A-1	s120509s1	O5s2LS	n16TTKO

AP121509Dcell1	sr3	s16-6cm1m	n19-cont02-sl1
AWa052709	TS052208s2c1	WT-AAV-GFP-06	n19TTKO
AWa052909	TS060607s2cell2	WT-AAV-GFP-07	N19ttwt
AWa70430	VN031605-0	WT-AAV-GFP-08	N1ttwt
AWab042309cell1	Vn05302006-0-B	WT-AAV-GFP-09	n20-cont02-sl1
AWb041109	Vn05302006-1-C	WT-AAV-GFP-14	N20ttwt
AWc021909	Vn05312006-1-B	WT-AAV-GFP-15	N22ttwt
AWd042009	Vn05312006-1-D	WT-AAV-GFP-17	n23TTKO
AWd101008cell1	Vn07142006-0-B	WT-AAV-GFP-18	n244
BE17A		WT-P270-05	N24ttwt
BE32B		WT-P270-12	n2TTKO
BE33G		WT-P270-16	N2ttwt
BE37A		WT-P30-02	n30-r01-03-sl3
BE37C		WT-P30-08	n3TTKO
BE55A		WT-P30-11	N3ttwt
BE62C		WT-P30-15	n4TTKO
BE91C		WT-P30-16	n5TTKO
BW042905-10		WT-P30-19	N5ttwt
C010600A2		WT-P30-22	n6TTKO
C011098A-I3		WT-P30-23	n7TTKO
C050600B1		WT-P30-25	N7ttwt
C050600B2		WT-P30-27	n8TTKO

Table 3.5 Names of the *outliers* identified in each cell class

Interneurons		Medium spiny		Granule
040809c6	JM092903-20-3	050803d_finaltrace	R93aAKE	3319201
081103a	MD070430-2-1	050803e_finaltrace	s1-4HCB4	3319202
09-06-01-4neurogliaform	MD070503-3	10-6vkd2m	s12-6cm11	4239204
092903a	MD070503-3-1	34-8s2LNS	s14-5cm1	4299202
14-L2int-na	MD070507-1	39-7b5AKE	s14-5vkd2	5199202
155-1-18b	MD07061-1	43-7b2AKE	s14-5vkd3	040726TTX2
156-2-17b	OLM	44-10s4LNS	s17-5cm1L	041014-vehicle
156-2-4b	P42-DEV384	44-9s1LNS	s18-4cm1	04n11023-TTX1
156-2-5b	P9-DEV304	44-9s2LNS	s18-4cm2	04n11023-TTX2
960824a	pairC3-C4-final-1	44-9s3LNS	s18-4vkd1	04n11024-APV
AWab092109cell2	pairC3-C4-final-2	45-6a4AKE	S6-8VD1	5-20-2011 cell1-GCcontrol
banke-06-03-07-01	R75C1	45-7b2AKE	s7-5cm4m	798-2
BE71B	Renshawcell1	46-7s4LNS	s7-6HCB1	B330886
C010700A	Renshawcell4	46-7s5LNS	S8-4CM1	DGC-657
C220798A-I	slm9	47-8a3AKE	WT-AAV-GFP-01	DGC660
C250300C2-ax	TS050207s1complete	A6b2AKE	WT-AAV-GFP-02	H906-2
dCH-cobalt	TS050907s5c1	B5a2AKE	WT-AAV-GFP-03	H910-1
DEV119T3	vCH-cobalt	B5a3AKE	WT-AAV-GFP-04	H910-2
DEV159T2	Vn07172006-0-D	C7a3AKE	WT-AAV-GFP-11	H917-2
DEV56T1	VS1-cobalt	C8a1AKE	WT-AAV-GFP-13	n08-cont03-sl2
DEV95T7	VS3-cobalt	D2OE-AAV-GFP-18	WT-AAV-GFP-20	n11-cont02-sl2
DIN-reconstruction	VS4-fluoro04	D2OE-AAV-GFP-23	WT-P270-01	n222

DR-int-July-24-08-686-R	VS4-fluoro07	D2OE-AAV-GFP-26	WT-P270-04	n225
GA01712c2anb-2	VS9-cobalt	D2OE-P30-02	WT-P270-20	n240
HSE-fluoro03		D2OE-P90-12	WT-P30-04	n241
HSE-fluoro13		D2OE-P90-16	WT-P30-06	n242
HSN-cobalt		D2OE-P90-19	WT-P30-09	n272
HSN-fluoro05		D7s1LS	WT-P90-15	n502
HSN-fluoro08		E7a1AKE	WT-P90-22	n503
HSN-fluoro09		E7a1LS	WT-P90-28	No39-TTX
HSN-fluoro10		E8a2AKE	WT-P90-32	No40-B-TTX
IaInt1		E8b3ake	WT-P90-34	No41-TTX
IaInt2		F9b5AKE	WT-P90-35	No44-APV
IaInt3		H6a2AKE		No50-Veh
IaInt4		J63aAKE		No51-pyramidal-Veh
IaInt5		M8a3AKE		No52-Veh
IaInt6		O6a1LS		No55-TTX
IaInt7		O6s1LS		No56-Veh
IaInt8		P6a2AKE		No60-TTX
JM091703-30-1		P7s1LS		No61-Veh

References

Andersen P, Morris R, Amaral D, Bliss T, O'Keefe J (2006) The hippocampus book, 1st edn. Oxford University Press, New York

Arisi GM, Garcia-Cairasco N (2007) Doublecortin-positive newly born granule cells of hippocampus have abnormal apical dendritic morphology in the pilocarpine model of temporal lobe epilepsy. Brain Res 1165:126–134

Ascoli GA (2002) Computational neuroanatomy: principles and methods. Humana Press, Totowa

Ascoli GA, Donohue DE, Halavi M (2007) Neuromorpho.org: a central resource for neuronal morphologies. J Neurosci 27(35):9247

Bausch SB, He S, Petrova Y, Wang XM, McNamara JO (2006) Plasticity of both excitatory and inhibitory synapses is associated with seizures induced by removal of chronic blockade of activity in cultured hippocampus. J Neurophysiol 96(4):2151–2167

Bishop GA, Chang HT, Kitai ST (1982) Morphological and physiological properties of neostriatal neurons: an intracellular horseradish peroxidase study in the rat. Neuroscience 7(1):179–191

Cannon RC, Turner DA, Pyapali GK, Wheal HV (1998) An on-line archive of reconstructed hippocampal neurons. J Neurosci Meth 84(1–2):49–54

Cannon RC, Wheal HV, Turner DA (1999) Dendrites of classes of hippocampal neurons differ in structural complexity and branching patterns. J Comp Neurol 413(4):619–633

Carim-Todd L, Bath KG, Fulgenzi G, Yanpallewar S, Jing D, Barrick CA, Becker J, Buckley H, Dorsey SG, Lee FS, Tessarollo L (2009) Endogenous truncated TrkB.T1 receptor regulates neuronal complexity and TrkB kinase receptor function in vivo. J Neurosci 29(3):678–685

Carnevale NT, Tsai KY, Claiborne BJ, Brown TH (1997) Comparative electrotonic analysis of three classes of rat hippocampal neurons. J Neurophysiol 78(2):703–720

Cazorla M, Shegda M, Ramesh B, Harrison NL, Kellendonk C (2012) Striatal d2 receptors regulate dendritic morphology of medium spiny neurons via kir2 channels. J Neurosci 32(7): 2398–2409

Chklovskii DB (2004) Synaptic connectivity and neuronal morphology: two sides of the same coin. Neuron 43(5):609–617

Costa LdF, Manoel ETM, Faucereau F, Chelly J, van Pelt J, Ramakers G (2002) A shape analysis framework for neuromorphometry. Network (Bristol, England) 13(3):283–310

Costa LDF, Zawadzki K, Miazaki M, Viana MP, Taraskin SN (2010) Unveiling the neuromorphological space. Front Comput Neurosci 4:150

Duda RO, Hart PE, Stork DG (2000) Pattern classification, 2nd edn. Wiley, New York

Gupta A, Elgammal FS, Proddutur A, Shah S, Santhakumar V (2012) Decrease in tonic inhibition contributes to increase in dentate semilunar granule cell excitability after brain injury. J Neurosci 32(7):2523–2537

Halavi M, Polavaram S, Donohue DE, Hamilton G, Hoyt J, Smith KP (2008) Neuromorpho.org implementation of digital neuroscience: dense coverage and integration with the nif. J Neuroinform 6(3):241–252

Hardle WK, Simar L (2012) Applied multivariate statistical analysis, 3rd edn. Springer, Berlin

Krichmar JL, Nasuto SJ, Scorcioni R, Washington SD, Ascoli GA (2002) Effects of dendritic morphology on CA3 pyramidal cell electrophysiology: a simulation study. Brain Res 941(1–2): 11–28

McDonald CG, Dailey VK, Bergstrom HC, Wheeler TL, Eppolito AK, Smith LN, Smith RF (2005) Periadolescent nicotine administration produces enduring changes in dendritic morphology of medium spiny neurons from nucleus accumbens. Neurosci Lett 385(2):163–167

Milosević NT, Ristanović D (2007) The sholl analysis of neuronal cell images: semi-log or log–log method? J Theor Biol 245(1):130–140

Montague PR, Friedlander MJ (1991) Morphogenesis and territorial coverage by isolated mammalian retinal ganglion cells. J Neurosci 11(5):1440–1457

Murphy BL, Pun RYK, Yin H, Faulkner CR, Loepke AW, Danzer SC (2011) Heterogeneous integration of adult-generated granule cells into the epileptic brain. J Neurosci 31(1):105–117

Pierce JP, McCloskey DP, Scharfman HE (2011) Morphometry of hilar ectopic granule cells in the rat. J Comp Neurol 519(6):1196–1218
Poirazi P, Brannon T, Mel BW (2003) Arithmetic of subthreshold synaptic summation in a model CA1 pyramidal cell. Neuron 37(6):977–987
Poznanski RR (1992) Modelling the electrotonic structure of starburst amacrine cells in the rabbit retina: a functional interpretation of dendritic morphology. Bull Math Biol 54(6):905–928
Preston R, Bishop G, Kitai S (1980) Medium spiny neuron projection from the rat striatum: an intracellular horseradish peroxidase study. Brain Res 183(2):253–263
Rihn LL, Claiborne BJ (1990) Dendritic growth and regression in rat dentate granule cells during late postnatal development. Brain Res Dev Brain Res 54(1):115–124
Ristanović D, Milosević NT, Stulić V (2006) Application of modified sholl analysis to neuronal dendritic arborization of the cat spinal cord. J Neurosci Methods 158(2):212–218
Rodrigues EP, Barbosa MS, Costa LdF (2005) Self-referred approach to lacunarity. Phys Rev E Stat Nonlin Soft Matter Phys 72(1 Pt 2):016707
Scorcioni R, Lazarewicz MT, Ascoli GA (2004) Quantitative morphometry of hippocampal pyramidal cells: differences between anatomical classes and reconstructing laboratories. J Comp Neurol 473(2):177–193
Scorcioni R, Polavaram S, Ascoli GA (2008a) L-measure: a web-accessible tool for the analysis, comparison and search of digital reconstructions of neuronal morphologies. Nat Protoc 3(5):866–876
Scorcioni R, Polavaram S, Ascoli GA (2008b) L-measure: a web-accessible tool for the analysis, comparison and search of digital reconstructions of neuronal morphologies. Nat Protoc 3:866–876
Sholl DA (1953) Dendritic organization in the neurons of the visual and motor cortices of the cat. J Anat 87(4):387–406
Tejada J, Arisi GM, Garcia-Cairasco N, Roque AC (2012) Morphological alterations in newly born dentate gyrus granule cells that emerge after status epilepticus contribute to make them less excitable. PLoS One 7(7):e40726
Toris CB, Eiesland JL, Miller RF (1995) Morphology of ganglion cells in the neotenous tiger salamander retina. J Comp Neurol 352(4):535–559
van Elburg RA, van Ooyen A (2010) Impact of dendritic size and dendritic topology on burst firing in pyramidal cells. PLoS Comput Biol 6:e1000,781
Vuksic M, Del Turco D, Bas Orth C, Burbach GJ, Feng G, Müller CM, Schwarzacher SW, Deller T (2008) 3D-reconstruction and functional properties of GFP-positive and GFP-negative granule cells in the fascia dentata of the thy1-GFP mouse. Hippocampus 18(4):364–375
Zhou D, Li S, Zhang Xh, Cai D (2013) Phenomenological incorporation of nonlinear dendritic integration using integrate-and-fire neuronal frameworks. PloS One 8(1):e53508

Chapter 4
Neuronal Arborizations, Spatial Innervation, and Emergent Network Connectivity

Jaap van Pelt, Harry B.M. Uylings, and Arjen van Ooyen

Abstract Neurons innervate space by their axonal and dendritic arborizations. Synapses can form when axons and dendrites are in close proximity. The geometry of neurons and their numerical densities in space are thus primary factors in the formation of synaptic connectivity of neuronal networks. A simulator of neuronal morphology based on principles of neural development (NETMORPH) has been used to show that realistic network connectivities emerge from realistic neuronal morphologies and simple proximity-based synapse formation rules.

4.1 Introduction

Activity dynamics underlying cognition depends crucially on the patterns and strengths of synaptic connections between neurons. Neurons innervate space by their axonal and dendritic arborizations as determined by their geometries. Axonal and dendritic arborizations are complex and show large variations in all their shape characteristics, such as in the number of segments and segment lengths, branching angles, tortuosity of branches, orientation of branches, thickness of branches, topology of their tree patterns, and their global 3D embedding, which may show some spherical symmetry or axial symmetry or even flat 2D patterns such as in Purkinje cells. Synapses can form only at those locations where axons and dendrites come

J. van Pelt (✉) • A. van Ooyen
Computational Neuroscience Group, Department of Integrative Neurophysiology, Center for Neurogenomics and Cognitive Research, VU University, Amsterdam, The Netherlands
e-mail: j.van.pelt@vu.nl

H.B.M. Uylings
Department of Anatomy and Neuroscience, VU University Medical Center, Amsterdam, The Netherlands

H. Cuntz et al. (eds.), *The Computing Dendrite: From Structure to Function*, Springer Series in Computational Neuroscience 11, DOI 10.1007/978-1-4614-8094-5_4,

sufficiently close to each other (Peters 1979). The geometry of neuronal arborizations and the numerical densities of neurons are therefore important determinants of synaptic connectivity of neuronal networks.

How neuronal morphology shapes neuronal network connectivity is, however, still poorly understood. From a computational point of view this question can be addressed by creating neuronal networks of computer generated or experimentally reconstructed neurons along with their axonal and dendritic arborizations in a given space and studying the emergent connectivity by applying Peters' rule (Peters 1979). The use of experimentally reconstructed neurons has the advantage of best morphological realism of their arborizations. For instance, NeuroConstruct (Gleeson et al. 2007) is a modeling tool for creating complex neuronal networks in 3D space using imported experimentally reconstructed neuronal morphologies.

The efforts of collecting accurate and complete reconstructions are time costly and large. Thanks to the initiative of creating the NeuroMorpho.Org database, an increasing number of reconstructions have become available (Ascoli 2006; Ascoli et al. 2007). For the construction of large networks, however, the finite datasets require the use of many copies of the same neuron. The use of computer generated neuronal arborizations has the advantage that neurons can be produced in any desired number and, depending on the generator, also with full and realistic variation between the individual neurons. A number of different neuronal morphological simulators are presently available in the field of computational neuromorphology, which differ amongst others in the algorithms used. Neurons may be generated by sampling experimental distributions of quantified neuronal shape parameters. These algorithms produce neurons with realistic variations in neuronal shapes. An important example of such an algorithm is L-Neuron (Senft and Ascoli 1999; Ascoli and Krichmar 2002; Samsonovich and Ascoli 2007) which was inspired by earlier work of Burke et al. (1992). Also the simulator NeuGen (Eberhard et al. 2006) uses this algorithmic approach. An algorithm explored by Luczak (2006) uses diffusion limited aggregation (DLA) for generating neuron-like arborizations. A recently developed algorithm by Cuntz et al. (2010) regards arborizations as locally optimized graphs and neuronal morphologies are generated by using the minimal spanning tree principle. Costa and Coelho (2005; 2008) generated 2D neuronal morphologies by statistically sampling a probabilistic model of neuronal geometry based on branch probabilities per branch level using a Monte Carlo approach, and form connections when neuronal trees overlap in 2D. Torben-Nielsen et al. (2008a; 2008b) developed an algorithm (KDE-Neuron and EvOL-Neuron) for generating arborizations *automatically* from a set of experimentally reconstructed neurons, without making a priori assumptions about the distributions generating the real data. Neuronal arborizations may also be produced by algorithms inspired by the developmental process of neurite outgrowth and branching. The simulator CX3D (Zubler and Douglas 2009) is a tool for simulating the growth of cortex in 3D space. It includes actions such as cell division, cell migration and neurite outgrowth. The simulated neurons have a physical implementation in space and may secrete and react to chemical components. CX3D combines mechanistic and stochastic rules for elongation and branching in the morphogenesis of single neurons. The simulator NETMORPH (Koene et al. 2009) is a simulation framework for generating realistic neuron morphologies also based on rules for neuronal development. It uses a phenomenological approach in modeling the development of neuronal morphology versus developmental time. Real neurons grow out by the actions of growth cones at the tips of outgrowing neurites. These actions include migration (elongation of the neurite), branching (into two daughter growth cones), turning (changing direction of outgrowth), and retraction (shortening of the neurite). The simulator NETMORPH uses the dendritic outgrowth model of Van Pelt et al. (e.g., 2001) which describes these actions in a stochastic way and which has been shown to generate realistic dendritic morphologies. The simulators NETMORPH and CX3D have recently been evaluated by Aćimović et al. (2011) and Mäki-Marttunen et al. (2011).

In this chapter, we show examples of networks created with NETMORPH, and emergent synaptic connectivities in a single layer of rat cortical L2/3 pyramidal neurons. This choice was based on the availability of experimental data on L2/3 pyramidal cell morphologies and pyramidal-pyramidal cell connectivity. The question was addressed whether model neurons, generated with the NETMORPH simulator, and placed in realistic densities in space resulted in networks with realistic synaptic connectivity patterns.

We found that networks formed by the realistic neuron morphologies generated with NETMORPH show connectivity patterns that are in very good agreement with the available experimental data. For a number of connectivity properties experimental data was not yet available and these outcomes must therefore be regarded as predictions.

4.2 NETMORPH Simulator

In the simulator NETMORPH, neurons, with their cell bodies initially placed in a 3D space, grow out independently from each other and fill the space with their axonal and dendritic arborizations. Candidate synaptic locations are searched for on the basis of the proximity of axonal and dendritic branches in conjunction with a "crossing criterion" for any pair of axonal and dendritic line-pieces that constitute the branches (Van Pelt et al. 2010). The synaptic locations are thus purely based on the 3D cell geometries.

A simulation run of NETMORPH is based on a script, a text file that contains parameter identifiers with associated values specifying the duration, the time step, and other parameters required for the growth process. These include the model specifications (parameters) and the various probability density functions from which random samples are to be drawn. The structure of the scripts and the parameters that can be used to specify the growth process are explained in the manual of NETMORPH (Koene et al. 2009). The NETMORPH simulator is used for both the optimization process in an iterative fashion, and for running the final simulations with the optimized growth parameter values. The optimization process aims at findings those parameter values that result in morphologies optimally matching experimental data of reconstructed neurons. A network simulation requires a network initialization step including the specification of the space, cell types, positions of the cell bodies, etc.

4.2.1 Model of Neurite Outgrowth in the NETMORPH Simulator

An important requirement of a simulator of neuronal morphology is that the simulated structures are realistic both in their shapes and in the variability in shapes. The outgrowth model used by NETMORPH is based on stochastic phenomenological rules for branching and elongation of neuritic segments (growth cones), formulated by Van Pelt et al., and has been shown to accurately describe the geometry of dendrites of various cell types (e.g., Van Pelt et al. 2001; Van Pelt and Uylings 2003; 2005). Note, that the current version of NETMORPH has also been used for generating axonal arborizations (see Sect. 4.6 for axonal outgrowth rules).

The growth model is a phenomenological one aiming at describing the growth actions, not a biophysical one that would describe the mechanisms underlying the growth actions. The model is based on a minimal set of growth rules formulated as close as possible to the observables in the neurite outgrowth process. These growth rules concern the branching process, producing topological trees, and the elongation process defining the lengthening of the segments. The modular structure allowed each growth rule to be validated separately on experimental data.

Neurite outgrowth starts with the formation of an initial segment emerging from the cell body and with a growth cone at the tip. The growth cone elongates the neurite, can change outgrowth direction, and can branch, creating two daughter growth cones. Note that on a smaller time scale growth cone behavior shows more complex dynamic behavior including phases of retraction.

4.2.2 Branching

During outgrowth, each growth cone j in a growing tree has a branching probability given at time t_i with time step Δt by

$$p_j\left(t_i\right) = \frac{B_\infty}{C_{n_i}} e^{-t_i/\tau}\left(e^{\Delta t/\tau} - 1\right) n_i^{-E} 2^{-S\gamma_j}. \tag{4.1}$$

with

$$C_{n_i} = \frac{1}{n_i}\sum_{k=1}^{n_i} 2^{-S\gamma_k},$$

(e.g., Van Pelt and Uylings 2002; 2003; 2005; 2007). The equation is the product of three factors: a factor $e^{-t_i/\tau}$ that decreases exponentially with time constant τ with developmental time; a factor $2^{-S\gamma_j}$, modulated by the parameter S, that changes with the growth cone's centrifugal order γ (the number of branch points between the tip and the root of the tree); and a factor n_i^{-E} that decreases with the momentary number (n_i) of growth cones in the tree. This last factor reflects competition between growth cones for resources. Parameter E denotes the strength of competition.

After each branching event, the branching probability thus decreases. The parameter B_∞ determines the expected number of terminal segments in the final tree for $E=0$. The coefficient C_{n_i} is a normalization term to be updated at each branching event. The branching process of a tree is thus fully defined by the parameters B_∞, E, S, τ, and the period of growth T. Examples of optimized values for these parameters for a variety of cell types are summarized in Van Pelt et al. (2001).

4.2.3 Elongation

The rate of elongation of a growth cone may vary considerably (e.g., Lamoureux et al. 1998; da Costa et al. 2002), also on the time scale of the chosen time steps Δt. In NETMORPH a more coarse grained approach is used by taking an averaged elongation rate for the period in which a terminal segment is elongating up to the occurrence of its next branching event. The daughter segments emerging from a branching event are then subsequently given an elongation rate up to their next branching events. NETMORPH assigns the elongation rates by randomly sampling a Gaussian distribution with mean *eri-mn* and standard deviation *eri-sd* (*eri* stands for elongation rate initialization). Alternatively, NETMORPH includes the option to select an elongation competition model. In addition to influencing branching probability, competition between growth cones for limited resources may also affect elongation rate (see also Van Ooyen et al. 2001; Van Ooyen 2011), which can be described as

$$\nu(t) = \nu_0 n(t)^{-F}, \tag{4.2}$$

with parameter F determining the strength of competition (Van Pelt and Uylings 2003). For $F=0$, terminal segments elongate with rate ν_0 independent of the number of terminal segments. For $F>0$, elongation rates depend on the momentary number of terminal segments. For $F=1$, elongation rates are inversely correlated with the number of terminal segments, implying that the total tree increase its length with rate ν_0. Strong experimental evidence for such dependencies is, however, not yet available. Therefore, this option has not been used in the simulations described in this chapter, i.e., $F=0$.

4.2.4 Neurite Turning and Direction of Outgrowth After Branching

The direction of outgrowth of growth cones is subject to different factors with intrinsic ones (such as cytoskeletal stiffness) and external ones (such as trophic and tropic factors, and irregularities of the surrounding tissue). NETMORPH uses a stiffness rule in combination with a stochastic factor to calculate momentary outgrowth directions of all dendritic growth cones (Koene et al. 2009).

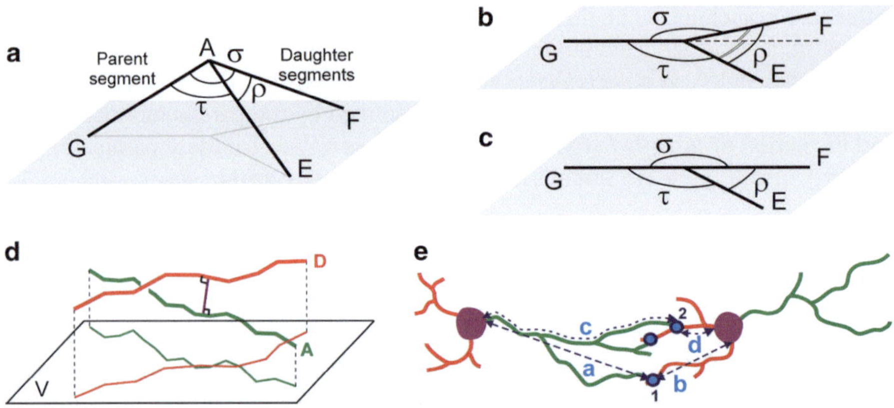

Fig. 4.1 (**a–c**) Flatness of bifurcations—(**a**) A schematic 3D bifurcation with parent segment GA and daughter segments AE and AF and it branching angles, with the intermediate angle ρ between the daughter segments, and the side angles σ and τ, between the parent and each of the daughter segments, respectively—(**b**) A flat bifurcation with the parent segment aligned to the daughters' bisector—(**c**) A flat bifurcation of a side branching bifurcation. Adapted from Van Pelt and Uylings (2012). (**d**) Shortest distance between an axonal and dendritic branch defined by the orthogonal distance between a pair of axonal and dendritic crossing line pieces. When this shortest distance is smaller than a given criterion value, the line piece pair marks the location of a candidate synapse. (**e**) Connectivity between axons (*green*) and dendrites (*red*) of two cells (*purple*) with three synaptic contacts, with presynaptic (*a*) and postsynaptic (*b*) Euclidean distances of synapse 1, and presynaptic (*c*) and postsynaptic (*d*) path distances of synapse 2. Adapted from Van Pelt et al. (2010)

4.2.5 Flatness of Bifurcations

An interesting property of dendritic bifurcations is that they are flat (parent and daughter segments lie predominantly in one plane) and that they are aligned with the bisector of the daughter segments oriented oppositely to the parent segment (Fig. 4.1a–c) (Uylings and Smit 1975; Van Pelt and Uylings 2012; Kim et al. 2012). Such flat geometries may arise when during neuronal development the segments at a newly formed bifurcation are subjected to elastic tensions, which force the bifurcation into an equilibrium planar shape. Such "balance of forces model" is used by NETMORPH to determine the geometry of a bifurcation and to determine the directions of outgrowth of daughter growth cones after a branching event (Koene et al. 2009).

4.2.6 Segment Diameters

Segment diameters are defined according to the power-law model by Rall (1959) based on the relation between the diameters between parent d_p and daughter segments d_1 and d_2.

$$d_p^e = d_1^e + d_2^e \tag{4.3}$$

Empirical values for the exponent *e* for different cell types were summarized in Van Pelt and Uylings (2005). NETMORPH determines the segment diameters after the outgrowth process has ended.

4.3 Neuronal Morphologies

Whether a simulator produces realistic neuronal shapes can only quantitatively be assessed by comparing shape properties of real and simulated neurons. Axonal and dendritic branching patterns are complex and require many different shape parameters to quantify their metrical and topological properties and their embedding in 3D space (e.g., Uylings and Van Pelt, 2002; Gillette et al. 2001). The shape parameters of the individual trees, used for the automatic optimization of the growth parameters, are the number of segments, the lengths of intermediate and terminal segments, the path distance of the terminal tips to the cell body, the tree asymmetry index (topological arrangement of the segments in the formation of the tree), and the mean centrifugal order (mean of the number of segments at successive centrifugal orders in the tree). In the optimization process the distributions of these shape parameters were compared with the experimental ones, both in their mean values and their standard deviations. The spatial innervation of the arborizations also depends on the curvature and orientation of the branches. These shape properties are in NETMORPH controlled by the stiffness parameters and the angular distributions for the outgrowth directions of growth cones. These parameters were optimized separately.

4.3.1 Parameter Optimization

The outgrowth rules in NETMORPH are determined by a main set of six growth parameters (E, S, B_∞, τ, *eri-mn*, and *eri-sd*) used for (semi-) automatic optimization and an additional set of parameters that are set separately. The outgrowth parameters for a given neuron type need to be optimized for an experimental morphological dataset of this type of neurons. Some growth parameter values have a direct relation with the generated morphological shapes. The topological properties of branching patterns emerge from the branching rules, while the metrical properties emerge from both the branching and the elongation rules. For instance, the parameters $B(T)$ and E directly determine the mean and standard deviation of the degree distribution of the generated trees at developmental time T and the user can find the best B-E values using for instance the graphs in Van Pelt et al. (2001) or in Van Pelt and Uylings (2002). The parameter B_∞ in (4.1) relates to $B(T)$ via $B_\infty = B(T)/(1 - e^{-T/\tau})$.

The parameter S in (4.1) relates to the mean topological asymmetry of the trees via the graphs in Van Pelt et al. (2001). The ratio of intermediate and terminal segment length depends on the decay constant τ in (4.1). The smaller this value the more rapidly in time the branching probability decreases and the longer the segments become (unless the elongation rate decreases as rapidly). Nevertheless, finding the optimal parameter values typically implicates an iterative process of selecting a parameter set, generating a sample of model neurons, analyzing their shapes and comparison with the shapes in the experimental dataset. Such optimization may proceed manually by selecting certain values for the parameters, but automated procedures are also available, such as genetic algorithms or maximum likelihood optimization methods. Note, that the six growth parameter values were optimized only for generating realistic neuronal morphologies, not for producing particular connectivity patterns.

4.4 Synaptic Connectivity

Axons and dendrites can only form connections at locations where they are in sufficiently close proximity (Peters 1979) which depends on the size of axonal boutons and dendritic spines and the "capture range" of growth cone filopodia. In NETMORPH such locations are searched, after having filled a space with axonal and dendritic ramifications, by using a preset criterion for the required proximity, typically in the range of a few microns. NETMORPH uses a recently developed algorithm for this search which combines a crossing requirement of a line piece pair (Fig. 4.1d), and the criterion for the orthogonal distance between these line pieces (Van Pelt et al. 2010). Note that these locations are based on geometrical considerations only and thus mark candidate sites for synapses. Whether synapses will actually occur depends on many other biological factors. In the following all these candidate sites are taken into account, no selection is thus made here.

4.4.1 Connectivity: Terminology and Measures

Two neurons are *connected* when they share at least one *synaptic contact* (Fig. 4.1e). The number of contacts per connection (*multiplicity*) is defined as the number of synapses from a given neuron projecting on the dendrites of another neuron. A synaptic contact is between the axon of a *presynaptic neuron* and a dendrite of a *postsynaptic neuron*. A synaptic contact is at certain distances from its presynaptic and its postsynaptic cell body, respectively. These distances can be measured as *Euclidean distances* (straight line between synapse location and soma center) or as *path distances* along the arborizations themselves. The Euclidean distance between the somata of two connected neurons is called the *connection length*.

4.4.2 Connectivity: Statistical Measures of Connectivity

The *connection probability* between two neurons is defined as the probability that two randomly selected neurons in a network are connected, i.e., have at least one synaptic contact from the axon of one of the neurons onto the dendrite of the other neuron.

The *connection probability versus Euclidean distance* between two neurons is defined as the probability that two randomly selected neurons in a network at a given Euclidean distance from each other are connected.

A distinction can be made between directed or undirected connections. The directed or one-way connection probability is defined as the probability that a given (presynaptic) neuron has at least one synaptic contact with another given (postsynaptic) neuron. Note that in a network of N neurons there are $N(N-1)$ directed neurons pairs but $N(N-1)/2$ undirected neuron pairs.

4.4.3 NETMORPH Example

For studying emergent network connectivity simulations were run mimicking a network of rat Layer 2/3 pyramidal neurons. A number of 250 cell bodies was uniform randomly placed within a cylinder with a height of 380 μm. This number was mainly restricted by the computational load of the simulation. A radius of 93 μm was adopted to obtain a realistic density of neurons (about 75,000 per mm^3, Hellwig 2000). A minimum distance of 20 μm was taken between the cell bodies. Each pyramidal cell was initialized with root points for an axon and an apical dendrite and a number of root points for basal dendrites uniform randomly selected between the numbers 4 and 8. The growth of the pyramidal cells has been simulated over a period of 18 days, a period in which these cells in the rat cortex obtain their mature total length (Uylings et al. 1994). Examples of simulated neurons are shown in Fig. 4.2b, while Fig. 4.3 illustrates the density of axonal and dendritic fibers in the network. Evidently, the axonal and dendritic arborizations covered a space much larger than the cylinder used for placing the cell bodies. Synapses thus could occur at all places where axonal and dendritic branches were present.

4.4.4 Validation of Morphological Shapes

The similarity between model and real neuronal shapes has been assessed quantitatively by comparing both the mean and standard deviation of the distributions of the quantitative shape parameters mentioned earlier for both axons and dendrites. Also visually do the NETMORPH examples shown in Fig. 4.2b agree very well with the experimental reconstructed neurons in Fig. 4.2a. Nevertheless, the visual

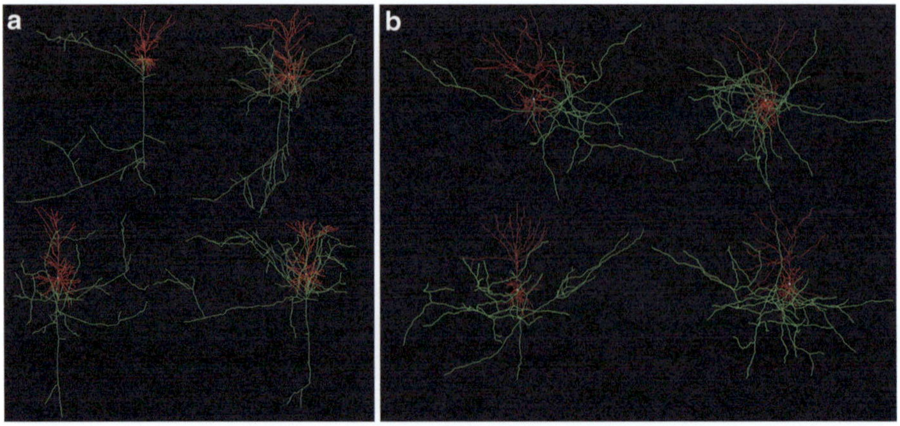

Fig. 4.2 (**a**) Four examples of rat neocortical Layer 2/3 pyramidal neurons, obtained from the Svoboda data set in NeuroMorpho.Org (Ascoli 2006; Ascoli et al. 2007). (**b**) Four examples of NETMORPH generated rat cortical L2/3 pyramidal neurons. Apical and basal dendrites in *red* and axons in *green*. Adapted from Van Pelt et al. (2010)

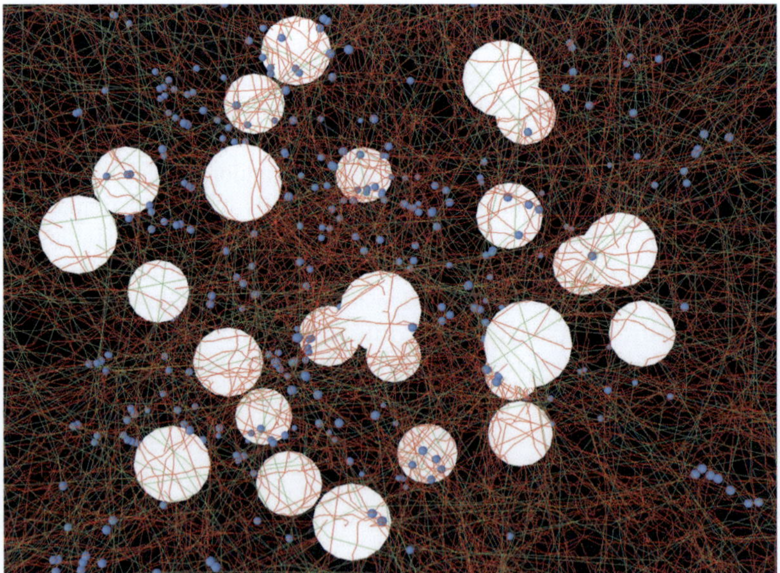

Fig. 4.3 Illustration of a NETMORPH generated network with the spatial distribution of candidate synaptic contacts. The 25 cell bodies are indicated as *white spheres*, axons are in *green*, dendrites are shown in *red*, and the synaptic contacts are indicated as *small blue spheres*. The simulation has been done with a time step parameter of dt = 20 s. Candidate synaptic locations are searched for with a distance criterion of 6 μm

comparison does show some differences in the pattern of axonal fibers which in the experimental data clearly show downward and sideward orientation preferences, which are not shown in the NETMORPH axons (see Sect. 4.6).

For illustration an example of a connected network of 25 neurons with a realistic numerical density is shown in Fig. 4.3. Axonal and dendritic fibers are indicated as green and red line pieces, respectively. In this example the candidate synapse locations were searched using a distance criterion of 6 μm, and are indicated as blue dots. Note that the actual simulations were done for networks of 250 neurons.

4.4.5 Connection Probability Versus Intersomal Euclidean Distance

With the set of identified synapses in the network of 250 neurons the one-way connection probability was determined by dividing the number of connected presynaptic and postsynaptic neuron pairs with a given intersomal distance by the total number of directed pairs of neurons with the given intersomal distance. The outcome in Fig. 4.4a shows that the connection probability decreases rather linearly with increasing intersomal distance.

4.4.6 Multiplicity of Connections

The number of synapse locations depends on the proximity criterion value used. The larger the criterion value, the more locations in space can be found where axonal and dendritic line pieces are within the required proximity. The effect of the proximity criterion value on the connectivity in a network is shown in Fig. 4.4b, both in the number of connected neurons pairs (n) and in the number of synaptic contacts per one-way connection (multiplicity). The frequency curves decrease monotonically, which is expected when the probability of a contact between two neurons is smaller than one. For a criterion value of 4 μm two neurons are on the average one-way connected by 2.5 synaptic contacts.

4.4.7 Path Distances and Euclidean Distances

For each synapse location its Euclidean and path distances were determined with respect to their presynaptic and postsynaptic cell body, resulting in the distance distributions in Fig. 4.4c, d. The Euclidean distances in Fig. 4.4c, d show that the cloud of synapses does not extend much beyond 500 μm from the presynaptic cell body, and does not extend much beyond 300 μm from the postsynaptic cell body. The figure also illustrates that pathlengths exceed Euclidean distances because of the tortuosity of the branches.

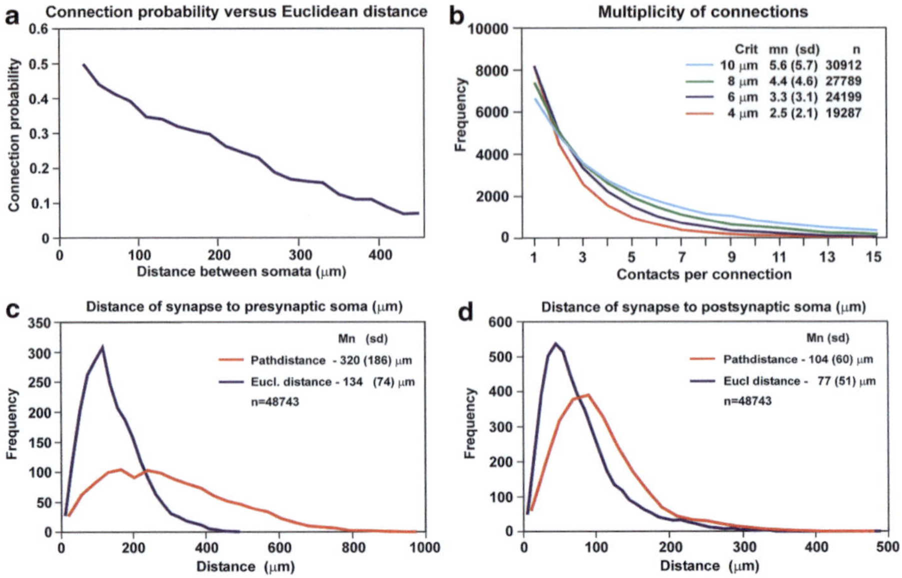

Fig. 4.4 (**a**) One-way connection probability between neurons versus intersomal distance. (**b**) Frequency distributions of the number of contacts per connection. The four distributions are obtained from a NETMORPH simulated network and for four proximity criterion values of 4, 6, 8, and 10 μm, respectively. The *inset* in the figure shows the mean number of contacts per connection for each of the criterion values. (**c**, **d**) Distribution of path distances and Euclidean distances of synapses to their (**c**) presynaptic and (**d**) postsynaptic somata in the model generated networks. Path distances exceed Euclidean distances because of the tortuosity of the branches. Presynaptic path distances have mean (sd) values of 320(186) μm; postsynaptic path distances of 104(60) μm; presynaptic Euclidean distances of 134(74) μm; and postsynaptic Euclidean distance of 77(51) μm

4.5 Comparison with Available Experimental Data

4.5.1 Connection Probability

Figure 4.4a shows connection probabilities gradually decreasing from about 0.5 down to less than 0.1 with increasing distance between the somata (30–440 μm). These values agree quite well in order of magnitude with a number of experimental findings.

Hellwig (2000) determined connectivity between reconstructed rat visual cortex L2/3 pyramidal neurons and found gradually decreasing connection probabilities down to less than 0.1 at a distance of 500 μm. Song et al. (2005) did 816 quadruple whole-cell recording attempts of thick tufted layer 5 pyramidal neurons in rat visual cortex and found functional connection probabilities of 11.6 % rather uniform over a short range of cell separations up to 100 μm, but with very few observations beyond a distance of 60 μm Perin et al. (2011) recorded simultaneously from up to 12 thick-tufted layer 5 pyramidal neurons in somatosensory cortical slices (300 μm thick) from Wistar rats (pn days 14–16). They found functional connectivity probabilities steadily decreasing with distance between cell bodies from about 0.21 at a distance of 15 μm down to about 0.03 at a distance of 300 μm. A direct comparison of these functional connectivity data and the anatomical connectivity in NETMORPH outcomes is difficult. The candidate synapses in NETMORPH emerge purely from the geometry of the neuronal arborizations. Whether they will become functional synapses depends on many other biological factors.

4.5.2 Multiplicity

For comparison we may use experimental results of Feldmeyer et al. (2002) showing a mean (sd) multiplicity value of 4.5 (0.5), based on a limited number of observations ($n=13$). This experimental outcome fits very well in the range of the model outcomes. The model outcomes further show decreasing frequencies with increasing multiplicity, which may be not surprising if the probability of having two contacts in a connection is smaller than having one contact, etc. The number of observations in the Feldmeyer data ($n=13$) is, however, too small to draw any conclusion about the shape of the experimental distribution.

4.5.3 Presynaptic and Postsynaptic Distances

Experimental data of Le Bé et al. (2007) shows a distribution of postsynaptic path distances ($n=55$) with mean (sd) of 130(138) μm, and thus with a SEM value of 18 μm. Although the synapses originate from layer 5 corticocallosal projecting neurons, their positions on the dendrites compare quite well with the model outcomes, also in the shape of the distribution, with few observations at higher distances. For comparison of postsynaptic Euclidean distances we may use experimental data of Feldmeyer et al. (2002) who measured connectivity between 13 pairs of L4 spiny neurons projecting onto L2/3 pyramidal neurons resulting in a mean (sd) of the geometric postsynaptic distances ($n=59$) of 67.2(33.6) μm. Although the synapses in this study originate from layer 4 spiny neurons, their positions on the dendrites compare quite well with the model outcomes, also in the shape of the distribution, with few observations at higher distance. Our NETMORPH outcomes on presynaptic and postsynaptic path distances also agree very well with qualitative experimental data from at least seven rat somatosensory cortical pyramidal neurons recently published by Hill et al. (2012). Experimental data of presynaptic path- and Euclidean distances was not yet available for comparison, making this distribution a prediction for future validation.

4.6 Discussion

4.6.1 Growth Model Assumptions

4.6.1.1 Stochastic Approach

The dendritic growth model, as earlier developed by Van Pelt et al. (e.g., Van Pelt et al. 2001; Van Pelt and Uylings 2005), described growth cone behavior in a stochastic way. This approach was based on the consideration that the actual behavior of growth cones within neural tissue depends on a multitude of mechanisms, interactions, and environmental structural, molecular, and chemical conditions, and that a statistical approach was most appropriate. Indeed, the variability in dendritic shape parameters could fully be described by the stochasticity in the modeled growth cone behavior.

4.6.1.2 Exponential Decreasing Baseline Branching Rate

An interesting outcome of the model-based analysis of the growth rate of outgrowing dendrites (Van Pelt and Uylings 2003) was that it required the assumption of an exponential decreasing baseline branching rate (4.1). This prediction may refer to intracellular processes during dendritic outgrowth that reduce the branching probability of individual growth cones with time.

4.6.1.3 Competition Between Growth Cones

For the dendritic growth model to produce realistic dendritic branching patterns, it was also necessary to include a dependence of the branching probability on the total number of growth cones in the dendritic tree. Such dependence may point to competition between growth cones during outgrowth. Such competition has indeed been observed in time-lapse movies of outgrowing neurites in tissue culture (Costa et al. 2002).

4.6.1.4 Axonal Outgrowth

The outgrowth model used by NETMORPH has initially been developed for and validated by dendritic arborizations of a large variety of cell types. The stochastic rules for dendritic branching and elongation did not include interactions of growth cones with their environment as it was assumed that this environment was homogeneous for the space of dendritic coverage. In the example the dendritic outgrowth model has also been applied for the generation of axonal arborizations. Although the dendritic shapes indeed appeared to agree very well with the experimental reconstructed neurons, the visual comparison of the pattern of axonal fibers did show some differences in the downward and sideward orientation preferences. As mentioned earlier, axonal growth cones migrate and branch in dependence of external cues and these interactions need to be included in the axonal outgrowth rules in NETMORPH which will further improve the realism of the 3D shape of axonal arborizations. In addition, axonal growth cones grow out over longer distances and in areas with different environmental cues that may influence their branching and elongation behavior. The assumption of spatial homogeneity is therefore not appropriate for axonal outgrowth. A neural simulator thus needs to specify the environment in which axons grow out, and the rules for interaction of axonal growth cones with the local environmental cues. Presently, NETMORPH is being extended with options for defining areas qualified by spatial scalar and vector fields. Scalar fields represent trophic actions on growth cones that modulate branching probability and elongation rate, and vector fields represent tropic actions on growth cones that modulate the direction of outgrowth of growth cones.

4.6.1.5 Independent Outgrowing Neurons

The simulator NETMORPH presently assumes that neurons grow out independently of each other and that synaptic connectivity fully depends on the geometry of the neurons. The independency assumption may be challenged when growth cone–environment interactions are being implemented for axon outgrowth, and in particular when the environment becomes influenced by the presence of other neurons. An interesting finding in this context was recently published by Hill et al. (2012). They compared presynaptic and postsynaptic path length distributions from experimental data with predictions from artificial networks composed of the reconstructed neurons and synapse formation based on geometrical considerations. They concluded that the predicted statistical connectivity was a good predictor of the experimentally observed functional connectivity. This finding may suggest that, even when neurite outgrowth proceeds by growth cones influenced by their local environments, the number of factors involved in such modulation is still so large that a stochastic approach is justified.

In the given example the outgrowth parameters are optimized on the morphologial properties of reconstructions, made available via the NeuroMorpho.Org database (Ascoli 2006; Ascoli et al. 2007). These reconstructions were all obtained from sliced tissue, containing only the parts of the axonal and dendritic arborizations within the slices (with thicknesses less than or equal to about 300 μm). By proper slicing and selecting only centrally placed neurons, the loss of cut branches can be minimized, but axons will certainly lose their long-distance projections. Therefore the growth rules have been optimized on the morphological properties of the dendritic and axonal parts within the slices. Consequently, the axons and dendrites generated by NETMORPH agree with these properties.

In the example discussed above the neuronal cell bodies were placed within a limited cylindrical space of height 380 μm and radius 93 μm. The space covered by the dendritic arborizations thus extends this cylinder by the dendritic Euclidean pathlengths, making synaptic connections only possible for the axonal arborizations within this dendritic space. Thus, long-distance parts of axonal arborizations were not required for these spatial dimensions.

4.6.2 Future Developments

Computational modeling is an essential tool towards understanding the brain. Cognition emerges from the activity dynamics in the neuronal networks comprising the brain. Modelers are thus challenged by the need of producing tools for reproducing the structural components of the brain and for simulating the activity dynamics on these structures. It depends on the questions addressed to what spatial and temporal details (granularity) these tools need to be developed.

The connectivity patterns in neuronal networks depend on the morphological characteristics of neuronal arborizations. This underscores the need for simulators able to generate neuronal arborizations with sufficient level of realism and detail. While the simulator NETMORPH has shown its functionality in generating dendritic arborizations, extensions are needed to reproduce axonal arborizations in more realistic detail. Such extensions include interactions of outgrowing neurites with local environmental cues. Preliminary work using scalar and vector fields for trophic and tropic actions indeed showed improved realism in the typical orientation preferences of axonal fibers and layer specific branching rates.

Long-range axonal connections, however, require also detailed qualifications of brain areas for axon guidance, and complex interaction rules to be included in the simulator. Finding a good balance between the need for a low-dimensional outgrowth model and a rich and realistic repertoire of emergent structures and connectivities will be a major challenge in further developments of the simulator.

Dendritic and axonal arborizations may also statistically be represented by so called density fields. The generation of connected networks from these density fields is still a challenge and a topic of ongoing research.

Acknowledgments This work was supported by grants from the Netherlands Organization for Scientific Research (NWO) through the Program Computational Life Sciences (grant number 635.100.005), the MC-RTN project NEURoVERS-it (grant number 019247) of the European Union, and the BIO-ICT project SECO (grant number 216593) of the Seventh Framework Programme of the European Union.

References

Aćimović J, Mäki-Marttunen T, Havela R, Teppola H, Linne M-L (2011) Modeling of neuronal growth in vitro: comparison of simulation tools NETMORPH and CX3D. EURASIP. J Bioinform Syst Biol 2011, 616382

Ascoli G, Krichmar J (2002) L-neuron: a modeling tool for the efficient generation and parsimonious description of dendritic morphology. Neurocomputing 32–33:1003–1011

Ascoli GA (2006) Mobilizing the base of neuroscience data: the case of neuronal morphologies. Nat Rev Neurosci 7:318–324

Ascoli GA, Donohue DE, Halavi M (2007) NeuroMorpho.Org: a central resource for neuronal morphologies. J Neurosci 27(35):9247–9251

Burke RE, Marks WB, Ulfhake B (1992) A parsimonious description of motoneuron dendritic morphology using computer simulation. J Neurosci 12:2403–2416

Cuntz H, Forstner F, Borst A, Häusser M (2010) One rule to grow them all: a general theory of neuronal branching and its practical application. PLoS Comput Biol 6(8):e1000877

Costa LF, Coelho RC (2005) Growth driven percolations: the dynamics of connectivity in neuronal systems. Eur Phys J B 47:571–581

Costa Lda F, Coelho RC (2008) Growth-driven percolations: the dynamics of community formation in neuronal systems. arXiv:q-bio/0411009v1;[q-bio.NC]

Costa LF, Manoel ETM, Faucereau F, Chelly J, Van Pelt J, Ramakers GJA (2002) A shape analysis framework for neuromorphometry. Network 13:283–310

Eberhard JP, Wanner A, Wittum G (2006) NeuGen: a tool for the generation of realistic morphology of cortical neurons and neuronal networks in 3D. Neurocomputing 70(1–3):327–342

Feldmeyer D, Lübke J, Silver RA, Sakmann B (2002) Synaptic connections between layer 4 spiny neurone- layer 2/3 pyramidal cell pairs in juvenile rat barrel cortex: physiology and anatomy of interlaminar signalling within a cortical column. J Physiol 538:803–822

Gillette T, Brown K, Ascoli G (2001) The DIADEM Metric: comparing multiple reconstructions of the same neuron. Neuroinformatics 9:233–245

Gleeson P, Steuber V, Silver RA (2007) neuroConstruct: a tool for modeling networks of neurons in 3D space. Neuron 54(2):219–235

Hellwig B (2000) A quantitative analysis of the local connectivity between pyramidal neurons in layers 2/3 of the rat visual cortex. Biol Cybern 82:111–121

Hill SL, Wang Y, Riachi I, Schurmann F, Markram H (2012) Statistical connectivity provides a sufficeint foundations for specific functional conenctivity in neocortical neural circuits. Proc Natl Acad Sci USA 109(42):E2885–E2894

Kim Y, Sinclair R, Chindapol N, Kaandorp JA, De Schutter E (2012) Geometric theory predicts bifurcations in minimal wiring cost trees in biology are flat. PLoS Comput Biol 8(4):e1002474

Koene RA, Tijms B, van Hees P, Postma F, de Ridder S, Ramakers G, van Pelt J, van Ooyen A (2009) NETMORPH: a framework for the stochastic generation of large scale neuronal networks with realistic neuron morphologies. Neuroinformatics 7:195–210

Lamoureux P, Buxbaum RE, Heidemann SR (1998) Axonal outgrowth of cultured neurons is not limited by growth cone competition. J Cell Sci 111:3245–3252

Le Bé JV, Silberberg G, Wang Y, Markram H (2007) Morphological, electrophysiological, and synaptic properties of corticocallosal pyramidal cells in the neonatal rat neocortex. Cereb Cortex 17:2204–2213

Luczak A (2006) Spatial embedding of neuronal trees modeled by diffusive growth. J Neurosci Methods 157:132–141

Mäki-Marttunen T, Aćimović J, Nykter M, Kesseli J, Ruohonen K, Yli-Harja O, Linne M-L (2011) Information diversity in structure and dynamics of simulated neuronal networks. Front Comput Neurosci

Perin R, Berger TK, Markram H (2011) A synaptic organization principle for cortical neuronal groups. Proc Natl Acad Sci USA 108(13):5419–5424

Peters A (1979) Thalamic input to the cerebral cortex. Trends Neurosci 2:1183–1185

Rall W (1959) Branching dendritic trees and motoneuron membrane resistivity. Exp Neurol 1:491–527

Senft S, Ascoli G (1999) Reconstruction of brain networks by algorithmic amplification of morphometry data. Lect Notes Comp Sci 1606:25–33

Samsonovich A, Ascoli G (2007) Computational models of dendritic morphology: From parsimonious description to biological insight. In: Laubichler MG, Müller G (eds) Modeling biology, structures, behaviors, evolution. MIT Press, Cambridge, MA, pp 91–113

Song S, Sjöström PJ, Reigl M, Nelson S, Chklovskii DB (2005) Highly nonrandom features of synaptic connectivity in local cortical circuits. PLoS Biol 3(3):e68

Torben-Nielsen B, Vanderlooy S, Postma EO (2008a) Non-parametric algorithmic generation of neuronal morphologies. Neuroinformatics 6:257–277

Torben-Nielsen B, Tuyls K, Postma EO (2008b) EvOL-Neuron: virtual neuron generation. Neurocomputing 71(4–6):963–972

Uylings H, Smit G (1975) Three dimensional branching structure of pyramidal cell dendrites. Brain Res 87:55–60

Uylings HBM, van Pelt J, Parnavelas JG, Ruiz-Marcos A (1994) Geometrical and topological characteristics in the dendritic development of cortical pyramidal and nonpyramidal neurons. In: van Pelt J, Corner MA, Uylings HBM, Lopes da Silva FH (eds) Progress in brain research (Vol 102), The self-organizing brain: from growth cones to functional networks. Elsevier, Amsterdam, pp 109–123

Uylings HBM, van Pelt J (2002) Measures for quantifying dendritic arborizations. Network 13:397–414

Van Ooyen A (2011) Using theoretical models to analyse neural development. Nat Rev Neurosci 12:311–326

Van Ooyen A, Graham BP, Ramakers GJA (2001) Competition for tubulin between growing neurites during development. Neurocomputing 38–40:73–78

Van Pelt J, Uylings HBM (2002) Branching rates and growth functions in the outgrowth of dendritic branching patterns. Network 13:261–281

Van Pelt J, Uylings HBM (2003) Growth functions in dendritic outgrowth. Brain and Mind 4:51–65

Van Pelt J, Uylings HBM (2005) Natural variability in the geometry of dendritic branching patterns. In: Reeke GN, Poznanski RR, Lindsay KA, Rosenberg JR, Sporns O (eds) Modeling in the neurosciences: from biological systems to neuromimetic robotics. CRC Press, Boca Raton, pp 89–115

Van Pelt J, Uylings HBM (2007) Modeling neuronal growth and shape. In: Laubichler MD, Müller GB (eds) Modeling biology – structures, behaviors, evolution. MIT Press, Cambridge, pp 195–215

Van Pelt J, Uylings HBM (2012) The flatness of bifurcations in 3D dendritic trees: an optimal design. Front Comput Neurosci 5:54

Van Pelt J, Van Ooyen A, Uylings HBM (2001) Modeling dendritic geometry and the development of nerve connections. In: de Schutter E, Cannon (CD-ROM) RC (eds) Computational neuroscience: realistic modeling for experimentalist. CRC Press, Boca Raton, pp 179–208

Van Pelt J, Carnell A, de Ridder S, Mansvelder HD, van Ooyen A (2010) An algorithm for finding candidate synaptic sites in computer generated networks of neurons with realistic morphologies. Front Comput Neurosci 4:148

Zubler F, Douglas RA (2009) Framework for modeling the growth and development of neurons and networks. Front Comput Neurosci 3:25

Chapter 5
Shaping of Neurons by Environmental Interaction

Artur Luczak

Abstract The geometry of dendritic trees plays an important role in determining the connectivity. However, the extent to which environmental factors shape dendritic geometry remains largely unknown. Recent development of computational models can help us to better understand it. This chapter provides a description of one such model (Luczak, J Neurosci Methods 157:132–41, 2006). It demonstrates that assuming only that neurons grow in the direction of a local gradient of a neurotrophic substance, and that dendrites compete for the same resources, it is possible to reproduce the spatial embedding of major types of cortical neurons. In addition, this model can be used to estimate environmental conditions which shape actual neurons, as proposed in Luczak, Front Comput Neurosci 4:135, 2010. In summary, the presented model suggests that basic environmental factors, and the simple rules of diffusive growth can adequately account for different types of axonal and dendritic shapes.

5.1 Introduction

Saying that shape determines function is especially true for neurons. The shape of dendritic trees to a large degree determines electrophysiological properties of neurons (Migliore et al. 1995; Mainen and Sejnowski 1996; Krichmar et al. 2002) and neuronal connectivity (Amirikian 2005; Stepanyants and Chklovskii 2005; Cuntz et al. 2010; Perin et al. 2011). Experimental data suggests that both intrinsic and

---

A. Luczak (✉)
Department of Neuroscience, Canadian Centre for Behavioural Neuroscience,
University of Lethbridge, 4401 University Drive, Lethbridge, AB, Canada, T1K 3M4
e-mail: luczak@uleth.ca

H. Cuntz et al. (eds.), *The Computing Dendrite: From Structure to Function*,
Springer Series in Computational Neuroscience 11, DOI 10.1007/978-1-4614-8094-5_5,

extrinsic factors influence dendritic geometry, but the relative importance of those factors is not well understood (Scott and Luo 2001). The complexity of interactions between different intrinsic and extrinsic factors during the development of neuronal arborization can make it very difficult to separate their contributions experimentally. One of the approaches to address this question is to use computational modeling. Nevertheless, even with advances in computational neuroscience, it is not a simple task, and it has taken decades to develop realistic models of neuronal growth. For example, in its earliest works, only dendrograms were modeled (i.e., connectivity among branches and their length and diameter) yet spatial embedding or environmental factors were not considered (Nowakowski et al. 1992; Van Pelt et al. 1997). In the later models of dendritic trees, 3D structures were included but without considering environment. In these models, several parameters measured from real neurons (e.g., the probability distribution of branching points as a function of the distance from a soma) were used and stochastic procedures were applied to recreate dendrites while disregarding influence of the environment (Ascoli 1999; Burke and Marks 2002; Samsonovich and Ascoli 2003; Samsonovich and Ascoli 2005; Lindsay et al. 2007; Torben-Nielsen et al. 2008; Koene et al. 2009; for review see Ascoli 2002).

In contrast to the above models based on statistical reconstruction of dendrites, the most recent models can simulate 3D neuronal growth with incorporating external factors. In these approaches, dendrite geometry parameters (e.g., number of segments, branching probability, orientation) are not built into the model, but rather geometry parameters emerge as a result of environmental factors such as the concentration of neurotropic factors, competition between neurons, and space limitations (Luczak 2006; Zubler and Douglas 2009). External cues are well known to play a significant role in shaping dendritic geometry (Horch and Katz 2002), and hence these models account for the important biological processes underlying neuronal geometry.

In this chapter, we focus on the model developed by Luczak (2006) which examines the effect of environmental factors on spatial embedding of neuronal trees. This model is based on diffusion-limited aggregation (DLA), which is a well-established physical model for the formation of structures controlled by diffusion processes (Witten and Sander 1981). Prior research has demonstrated that DLA can provide a good description of a variety of natural processes, such as electrical discharge in gas (lightning) (Niemeyer et al. 1984), electrochemical deposition (Halsey 1990; Brady and Ball 1984), or the growth of snowflakes (Family et al. 1987). The form of a typical DLA structure is illustrated in the insert of Fig. 5.1.

Previously, diffusive processes were invoked to explain the origin of dendritic arbors by Hentschel and Fine (1996), who proposed a two-dimensional diffusion-regulated model of dendritic morphogenesis in which cell growth depended upon the local concentration of calcium. However, their model was restricted to only the early stage of neuronal growth, and was based on in vitro dissociated cultures. In order to generate a 3D embedding of fully developed dendritic trees, the present model operates at a coarser level. It takes into account the local concentrations of neurotrophic factors, but without including such details as changes in the concentrations of ions along the dendrite membrane.

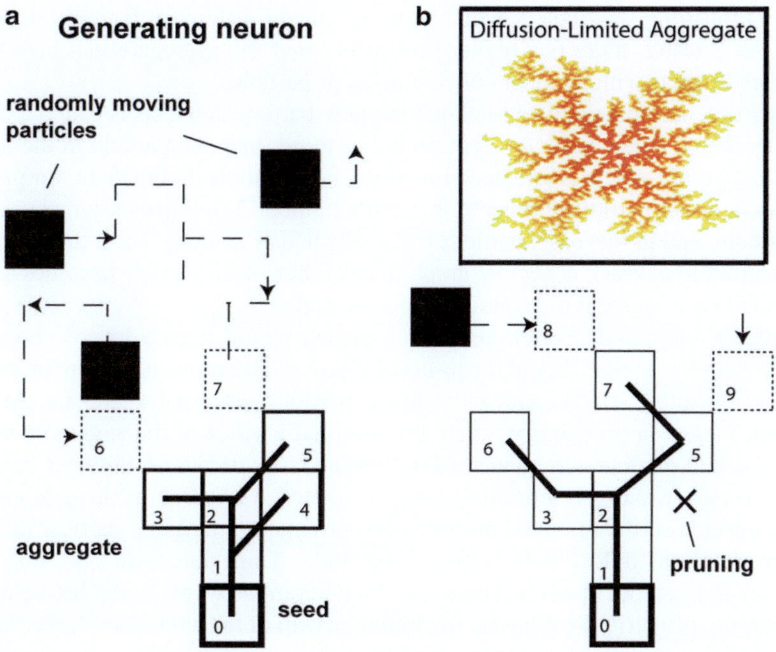

Fig. 5.1 Illustration of the DLA algorithm. (**a**) Randomly moving particles (*black*) stick irreversibly at their point of first contact with the aggregate (composed of particles 0–5). To each newly jointed particle, a parent particle is assigned and both become connected by a line segment. (**b**) While the aggregate grows, the particles at the terminals are randomly deleted from the aggregate (pruning) during a specified time window. Insert: Example of a two-dimensional DLA comprising 6,000 particles. The color intensity decreases in the order in which particles connected to the aggregate (reproduced from Luczak 2006 with permission from Elsevier)

5.2 Model Description

As described in details in (Luczak 2006), the growth rule for DLA can be defined inductively as follows: introduce a randomly moving particle at a large distance from an *n*-particle aggregate, which sticks irreversibly at its first point of contact with the aggregate, thereby forming an *n*+1-particle aggregate. Figure 5.1a illustrates a sample trajectory of particles that stick to an aggregate composed of five particles (each particle is numbered in the order in which it contacts the aggregate; the seed particle is numbered 0). Stated differently, the aggregate grows by one step at the point of contact with a particle, thus prominent branches screen internal regions of the aggregate, preventing them from growing further (Halsey 1997). For computational efficiency, instead of one moving particle, *m* simultaneously moving particles were introduced (Voss 1984). For computational convenience, particles leaving box on one side, enter the box on the opposite side. In the presented model, the initial distribution of particles is a model parameter and thus particles are not always uniformly distributed, which is a significant difference from the classical DLA. As a result, there is a higher probability that the aggregate will have more branches in areas with a higher concentration of particles.

A particle in the aggregate to which the new particle connects is called a parent particle. When a new particle is connected to more than one particle in the aggregate, the parent particle is selected at random. For example, in Fig. 5.1a, for particle number 4, either particle number 1 or particle number 2 could be assigned as a parent particle, and in this case, particle 1 was selected at random. Thus, the aggregate is converted to a directed, acyclic graph (tree), where each particle becomes a node connected by a segment to an assigned parent node.

Without additional restrictions, the aggregate would form a heavily branched structure similar to the DLA in the insert of Fig. 5.1. Therefore, a pruning procedure was implemented, which removes terminal particles from the aggregate. At each iteration there is a probability p that any terminal particle of the aggregate can be deleted if that particle was connected within the last s iterations, but later than five iterations ago, where s is a pruning span parameter. As a result of the deletion, the parent particle of the removed particle becomes again a terminal particle (eligible for the deletion) unless it is a branching node. Thus, increasing pruning span increases the number of deleted particles. Five iterations were chosen before applying pruning, primarily to allow for the initial growth of the aggregate. Nevertheless, this parameter has a very minor effect on the geometry of a dendrite as compared to pruning span. The removed particles do not return to the pool of particles, and the seed particle cannot be removed by definition. The algorithm stops when no new particle is connected for 100 iterations.

Illustration of the initial spatial distribution of particles for the granule cells is presented in Fig. 5.2a. To eliminate tree variability due to seed position, the seed is always located in the center at the bottom of the box, or when generating multiple trees simultaneously, seeds are uniformly distributed at the bottom of the box. The number of seed particles placed inside a box determined the number of aggregates.

The MATLAB code used to produce the described simulations is available upon request, or it can be downloaded from: http://lethbridgebraindynamics.com/artur_luczak.

5.3 Generating Different Neuronal Tree Types

The above description of how branching structures grow could be seen as an "innate" set of rules which is the same for all neuronal types generated with this model. What differentiates between growing a granule cell and a pyramidal cell in our model are mostly differences in parameters defining environment. Interestingly, only two environment factors were enough to define and reproduce diverse types of neuronal trees. These factors were: the distribution of "neurotrophic particles" (NPs), and space available to grow. Below we will describe in more detail the effect of both factors on neuronal shape, beginning with the space constraints.

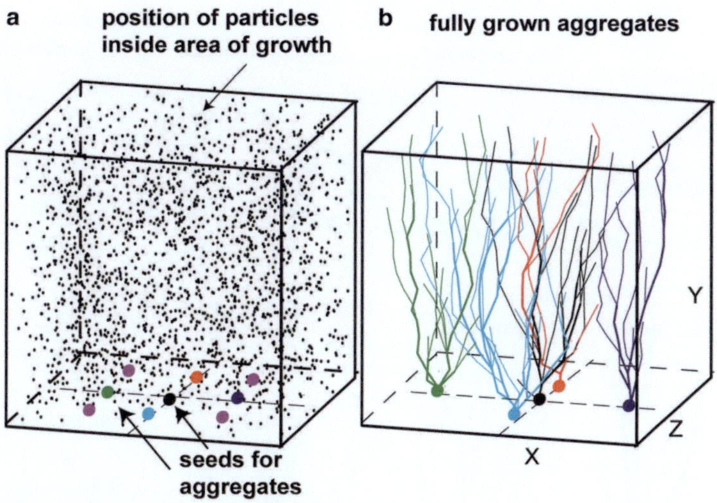

Fig. 5.2 Generating neurons in ensemble. (**a**) Illustration of the initial condition for generating nine aggregates. *Small black dots* denote initial location of NP particles. For illustration clarity only centers of NPs are marked, as the full size of NP used in this simulation (~10 μm) would obstruct view of particles in the back. (**b**) Generated granule cells (cells in corners are not shown for visualization clarity). *Rectangular box* represents a space limitation imposed on the growth of aggregates (reproduced from Luczak 2006 with permission from Elsevier)

In our model, the shape of generated dendrites directly depends on the size of the rectangular box in which the neuron is growing (Fig. 5.2). For example, increasing the height–width ratio of the box changes the dendrite shape from that of a basal dendrite to a granule cell and, ultimately, to an apical dendrite. Decreasing relative width only in the Z-coordinate direction changes it from a basal dendrite to a Purkinje cell.

The use of a rectangular box to limit neuron growth may, at first sight, appear to impose an artificial constraint, whereas this actually simulates the space limitations imposed by the thickness of the cortical layers (height of the box), and by neighboring neurons growing simultaneously and competing for space and neurotrophic substances (length and width of the box) (Devries and Baylor 1997). For example, generating a neuron surrounded by other simultaneously growing cells resulted in a shape almost identical to growing this neuron in an isolated box with appropriately reduced size (Fig. 5.2). This is because the neighboring aggregates competed for available space and access to NPs, which limited the sideways growth of the neuron. Thus, sideways space limitation imposed in the model could also be viewed as another manifestation of constrains imposed by the spatial distribution of neurotrophic factors, where a neuron has a higher chance of getting access to NPs by extending its processes up rather than to the sides due to competition from other cells.

We found that it was also important to vary the distribution of NPs along the vertical direction to accurately reflect branching patterns of reproduced neurons.

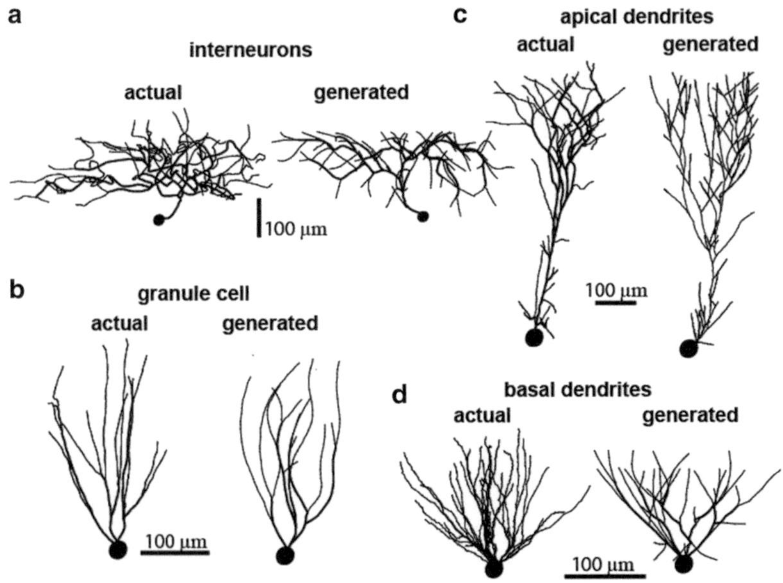

Fig. 5.3 Examples of real and generated neurons. (**a**) Examples of real and generated axonal trees of interneurons. (**b**) Examples of real and generated granule cells. (**c**) Examples of real and generated apical dendrites of pyramidal cells. (**d**) Examples of real and generated basal dendrites. For all neurons, the cell bodies are depicted by spheres (modified from Luczak 2006)

This at first was a surprise, but after examining the required patterns of NPs distribution, we found a striking similarity to the laminar structure of cortical layers. Thus, the difference in NPs concentration at different heights in the box could be a reflection of a different concentration of actual neurotrophic factors across different cortical layers. A direct consequence of "innate" properties of the DLA-based model is that increasing the concentration of NPs increases the density of branches. For example, increasing the concentration of NPs in the upper 30 % section of the box, while reducing it to almost zero elsewhere, produces an aggregate with the appearance of an interneuron (Fig. 5.3a) rather than of a granule cell or basal dendrite (Fig. 5.3b, d). As mentioned before, such changes in particle density along the vertical axis may be biologically justified as reflecting different cortical layers. In the model, the initial distribution of particle densities along the vertical axis exhibits a sharp transition between two regions with different concentrations. However, after a few iterations, the diffusive motion of NPs creates a smooth concentration gradient between the layers which is closer to real biological conditions. Due to much faster growth of the aggregate than diffusion of NPs between layers, the difference in concentration between layers is maintained throughout simulation. Thus, by changing only the space available for growth, the spatial distribution of NPs (and the pruning span to a smaller degree, Luczak 2006), the DLA-based model makes it possible to generate 3D structures similar to different types of dendritic and axonal trees (Fig. 5.3; Table 2 in Luczak 2006 provides detailed comparison of dendrite geometry parameters for generated and actual neurons which were obtained from http://neuromorpho.org, Ascoli et al. 2001).

5.4 Estimating the Effect of Environmental Variables on Neuronal Shape

Modeling neuronal growth can provide information on which environmental factors may be influencing neuronal shape. Moreover, it was proposed in (Luczak 2010), that modeling can also be used to infer what environmental conditions for any actual reconstructed neuron were. By measuring "how easy" it is to reproduce the shape of the actual neuron with the appropriate model, it can be estimated how "typical" the neuron is, and which environmental variables have to be adjusted to better reproduce it. For example, if increasing the concentration of NPs in the upper part of the box results in a model that can "easier" or "better" reproduce a given neuronal tree, then it may indicate that this neuron experienced more neurotrophic factors in upper cortical layers.

As an example of how to measure "how easy" it is to reproduce a neuron with a DLA-based model, let's use the DLA algorithm with uniform distribution of NPs as described in details in (Luczak 2010). The "reproduction" rules are similar to rules generating DLA. The algorithm starts by placing the seed particle at the location of the cell body of the reproduced neuron (Fig. 5.4a). Randomly moving particles (NPs) stick irreversibly at the first point of contact with the aggregate only if the point of that contact overlaps with a reproduced object (Fig. 5.4b). By definition, every particle at the time of connecting to the aggregate has a hit value equal to 1. Moving particles which contact the aggregate at the point not overlapping with the object are not connected (Fig. 5.1c, particle on the right side). If such a moving particle in the next step moves to a place already occupied by the aggregate, then that particle is deleted, and the aggregate at that point registers a new hit (Fig. 5.4d). In summary, the aggregate grows at the points of contact with randomly moving particles, covering the reproduced neuron particle-by-particle.

Thus, a randomly moving particle after hitting the aggregate will either become a new particle of the aggregate, or it will be deleted depending on whether or not the place of contact overlaps with the reproduced shape. To illustrate how many times a particle of the reproduced neuron was hit by moving particles during the growth process, a sample pyramidal neuron and Purkinje cell are color-coded for the total number of hits (Fig. 5.4e, f). As can be seen just by visual inspection, the distribution of hits will be different for both cell types. For pyramidal cell most branches were hit a similar number of times. In contrast, for Purkinje cell the number of hits decreased gradually with distance from the soma. One could measure distance between hit distributions for analyzed objects to evaluate similarity between them. For example, comparing the hit distribution of an analyzed structure to the

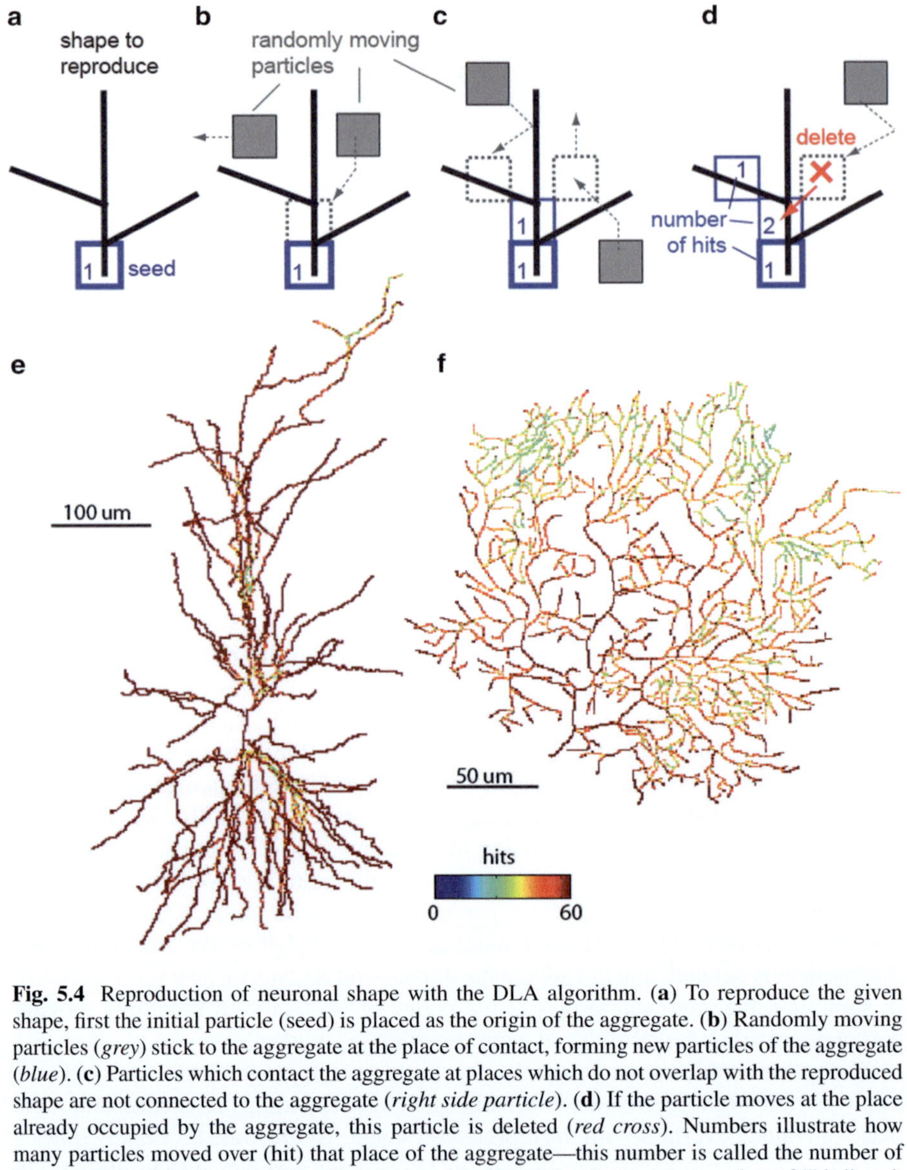

Fig. 5.4 Reproduction of neuronal shape with the DLA algorithm. (**a**) To reproduce the given shape, first the initial particle (seed) is placed as the origin of the aggregate. (**b**) Randomly moving particles (*grey*) stick to the aggregate at the place of contact, forming new particles of the aggregate (*blue*). (**c**) Particles which contact the aggregate at places which do not overlap with the reproduced shape are not connected to the aggregate (*right side particle*). (**d**) If the particle moves at the place already occupied by the aggregate, this particle is deleted (*red cross*). Numbers illustrate how many particles moved over (hit) that place of the aggregate—this number is called the number of hits. (**e**) Examples of pyramidal cell reproduced in 3D. *Color* denotes the number of "hits" each part of the dendrite received. (**f**) Reproduced Purkinje cell (modified from Luczak 2010)

distribution of hits for a structure generated with a DLA algorithm, it could provide a reliable measure for how similar to a typical DLA structure ("DLA-like" or "tree-like") is a given shape (Luczak 2010). Likewise, using a DLA-based model optimized to generate specific type of neurons (as described in Sect. 5.3) could be used to measure, for example, how similar to typical Purkinje cell is an analyzed neuron, and it can be further used to infer what parameters reflecting environmental conditions have to be adjusted to improve the reproduction.

5.5 Discussion

The main significance of the presented model is that it illustrates that the creation of complex dendritic trees does not require precise guidance or an intrinsic plan of the neuron geometry, but rather, that external factors can account for the spatial embedding of the major types of dendrites observed in the cortex. In this model, the number of terminal branches, the mean and maximum branch orders, the fractal dimension, and other parameters of dendrite geometry are all controlled by a few basic environmental factors (Luczak 2006). The most important factor in determining the shape of generated neurons is the space available for growth and the spatial distribution of NPs. Thus, the presented DLA-based model reveals that a simple, diffusive growth mechanism is capable of creating complex and diverse 3D trees strictly similar to observed neuronal shapes.

The main criticism of this model is that although it is able to generate diverse neuronal shapes, a direct translation to biological processes is difficult, as real dendrites do not grow directly by aggregating particles from their environment (Van Ooyen 2011). This is a very important point, but with a closer examination it can be found that the presented model is much less artificial than it would appear. This is because in the DLA model, connecting a new particle to the aggregate is computationally equivalent to the growth in the direction of that particle. DLA is an approximation of Laplacian growth where the probability of growth at any point on the boundary of the growing object is determined by Laplace's equation, which describes the "attraction" field around the object (Hastings and Levitov 1998). Therefore, the growth in the direction of a local gradient is computationally equivalent to the DLA model which connects particles to the aggregate with higher probability in places which have higher local concentration of NPs. Thus, DLA is used as a computationally convenient tool to model (1) the growth of a dendrite *toward* a higher concentration of NPs, (2) diffusive motion of NPs, and (3) competition between dendrites for access to NPs.

The real dendrites grow by elongation and can branch either via bifurcation of growth cone-like tips or through interstitial sprouting of new branches from an existing dendritic branch. These new branches extend and retract to undergo constant remodeling. Only a subset is eventually stabilized (Jan and Jan 2010). This phenomenon of constant pruning of dendritic branches during neuron development is modeled here by probabilistic deleting the terminals. Parts of the neuron, which were not deleted during a specified number of iteration (pruning span), become "stabilized" by being excluded from any further pruning. The growth and pruning of real cortical neurons is strongly influenced by the excess or deficit of extrinsic factors, which includes for example: neurotrophin 3, brain-derived neurotrophic factor (BDNF), and nerve growth factor (McAlister et al. 1997). For instance, BDNF released from an individual cell alters the structure of nearby dendrites on an exquisitely local scale (Horch and Katz 2002). The intrinsic factors have an effect on stability rather than the directionality of the dendrite by affecting the dynamics of the structural components of dendrites (Scott and Luo 2001). The NPs in the presented model do not refer to any concrete neurotrophic substance. We chose to call these particles "neurotrophic" to suggest a biological interpretation of the model, which is that a new dendrite branch sprouts at the point of contact with neurotrophic particles. Stated differently, connecting NP to the aggregate can be seen as equivalent to the process where a new part of a dendrite comes from the cell itself at the location where the NP was detected. Also, a decrease in the number of freely moving NPs after contacting the aggregate has a biological justification, namely, that the neurotrophic molecules are commonly uptaken by neurons and transported to the cell body (Purves 1988; von Bartheld et al. 1996). As mentioned above, the neuron development is a very complicated process and the model presented here cannot account for all possible phenomena affecting neuron shape. For example, the morphology of axons and dendrites can be affected by mechanical tensions during brain development (Van Essen 1997). Additional model parameters could improve the model's accuracy, but would also increase its complexity. Thus, in light of the fact that the existing model performs well, at the goal of keeping the model simple, we believe the model's current level of complexity and accuracy are appropriately balanced.

The presented DLA-based model can also be used to infer parameters describing environmental factors for a given actual neuron. It is a conceptually new approach based on measuring "how easy" it is to reproduce neuronal shape by using a tree-generating algorithm. Thus, by finding what values of environmental variables allow the model to "best" reproduce of the analyzed neuron, it could provide information about conditions shaping this neuron. The performance of DLA algorithms could be measured in a variety of ways: for example, how quickly it can cover a shape, how completely it covers, how broad the distribution of hits is. From all of the different measures tried, the distance to the hit distribution of DLA provided the most reliable measure of similarity to modeled shape. In addition, taking for example the model optimized for generating Purkinje cells, and measuring how well this model can reproduce a given shape, this can provide a new measure of shape, i.e., how "Purkinje-like" is that object. There is still no single measure which quantifies our intuitive perception of how much this cell resembles a Purkinje neuron. This type of question is easily answered by humans, but it is very difficult to quantify using computers. The reason is that perception of a tree-like shape requires simultaneously combining a multitude of global and local measures like spatial distribution of segments, relative lengths and direction, connectivity, symmetry, space filling, etc. For example, we would consider as a tree only shapes with a particular type of connectivity pattern, and with a particular spatial distribution of segments, branching angles, relative lengths, orientation, etc. By using the DLA model to reproduce analyzed objects, we can quantify the tree-like resemblance of an object by simply measuring performance of the DLA algorithm. Thus, this approach presents a new conceptual advancement where the use of a computational model allows one to assess complex properties of an object, which otherwise would be very difficult to quantify with any other existing measures.

In conclusion, this chapter describes a realistic model of the formation of diverse neuron shapes. The results demonstrate that simultaneously grown diffusion-limited aggregates competing for available resources create reproducible, self-organized structures that are strikingly similar to real neurons (Fig. 5.3). This is the first model to simulate 3D neuronal growth accounting for external factors such as the NP concentration, competition between neurons, and space limitations. Moreover, it advances DLA-based models by incorporating pruning and space limitations. Analysis of the discrepancies between generated and real neurons may also elucidate the relative contribution of different environmental factors on neuronal outgrowth.

Acknowledgments This work was partly supported by grants from NSERC and AHFMR.

References

Amirikian B (2005) A phenomenological theory of spatially structured local synaptic connectivity. PLoS Comput Biol 1:74–85

Ascoli GA (1999) Progress and perspectives in computational neuroanatomy. Anat Rec 257:195–207

Ascoli GA (2002) Neuroanatomical algorithms for dendritic modeling. Network 13:247–260

Ascoli GA, Krichmar J, Nasuto S, Senft S (2001) Generation, description and storage of dendritic morphology data. Philos Trans R Soc Lond B Biol Sci 356:1131–1145

Brady R, Ball R (1984) Fractal growth of copper electrodeposits. Nature 309:225–229

Burke RE, Marks WB (2002) Some approaches to quantitative dendritic morphology. In: Ascoli GA (ed) Computational neuroanatomy: principles and methods. Humana, Totowa, NJ, pp 27–48

Cuntz H, Forstner F, Borst A, Häusser M (2010) One rule to grow them all: a general theory of neuronal branching and its practical application. PLoS Comput Biol 6:e1000877

Devries SH, Baylor DA (1997) Mosaic arrangement of ganglion cell receptive fields in rabbit retina. J Neurophysiol 78:2048–2060

Family F, Platt DE, Vicsek T (1987) Deterministic growth model of pattern formation in dendritic solidification. J Phys A 20:1177–1183

Halsey TC (1990) Electrodeposition and diffusion-limited aggregation. J Chem Phys 92:3756–3767

Halsey TC (1997) The branching structure of diffusion limited aggregates. Europhys Lett 39:43–48

Hastings MB, Levitov LS (1998) Laplacian growth as one-dimensional turbulence. Physica D 116:244–250

Hentschel HGE, Fine A (1996) Diffusion-regulated control of cellular dendritic morphogenesis. Proc R Soc Lond B 263:1–8

Horch HW, Katz LC (2002) BDNF release from single cells elicits local dendritic growth in nearby neurons. Nat Neurosci 5:1177–1184

Jan YN, Jan LY (2010) Branching out: mechanisms of dendritic arborization. Nat Rev Neurosci 11:316–328

Koene RA, Tijms B, van Hees P, Postma F, de Ridder A, Ramakers GJ, van Pelt J, van Ooyen A (2009) NETMORPH: a framework for the stochastic generation of large scale neuronal networks with realistic neuron morphologies. Neuroinformatics 7(3):195–210

Krichmar JL, Nasuto SJ, Scorcioni R, Washington SD, Ascoli GA (2002) Effects of dendritic morphology on CA3 pyramidal cell electrophysiology: a simulation study. Brain Res 941:11–28

Lindsay KA, Maxwell DJ, Rosenberg JR, Tucker G (2007) A new approach to reconstruction models of dendritic branching patterns. Math Biosci 205:271–296

Luczak A (2006) Spatial embedding of neuronal trees modeled by diffusive growth. J Neurosci Methods 157(1):132–141

Luczak A (2010) Measuring neuronal branching patterns using model-based approach. Front Comput Neurosci 4:135

Mainen ZF, Sejnowski TJ (1996) Influence of dendritic structure on firing pattern in model neocortical neurons. Nature 382:363–366

McAlister AK, Katz LC, Lo DC (1997) Opposing roles for endogenous BNDF and NT-3in regulating cortical dendritic growth. Neuron 18:767–778

Migliore M, Cook EP, Jaffe DB, Turner DA, Johnston D (1995) Computer simulations of morphologically reconstructed CA3 hippocampal neurons. J Neurophysiol 73:1157–1168

Niemeyer L, Pietronero L, Wiesmann HJ (1984) Fractal dimension of dielectric breakdown. Phys Rev Lett 52:1033–1036

Nowakowski RS, Hayes NL, Egger MD (1992) Competitive interactions during dendritic growth: a simple stochastic growth algorithm. Brain Res 576:152–156

Perin R, Berger TK, Markram H (2011) A synaptic organizing principle for cortical neuronal groups. Proc Natl Acad Sci USA 108:5419–5424

Purves D (1988) Body and brain. A trophic theory of neural connections. Harvard University Press, Cambridge, MA

Samsonovich AV, Ascoli GA (2003) Statistical morphological analysis of hippocampal principal neurons indicates cell-specific repulsion of dendrites from their own cells. J Neurosci Res 71:173–187

Samsonovich AV, Ascoli GA (2005) Statistical determinants of dendritic morphology in hippocampal pyramidal neurons: a hidden Markov model. Hippocampus 15:166–183

Scott E, Luo L (2001) How do dendrites take their shape? Nat Neurosci 4:353–359

Stepanyants A, Chklovskii DB (2005) Neurogeometry and potential synaptic connectivity. Trends Neurosci 28:387–394

Torben-Nielsen B, Vanderlooy S, Postma EO (2008) Non-parametric algorithmic generation of neuronal morphologies. Neuroinformatics 6(4):257–277

Van Essen DC (1997) A tension-based theory of morphogenesis and compact wiring in the central nervous system. Nature 385:313–318

Van Ooyen A (2011) Using theoretical models to analyse neural development. Nat Rev Neurosci, 12(6):311–326

Van Pelt J, Dityatev A, Uylings HBM (1997) Natural variability in the number of dendritic segments: model-based inferences about branching during neurite outgrowth. J Comp Neurol 387:325–340

Von Bartheld CS, Williams R, Lefcort F, Clary DO, Reichardt LF, Bothwell M (1996) Retrograde transport of neurotrophins from the eye to the brain in chick embryos: roles of the p75NTR and trkB receptors. J Neurosci 16:2995–3008

Voss RF (1984) Multiparticle fractal aggregation. J Stat Phys 36:861–872

Witten TA, Sander LM (1981) Diffusion limited aggregation, a kinetic critical phenomena. Phys Rev Lett 47:1400–1408

Zubler F, Douglas R (2009) A framework for modeling the growth and development of neurons and networks. Front Comput Neurosci 3:25

Chapter 6
Modelling Dendrite Shape from Wiring Principles

Hermann Cuntz

Abstract The primary function of a dendrite is to connect a neuron to its inputs. In this chapter, I describe a model that captures the general features of dendritic trees as a function of the connectivity they implement. This model is based on locally optimising connections by weighing costs for total wiring length and conduction times. The model was used to generate synthetic dendrites that are visually indistinguishable from their real counterparts for all dendrite types tested so far. Dendrites of different cell types vary only in the shape of the volume that they span and in the weight between costs for wiring length versus conduction times. Using the model, an equation was derived that relates total dendrite length, number of branch points, spanning volume and the number of synapses, measures that are among the most commonly employed in the study of the molecular and genetic background of dendrite morphology and growth. This equation holds true for all neurons measured so far and confines the possible computations a dendrite is capable of. Finally, beyond the consequences for neuronal morphology and computation, an outlook is given on a number of ways to scale up the single cell model to study the formation of larger neural circuits.

The large network of finely branched neurons in the nervous system is suggestive of its exquisitely complex underlying connectivity. Among these branched processes, the dendritic input trees exist in the most striking variety of shapes and sizes (DeFelipe 2010; Ramón y Cajal 1995; Stuart et al. 2007). This variety most likely reflects the different computations required throughout the nervous system. Indeed, the anatomy of the dendritic tree is known to shape synaptic integration (Gulledge et al. 2005;

H. Cuntz (✉)
Ernst Strüngmann Institute (ESI) for Neuroscience in Cooperation with Max Planck Society, Frankfurt/Main, Germany

Institute of Clinical Neuroanatomy, Goethe University, Theodor-Stern-Kai 7, Frankfurt/Main, D-60590, Germany
e-mail: hermann.neuro@gmail.com

H. Cuntz et al. (eds.), *The Computing Dendrite: From Structure to Function*, Springer Series in Computational Neuroscience 11, DOI 10.1007/978-1-4614-8094-5_6,

London and Häusser 2005; Rall et al. 1967; Rall 1959; Rall and Rinzel 1973; Segev and London 2000) and the corresponding neural firing behaviour (e.g. Mainen and Sejnowski 1996). Understanding the principles governing dendritic branching is therefore essential for unravelling the functionality of single neurons with respect to computation and to the way in which they connect. However, so far no unifying theory exists that links structure of dendrites to their particular function. Very little is known about any general rules leading to the distinct appearance of dendrites partly due precisely to the wide variety of shapes among different neuron types. On the other hand, a vast amount of recent research has expanded our knowledge on the molecular and genetic factors governing dendrite growth in recent years (Acebes and Ferrús 2000; Cline 2001; Grueber and Jan 2004; Jan and Jan 2003, 2010; Parrish et al. 2007). But in most of these cases the quantitative analysis of dendritic arborisations is restricted to descriptions of local and global branching parameters (overview: Uylings and Van Pelt 2002; see Chap. 1). Measures that are most commonly used are the total dendrite length, the number of branch points as well as the classical Sholl analysis that measures the number of intersections between the branches of the dendrite and an imaginary sphere of increasing diameter centred on the dendritic root (Sholl 1959). However, these measures have little to do with dendrite function. Moreover, classifying individual neurons into different neuron types solely on the basis of morphological parameters has been generally unsuccessful although it has worked in very specific cases (e.g. Cannon et al. 1999; Oberlaender et al. 2011). More recently, computational techniques have appeared that make it possible to construct synthetic neuronal structures *in silico* governed by the simulation of physical and biological constraints (Koene et al. 2009; van Pelt et al. 2001; van Pelt and Uylings 2002) or by the implementation of simple geometric rules (Ascoli 1999; Hillman 1979; Luczak 2006; Torben-Nielsen et al. 2008). Modelling in the field of neuroscience has become an established tool in identifying underlying principles in the complex and often dynamical systems of neural circuits (since e.g. Hodgkin and Huxley 1952; Rall 1959). This chapter summarises the work of several papers (Cuntz 2012; Cuntz et al. 2007, 2008, 2010, 2012) for which the goal was to develop a model that incorporates known functional constraints directly into the generative process of building synthetic dendritic morphologies in order to interpret the consequences of these known constraints onto the neural circuit.

6.1 Morphological Modelling: Synthetic Dendrites Shaped by Wiring Constraints

Because dendrites are the main receptive region of neurons, one requirement common to all dendrites is that they need to connect with often widespread input sources. More than a century ago, Ramón y Cajal was able to make sense of the neuroanatomical complexity by formulating fundamental anatomical principles of nerve cell organisation (Ramón y Cajal 1995). He described three biological laws of neuronal architecture (Chapter V, p.115–125, in Ramón y Cajal 1995). These were optimisation principles for the conservation of (1) space, (2) cytoplasm and (3) conduction times in the neural circuitry. These principles were instrumental for him to interpret his anatomical observations in terms of wiring and allowed him to postulate a wide variety of theories of functionality and directionality of signal flow in various brain areas. In the meantime, the principles of optimal wiring that he suggested have been substantiated by more rigorous statistical analyses: circuitry and connectivity considerations as well as simple wire-packing constraints have been shown to describe statistics of dendritic and axonal morphology (Cherniak 1992; Chklovskii 2004; Shepherd et al. 2005). It has also been shown mathematically that the specific organisation and architecture of many parts of the brain reflect the selection pressure to reduce wiring costs for the circuitry (Chklovskii and Koulakov 2004; Klyachko and Stevens 2003; Wen et al. 2009; Wen and Chklovskii 2008). In a separate chapter of this book, Quan Wen describes these findings in great detail (see Chap. 7). Are dendrites and axons then ideally matched to anatomically optimise connections to their potential connection partners in terms of Cajal's rules? In order to study the relation between dendrite shape and connectivity we derived a growth algorithm for building dendritic arborisations following closely the constraints described by Ramón y Cajal. The algorithm was introduced in (Cuntz et al. 2007, 2010). Implementing optimisation procedures known from graph theory the algorithm builds tree structures that minimise the total amount of wiring and the path from the root to all points on the tree, corresponding to the conservation of cytoplasm and of conduction times respectively.

Figure 6.1a exemplifies the general approach of assembling a set of input target points to such an optimised graph. A greedy algorithm based on the minimum spanning tree algorithm (Prim 1957) starts at the root (large black dot) with an empty tree and connects unconnected target points (dark circles) one by one to the nodes of the tree (light circles). At each step, the unconnected target point which is the point closest to the tree according to a cost function is connected to the node in the tree to which it is nearest. The distance cost is composed of two factors: (1) A wiring cost represented by the Euclidean distance between the target point and the node in the tree (dashed lines show three sample segment distances for target point *P*); this quantity loosely corresponds to the material or cytoplasm conservation constraint by Cajal; (2) A path length cost of the path along the tree from the root to the target point; this quantity is consistent with the conduction time conservation constraint by Cajal. In the example in Fig. 6.1a, even though *P* is closer to node 5 in Euclidean terms, the additional cost of path length through node 5 might tip the balance in favour of connecting *P* to node 4 or 3, etc. A balancing factor *bf*, which weighs these two costs against each other in the cost function, represents the one and only parameter of the model once the target points are defined. Figure 6.1b illustrates the results for synthetic trees grown on homogeneously distributed random target points in a circular envelope when the root is located at its centre. Since the two constraints (minimising wiring and minimising path length to the root) are weighted according to the balancing factor *bf* determining the contribution of the second constraint, the synthetic trees range along the dimension of that parameter from a pure minimum spanning tree, which grows in wide spirals, to a more stellate architecture (Fig. 6.1b, from left to right).

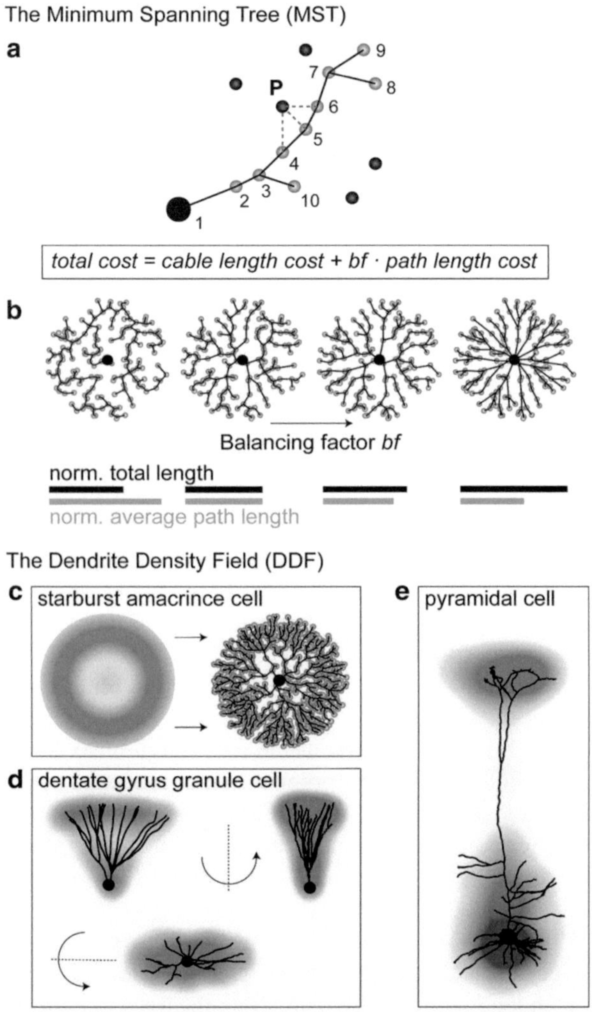

Fig. 6.1 Morphological modelling of dendrite structure. (**a**) Dendrite morphology as described by an extended minimum spanning tree algorithm (see text). Unconnected target points (*dark circles*) are connected one by one to the nodes of a tree (*light circles*) starting with a dendrite root (*large black dot*). *Dashed lines* indicate three sample Euclidean distances to the nodes of the tree for sample point *P*. (**b**) From left to right the same set of target points (*light circles*) were connected to a starting node (*large black dot*) according to the minimum spanning tree algorithm with increasing balancing factor *bf* between the two wiring costs: total cable length and sum of all paths. Below the trees *dark* and *light bars* indicate total length and average path length respectively for each tree. (**c**) Geometric description of the dendritic density field (DDF) of a starburst amacrine cell (*left*). By randomly selecting target points according to the DDF and subsequently connecting them using the minimum spanning tree algorithm (*right*) to a starting node in the middle (*large black dot*), a synthetic starburst amacrine cell dendrite is generated. (**d**) The DDF obtained directly from reconstructions of dentate gyrus granule cells with its characteristic cone-like shape. Views from three different angles; A representative real morphology is shown. (**e**) Separate DDFs for L5 cortical pyramidal cell basal trees (*dark grey*) and apical trees (*light grey*). A representative real morphology is shown. The figure was adapted from previous work (Cuntz et al. 2010)

The dendritic branching structure is therefore fully determined by the proper selection of target points and the one parameter of the model, *bf*. For example, in order to generate a synthetic starburst amacrine cell we designed a geometrical arrangement of densities (Fig. 6.1c, a ring shaped density function) according to which target points were distributed stochastically. After distributing target points in such a way, the resulting synthetic tree structure accurately reproduced the architecture of a real starburst amacrine cell. This type of approach yields good results for a large number of dendrites, e.g. for fly interneurons (Cuntz et al. 2008), dentate gyrus granule cells, cortical pyramidal cells from different layers and at different developmental stages (Cuntz et al. 2010). In all cases also the branching statistics such as branch order distributions and the aforementioned Sholl analysis were comparable with available reconstructions from real dendrites (see Cuntz et al. 2010). *bf* values between 0.1 and 0.85 were shown to reproduce realistic dendritic morphologies (Cuntz et al. 2010). This simple procedure highlights the usefulness of a dendritic density field (DDF), here representing the density of target points per area, to describe a dendrite's morphology. The DDF and the one parameter balancing the two costs for wiring mentioned above are thereby sufficient to describe the dendrite type in certain cases. In fact, we have shown that for fly interneurons the spanning field is the most informative element for classifying dendritic trees into their respective cell types (Cuntz et al. 2008) using this model. In some cases, DDFs can be obtained directly from reconstructions of real neurons. By superimposing their soma locations and rotating them such that they lie symmetrically around each axis and then scaling their width, height and depth, a set of reconstructions of dentate gyrus granule cell dendrites (data from Rihn and Claiborne 1990) was brought into a common context. An approximation of the DDF was then obtained directly from the density of branch and termination points from the reconstructions (Fig. 6.1d). Note that this is not identical to the distribution of target points. Also, DDFs might be calculated as density functions of dendrite length, dendrite membrane or dendrite volume. When apical and basal dendrites were considered separately, the DDFs of pyramidal cells were calculated in a similar way (L5 PCs, Fig. 6.1d; data from Wang et al. 2002). Using similar approaches to design sets of target points we have successfully first modelled fly interneuron dendrites (Cuntz et al. 2007, 2008) and a large number of other dendrite types (Cuntz et al. 2010). The approach has meanwhile also shown to be suitable for neocortical axons (Budd et al. 2010). This indicates that wiring constraints as proposed by Ramón y Cajal play a major role in determining dendrite and axon morphology.

6.2 Consequences for Branching Statistics and Dendritic Computation: A Scaling Law for Dendritic Trees

Since the wiring rules described above constrain the space of possible real dendrites they have consequences on fundamental morphological parameters of dendrites and on dendritic computation. This section summarises the work in Cuntz et al. 2012 that showed one such relation arising from studying the morphological model. Let us assume that a dendritic tree of total length L connects n target points distributed over a volume V. Each target point occupies an average volume V/n. A tree that optimises the total amount of wiring will tend to connect the target points to their nearest neighbours, which are located at distances proportional to the cubic root of this average volume or $(V/n)^{1/3}$—since the dendritic segments are one-dimensional compared to the volume surrounding them that is three-dimensional. Exactly n such dendritic sections are required to make up the tree. The total length L of these sections sums up to:

$$L = c \cdot n \cdot \left(V / n\right)^{1/3} = c \cdot V^{1/3} \cdot n^{2/3}.$$

This shows that a 2/3 power law relationship between L and n (Beardwood et al. 1958) provides a lower bound for the total dendritic length, where c is a proportionality constant. Approximating the volume around each target point by a sphere of the same volume, the length of the dendritic section required to leave the volume corresponds to the radius of that sphere. Then $c = (3/4\pi)^{1/3}$, giving:

$$L = \sqrt[3]{3/4\pi} \times \sqrt[3]{V} n^{2/3}. \tag{6.1}$$

To compare this scaling property between L and n with the one observed in neuronal dendritic trees, we took advantage of our morphological model which we have shown can reproduce a wide range of dendrite morphologies. As mentioned above, in addition to minimising the total length L to connect a set of n target points to a tree, this procedure introduces the additional cost to minimise all path lengths from any target point toward the root along the tree, parameterised with the balancing factor *bf*. It was therefore important to use the model to check how tightly (6.1) matches the model. When target points were randomly distributed within a spherical volume and connected to a tree to minimise these costs we found that (6.1) provided a tight lower bound for the total dendritic length, particularly for low *bf* (Fig. 6.2a). With increasing *bf* the exponent in the power law increased from precisely 2/3 i.e. 0.66 (root mean square error of 1.0%) up to 0.72 (root mean square error 5.2%) for *bf*=0.9, the maximal realistic balancing factor. Interestingly, in the model the number of branch points *bp* grew proportionally to the number of target points n leaving *bp*/n constant. Such a proportional relationship has been well described previously (Janson 2005; Steele et al. 1987) but no analytical formula exists to compute the precise relation and it must therefore be derived empirically. We found that both the scaling behaviour of L with n and the ratio *bp*/n were

Fig. 6.2 (continued) database NeuroMorpho.Org (Ascoli 2006). All data strictly follow the predicted equation in all cases, while different datasets cover their respective range of length and complexity. Both measures were normalised by the volume $V = 1{,}000{,}000\ \mu m^3$ which the dendrite covers since the volume covered by the spanning fields of the dendrites was variable. Differences in balancing factor cause a vertical offset while conserving the characteristic 2/3 power law trend shown here (*black line*) for $\alpha = bp/n = 0.29$ corresponding to $bf = 0.5$. Each datapoint is one neuron, with the grey shades corresponding to neuronal types. Best fit (*dashed*) has a power of 0.70. The figure was adapted from previous work (Cuntz et al. 2012)

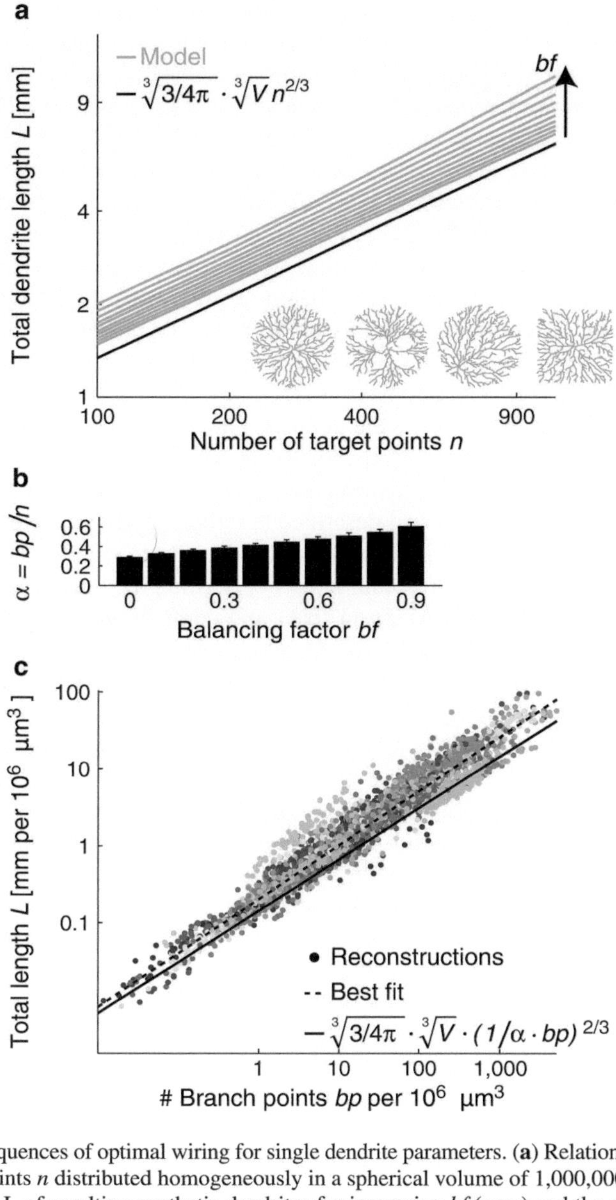

Fig. 6.2 Consequences of optimal wiring for single dendrite parameters. (**a**) Relation between number of target points *n* distributed homogeneously in a spherical volume of 1,000,000 μm³ and total dendrite length L of resulting synthetic dendrites for increasing *bf* (*grey*) and the predicted lower bound (*black*). The results were largely independent of arrangements of target points (see *inset* from left to right: spherical boundaries with homogeneous distribution; inhomogeneous distribution; root offset from the centre; cubic boundaries) (**b**) Ratios between number of branch points *bp* and number of target points *n* were independent of arrangements (see *inset* in **a**) or number of target points, but depended on the balancing factor *bf* between the two wiring costs. *Error bars* are for shuffled trees of all conditions. 100 trees were grown for each individual condition in (**a**) and (**b**). (**c**) General applicability of optimal space-filling power law relations. Relation between total dendrite length *L* and number of branch points *bp* for all reconstructions from the morphological

independent of the geometrical arrangement of target point distributions. Specifically, they did not change regardless of whether target points were distributed inhomogeneously, the root was displaced from the centre of the sphere, or the physical boundaries of the sphere were replaced by those of a cube (Fig. 6.2a, inset). However, *bp*/*n* depended in a linear manner on the model parameter *bf* (Fig. 6.2b), which we know varies between different classes of dendrites. Therefore, when two of the three quantities *bp*, *n* or *bf* are known, the third can be inferred independently of tree conditions or metric scale. The 2/3 power was therefore equally present between *bp* and *L* in the model (not shown, see Cuntz et al. 2012), with powers ranging from 0.66 to 0.72 and a mean square error below 2% between the prediction and the model for all *bf*. To test for the presence of our predicted power law in real neuronal morphologies, we analysed all available dendritic trees of all different cell types from the NeuroMorpho.Org database (Ascoli 2006). In these reconstructions, it is unclear what the target points of the dendrites are, but branch points can unambiguously be counted. We therefore plotted the wiring length of these cells against the number of branch points after normalising these values to a volume of $V = 10^6 \mu m^3$ (Fig. 6.2c). The power which best described the trends between the dendrite length and the number of branch points obtained for all cell types individually was very close to 2/3: 0.72 ± 0.10 and residual norms confirmed the goodness of fit (Cuntz et al. 2012). We replaced *n* by *bp* in (6.1) and compensated for the decrease in points by using the $\alpha = bp/n$ ratio derived from the simulations in Fig. 6.2b to obtain our lower bound for dendrite length (Fig. 6.2c, straight line). Although providing a lower bound, the resulting equation well described the data (with a root mean square error of 6.4%) and was very similar to the best fit (Fig. 6.2c, dashed line) with a power of 0.70.

Taking advantage of the morphological model, we have therefore derived from first principles a simple equation that relates the most fundamental measures of dendrite branching: the total length, the number of branch points, the spanning volume and the number of target points. Supporting the intuitive derivations of power laws described here, there have been several proofs (Beardwood et al. 1958; Steele et al. 1987) that in a minimum spanning tree in *d* dimensions—the canonical model of a tree constructed to minimise wiring length—*L* scales as a $(d-1)/d$ power of the number of target points or branch points. In summary, the wiring minimisation hypothesis predicts a 2/3 power law between *L* and *bp*, and a 2/3 power law between *L* and *n*. By contrast, a process that randomly connects target points without optimising wiring yields a power law with exponent 1. We showed that the equation that we derived holds in our model of dendrite branching. When realistic dendritic diameters were mapped on the synthetic dendrites to optimise synaptic democracy (Cuntz et al. 2007), we found that the passive electrotonic properties of the dendritic trees scaled in a similar manner as the dendrite length *L*. The number of independent electrical compartments and the total electrotonic length were therefore found to scale with the number of target points governed by a 2/3 power law when $bf > 0$ (Cuntz et al. 2012). This indicates a functional consequence of the scaling law on the integration properties of dendrites. Taking advantage of the measured ratio *bp*/*n* in the model we show that the equation holds for all dendrite and axon reconstructions available in the NeuroMorpho.Org database, including over 6000 reconstructions from 140 datasets. For one dataset consisting of newborn neurons in the adult olfactory bulb (Livneh et al. 2009; Mizrahi 2007) for which both faithful reconstructions of dendrites and the synapse locations were known (synapses were targeted with PSD95-GFP) we found that synapse locations corresponded to the target points used in our morphological model (Cuntz et al. 2012). Both the scaling of L with n and the bp/n ratio were comparable to the model, where n in this case was the number of synapses. Also the location of the synapses when used as target points for the minimum spanning tree algorithm were sufficient to predict more than 80 % of the tree segments. This indicates that in some cases the target points used in the morphological model can be considered to be the actual connection points between the neurons. An interesting consequence of minimising the wiring to target synapses is that synapses that are close together are more likely to be linked by the same stretch of dendrite and therefore involved in a local computation (Poirazi et al. 2003; Polsky et al. 2004; Williams and Stuart 2002), such that geometry is a key determinant of information processing. There is increasing evidence that such sophisticated local processing may be carried out within the dendritic tree (Häusser and Mel 2003; London and Häusser 2005), with nonlinear interactions between synaptic inputs shaping the output of the neuron (Branco et al. 2010; Branco and Häusser 2010; Koch et al. 1983). It will therefore be fruitful to study in detail how the scaling laws of wiring and branching place constraints on the overall computational power of single neurons.

6.3 Dendrite Types as a Function of Local Connectivity

After showing that the morphological model is useful to infer morphological and functional properties and constraints for single cells this section describes the particular relation between dendrite synaptic connectivity and morphology. In the beginning of this chapter it was shown that the dendrite density field (DDF) is useful for describing dendrite morphology. The next step is to identify useful DDFs based on underlying potential connectivity schemes. While the dendrites of starburst amacrine cells, dentate gyrus granule cells and many other neurons exhibit complex DDFs, some simpler ones such as pyramidal cell basal dendrites observe basic principles that can be characterised by studying the statistical moments of their density distributions (Snider et al. 2010; Teeter and Stevens 2011). Stevens and colleagues have shown that the cable density distributions, i.e. the cable length in a given volume, in dendrites roughly follow separate Gaussian distributions for each dimension in space. These are cut off at about two standard deviations. What determines this particular density profile? In synthetic dendrites generated using the morphological model, homogenously distributed target points lead to a homogeneous cable density. The criteria for optimal wiring alone can therefore not be responsible for the observed Gaussian cable density distribution. Intuitively, the shape of a DDF should be determined by the set of axons that are a dendrite's potential, i.e. anatomical, connection partners. This simple intuition can explain some basic features of DDFs as

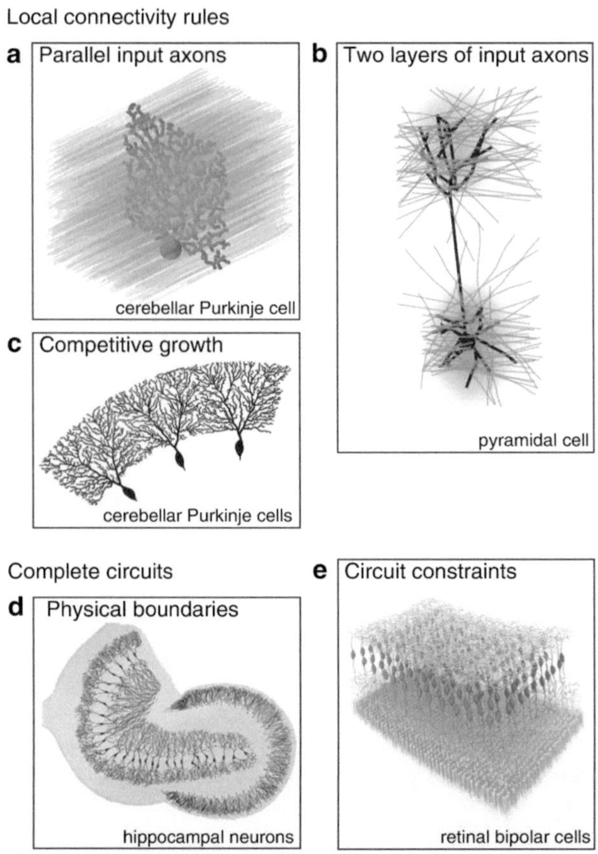

Fig. 6.3 Studying circuits using the morphological model. (**a**) Synthetic dendrite (*black*) connecting a starting point (*large black dot*) in 3D space to a set of parallel axons (*grey*): the resulting synthetic dendrite is flat, similar to a Purkinje cell that connects to a set of parallel fibres. (**b**) Two target axon sets (*grey*) are arranged in two parallel layers and the axons in each layer are isotropically oriented within the two parallel layers. The synthetic dendrite (*black*) connecting a starting point (*large black dot*) to these two sets of target axons displays a characteristic pyramidal cell shape with an apical and a basal dendrite. Average DDF of 100 such synthetic dendrites in *inset*; Compare with Fig. 6.1e. (**c**) Three out of 16 trees grown competitively on random target points homogeneously distributed on a ring. The resulting sharp boundaries (*tiling*) are reminiscent of Purkinje cells. (**d**) Hippocampal contours obtained from a sketch by Camillo Golgi (translated in Bentivoglio and Swanson 2001) with synthetic dentate gyrus granule cells and CA3 hippocampal pyramidal cells observing the limits from the template (see text). (**e**) Bipolar cells (*black*) in the retina were grown competitively to connect an array of photoreceptors (*grey*, *bottom*) to an array of starburst amacrine cells (*grey*, *top*; obtained as in Fig. 6.1c). In such a case the full morphology of bipolar cells is determined by the context of the circuitry, after prescribing soma locations of the bipolar cells. Precise scale bars were omitted in all figures since they depend on the system to be modelled. The figure was adapted from previous work (Cuntz 2012; Cuntz et al. 2010)

was demonstrated in (Cuntz 2012). After using target axons instead of input target points (e.g. Fig. 6.3a) the morphological model was set to connect to each target axon only once. To do this, target points were distributed every 50 μm along the axon and all target points belonging to one axon were removed once one of them was selected by the iterative growth algorithm. A synthetic dendrite connecting to a set of parallel axons traversing space was flat and perpendicular to the axons (Fig. 6.3a). Beyond the obvious intuition that direct connections lead to perpendicular angles between the dendrite and the axons, a flat dendrite that lies perpendicular to a set of parallel axons was shown to be the overall ideal geometrical arrangement to connect a set of axons with a dendrite to optimise wiring costs (Wen and Chklovskii 2008). It is not surprising that the morphological model reproduced these findings. Such a planar dendritic structure can for example be observed in cerebellar Purkinje cells which reach out to the molecular layer in the cerebellum to collect their parallel fibre inputs. These being arranged entirely in parallel the dendrite must grow in a planar way perpendicular to the parallel fibres to connect to them most efficiently. Most neural systems however are not entirely optimised for the connection between only one set of axons and one dendrite type. Axon distributions are typically more complex. In order to illustrate the potential of this constructive approach, a sample configuration is shown in which two separate layers of target axons were both connected to a single starting point located at the centre of the lower layer (Fig. 6.3b). Under these conditions, the natural shape of a pyramidal cell, including its apical dendrite and its two separate DDFs (Fig. 6.3b, shading) were a natural consequence. Interestingly, the toy model of a pyramidal cell presented here does not build on any functional differences between the two layers that it connects. The location of the starting point alone determines which of the layers will yield an apical dendrite and which one a basal dendrite. This dendrite root corresponds to the integration point of the input signals but more importantly it is the coordinate from which the axon, the output of the neuron, exits the cell. The fact that the location of the exiting axon determines the neuron's shape was at the centre of Cajal's axopetal polarity theory. In the cortex, most long-range axons project toward the white matter away from the pial surface. The axonal output being on one side of the cortex, we would predict a stereotypic polarity of pyramidal cell morphology: Basal dendrites of pyramidal cells should be on the inside closer to the white matter where the output of the cell is located, whereas the apical dendrites should be on the pial side. This indeed is the case. At the example of the pyramidal cell dendrite we showed that it is possible to predict both their DDF and their precise branching structure by making assumptions about the input axon distribution. In this way it might be possible to link regionalised specialisations in pyramidal cell dendrite shape with differences in local connectivity. To summarise, the DDF is a promising attribute of dendrite shape which enables direct conclusions on both the morphology of a dendrite and on the arrangement of a neuron's axonal inputs, thereby linking morphology to connectivity directly.

6.4 Formation of Neural Circuits

In the final section of this chapter the morphological model of single dendrites is expanded to larger circuits as sketched previously in Cuntz et al. 2010. The local circuitry ultimately determines the context in which neuronal trees grow. There are global boundaries given by the neural tissue such as layers, topography or physical borders. However, competition for input space between neighbouring neurons also seems to play a role (e.g. Samsonovich and Ascoli 2003). Competition is easily implementable in the greedy minimum spanning tree algorithm that underlies the morphological model because of the iterative nature of the algorithm. When grown under competitive conditions in which trees connect to a target point one after the other, the immediate consequence is spatial tiling. In fact, both Cajal's material cost and his conduction time cost independently lead to this type of tiling. This is not the case for random wiring. Competitive dendrite growth can for example directly reproduce the sharp borders observed in Purkinje cell dendrites of the cerebellum. To demonstrate this, 16 cells were grown in a competitive manner on random target points distributed homogeneously in a ring-shaped area. Purkinje cell dendrites were required to be grown in two stages: first the thick primary dendrites and then the thinner ones covered in spines (three sample cells are displayed in Fig. 6.3c). The sharp borders of Purkinje cell dendrites were well reproduced. However, whether these sharp borders are actually a result of tiling in sagittal planes of the cerebellum remains to be determined experimentally. Cajal's laws can therefore explain more than just the inner branching rule: tiling between cells can emerge directly from his suggested optimisation principles applied at the network level. The broader network context also plays a major role in governing neuronal spanning fields and their density profiles. Arranging synthetic hippocampal granule cells obtained from the DDF shown in Fig. 6.1d onto the contours of the dentate gyrus demonstrates the effect of the physical boundaries onto single cell variability (Fig. 6.3d). Growing CA3 hippocampal pyramidal cells in a context-dependent manner (here in a competitive growth process bounded by the CA3 contours of the hippocampus, obtained from Golgi; Bentivoglio and Swanson 2001) reproduces the variability in neuronal branching seen in the reconstructions. The idea that the network context determines a neuron's branching can be followed further: both input and output locations can serve as direct constraints for the cell morphology, as is the case when an array of photoreceptors (Fig. 6.3e, grey, bottom layer) in the retina connects to an array of starburst amacrine cells (Fig. 6.3e, grey, top layer; cells were obtained as in Fig. 6.1c) via a set of bipolar cells. In such a case, the input–output topography of the network determines the morphology of bipolar cells given that these grow in a competitive manner (Fig. 6.3e, black). Extending this idea, it might be possible to obtain the complete morphology of an entire circuitry based solely on the input–output topography of the underlying network, assuming that Cajal's optimisation principles hold. To then simulate the complete spatial packing of the circuit will require the inclusion of dendrite diameters in the growth algorithm.

6.5 Conclusions and Outlook

I showed here that the laws of conservation of cytoplasm and conduction time formulated by Ramón y Cajal from simple observation represent a strict constraint on neuronal branching. These were sufficient to constrain a morphological model that is based on the minimum spanning tree to generate synthetic dendrites of a wide variety of dendrite types (Cuntz et al. 2007, 2008, 2010). The resulting dendrites were indistinguishable from real reconstructions by eye and by a variety of dendrite branching statistics. The optimal wiring considerations led to a unifying equation for dendrite length, number of branch points, spanning volume and number of synapses that the morphological model and all real reconstructions follow (Cuntz et al. 2012). This represents a direct validation of the fundamental constraints on neuronal circuit organisation described originally by Cajal. It provides direct evidence that the major determinant for a dendrite's morphology is its role in the connectivity. The same wiring constraints are further applicable to local and global network considerations and directly lead to dendrite tiling, particular dendrite density fields and to complete circuits (Cuntz 2012; Cuntz et al. 2010). These results are in close agreement with complementary statistical calculations based on similar connectivity requirements (Wen et al. 2009; Wen and Chklovskii 2008).

Does this mean that dendritic computation or other requirements do not play a role in the formation of dendritic morphology? While operating within the strict constraints described above, a dendrite has a number of remaining degrees of freedom. The weighting of both wiring costs, the balancing factor *bf*, is an adjustable parameter and differs from one cell type to the next. By adjusting the balance between the two wiring costs, a dendrite can efficiently set its electrotonic compartmentalisation and conduction speeds, quantities attributable to computation. These are then malleable by specialised ion channel distributions to implement the computations known from dendrite structures. Furthermore, the density profile of the spanning field in which a dendrite grows determines its shape dramatically. A few weaker constraints such as the suppression of multifurcations, the addition of spatial jitter or the sequential growth of sub-regions of a dendrite are helpful for reproducing the dendritic branching patterns of particular preparations. These additional constraints might shed light on further functional, computational, developmental or network determinants for certain dendritic structures, and more of these will follow when applying our method to many more preparations.

Other methods exist to generate synthetic dendrite morphology in models based for example on mechanisms of dendrite growth (Koene et al. 2009), the fractal organisation of the brain (Ascoli et al. 2001; Luczak 2006), specific computations (Torben-Nielsen and Stiefel 2010), or simply on branching statistics (Hillman 1979). Some of these approaches are described in other chapters in this book. It will be useful to combine these approaches and to find out how biology really implements the optimal wiring constraints that define dendrite structure. I have shown here that morphological models based on wiring constraints provide more than just anatomically realistic profiles that are useful for neural network simulations. They can be a tool to understand how dendrite morphology comes about and a tool to test our knowledge about the local connectivity in the brain. They can further help elucidate constraints on morphological statistics, on dendritic computation, on the process of dendrite growth and on large-scale network connectivity principles.

Acknowledgments The work presented here was a collaborative effort together with Alexander Borst, Friedrich Förstner, Jürgen Haag, Michael Häusser, Alexandre Mathy and Idan Segev. I would like to thank Peter Jedlicka for constructive criticism on this manuscript.

References

Acebes A, Ferrús A (2000) Cellular and molecular features of axon collaterals and dendrites. Trends Neurosci 23:557–565
Ascoli GA (1999) Progress and perspectives in computational neuroanatomy. Anat Rec 257:195–207
Ascoli GA (2006) Mobilizing the base of neuroscience data: the case of neuronal morphologies. Nat Rev Neurosci 7:318–324
Ascoli GA, Krichmar JL, Nasuto SJ, Senft SL (2001) Generation, description and storage of dendritic morphology data. Philos Trans R Soc Lond B Biol Sci 356:1131–1145
Beardwood J, Halton JH, Hammersley JM (1958) The shortest path through many points. Math Proc Cambridge 55:299–327
Bentivoglio M, Swanson LW (2001) On the fine structure of the pes Hippocampi major (with plates XIII–XXIII) by Camillo Golgi. Brain Res Bull 54:461–483
Branco T, Clark BA, Häusser M (2010) Dendritic discrimination of temporal input sequences in cortical neurons. Science 329:1671–1675
Branco T, Häusser M (2010) The single dendritic branch as a fundamental functional unit in the nervous system. Curr Opin Neurobiol 20:494–502
Budd JML, Kovács K, Ferecskó AS et al (2010) Neocortical axon arbors trade-off material and conduction delay conservation. PLoS Comput Biol 6:e1000711
Cannon RC, Wheal HV, Turner DA (1999) Dendrites of classes of hippocampal neurons differ in structural complexity and branching patterns. J Comp Neurol 633:619–633
Cherniak C (1992) Local optimization of neuron arbors. Biol Cybern 66:503–510
Chklovskii DB (2004) Synaptic connectivity and neuronal morphology: two sides of the same coin. Neuron 43:609–617
Chklovskii DB, Koulakov AA (2004) Maps in the brain: what can we learn from them? Annu Rev Neurosci 27:369–392
Cline HT (2001) Dendritic arbor development and synaptogenesis. Curr Opin Neurobiol 11:118–126
Cuntz H (2012) The dendritic density field of a cortical pyramidal cell. Front Neuroanat 6:2
Cuntz H, Borst A, Segev I (2007) Optimization principles of dendritic structure. Theor Biol Med Model 4:21
Cuntz H, Forstner F, Borst A, Häusser M (2010) One rule to grow them all: a general theory of neuronal branching and its practical application. PLoS Comput Biol 6:e1000877
Cuntz H, Forstner F, Haag J, Borst A (2008) The morphological identity of insect dendrites. PLoS Comput Biol 4:e1000251
Cuntz H, Mathy A, Häusser M (2012) A scaling law derived from optimal dendritic wiring. Proc Natl Acad Sci USA 109:11014–11018
DeFelipe J (2010) Cajal's butterflies of the soul: science and art. Oxford University Press, New York
Grueber WB, Jan Y-N (2004) Dendritic development: lessons from Drosophila and related branches. Curr Opin Neurobiol 14:74–82
Gulledge AT, Kampa BM, Stuart GJ (2005) Synaptic integration in dendritic trees. J Neurobiol 64:75–90
Häusser M, Mel BW (2003) Dendrites: bug or feature? Curr Opin Neurobiol 13:372–383
Hillman DE (1979) Neuronal shape parameters and substructures as a basis of neuronal form. In: S F (ed) The neuroscience, fourth study program. The MIT Press, Cambridge, MA, pp 477–498
Hodgkin AL, Huxley AF (1952) A quantitative description of membrane current and its application to conduction and ecitation in nerve. J Physiol 117:500–544
Jan Y-N, Jan LY (2010) Branching out: mechanisms of dendritic arborization. Nat Rev Neurosci 11:316–328
Jan Y-N, Jan LY (2003) The control of dendrite development. Neuron 40:229–242

Janson S (2005) Asymptotic degree distribution in random recursive trees. Random Struct Algor 26:69–83

Klyachko VA, Stevens CF (2003) Connectivity optimization and the positioning of cortical areas. Proc Natl Acad Sci USA 100:7937–7941

Koch C, Poggio TA, Torre V (1983) Nonlinear interactions in a dendritic tree: localization, timing, and role in information processing. Proc Natl Acad Sci USA 80:2799–2802

Koene RA, Tijms B, Van Hees P et al (2009) NETMORPH: a framework for the stochastic generation of large scale neuronal networks with realistic neuron morphologies. Neuroinformatics 7:195–210

Livneh Y, Feinstein N, Klein M, Mizrahi A (2009) Sensory input enhances synaptogenesis of adult-born neurons. J Neurosci 29:86–97

London M, Häusser M (2005) Dendritic computation. Annu Rev Neurosci 28:503–532

Luczak A (2006) Spatial embedding of neuronal trees modeled by diffusive growth. J Neurosci Methods 157:132–141

Mainen ZF, Sejnowski TJ (1996) Influence of dendritic structure on firing pattern in model neocortical neurons. Nature 382:363–366

Mizrahi A (2007) Dendritic development and plasticity of adult-born neurons in the mouse olfactory bulb. Nat Neurosci 10:444–452

Oberlaender M, De Kock CPJ, Bruno RM et al (2011) Cell type-specific three-dimensional structure of thalamocortical circuits in a column of rat vibrissal cortex. Cereb Cortex 22:2375–2391

Parrish JZ, Emoto K, Kim MD, Jan Y-N (2007) Mechanisms that regulate establishment, maintenance, and remodeling of dendritic fields. Annu Rev Neurosci 30:399–423

Van Pelt J, Van Ooyen A, Uylings HBM (2001) The need for integrating neuronal morphology databases and computational environments in exploring neuronal structure and function. Anat Embryol 204:255–265

Van Pelt J, Uylings HBM (2002) Branching rates and growth functions in the outgrowth of dendritic branching patterns. Network 13:261–281

Poirazi P, Brannon T, Mel BW (2003) Pyramidal neuron as two-layer neural network. Neuron 37:989–999

Polsky A, Mel BW, Schiller J (2004) Computational subunits in thin dendrites of pyramidal cells. Nat Neurosci 7:621–627

Prim RC (1957) Shortest connection networks and some generalizations. Bell Syst Tech J 36:1389–1401

Rall W (1959) Branching dendritic trees and motoneuron membrane resistivity. Exp Neurol 527:491–527

Rall W, Burke RE, Smith TG et al (1967) Dendritic location of synapses and possible mechanisms for the monosynaptic EPSP in motoneurons. J Neurophysiol 30:1169–1193

Rall W, Rinzel J (1973) Branch input resistance and steady attenuation for input to one branch of a dendritic neuron model. Biophys J 13:648–687

Ramón y Cajal S (1995) Histology of the nervous system of man and vertebrates. Oxford University Press, New York

Rihn LL, Claiborne BJ (1990) Dendritic growth and regression in rat dentate granule cells during late postnatal development. Brain Res Dev Brain Res 54:115–124

Samsonovich AV, Ascoli GA (2003) Statistical morphological analysis of hippocampal principal neurons indicates cell-specific repulsion of dendrites from their own cell. J Neurosci Res 71:173–187

Segev I, London M (2000) Untangling dendrites with quantitative models. Science 290:744–750

Shepherd GMG, Stepanyants A, Bureau I et al (2005) Geometric and functional organization of cortical circuits. Nat Neurosci 8:782–790

Sholl DA (1959) A comparative study of the neuronal packing density in the cerebral cortex. J Anat 93:143–158

Snider J, Pillai A, Stevens CF (2010) A universal property of axonal and dendritic arbors. Neuron 66:45–56

Steele JM, Shepp LA, Eddy WF (1987) On the number of leaves of a Euclidean minimal spanning tree. J Appl Probab 24:809–826
Stuart GJ, Spruston N, Häusser M (2007) Dendrites. Oxford University Press, USA
Teeter CM, Stevens CF (2011) A general principle of neural arbor branch density. Curr Biol 21:2105–2108
Torben-Nielsen B, Stiefel KM (2010) An inverse approach for elucidating dendritic function. Front Comput Neurosci 4:128
Torben-Nielsen B, Vanderlooy S, Postma EO (2008) Non-parametric algorithmic generation of neuronal morphologies. Neuroinformatics 6:257–277
Uylings HBM, Van Pelt J (2002) Measures for quantifying dendritic arborizations. Network 13:397–414
Wang Y, Gupta A, Toledo-rodriguez M, Wu CZ (2002) Anatomical, physiological, molecular and circuit properties of nest basket cells in the developing somatosensory cortex. Cereb Cortex 395–410
Wen Q, Chklovskii DB (2008) A cost-benefit analysis of neuronal morphology. J Neurophys 99:2320–2328
Wen Q, Stepanyants A, Elston GN et al (2009) Maximization of the connectivity repertoire as a statistical principle governing the shapes of dendritic arbors. Proc Natl Acad Sci USA 106:12536–12541
Williams SR, Stuart GJ (2002) Dependence of EPSP efficacy on synapse location in neocortical pyramidal neurons. Science 295:1907–1910

Chapter 7
A Statistical Theory of Dendritic Morphology

Quan Wen

Abstract Since Santiago Ramon y Cajal, neuroscientists have been fascinated by the shapes of dendritic arbors for more than 100 years. However, the principle underlying these complex and diverse structures remains elusive. Here we propose that evolution has tinkered with brain design to maximize its functionality while minimizing the cost associated with building and maintaining it. We hypothesize that the functionality of a neuron benefits from a larger repertoire of connectivity patterns between dendrites and surrounding axons, and the cost of a dendritic arbor increases with its total length and path length from synapses to soma. We solved this optimization problem by drawing an analogy with maximization of the entropy for a given energy in statistical physics. The solution predicts several scaling relationships between arbor dimensions and closely fits with experimental data. Moreover, our theory may explain why basal dendrites of pyramidal cells and Purkinje cells, the two major cell types in the mammalian brains, exhibit distinct morphologies.

7.1 Introduction

Studying the structure of an organ helps unravel its function. More than 100 years ago, by applying the histological staining technique to study the anatomy of nervous systems, Santiago Ramon y Cajal formulated the "neuron doctrine." He proposed that neurons are interconnected with each other by two polarized components, axons and dendrites. The function of dendrites and axons is to conduct electrical signals from postsynaptic terminals to the integration site, which often is the cell body, and from the integration site to the presynaptic terminals, respectively.

Q. Wen (✉)
Department of Physics and Center for Brain Science, Harvard University,
17 Oxford Street, Cambridge, MA 02138, USA
e-mail: vineygeyser@gmail.com

H. Cuntz et al. (eds.), *The Computing Dendrite: From Structure to Function*, Springer Series in Computational Neuroscience 11, DOI 10.1007/978-1-4614-8094-5_7,

One hundred years later, neurobiologists, armed with advanced tools, are embracing a great opportunity to dissect the complex structure of the neuronal circuitry. For instance, cell-labeling methods such as those based on biocytin or green fluorescent protein make it possible to describe neuronal arbors in a comprehensive and quantitative way (Ascoli 2006). Reconstructing the wiring diagram of an entire brain circuitry may become an attainable goal with the advances of high-throughput electron microscopy and computer-aided image analysis (Lichtman et al. 2008; Denk and Horstmann 2004; Chklovskii et al. 2010; Kleinfeld et al. 2011). However, a quantitative theory of neuronal arbor shape does not exist. By establishing a direct relationship between neuronal structure and function, such a theory could elucidate the design principles of neural circuit.

Insight in this direction was also contributed by Cajal, as he wrote (Ramón y Cajal 1899), "After the many shapes assumed by neurons, we are now in a position to ask whether this diversity … has been left to chance and is insignificant, or whether it is tightly regulated and provides an advantage to the organism. … we realized that all of the various conformations of the neuron and its various components are simply morphological adaptations governed by laws of conservation for time, space, and material."

Echoing Cajal's thoughts, we propose that evolution has "tinkered" with the brain design to maximize the functionality of a neuronal arbor while minimizing the cost associated with building and maintaining it. Specifically, brain functionality must benefit from high synaptic connectivity and large plasticity, because synaptic connections are central to information processing as well as learning and memory. However, increasing connectivity requires adding wiring to the network, which comes at a cost. The cost of wiring could be accrued by various things, including conduction delays, signal attenuation, and wiring volume. Below, I will show that these ideas can be formulated in the framework of constrained optimization. By solving this mathematical problem, we can account for several salient features of the shape of pyramidal and Purkinje dendritic arbors, the two major cell types in mammalian brains. This chapter is based on two papers that I have published earlier (Wen and Chklovskii 2008; Wen et al. 2009).

7.2 Dendritic Cost Function and Potential Connectivity Constraints

To start, let us consider the cost of wiring. We assume that the cost of dendritic arbors E grows with the total dendritic length L, i.e., $\partial E/\partial L>0$, and with the average path length from a synapse to the soma l, i.e., $\partial E/\partial l>0$ for all L and l. The path length cost may be contributed by several factors such as longer delays and greater attenuation of synaptic signals and higher metabolic costs for intracellular transport (Cuntz et al. 2007; Wen and Chklovskii 2008). These two quite general assumptions will be sufficient to make predictions about optimal dendritic shape, which I will then compare with the shape of purkinje dendritic arbors in the cerebellum.

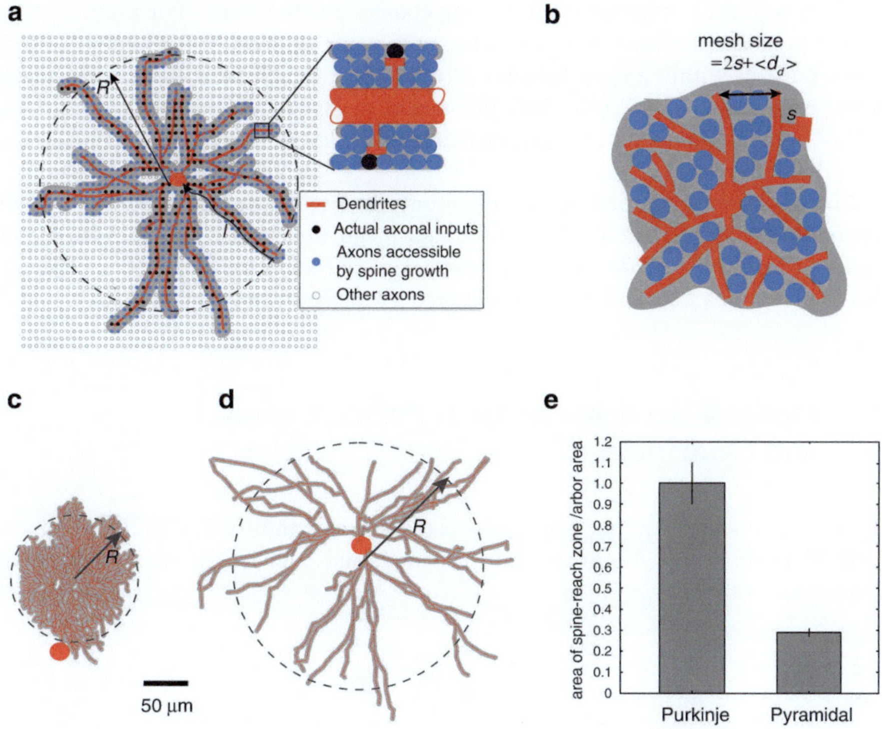

Fig. 7.1 The shape of a dendritic arbor determines which axons are accessible to a neuron. (**a**) Schematic illustration of a 3D dendritic arbor projected onto a plane and nearby axons that are labelled based on their relations to the arbor. For illustration purposes, we have shown all axons as running orthogonally to the plane of the drawing. Actual synapses (*black dots*) are a subset of potential synapses (*solid blue circles*), which in turn are chosen out of a larger set of axons (*circles*). Potential synapses are located within the spine-reach zone of a dendrite (*gray*). (**b**) A compact branching arbor makes on average one potential synapse with each axon (*blue*) passing through the arbor. The mesh size is defined as the arbor area divided by the total dendritic length. The mesh size is $2s + \langle d_d \rangle$ for a compact planar arbor and is the same, up to a numerical factor of order one, for a compact 3D arbor. (**c**) Dendritic arbor of a Purkinje cell from mouse cerebellum and its spine-reach zone. (**d**) Basal dendritic arbor (*red*) of a pyramidal cell from rat cortex (Shepherd and Svoboda 2005) and its spine-reach zone (*gray*). The arbor territory is approximated by the *dashed circle*. (**e**) Ratio of the area of the spine-reach zone to the area of the arbor territory on a 2D projection [$n = 10$ for both pyramidal and Purkinje cells (Kisvarday and Eysel 1992; Martone et al. 2003; Ascoli 2006; Rapp et al. 1994; Vetter et al. 2001)]. The spine-reach zone area is calculated as $2sL$, where $s = 2$ μm. Pyramidal basal dendrites are sparser than Purkinje dendrites ($p < 10^{-5}$, *t*-test with unequal variances) and, therefore, have lower probability to form a potential synapse with an axon passing through its territory. Error bars are s.e.m. These digitally reconstructed arbors are publically accessible from http://NeuroMorpho.Org [see also (Ascoli 2006)]. This figure is modified from Wen et al. (2009)

Next, we switch our attention to the constraints. We consider wiring up a neural circuit for a fixed number of potential synapses per neuron. We define *potential synapse* as a location where an axon passes within a spine length of a dendrite (Fig. 7.1a and Stepanyants et al. 2002). A potential synapse can be converted into an actual one by growing a spine (Trachtenberg et al. 2002). Requiring a potential convergence factor imposes the following constraints on L and l. For simplicity, let us first consider a planar dendritic arbor and axons running orthogonally to it. C axons must fit within a spine length s of a dendrite, which we call the spine-reach zone (gray area in Fig. 7.1a). Then, the area of the spine-reach zone ($2sL$) must be at least equal to the total cross-sectional area of the axons ($\pi/4Cd_a^2$), where d_a is the axon diameter.

Likewise, the mean path length l can only be greater or on the same order of linear arbor radius R. We define R as the root-mean-square distance from the center mass of the arbor to all dendritic segments. A planar arbor must at least encompass all the potential axons and the area occupied by the dendrites on the plane.

7.3 Optimal Dendritic Arbor Is Planar, Compact, and Centripetal

By minimizing the cost function E subject to the constraints of L and l, it is straightforward to show that the total dendritic length and path length satisfy (Wen and Chklovskii 2008)

$$L = \frac{\pi}{4} C d_a^2 / 2s. \tag{7.1}$$

$$l \sim R. \tag{7.2}$$

Here approximately (~) suggests that l and R only differ by a numerical factor that is close to one. In addition, in an optimal arbor, the total arbor area A should satisfy

$$A = \frac{\pi}{4} C d_a^2 \left(1 + d_d / 2s\right). \tag{7.3}$$

In (7.3), $\langle d_d \rangle$ is the mean dendritic diameter. Dendritic arbors satisfying (7.1)–(7.3) have the following properties. First, minimizing total dendritic length ((7.1)) demands a spatial organization of the neuropil, in which adjacent dendrites from different neurons are excluded from each other's spine-reach zone (Fig. 7.1a, inset). If dendrites penetrated each other's spine-reach zones, they would add to the excluded volume of axons and would increase the total dendritic length.

Second, to achieve the minimum path length l in (7.2), each segment of the dendrite should be directed towards the soma. We call such arbor design centripetal (Fig. 7.1a). If the total dendritic length is greater than the dendritic arbor span, the centripetal arbor must branch. Therefore, branching of dendrites is a trivial consequence of minimizing the mean path length.

In suboptimal dendritic arbors, the dependence of l on R does not have to be linear as suggested by (7.2). For example, if dendrites consisted of randomly oriented segments, like in a random walk, the path length would be $l \sim R^2$.

Third, we calculate the arbor mesh size—a parameter that quantifies the sparseness of an arbor—by dividing the arbor area by the total length, A/L. By combining (7.1) and (7.3), we find that

$$\frac{A}{L} = 2s + d_d. \tag{7.4}$$

We call an arbor satisfying (7.4) compact (Fig. 7.1b). One property of a compact arbor is that it forms on average one potential synapse with each axon passing through the arbor.

A compact branching arbor is less costly than other branching arbors with the same potential convergence. Consider a sparse arbor with a mesh size that is much larger than $2s+\langle d_d \rangle$ (Fig. 7.1a) and which does not form potential synapses with every axon passing through the arbor (Fig. 7.1a). A compact arbor is less costly because it has smaller span than a sparse branching arbor.

How does this analysis of a planar dendritic arbor generalize to 3D dendritic arbors and axons running in different directions? Note that we can always project a 3D dendritic arbor onto a plane that is orthogonal to the running direction of some axons. One can then calculate the spine-reach zone area that captures these axons. The same procedure can be performed on different arbor projections, and the sum of these spine-reach zone areas will determine the total dendritic length and the arbor size. Direct calculation shows that in the 3D case, the optimality conditions still hold if numerical factors close to one are ignored. Yet when these numerical factors are included, a planar arbor is preferable because a 2D projection of a 3D arbor is always shorter than the original. Thus, both the minimum path length, l, and the minimum total dendritic length, L, in a planar arbor are shorter than those in a 3D arbor.

7.4 Purkinje Dendritic Arbors Are Planar, Compact, and Centripetal

In the previous section, we derived the properties of optimal dendritic arbors that minimize cost for a given potential connectivity. Next, we compare our predictions with the experimental measurements for Purkinje dendritic arbors. As Purkinje arbors are evidently planar, we demonstrate that they are compact and centripetal.

To prove that Purkinje dendritic arbors are compact, we refer to (7.4) and show that the mesh size of a dendritic arbor is $2s+\langle d_d \rangle$. Napper and Harvey (1988b) reported that $s = 1.4$ μm as measured from the surface of the dendrites to the tip of the spine. They also found the diameter of spiny dendrites receiving parallel fiber inputs to be $\langle d_d \rangle = 1.5$ μm. Dividing the arbor area by the total dendritic length yields a value close to 4 μm (Fig. 7.1c, e), which is consistent with (7.4).

To demonstrate that Purkinje cell dendrites are centripetal, we calculated the distribution of dendritic segment orientation angle θ (Fig. 7.2a), which is defined as the angle between the vector of the signal flowing along a dendritic segment and the vector pointing centripetally from the segment to the soma. Figure 7.2b shows a Purkinje dendritic arbor where each dendritic segment is colored according to the

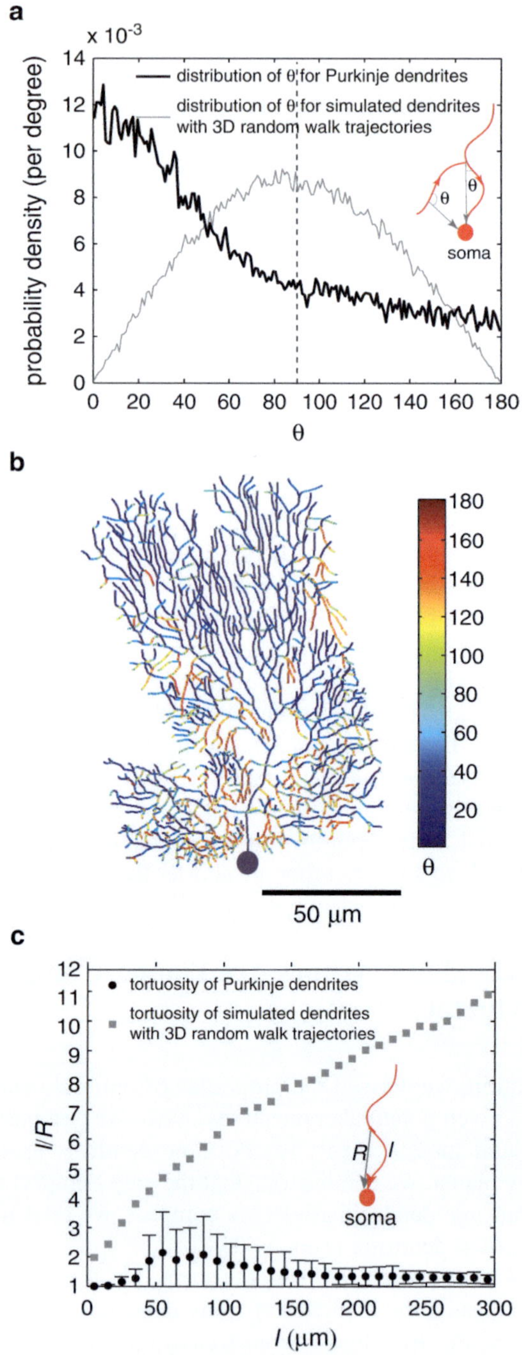

Fig. 7.2 Purkinje dendritic arbors are centripetal. (**a**) The probability distribution of the orientation angle θ between the vector associated with a Purkinje dendritic segment and the vector pointing centripetally from the segment to the soma (*black line*). Seventy percent of the segments have angles less than 90° (the total area under the *black curve* from 0 to 90°). *Gray line* is the probability distribution of angles for dendritic branches generated by 3D random walk trajectories.

value of θ. We found that 70 % of the segments have orientation angles less than 90° (Fig. 7.2a), which suggests that Purkinje dendritic segments are predominately centripetal. This observation is consistent with the measurements of pyramidal cell dendrites in hippocampus (Samsonovich and Ascoli 2003).

Many trajectories observed in nature are not centripetal, and one of the classical examples is a random walk. For comparison, we simulated a random walk comprising rigid segments 5 μm in length with random orientations and plotted the probability distribution of orientation angles (gray line in Fig. 7.2a). In a 2D random walk, the orientation angle would have a uniform distribution. However, in a 3D random walk, the most likely orientation angle is near 90° (Fig. 7.2a).

In optimal dendrites, the typical path length from a dendritic segment to the soma must be close to the Euclidean distance between them (7.2), i.e., the tortuosity index, defined as the ratio of the path length from a dendritic segment to the soma to the Euclidean distance between the two locations, is close to one. To verify that dendrites are optimal, we plot the tortuosity in Purkinje cell dendrites as a function of path length (Fig. 7.2c). Unlike the simulated dendrites with random walk trajectories, the tortuosity of real dendrites is close to one (Fig. 7.2c), consistent with optimality.

7.5 Micro-Architecture of the Cerebellum Molecular Layer

To verify that Purkinje cell dendrites from different neurons are excluded from each other's spine-reach zone (Fig. 7.1a, inset), we shall estimate the interval b between potential synapses along a parallel fiber in the molecular layer and show that $b = 2s + \langle d_d \rangle$ (Napper and Harvey 1988a).

We can view the interval b as the distance a parallel fiber travels before encountering a different Purkinje dendrite. Along its course, a parallel fiber with length L_a will encounter totally D different dendrites and $b = L_a/D$. Below I will call D the potential divergence factor of an axon, and it can be calculated as follows. Because Purkinje cell dendrites are compact, an axon can potentially connect with all the dendrites in the volume $L_a wh$, where w is the width of the dendritic arbor and h is the height of the arbor. Therefore, we have

$$D = \sigma L_a wh, \tag{7.5}$$

Fig. 7.2 (continued) Random walk is not centripetal as the distribution has a peak near 90°. (**b**) A digitized Purkinje dendritic arbor, where the color of each dendritic segment represents the value of the orientation angle. (**c**) The tortuosity of Purkinje cell dendrites, which is defined as the ratio of the path length from the soma to a dendritic segment to the Euclidian distance between the two locations, is close to one for different path lengths from dendritic segments to the soma. The tortuosity of dendrites generated by random walk trajectories is much higher than that of Purkinje cell dendrites, and it scales with the square root of the path length. All error bars are standard deviations. The above analysis is done on ten digitally reconstructed Purkinje dendritic arbors (Rapp et al. 1994; Vetter et al. 2001; Martone et al. 2003) available from http://neuromorpho.org. The arbors were projected onto the plane perpendicular to the parallel fibers (*sagittal plane*). This figure is modified from Wen and Chklovskii (2008) where ρ is the neuronal density. Because Purkinje cell bodies are arranged uniformly in a layer, we may rewrite (7.5) as a function of the neuronal density per unit area σ,

$$D = \sigma L_a w. \tag{7.6}$$

As a result, the interval of potential synapses on an axon b is given by

$$b = L_a / D = \frac{1}{\sigma w}. \tag{7.7}$$

By substituting the values from the rat cerebellum $\sigma = 1{,}018\ \text{mm}^{-2}$, $w = 250\ \mu\text{m}$, we obtain $b = 4\ \mu\text{m}$. Recalling that $s = 1.4\ \mu\text{m}$ and $d_d = 1.5\ \mu\text{m}$ (Napper and Harvey 1988b), we find that the relation $b = 2s + d_d$ is satisfied and adjacent Purkinje cell dendrites are on average excluded from each other's spine-reach zone. We hope this calculation can be verified directly by electron microscopic reconstructions.

7.6 Purkinje and Pyramidal Dendritic Arbors Exhibit Distinct Morphologies

Although we were able to explain the shape of Purkinje cell dendrites, neuronal arbor shape varies among cell classes. In particular, cortical pyramidal cell dendrites in the neocortex have 3D shape and are sparser than Purkinje cell dendrites on a projection: the distance between adjacent branches is much greater than spine length (Fig. 7.1d). How can we understand the shape of such dendrites?

One difference between the Purkinje cells in the cerebellum and the pyramidal cells in the cortex is the geometry of axons representing their dominant input. Unlike parallel fibers in the cerebellum, cortical axons run in different directions. Thus, a flat dendritic arbor can effectively capture only those axons that are oriented near orthogonally to the dendritic plane while a 3D dendritic arbor can effectively capture axons from all directions. However, the sparseness of pyramidal cells suggests that they are not minimizing their dendritic cost. Instead, understanding the shape of pyramidal dendrites requires the introduction of a new principle.

7.7 Maximizing the Connectivity Repertoire as a Statistical Principle

Mammalian neocortex is a highly plastic brain region vital for learning and memory. Because the shape of a dendritic arbor determines which axons are accessible to which dendrites, we hypothesize that the sparseness of a pyramidal dendritic arbor may reflect its flexibility to find appropriate presynaptic neurons. In an adult brain, a dendritic arbor connects with a combination of appropriate neurons, whose axons are sparsely distributed [constituting less than 10 % of all axons passing through the arbor territory (Stepanyants et al. 2002; Markram et al. 1997; Thomson and Bannister 2003; Sjöström et al. 2001)] and are not known to the dendrites prior to arbor growth. In the course of neuronal development, the arbor must find this combination in the ensemble of accessible input neuron combinations (via physical contacts between axons and dendrites) defined by arbor dimensions. Then, finding a more appropriate combination would require choosing arbor dimensions that have a larger ensemble. Furthermore, if the developmental search strategy is sufficiently good, the larger the ensemble of input neuron combinations available to the arbor of given dimensions, the more appropriate the input neuron combination can be found.

Therefore, we can quantify arbor functionality by the number of different combinations of input neurons available to an arbor of given dimensions. To reduce the arbor functionality to an additive and extensive quantity (i.e., it doubles if L is doubled), we consider the logarithm of the number of accessible input neuron combinations for given arbor dimensions, which we call the *connectivity repertoire*. Thus, maximizing arbor functionality reduces to finding arbor dimensions that maximize the connectivity repertoire.

We computed the connectivity repertoire by adding the following two contributions. First, for given arbor dimensions, we counted all possible different shapes of a dendritic arbor. Each arbor shape selects a different subset of axons that pass within a spine length of a dendrite (Fig. 7.1a). Second, for a given arbor shape, we enumerated the number of combinations of choosing actual axonal inputs out of potential synapses (Fig. 7.1a).

Whereas the second contribution is straightforward to calculate (Stepanyants et al. 2002), the first contribution is challenging because locations of potential synapses are strongly correlated with each other due to the contiguous nature of an arbor. Fortunately, a similar problem has been solved in statistical physics by expressing the astronomical number of different conformations of a branched polymer in terms of its dimensions (Gutin et al. 1993; Rubinstein and Colby 2003). Using this analogy, we derived an expression for the connectivity repertoire, S, in terms of three arbor dimensions (Wen et al. 2009): L, R, and the average distance along the path from the tip of a branch to the cell body, l (Fig. 7.1a),

$$S \sim S_0 + \frac{L}{a}\ln\left(1 - R/l\right) - \frac{l^2}{La} - \frac{L^2}{R^2}. \tag{7.8}$$

Prior to maximizing S in (7.8) with respect to R and l, let us explain the biological meaning of each term. The first term contains contributions independent of R and l (both from variations in arbor shape and from selecting actual synapses out of potential ones) and, thus, will not be considered further.

The second and third terms are R- and l-dependent corrections to the number of different arbor shapes. The second term reflects the fact that straighter branches can come in fewer different shapes (Rubinstein and Colby 2003) and is always negative. As dendrites become straighter, the path length l approaches from above the

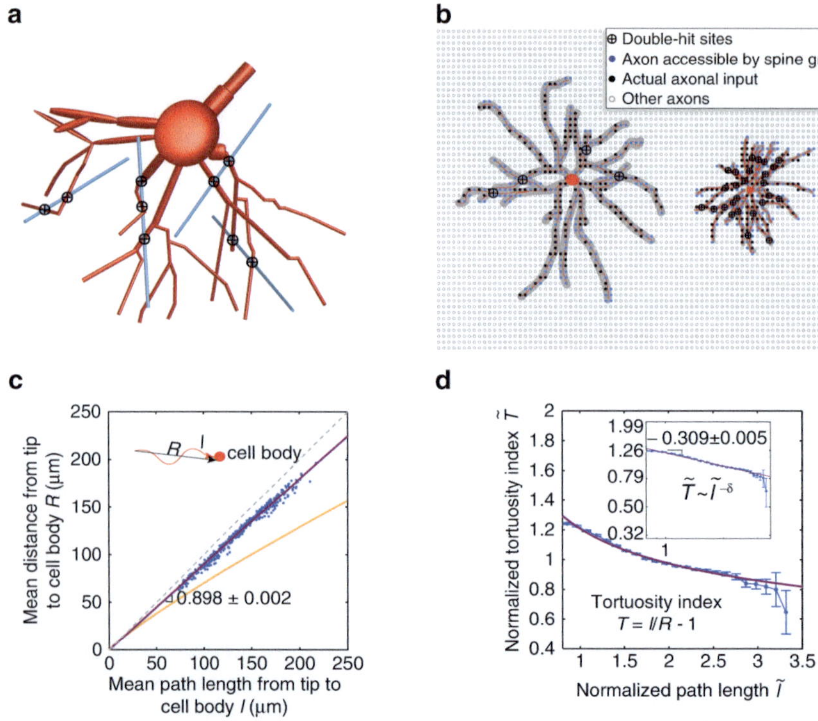

Fig. 7.3 Dendritic arbor shapes reflect maximization of the connectivity repertoire for a given dendritic cost. (**a**) Some axons form multiple potential synapses with a 3D dendritic arbor (*crossed black circles*). (**b**) A sparse arbor, which has lower dendrite density for a given total dendritic length, has fewer multiple potential synapses and, hence, maximizes the connectivity repertoire. (**c**) Tortuosity measured as a function of the average Euclidean distance from the tip of a branch to the cell body on l. The best fit (*magenta line*) suggests that basal pyramidal dendrites are approximately straight. Orange line shows tortuous dendrites predicted from maximizing the connectivity repertoire alone, $R \sim a^{1/8} l^{7/8}$. Each point represents a different cell. (**d**) Normalized tortuosity index as a function of the normalized path length from a dendritic segment to cell body. The tortuosity index T is defined as the ratio of the length along a path to the Euclidean distance between its ends minus one. In rescaled coordinates, the tortuosity was measured in units of the average index value over all paths within an arbor. The path length was measured in units of the average length over all paths within an arbor as well. *Magenta line* is a power law fit. *Inset* shows the same data on log–log scale. Error bars show s.e.m. This figure is reproduced from Wen et al. (2009)

Euclidian distance between the tip of a branch and the cell body (approximated by R), and the second term decreases dramatically. Therefore, maximizing the number of arbor shapes favors $R \ll l$, that is, tortuous branches. The third term reflects the fact that maximizing the number of arbor shapes favors branchy dendrites ($l \ll L$) (Gutin et al. 1993). In these terms, a is the persistence length, below which a dendrite cannot bend (Rubinstein and Colby 2003).

The last term in equation is an R-dependent correction to the contribution to S arising from choosing actual synapses out of potential ones. Some axons could establish more than one potential synapse in different locations on a dendritic arbor (Fig. 7.3a).

If the entire dendritic arbor acts as a unit, differences in the locations of actual synapses from the same axon do not affect the function (Magee and Cook 2000; Chklovskii et al. 2004; Cash and Yuste 1999; Nevian et al. 2007). As a result, additional potential synapses from the same axon are redundant and do not make a contribution to S. Thus, in calculating S, we should subtract the over-counting due to multiple potential synapses, which, for isotropically distributed straight axons (Braitenberg and Schuz 1998; Binzegger et al. 2005; Kisvarday and Eysel 1992), is given by the last term.

Maximizing S favors fewer multiple potential synapses from the same axon and hence smaller last term. Reducing the last term calls for a larger R for a given L or lower dendrite density (Fig. 7.3b). Thus, avoidance of multiple potential synapses leads to a statistical preference for arbors with sparsely distributed branches (Fig. 7.3b).

However, maximization of S alone resulted in more tortuous dendrites than those observed in a pyramidal cell (Fig. 7.3c, yellow line). This fact indicates that the straightness of dendrites (Kawaguchi et al. 2006; Samsonovich and Ascoli 2003; Stepanyants et al. 2004) is driven by reducing the dendritic cost. By taking into account the cost contributed by path length and total wiring length as discussed in the previous section, we derived a scaling form of the cost function $E \sim Ll^{\delta}$, where δ is model dependent (see Appendix).

7.8 Predictions from Functionality–Cost Optimization and Experimental Measurements

Maximizing the connectivity repertoire for a given dendritic cost reduces to maximizing their combination: $S-\beta E$, where β is the weight of E relative to S. We solved this optimization problem (analogous to maximizing entropy for a given energy in statistical physics) and derived analytical expressions relating arbor dimensions l, R, and L. Next, we compare these expressions with experimental measurements.

First, optimal l and R satisfy the relation $l/R-1 \sim 1/l^{\delta}$, where $l/R-1$ is defined as the tortuosity index. This prediction is consistent with measurements from basal dendrites of pyramidal cells. The tortuosity index of a path from a dendritic segment to cell body decreases with the path length, and their relationship can be well fit by a power law with exponent $\delta = 0.309 \pm 0.005$ (Fig. 7.3d).

Second, our optimization framework predicts that R and L should satisfy a power law relation $R \sim L^{\nu}$ with the exponent $\nu = 1/(2+\delta)$. Substituting the empirically measured value for δ, we find $\nu = 0.43$. Experimentally, by measuring the arbor dimensions of 2,171 basal pyramidal dendritic arbors, we found that the relation between R and L can be fit via a power law with an exponent, $\nu = 0.44 \pm 0.01$ (Fig. 7.4a), which is not significantly different from our theoretical prediction.

Third, we explored statistical properties of branching by analyzing the spatial correlations among branches within an arbor. To this end, we counted the number of

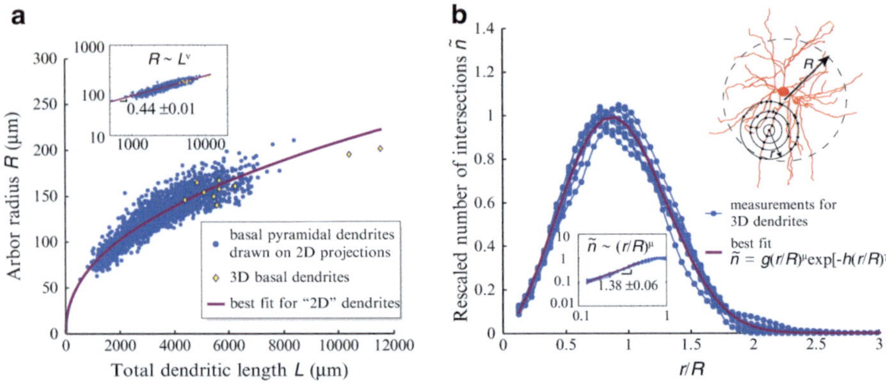

Fig. 7.4 Scaling, universality, and self-similarity of basal dendritic arbors of pyramidal neurons. (**a**) Dendritic arbor radius scales as a power of the total dendritic length. For 2,161 shrinkage-corrected 2D reconstructions of basal dendrites of pyramidal cells from different cortical areas, the power is 0.44 ± 0.01. The ten available shrinkage-corrected 3D reconstructions (Kisvarday and Eysel 1992) are consistent with this relationship. Inset shows the same data on log–log scale. (**b**) Spatial pairwise correlation between dendritic segments for each neuron shown in rescaled coordinates. *Inset* at the *upper right* illustrates how the measurement was performed: counts of intersections between concentric spheres and dendritic branches were averaged over random sphere center locations in the central part of the dendritic arbor. *Curves* corresponding to different neurons collapse onto a master curve, fit by a universal function (*magenta line*), in which μ and γ are the only fitting parameters; coefficients g and h are fully determined by the normalization conditions. *Inset* under the *main curves* shows the rising part of the plot on log–log scale, indicating a power law relationship and self-similarity of arbor shape. This figure is reproduced from Wen et al. (2009)

pairs of dendritic segments separated by distance r. We expressed the distance in units of R and rescaled the counts, $\tilde{n}$, so that the area under the $\tilde{n}(r/R)$ curve is normalized to one (Fig. 7.4b). The dependence of $\tilde{n}$ on r/R for different arbors collapses onto a single curve (Fig. 7.4b).

The universality of the pairwise correlations in the locations of dendritic segments suggests that dendritic arbors of pyramidal cells are built by statistically similar processes. Indeed, the analogy between our optimization framework and the statistical physics of polymers (Rubinstein and Colby 2003) suggests a functional form of $\tilde{n}$. In particular, the rising part of $\tilde{n}$ would follow a power law $\tilde{n} \sim (r/R)^{\mu}$ (Fig. 7.4b, inset), indicating that a fragment of an arbor is statistically similar to the scaled-down version of the whole arbor (excluding the periphery) (Stevens 2008; Smith et al. 1989; Jelinek et al. 2005; Caserta et al. 1990; Rothnie et al. 2006; Jelinek and Fernandez 1998; Milosevic et al. 2005). These conclusions are further supported by the fact that the measured exponent ν, characterizing the scaling of arbor dimensions across different cells (Fig. 7.4a), and the self-similarity exponent μ (1.38 ± 0.06 and Fig. 7.4b) are consistent with $\mu = 1/\nu - 1$, a relation that results from statistical similarity through a standard scaling argument (see Appendix).

7.9 Why Pyramidal Dendritic Arbors Are Sparser than Purkinje Dendritic Arbors?

We have showed that, in the case of isotropically distributed axons, avoiding multiple potential synapses leads to an arbor territory greater than the total area of the spine-reach zone (Fig. 7.1a). This accounts for the sparseness of basal dendritic arbors of pyramidal cells, which form potential synapses with only a small fraction of axons passing through the arbor territory (Fig. 7.1d, e).

In the case of anisotropic distribution of axons and dendrites, avoidance of multiple potential synapses may not lead to a sparse arbor. For example, if all axons were oriented orthogonally to a planar dendritic arbor, such as parallel fibers and Purkinje cell dendrites in the cerebellum (Llinás et al. 2004), multiple potential synapses could be avoided by arranging dendritic branches so that their spine-reach zones do not overlap. Thus, to minimize the cost of dendrites, the arbor must contract until the spine-reach zone covers most of the arbor territory, just as in Purkinje cell dendrites (Fig. 7.1c). Although a projection of Purkinje dendritic arbor on the direction other than that of parallel fibers would contain many overlaps of the spine-reach zone, this would not generate multiple potential synapses for parallel fibers.

7.10 Discussion

Are our results applicable to cell classes other than pyramidal and Purkinje cells? We note that the basic aspects of our theory still hold even if some assumptions were relaxed. (a) If presynaptic axons are tortuous and/or branchy rather than straight on the scale of a dendritic arbor, our theory would still hold but predict different scaling exponents. (b) If an individual dendritic branch rather than the whole arbor acts as a unit (Losonczy and Magee 2006; Poirazi et al. 2003a, b), actual synapses from the same axon lead to the same functionality only when they occur on the same branch. In this case, calculations would be different but the general framework would still hold.

However, arbor shapes would fall outside the current theoretical framework in the following cases. (a) Instead of maximizing the connectivity repertoire, the objective of some dendrites could be to maximize the number of synaptic contacts with the same presynaptic neuron. An example includes large monopolar cells in the insect optic lobe innervated by photoreceptors (Nicol and Meinertzhagen 1982). (b) Spatial distribution of axons from appropriate input neurons may not be a priori unknown to the arbor but partially or fully predetermined. For example, layer V apical dendrites reach out to upper cortical layers I–III containing appropriate inputs, which is why we restricted our analysis to basal dendritic arbors. To determine whether our theory is applicable to various cell types, large data sets of high-quality arbor reconstructions would be desirable.

Why do we believe that the distribution of axons presynaptic to basal pyramidal dendrites is not predetermined? (a) Specificity in the location of pyramidal axons has been demonstrated only on the scale of cortical layers and columns, which is greater than the radius of a basal dendritic arbor (~100 μm) (Gilbert and Wiesel 1989; Mooser et al. 2004). (b) On the smaller scale (<100 μm), pyramidal axons follow relatively straight trajectories in various directions (Binzegger et al. 2005; Stepanyants et al. 2004; Fujita and Fujita 1996; Lund et al. 1993; Anderson et al. 2002; Kisvarday and Eysel 1992) and, thus, are unlikely to organize into specific patterns. (c) As synapses on a pyramidal dendrite are made about every half micron, it is unclear how axons from appropriate input neurons could align along dendrites in such a precise manner. Even if this arrangement could be achieved for one dendritic arbor, thousands of other arbors present in the dendritic territory would require a rearrangement of axons from one dendrite to the next, a scenario that is highly improbable.

One may question the validity of our assumption that pyramidal cell dendrites avoid multiple potential synapses because experiments indicate that synaptically coupled cortical neurons share multiple synaptic contacts (Markram et al. 1997; Silver et al. 2003). However, this observation does not contradict our theory for two reasons. First, electrophysiological recordings are heavily biased towards connected and hence nearby neurons. Although nearby neurons may make many potential synapses, these inputs are in a minority. Indeed, in V1, 83 % of synapses on a dendrite originate from neurons located farther than 200 μm in the cortical plane (Stepanyants et al. 2009). For these presynaptic neurons, typically only one axonal branch courses through the dendritic arbor territory and hence is unlikely to make multiple potential synapses. Second, it is possible that various synaptic contacts form independently according to some local rules. Therefore, development can only strive to avoid multiple potential synapses but cannot guarantee their complete absence.

The cornerstone of our theory is the idea that a dendritic arbor avoids multiple hits with the same axon. Statistically, such avoidance results in an apparent repulsion between branches. To go beyond our teleological theory and speculate developmental mechanisms responsible for repulsion, we must distinguish the following two logical possibilities. First, growing dendrites may experience some kind of interaction, which forces them to become more extended, thus leading to reduction of multiple hits. Second, arbors somehow detect multiple hits with the same axon and rearrange their shapes in response. We think that the latter possibility is less realistic, not only because it would be difficult for a cell to detect multiple hits from the same axon on the background of simultaneous signals from thousands of other inputs, but also mostly because dendrites should avoid multiplying even potential synapses which generate no electrical signal in dendrites. We therefore think that the first possibility—avoidance of multiple hits as a consequence of interaction between branches—is more realistic.

Although the mechanism for such interaction among dendrites is unknown in vertebrates, it has been discovered in invertebrates in the form of homophilic Dscam interaction (Samsonovich and Ascoli 2003; Hattori et al. 2007; Schmucker et al.

2000; Hughes et al. 2007). Due to a large number of stochastically chosen splice isoforms, this molecule can mediate self-repulsion among branches of the same neuron without inducing interactions among different neurons. However, Dscams are thought to be nondiffusible and hence incapable of acting over long distances. Moreover, a vertebrate homolog of invertebrate Dscam does not have a large number of splice isoforms (Fuerst et al. 2008; Yamagata and Sanes 2008). We conjecture that a diffusible molecule with a large number of isoforms mediates self-repulsion in vertebrates.

Prior works on understanding arbor shape have involved computer simulations (Samsonovich and Ascoli 2003, 2005; van Pelt and Uylings 2002) using specific rules of branching and orientation of dendrites. Interestingly, it was suggested that reproducing realistic arbor shape requires effective repulsion between the cell body and dendrites (Samsonovich and Ascoli 2003). Although the proposed scheme can explain the straightness of dendritic branches and their centripetal orientation (Samsonovich and Ascoli 2003; Marks and Burke 2007a), it does not account for the distribution of the nearest neighbor distance among branches (Marks and Burke 2007b). Here, based on the analysis of dendritic functionality, we proposed that effective repulsion exists among all segments of a dendritic arbor, not just between the cell body and dendrites.

More recently, Hermann Cuntz and colleagues (Cuntz 2012; Cuntz et al. 2010, 2012) have used a greedy optimization algorithm to generate realistic dendritic arbor shapes with a variability that matches different neuronal types in the insect and mammalian brains (see Chap. 6). In his approach, an in silico arbor must calculate a strategy to minimize both the total wire length and path length cost while connecting given target points in the neuropil with defined spatial distribution. Our complementary approaches reach similar conclusions that total wire length and path length costs are important for determining the shape of an arbor. In the future, however, it is important to understand how these target points (or seeds) are chosen. Do these target points reflect the location of the most appropriate axonal inputs? If so, what determines their sparseness and spatial distribution? These questions would be answered if we have a better understanding of the organization and connectivity of a neuropil on the size of a dendritic arbor.

Finally, the idea that avoidance of multiple hits affects arbor shape may not be limited to neurons and apply to other biological objects. Consider, for example, the spatial arrangement of tree branches. As a tree strives to maximize its exposure to sunlight, its leaves must avoid shading each other (Mauseth 2003; Thomas 2001). This means that the branches should minimize multiple hits with straight lines directed from the sun. Because light rays come from varying directions, they induce repulsion among branches, similarly to self-repulsion in dendritic arbors of pyramidal cells. As the cost of branches is likely to grow with both the total length and the path length, our theory should apply to the spatial distribution of tree branches as well. Thus the similarity of shapes among neurons and trees may not be a coincident but arises from similar evolutionary objectives.

Appendix

A.1 Relation Between Scaling Exponents

If the statistical properties of different dendritic arbors were universal and dendritic arbors were indeed self-similar, the scaling exponent, ν, from the power law between R and L, would be related to the exponent from the rising part of spatial pairwise correlation function between dendritic segments, μ, as shown below.

For self-similar dendritic arbors built by common rules, the rising part of the pairwise correlation function on the regular scale, $n(r)$, would be described by the power law function

$$n \sim r^{\mu}, r < R. \tag{7.9}$$

In addition, the pairwise correlation function would be characterized by a single length scale, R. Specifically, if $n(r)$ decays fast for $r>R$ and the rising part of $n(r)$ remains a power law until r is on the order of R (as in agreement with our data), then the normalization condition $\int_0^{\infty} n(r)\,dr = L$ can be well approximated by introducing a cutoff R:

$$\int_0^{R} n(r)\,dr \approx L. \tag{7.10}$$

By using (7.9) for $n(r)$ in (7.10), we obtain

$$L \sim R^{\mu+1}. \tag{7.11}$$

Recall that our measurement for arbor dimensions shows $R \sim L^{\nu}$ (Fig. 7.4a). Therefore, we arrive at a relation between exponents μ and ν:

$$\mu = \frac{1}{\nu} - 1. \tag{7.12}$$

In addition, given (7.9) and (7.12), the rising part of the rescaled pairwise correlation function should follow a similar power law $\tilde{n} \sim (r/R)^{\mu}$. This can be derived straightforwardly from the relation $\tilde{n} = nR/L$, which is a consequence of the normalization conditions.

A.2 A Cost Model for Dendritic Arbors

As maximizing the connectivity repertoire alone leads to more tortuous branches than observed, we must include the dendritic cost in the optimization framework.

After considering several simplified models based on the above reasoning (Wen and Chklovskii 2008), we propose that the cost scales with L and l as

$$E \sim Ll^{\delta}. \tag{7.13}$$

The value of exponent δ is model dependent. Below, we consider one simplified model that yields a specific value of δ.

We assume that E is proportional to the total area of dendritic surface for the following reason. To maintain the resting potential of a neuron or low [Na+] inside a cell, sodium ions permeating into a neuron should be actively extruded by Na+/K+ pumps (Attwell and Laughlin 2001). The amount of ATP expended is proportional to the membrane capacitance for action potential propagation or membrane input conductance in the absence of spikes (Attwell and Laughlin 2001). Hence, in both considerations, E is proportional to the membrane surface area,

$$E \sim Ld, \tag{7.14}$$

where d is the mean dendritic diameter.

Although E decreases as the mean dendritic diameter d is reduced, a small diameter would detrimentally affect dendritic function. One possible reason is that proteins must be transported from the cell body to synapses and the rate of transport is proportional to the number of microtubules at a given cross section of a dendrite (Hillman 1979). The rate of protein transport through a cross section must be proportional to the number of synapses downstream of that cross section. If the density of microtubules (per cross-sectional area) is roughly invariant, the cross-sectional area must be proportional to the number of microtubules and hence to the number of synapses downstream (Hsu et al. 1998; Hillman 1979; Wittenberg and Wang 2007). If the typical branch has length l, the number of synapses on it is Nl/L. Therefore, a typical dendritic diameter is

$$d^2 \sim ANl / L, \tag{7.15}$$

where A is a cross-sectional area needed to support one synapse. By substituting (7.15) with (7.14), we find that

$$E \sim N^{1/2}L^{1/2}l^{1/2}. \tag{7.16}$$

Given that $N \sim L$ (Larkman 1991), we obtain

$$E \sim Ll^{1/2}. \tag{7.17}$$

References

Anderson JC, Binzegger T, Douglas RJ, Martin KA (2002) Chance or design? Some specific considerations concerning synaptic boutons in cat visual cortex. J Neurocytol 31(3–5): 211–229

Ascoli GA (2006) Mobilizing the base of neuroscience data: the case of neuronal morphologies. Nat Rev Neurosci 7(4):318–324

Attwell D, Laughlin SB (2001) An energy budget for signaling in the grey matter of the brain. J Cereb Blood Flow Metab 21(10):1133–1145

Binzegger T, Douglas RJ, Martin KA (2005) Axons in cat visual cortex are topologically self-similar. Cereb Cortex 15(2):152–165

Braitenberg V, Schuz A (1998) Cortex: statistics and geometry of neuronal connectivity. 2nd thoroughly rev. edn. Springer, New York

Caserta F, Stanley HE, Eldred WD, Daccord G, Hausman RE, Nittmann J (1990) Physical mechanisms underlying neurite outgrowth: a quantitative analysis of neuronal shape. Phys Rev Lett 64(1):95–98

Cash S, Yuste R (1999) Linear summation of excitatory inputs by CA1 pyramidal neurons. Neuron 22(2):383–394

Chklovskii DB, Mel BW, Svoboda K (2004) Cortical rewiring and information storage. Nature 431(7010):782–788

Chklovskii DB, Vitaladevuni S, Scheffer LK (2010) Semi-automated reconstruction of neural circuits using electron microscopy. Curr Opin Neurobiol 20(5):667–675

Cuntz H (2012) The dendritic density field of a cortical pyramidal cell. Front Neuroanat 6:2

Cuntz H, Borst A, Segev I (2007) Optimization principles of dendritic structure. Theor Biol Med Model 4:21

Cuntz H, Forstner F, Borst A, Häusser M (2010) One rule to grow them all: a general theory of neuronal branching and its practical application. PLoS Comput Biol 6(8)

Cuntz H, Mathy A, Häusser M (2012) A scaling law derived from optimal dendritic wiring. Proc Natl Acad Sci USA 109(27):11014–11018

Denk W, Horstmann H (2004) Serial block-face scanning electron microscopy to reconstruct three-dimensional tissue nanostructure. PLoS Biol 2(11):e329

Fuerst PG, Koizumi A, Masland RH, Burgess RW (2008) Neurite arborization and mosaic spacing in the mouse retina require DSCAM. Nature 451(7177):470–474

Fujita I, Fujita T (1996) Intrinsic connections in the macaque inferior temporal cortex. J Comp Neurol 368:467–486

Gilbert CD, Wiesel TN (1989) Columnar specificity of intrinsic horizontal and corticocortical connections in cat visual cortex. J Neurosci 9(7):2432–2442

Gutin AM, Grosberg AY, Shakhnovich EI (1993) Polymers with annealed and quenched branchings belong to different universality classes. Macromolecules 26(6):1293–1295

Hattori D, Demir E, Kim HW, Viragh E, Zipursky SL, Dickson BJ (2007) Dscam diversity is essential for neuronal wiring and self-recognition. Nature 449(7159):223–227

Hillman D (1979) Neuronal shape parameters and substructures as a basis of neuronal form. In: The Neurosciences: fourth study program. MIT Press, Cambridge, MA, pp vii, 1185 p

Hsu A, Tsukamoto Y, Smith RG, Sterling P (1998) Functional architecture of primate cone and rod axons. Vision Res 38(17):2539–2549

Hughes ME, Bortnick R, Tsubouchi A, Baumer P, Kondo M, Uemura T, Schmucker D (2007) Homophilic Dscam interactions control complex dendrite morphogenesis. Neuron 54(3):417–427

Jelinek H, Elston GN, Zietsch B (2005) Fractal analysis: pitfalls and revelations in neuroscience. In: Losa GA, Merlini D, Nonnenmacher TF, Weibel ER (eds) Fractals in biology and medicine, vol 4. Birkhauser Verlag AG, Switzerland, pp 85–94

Jelinek HF, Fernandez E (1998) Neurons and fractals: how reliable and useful are calculations of fractal dimensions? J Neurosci Methods 81(1–2):9–18

Kawaguchi Y, Karube F, Kubota Y (2006) Dendritic branch typing and spine expression patterns in cortical nonpyramidal cells. Cereb Cortex 16(5):696–711
Kisvarday ZF, Eysel UT (1992) Cellular organization of reciprocal patchy networks in layer III of cat visual cortex (area 17). Neuroscience 46(2):275–286
Kleinfeld D, Bharioke A, Blinder P, Bock DD, Briggman KL, Chklovskii DB, Denk W, Helmstaedter M, Kaufhold JP, Lee WC, Meyer HS, Micheva KD, Oberlaender M, Prohaska S, Reid RC, Smith SJ, Takemura S, Tsai PS, Sakmann B (2011) Large-scale automated histology in the pursuit of connectomes. J Neurosci 31(45):16125–16138
Larkman AU (1991) Dendritic morphology of pyramidal neurones of the visual cortex of the rat: III. Spine distributions. J Comp Neurol 306(2):332–343
Lichtman JW, Livet J, Sanes JR (2008) A technicolour approach to the connectome. Nat Rev Neurosci 9(6):417–422
Llinás RR, Walton KD, Lang EJ (2004) Cerebellum. In: Shepherd GM (ed) The synaptic organization of the brain, 5th edn. Oxford University Press, New York, p xiv, 719 p
Losonczy A, Magee JC (2006) Integrative properties of radial oblique dendrites in hippocampal CA1 pyramidal neurons. Neuron 50(2):291–307
Lund JS, Yoshioka T, Levitt JB (1993) Comparison of intrinsic connectivity in different areas of macaque monkey cerebral cortex. Cereb Cortex 3:148–162
Magee JC, Cook EP (2000) Somatic EPSP amplitude is independent of synapse location in hippocampal pyramidal neurons. Nat Neurosci 3(9):895–903
Markram H, Lubke J, Frotscher M, Roth A, Sakmann B (1997) Physiology and anatomy of synaptic connections between thick tufted pyramidal neurones in the developing rat neocortex. J Physiol 500(Pt 2):409–440
Marks WB, Burke RE (2007a) Simulation of motoneuron morphology in three dimensions. I. Building individual dendritic trees. J Comp Neurol 503(5):685–700
Marks WB, Burke RE (2007b) Simulation of motoneuron morphology in three dimensions. II. Building complete neurons. J Comp Neurol 503(5):701–716
Martone ME, Zhang S, Gupta A, Qian X, He H, Price DL, Wong M, Santini S, Ellisman MH (2003) The cell-centered database: a database for multiscale structural and protein localization data from light and electron microscopy. Neuroinformatics 1(4):379–395
Mauseth JD (2003) Botany : an introduction to plant biology, 3rd edn. Jones and Bartlett Publishers, Boston
Milosevic NT, Ristanovic D, Stankovic JB (2005) Fractal analysis of the laminar organization of spinal cord neurons. J Neurosci Methods 146(2):198–204
Mooser F, Bosking WH, Fitzpatrick D (2004) A morphological basis for orientation tuning in primary visual cortex. Nat Neurosci 7(8):872–879
Napper RM, Harvey RJ (1988a) Number of parallel fiber synapses on an individual Purkinje cell in the cerebellum of the rat. J Comp Neurol 274(2):168–177
Napper RM, Harvey RJ (1988b) Quantitative study of the Purkinje cell dendritic spines in the rat cerebellum. J Comp Neurol 274(2):158–167
Nevian T, Larkum ME, Polsky A, Schiller J (2007) Properties of basal dendrites of layer 5 pyramidal neurons: a direct patch-clamp recording study. Nat Neurosci 10(2):206–214
Nicol D, Meinertzhagen IA (1982) An analysis of the number and composition of the synaptic populations formed by photoreceptors of the fly. J Comp Neurol 207(1):29–44
Poirazi P, Brannon T, Mel BW (2003a) Arithmetic of subthreshold synaptic summation in a model CA1 pyramidal cell. Neuron 37(6):977–987
Poirazi P, Brannon T, Mel BW (2003b) Pyramidal neuron as two-layer neural network. Neuron 37(6):989–999
Ramón y Cajal S (1899) Textura del Sistema Nervioso del Hombre y de los Vertebrados (Texture of the nervous system of man and the vertebrates), vol 1. Springer, New York
Rapp M, Segev I, Yarom Y (1994) Physiology, morphology and detailed passive models of guinea-pig cerebellar Purkinje cells. J Physiol 474(1):101–118
Rothnie P, Kabaso D, Hof PR, Henry BI, Wearne SL (2006) Functionally relevant measures of spatial complexity in neuronal dendritic arbors. J Theor Biol 238(3):505–526

Rubinstein M, Colby RH (2003) Polymer physics. Oxford University Press, Oxford, NY
Samsonovich AV, Ascoli GA (2003) Statistical morphological analysis of hippocampal principal neurons indicates cell-specific repulsion of dendrites from their own cell. J Neurosci Res 71(2):173–187
Samsonovich AV, Ascoli GA (2005) Statistical determinants of dendritic morphology in hippocampal pyramidal neurons: a hidden Markov model. Hippocampus 15(2):166–183
Schmucker D, Clemens JC, Shu H, Worby CA, Xiao J, Muda M, Dixon JE, Zipursky SL (2000) Drosophila Dscam is an axon guidance receptor exhibiting extraordinary molecular diversity. Cell 101(6):671–684
Shepherd GM, Svoboda K (2005) Laminar and columnar organization of ascending excitatory projections to layer 2/3 pyramidal neurons in rat barrel cortex. J Neurosci 25(24):5670–5679
Silver RA, Lubke J, Sakmann B, Feldmeyer D (2003) High-probability uniquantal transmission at excitatory synapses in barrel cortex. Science 302(5652):1981–1984
Sjöström PJ, Turrigiano GG, Nelson SB (2001) Rate, timing, and cooperativity jointly determine cortical synaptic plasticity. Neuron 32(6):1149–1164
Smith TG Jr, Marks WB, Lange GD, Sheriff WH Jr, Neale EA (1989) A fractal analysis of cell images. J Neurosci Methods 27(2):173–180
Stepanyants A, Hof PR, Chklovskii DB (2002) Geometry and structural plasticity of synaptic connectivity. Neuron 34(2):275–288
Stepanyants A, Martinez LM, Ferecsko AS, Kisvarday ZF (2009) The fractions of short- and long-range connections in the visual cortex. Proc Natl Acad Sci USA 106(9):3555–3560
Stepanyants A, Tamas G, Chklovskii DB (2004) Class-specific features of neuronal wiring. Neuron 43(2):251–259
Stevens CF (2008) All arbors have the same shape. In: Neuronal circuits: from structure to function, CSHL and private communication, 2008
Thomas PA (2001) Trees: their natural history, 1st edn. Cambridge University Press, UK
Thomson AM, Bannister AP (2003) Interlaminar connections in the neocortex. Cereb Cortex 13(1):5–14
Trachtenberg JT, Chen BE, Knott GW, Feng G, Sanes JR, Welker E, Svoboda K (2002) Long-term in vivo imaging of experience-dependent synaptic plasticity in adult cortex. Nature 420(6917):788–794
van Pelt J, Uylings HB (2002) Branching rates and growth functions in the outgrowth of dendritic branching patterns. Network 13(3):261–281
Vetter P, Roth A, Häusser M (2001) Propagation of action potentials in dendrites depends on dendritic morphology. J Neurophysiol 85(2):926–937
Wen Q, Chklovskii DB (2008) A cost-benefit analysis of neuronal morphology. J Neurophysiol 99(5):2320–2328
Wen Q, Stepanyants A, Elston GN, Grosberg AY, Chklovskii DB (2009) Maximization of the connectivity repertoire as a statistical principle governing the shapes of dendritic arbors. Proc Natl Acad Sci USA 106(30):12536–12541
Wittenberg GM, Wang SS-H (2007) Evolution and scaling of dendrites. In: Stuart G, Spruston N, Häusser M (eds) Dendrites, 2nd edn. Oxford University Press, USA, pp 43–67
Yamagata M, Sanes JR (2008) Dscam and Sidekick proteins direct lamina-specific synaptic connections in vertebrate retina. Nature 451(7177):465–469

Chapter 8
Reverse Engineering the 3D Structure and Sensory-Evoked Signal Flow of Rat Vibrissal Cortex

Robert Egger, Vincent J. Dercksen, Christiaan P.J. de Kock, and Marcel Oberlaender

Abstract Soma location, dendrite morphology, and synaptic innervation are key determinants of neuronal function. Unfortunately, conventional functional measurements of sensory-evoked activity in vivo yield limited structural information. In particular, when trying to infer mechanistic principles that underlie perception and behavior, interpretations from functional recordings of individual or small groups of neurons often remain ambiguous without detailed knowledge of the underlying network structures. Here we review a novel reverse engineering approach that allows investigating sensory-evoked signal flow through individual and ensembles of neurons within the context of their surrounding neural networks. To do so, spontaneous and sensory-evoked activity patterns are recorded from individual neurons in vivo. In addition, the complete 3D dendrite and axon projection patterns of such in vivo-characterized neurons are reconstructed and integrated into an anatomically realistic

---

R. Egger
Computational Neuroanatomy Group, Max Planck Institute for Biological Cybernetics, Spemannstraße 38-44, Tuebingen 72076, Germany

Graduate School of Neural Information Processing, University of Tuebingen, Tuebingen, Germany

V.J. Dercksen
Department of Visualization and Data Analysis, Zuse Institute Berlin, Berlin, Germany

C.P.J. de Kock
Center for Neurogenomics and Cognitive Research, Neuroscience Campus Amsterdam, VU University Amsterdam, Amsterdam, The Netherlands

M. Oberlaender (✉)
Computational Neuroanatomy Group, Max Planck Institute for Biological Cybernetics, Spemannstraße 38-44, Tuebingen 72076, Germany

Bernstein Center for Computational Neuroscience, Tuebingen, Germany

Digital Neuroanatomy, Max Planck Florida Institute for Neuroscience, Jupiter, FL, USA
e-mail: marcel.oberlaender@tuebingen.mpg.de

H. Cuntz et al. (eds.), *The Computing Dendrite: From Structure to Function*, Springer Series in Computational Neuroscience 11, DOI 10.1007/978-1-4614-8094-5_8,

model of the rat vibrissal cortex. This model allows estimating the number and cell type-specific subcellular distribution of synapses on these neurons with 50 μm precision. As a result, each neuron can be described by a rich set of parameters that allows investigating structure–function relationships and simulation experiments at single-neuron and network levels.

8.1 Introduction

A principal challenge in neuroscience research is to understand how the brain processes sensory information and how the diverse stimuli give rise to behavioral responses. Despite a long history in recording the activity of individual or small groups of neurons by using electrical or optical methods (Armstrong-James and Fox 1987; Simons et al. 1992; Brecht et al. 2003; Martinez et al. 2005; Bruno and Sakmann 2006; de Kock et al. 2007; Kerr et al. 2007; Poulet and Petersen 2008; Wallace and Sakmann 2008; Sakata and Harris 2009; O'Connor et al. 2010a; O'Connor et al. 2010b), our understanding of mechanistic principles underlying sensory perception and behavior remains limited. This is mainly due to the fact that functional recordings often lack structural information about the underlying neural networks, such as the morphology of its constituent neurons and the complex connectivity patterns established among them. Consequently, to understand how complex spatiotemporal patterns of synaptic input drive neuronal output during perception and behavior, one has to (1) investigate the synaptic "wiring" of functionally characterized neurons, (2) identify the microcircuits that they form, and (3) determine how these microcircuits participate in neural networks during different stimulus conditions.

At present, methods to monitor sensory-evoked streams of excitation through the entire neural networks, combined with quantitative reconstructions of the underlying network structure, are scarce. However, recent studies using functional two-photon calcium imaging (Svoboda et al. 1997) combined with high-throughput electron microscopy techniques (Denk and Horstmann 2004) allowed investigating structure–function relationships in mouse retina (Briggman et al. 2011) or primary visual cortex (Bock et al. 2011). Rendering important steps towards a deeper understanding of visual information processing, at present, these methods remain however limited to relatively small brain volumes of approximately 200 μm × 200 μm × 200 μm cubes (Helmstaedter et al. 2008). Neural networks that include long-range projections, for example such as those between different cortical layers or even between different cortical fields and subcortical neuron ensembles, may not be accessible with electron microscopic approaches. Furthermore, the limited temporal resolution of two-photon calcium imaging (Grewe and Helmchen 2009) and the lack of subthreshold information may prevent from describing important aspects of structure–function relationships using this technique.

In this chapter, we review an alternative, reverse engineering approach that is based on electrical recordings from individual neurons in vivo, combined with a conventional neuron filling method (Pinault 1996; Margrie et al. 2002; Horikawa and Armstrong 1988). By recording the functional responses to different sensory stimuli in large numbers of individual neurons and with high temporal resolution, followed by reconstructing the complete three-dimensional (3D) dendritic and axonal projection patterns, our approach allows investigating how anatomical parameters of individual neurons may influence functional responses. To do so, structural data from a large number of animals is registered into a common 3D reference frame to reconstruct average anatomically realistic neural network models. These networks are then populated with in vivo-measured activities, and responses in silico are compared with responses measured in the respective in vivo network (here, in silico refers to numerical simulations of neuronal activity; see Lang et al. 2011 for details).

We illustrate our approach by reconstructing the vibrissal cortex in rodents, such as rats and mice, which is particularly suited to reverse engineer the structure and function of neural circuits that underlie sensory perception and behavior. First, sensory input from a single principal whisker (PW) located on the animal's snout is capable of eliciting simple behaviors, such as gap crossing (Celikel and Sakmann 2007; Hutson and Masterton 1986). Second, in the vibrissal area of rodent somatosensory cortex (S1), a cytoarchitectonic equivalent to the elementary functional unit of sensory cortices (Mountcastle 1957; Hubel and Wiesel 1959), designated as a barrel column, has been described (Woolsey and Van der Loos 1970). Barrel columns are somatotopically organized, resembling the layout of the facial whiskers (Fig. 8.1). Furthermore, a barrel column primarily processes sensory input from a related PW (Welker 1976). Specifically, sensory-evoked excitation after whisker touch and presumably during whisker motion (Yu et al. 2006) is mediated by neurons located in the ventral posterior medial (VPM) division of the thalamus. The VPM, like the vibrissal cortex, is subdivided into whisker-specific areas, referred to as barreloids (Land et al. 1995). Third, the vibrissal system in rodents is an active sensory system. Thus, unlike for example in cats, rodents actively move their whiskers back and forth to explore their environment.

8.2 Reverse Engineering the Structure and Function of Neural Networks

8.2.1 Cell Type-Specific Representation of Sensory Stimuli in Rat Vibrissal Cortex

Cell-attached (juxtasomal) and whole-cell recording techniques allow electrophysiological measurements on individual neurons. The high temporal resolution and the wide range of applications render single-neuron recordings as one of the primary sources of our current understanding of neuron and network function (Barth and Poulet 2012).

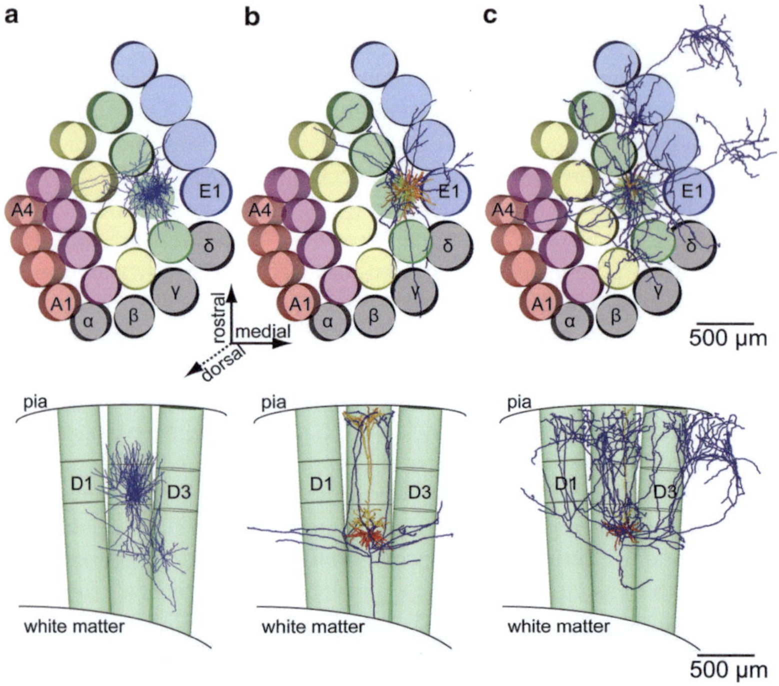

Fig. 8.1 Cell type-specific axonal projection patterns in rat vibrissal cortex. (**a**) *Top*: Tangential view of the barrel field in rat vibrissal cortex. An axon from the ventral posterior medial (VPM) division of the thalamus projecting to the D2 barrel is shown in *blue*. *Bottom*: Semi-coronal view of the same axon with respect to the D2 and two surrounding columns. (**b**) *Top*: As in (**a**). Layer 5 thick tufted (L5tt) neuron with apical (*orange*) and basal (*red*) dendrites and axonal projections (*blue*). *Bottom*: As in (**a**). (**c**) *Top*: As in (**b**). Layer 5 slender tufted (L5st) neuron. *Bottom*: As in (**b**). Panels in (**b**) and (**c**) modified from Oberlaender et al. (2011)

In order to combine these functional measurements with structural information about the underlying circuitry, the recorded neuron is labeled with a tracer, for example biocytin (Horikawa and Armstrong 1988), which allows reconstructing its 3D morphology. This approach has been largely used for preparations in vitro (Helmstaedter et al. 2009; Staiger et al. 2004), for example resulting in various detailed computer models that relate the structure of an individual neuron to its measured function (Hay et al. 2011; Sarid et al. 2007). Unfortunately, the typical thickness of a brain slice is 300 μm. Consequently, in vitro tracings usually suffer from cutoff dendrites and axons (Oberlaender et al. 2012a). Thus, in vitro approaches may be sufficient to investigate structure–function relationships at the single-neuron level, but insights into a neuron's function with respect to the underlying neural circuits and stimulus conditions are limited.

To compensate for these limitations, labeling of individual neurons in vivo allows studying the relationships between sensory-evoked activity and the structure of the intact network. The difficulty in recovering well-labeled dendrites and axons from recordings in vivo and the enormous extent and complexity of axons previously prevented from manually reconstructing large numbers of complete 3D morphologies. We thus developed a semiautomated imaging and tracing pipeline that rivals manual tracings by human expert users in accuracy and reduces the time required for large-scale reconstruction of 3D neuron morphology (Dercksen et al. 2009, 2012b; Oberlaender et al. 2007, 2009a).

Figure 8.1 shows three examples of completely reconstructed dendrite and axon morphologies of the major cell types that provide input to (VPM, Fig. 8.1a) and output from the vibrissal cortex (L5 thick-tufted pyramid (L5tt), L5 slender-tufted pyramid (L5st); Fig. 8.1b, c, respectively). Slender-tufted pyramidal neurons project to the striatum and are commonly referred to as corticostriatal neurons. Thick-tufted pyramidal neurons project to the posterior nucleus of the thalamus, brainstem, superior colliculus, and pons (Alloway 2008; Aronoff et al. 2010; Groh et al. 2008). The two neuron types have been characterized across cortical areas, including somatosensory and visual cortices, and therefore represent canonical elements of the cortical microcircuitry (Groh et al. 2010; Binzegger et al. 2004).

All neurons were reconstructed with respect to anatomical landmarks, such as the pia and white matter (WM) surfaces, the local blood vessel pattern, and the barrels in L4 (Egger et al. 2012). Thus, our reconstruction pipeline allows determining the complete dendrite morphology, axon projection pattern, and 3D location within the vibrissal cortex for large numbers of in vivo-recorded neurons.

Reconstructing approximately 150 neurons across L2–6, we found that dendritic and axonal projection patterns are highly cell type specific. For example, VPM axons (see Oberlaender et al. 2012b for more details and a gallery of cells) delineate the vertical barrel column boundaries, display the densest innervation within L4, and show a second, less pronounced innervation zone at the border between L5 and L6. L5tt neurons display widespread apical tuft dendrites that usually reach into the septum surrounding the principal column (PC, i.e., containing the neuron's soma) or even into surrounding barrel columns (Oberlaender et al. 2012a). Axons of L5tt neurons are usually sparse within the vibrissal cortex, confined to the vertical barrel column borders in L1–4, and project to surrounding barrel columns in L5 (see Oberlaender et al. 2011 for more details and a gallery of cells). Somata of L5st neurons intermingle with those of L5tt cells but have very different dendritic and axonal projection patterns. The apical tuft dendrite of L5st cells is usually confined to the vertical boundaries of the respective PC. L5st axons within the vibrissal cortex are by a factor of 3 more elaborate than L5tt neurons, display widespread projections within L1–3, and remain largely confined to the PC in L5 (see Oberlaender et al. 2011 for more details and a gallery of cells).

The anatomically defined cell types show also cell type-specific functional properties. Figure 8.2a summarizes the average spiking activity for VPM neurons after passive whisker deflection ("passive touch") (Brecht and Sakmann 2002) and the evoked activity of L5tt and L5st neurons (de Kock et al. 2007). VPM neurons reliably, and almost exclusively, respond to the deflection of the respective PW. In contrast, L5tt neurons show strong responses to deflection of the principal and

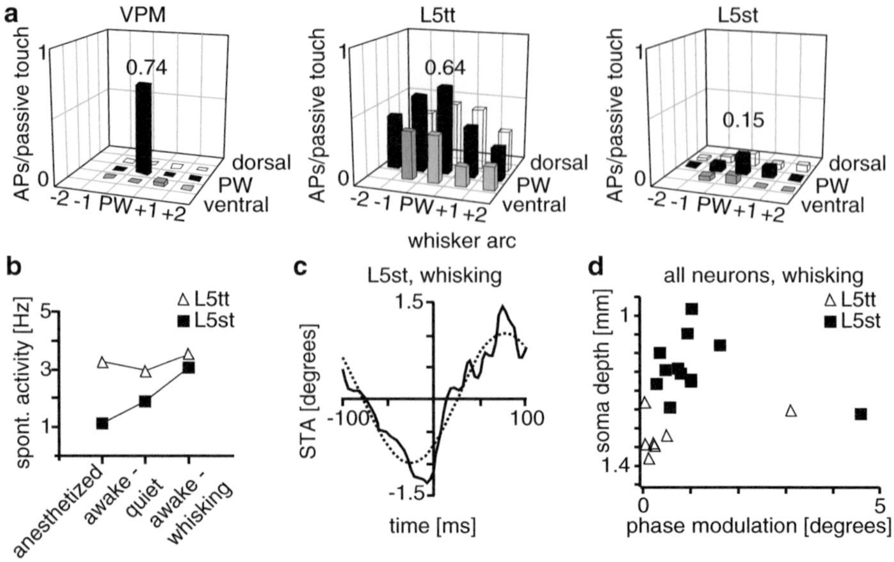

Fig. 8.2 Functional responses of neurons depend on cell type and behavioral state. (**a**) Suprathreshold receptive fields of the three cell types in Fig. 8.1 after passive whisker deflection. (**b**) Spiking activity of L5st and L5tt neurons during different behavioral states. (**c**) Phase-modulated spiking of one L5st neuron. *STA* spike-triggered average. (**d**) Phase modulation of L5st and L5tt neurons. Panels in (**a**) are modified from Brecht and Sakmann (2002) and de Kock et al. (2007), panels in (**b**) and (**c**) are modified from de Kock and Sakmann (2009), and panel in (**d**) is modified from Oberlaender et al. (2012a)

the eight surrounding whiskers (Su8W), while L5st neurons show almost no response to passive touch at all.

The situation changes dramatically when the animal is awake and actively moving its whiskers back and forth in a periodic manner (Fig. 8.2b de Kock and Sakmann 2009). The spiking activity of L5tt neurons remains largely the same compared to the anesthetized state or periods without whisker movement (i.e., quiet-awake). In contrast, L5st neurons increase their activity during whisking. More importantly, L5st neurons carry information about whisker location, which is represented as the instantaneous phase of the periodic whisker movement (Hill et al. 2011). Neurons of all other excitatory cell types, including L5tt neurons, carry only little or no phase information at all (de Kock and Sakmann 2009; Oberlaender et al. 2011).

In summary, single-neuron recordings and fillings in vivo allow (1) determining the neuron's location with respect to anatomical landmarks, (2) reconstructing the complete 3D dendritic and axonal projection patterns, (3) classifying anatomical cell types, and (4) determining cell type-specific functional properties during different behavioral states and sensory stimuli.

8.2.2 Reconstruction of 3D Anatomically Realistic Neural Networks in Rat Vibrissal Cortex

Correlating the 3D morphology of individual neurons with their spontaneous or sensory-evoked activity in vivo allows specifying cell type-specific structure–function relationships. However, to relate such relationships to the overall stream of sensory-evoked excitation and to ultimately infer mechanistic principles of perception and behavior, the individual neurons have to be embedded and investigated with respect to the underlying neural network structure. To do so, we developed a set of tools that allow reconstructing an anatomically realistic average model of rat vibrissal cortex (Dercksen et al. 2012a; Lang et al. 2011). Using this network model we are able to estimate the number, cell type-specific origin, and 3D distribution of synapses along the dendrites of each reconstructed neuron. Subsequently, we review the custom-designed tools and anatomical data necessary to reconstruct an average model of the vibrissal cortex.

First, to generate an average anatomically realistic model, we needed to investigate whether the 3D layout of the vibrissal cortex is preserved across animals (see Egger et al. 2012 for a detailed description). This step yields the maximal possible precision of the average vibrissal cortex and allows registering each reconstructed neuron to its respective location within the cortex model. We defined five anatomical parameters to describe the 3D structure of each barrel column: (1) the *barrel area*, defined as the maximal circumference of the L4 barrel in the tangential plane; (2) the *barrel top*, defined as the closest point of the barrel to the pia in the coronal plane; (3) the *barrel bottom*, defined as the closest point of the barrel to the WM; (4) the *barrel column orientation*, defined as the perpendicular axis from the barrel center (BC) to the pia; and (5) the *barrel column height*, defined by extrapolating the barrel circumference along the BC axis towards the WM and pia (i.e., pia–WM distance). We measured these parameters for almost 1,000 barrel columns in more than 100 different rats. We found that the standard deviations (SD) of the five parameters across animals were small for corresponding barrel columns, usually around 5 % of the mean. Furthermore, we generated a standard model of the vibrissal cortex by superimposing the respective BC locations of 12 reconstructed vibrissal cortices using only rigid transformations. We found that the standard model (Fig. 8.1) captured the average 3D layout of the vibrissal cortex with approximately 50 μm precision. Thus, by reconstructing each neuron morphology with respect to the anatomical reference landmarks that were used to generate the standard model (i.e., pia, WM, barrels), rigid transformations can be used to register (i.e., place) each reconstruction into the standard model with 50 μm accuracy.

Second, in order to obtain anatomically realistic estimates of synaptic innervation, we needed to determine the number and 3D distribution of neurons within the vibrissal cortex and thalamus. To do so, (1) we stained consecutive brain sections with NeuN to label all neuron somata (Meyer et al. 2010b), (2) counterstained the sections with GAD67 to distinguish between excitatory and inhibitory cells (Meyer et al. 2011) and to reveal the L4 barrels and VPM barreloids, and (3) developed an automated pipeline to detect the center location of each labeled soma (Oberlaender et al. 2009b). Using this method, we determined the number of excitatory and inhibitory neurons for each barrel column, barreloid, and the septa between them in three different rats (Meyer et al. in press). We found that the number and 3D distribution of neurons per barrel column and barreloid were preserved across animals. As for the five parameters describing the geometry of the respective barrel columns, the SDs of the neuron densities were usually around 5 % of the mean. Further, the number of cells per barrel column correlated strongly with the number of neurons in the respective barreloid ($p < 10^{-5}$, two-tailed t-test). For example the average numbers of neurons in the A2 barrel column and barreloid were 11,978 and 198, and the respective values for C2 were 19,707 and 350, yielding ratios of 61 and 56, respectively. Finally, again using the pia, WM, and barrels as reference landmarks, we generated an average 3D distribution of excitatory neuron somata and registered this distribution to the standard model of the vibrissal cortex (Fig. 8.3a).

Third, in order to assign a cell type to each soma in the average 3D soma distribution, we needed to determine the number and location of excitatory cell types in a barrel column. We thus registered 95 reconstructed dendrite morphologies, obtained across L2–6, to the standard cortex and extracted 21 dendritic features for each cell (see Oberlaender et al. 2012a for a full list of features). Using these parameters, we performed a supervised cluster analysis using the OPTICS algorithm (Ankerst et al. 1999) implemented in the WEKA software package (http://sourceforge.net/projects/weka). The clustering yielded nine different dendritic cell types. Since the soma location of each registered neuron can be determined with 50 μm accuracy, the cluster analysis allowed determining the minimal and maximal depth locations of each cell type as well as the overlap ratio between cell types (Oberlaender et al. 2012a). Using these values, each neuron in the soma distribution can be assigned to a unique cell type (Fig. 8.3a).

Fourth, as a final step to reconstruct an anatomically realistic average model of the vibrissal cortex, we replace each soma in the average soma distribution by a dendrite/axon tracing of the respective cell type. Because the number of reconstructed cells is much smaller than the number of neurons in the vibrissal cortex, we need to scale our sample up. During the up-scaling process, a custom-designed network assembly pipeline automatically determines a neuron from the pool of reconstructed morphologies (Dercksen et al. 2012a; Lang et al. 2011). It is assured that the neurons in the soma distribution and the morphology are of the same cell type and located within the same barrel column, at approximately the same depth. The morphology is then registered to the new location, preserving dendrite/axon orientation and scaling relative to the PC. This step yields the 3D distribution of cell type-specific dendrites and axons within the vibrissal cortex, as exemplarily shown in Fig. 8.3a for the ensemble of L5tt dendrites and VPM axons in different columns, respectively. Finally the dendrites and the axon of each neuron in the network model are converted into 3D density distributions with 50 μm voxel resolution (i.e., the approximate precision of the average cortex model), where each voxel is then

Fig. 8.3 Reverse engineering synaptic connectivity and sensory-evoked signal flow. (**a**) The reverse engineering pipeline. From *left* to *right*: Assignment of neuron somata to different cell types, replacement of each neuron soma with a dendrite tracing of the same cell type, all L5tt dendrites of one column, all VPM axons of one column, an individual L5tt neuron registered to its location in the average network model. (**b**) Subcellular distribution of VPM synapses from the principal whisker (PW) and the eight surrounding whiskers (Su8W) on the individual L5tt neuron from panel (**a**). Scale bar applies to (**b**) and (**c**). (**c**) Subcellular distribution of L5st synapses from the PW and Su8W on the individual L5tt neuron from (**a**). (**d**) Active VPM synapses after passive touch. Scale bar applies to (**d**)–(**g**). (**e**) Spread of depolarization along the neuron 12 and 17 ms after passive touch (**f**) L5st synapses with increased activity during whisking. (**g**) Spread of depolarization along the neuron 125 and 175 ms after beginning of the whisking cycle

multiplied with a cell type-specific spine and bouton density value, respectively (e.g., 1 spine per μm dendrite; see Lang et al. 2011 for more details).

In summary, the network assembly process is based on (1) a 3D standard model that captures the average geometry of the vibrissal cortex with 50 μm accuracy, (2) the average number and 3D distribution of excitatory and inhibitory somata, (3) more than 150 complete dendrite and/or axon reconstructions obtained from in vivo recordings in VPM and across all cortical layers, and (4) spine and bouton distributions measured for each cell type.

8.2.3 Mapping of Cell Type-Specific Subcellular Synapse Distributions in Rat Vibrissal Cortex

The network assembly process yields an average model of rat vibrissal cortex which contains approximately 500,000 neurons of ten different cell types. Each of these neurons is represented as a 3D spine density distribution with 50 μm voxel resolution. Specifically, each voxel contains the number of spines a respective neuron contributes to the total number of spines within the voxel. Based on this network model we can estimate the number and subcellular distribution of synapses using structural overlap of spine and bouton distributions.

This way of estimating synaptic innervation is commonly referred to as Peter's rule (White 1979). While it has been shown that structural overlap between axons and dendrites is not sufficient to predict connectivity at the (sub)micron level (Mishchenko et al. 2010), estimates at lower resolution (e.g., 50 μm) yielded reasonable first-order approximations (Binzegger et al. 2004; Meyer et al. 2010a). Furthermore, in case of the present network assembly process we argue that connectivity mapping at resolutions lower than 50 μm, for example by detecting intersecting dendrites and axons, as suggested previously (Kozloski et al. 2008), is not appropriate for such average cortex models. The variability of cortex geometry limits registering of anatomical data to a precision of 50 μm, which can thus be regarded as the lowest possible resolution to predict synaptic innervation by structural overlap in the vibrissal cortex.

Consequently, during the present connectivity mapping approach we determine the number of spines each cell contributes to the total number of spines within 50 μm × 50 μm × 50 μm voxels. To estimate the number of synapses each neuron receives from any presynaptic cell type and in each voxel of its spine distribution, the number of boutons within each voxel is divided by the total number of spines and multiplied with the number of spines from the respective neuron in the same voxel. For example, the total number of spines within a voxel may be 1,000 originating from various neurons of multiple cell types overlapping in this voxel. The total number of boutons from a single presynaptic axon distribution may be 100 in the same voxel. One neuron may contribute ten spines to the total number of spines (i.e., 1 %). Thus, one bouton from this presynaptic cell type is assigned to this cell and to this voxel. Within a voxel, potential synapse locations are randomly assigned (i.e., for the described example, one synapse is placed randomly on a piece of dendrite within the respective voxel). See (Lang et al. 2011) for a detailed description of the synapse mapping calculations.

This connectivity mapping approach is illustrated on the example of one L5tt neuron and its innervation by the VPM (Fig. 8.3b) and by L5st neurons (Fig. 8.3c). The L5tt neuron has been functionally characterized in vivo (Table 8.1). The dendrite and axon projection patterns have been completely reconstructed, and it has been registered to the standard model of the vibrissal cortex. This allows estimating the cell type-specific number (20,658 synapses in total) and subcellular distribution of synapses this particular L5tt neuron may receive (Table 8.1). Specifically, the synapse

Table 8.1 Structural and functional characterization of the L5tt neuron used for simulations (Fig. 8.3)

Anatomical parameters		Functional parameters	
Dendrite length	18,747 μm	Spontaneous spiking (anesthetized)	2.59 Hz
Axon length	26,920 μm	Spiking 0–50 ms after passive touch (anesthetized)	0.99 Hz
Soma location: cortical depth	1,285 μm	*Parameters used in simulation*	
Soma location: distance to vertical column axis	147 μm	R_m/E_{pas}	12,000 Ωcm²/–70 mV
Nr. of synapses	20,658	C_m	0.75 μF/cm²
Nr. of VPM PW synapses		R_a	150 Ωcm
Proximal	168	Synapses	
Distal	77	Peak conductance	0.6 nS
Nr. of L5st PW synapses		Transmitter binding	1.885/ms
Proximal	435	Transmitter unbinding	0.89/ms
Distal	878	VPM PW activity	
Nr. of VPM Su8W synapses		Nr. of neurons	311
Proximal	194	Active fraction	0.65
Distal	21	Convergence	0.43
Nr. of L5st Su8W synapses		Nr. of spikes	1
Proximal	884	Avg. spike time	10 ms
Distal	1,171	Spike time dispersion	1.7 ms
Distance of VPM PW synapses		VPM Su8W activity	
Proximal	151 ± 88 μm	Nr. of neurons	2,560
Distal	821 ± 182 μm	Active fraction	0.058
Distance of L5st PW synapses		Convergence	0.43
Proximal	205 ± 94 μm	Nr. of spikes	1
Distal	1,208 ± 146 μm	Avg. spike time	15 ms
Distance of VPM Su8W synapses		Spike time dispersion	1.0 ms
Proximal	180 ± 92 μm	L5st PW/Su8W activity	
Distal	778 ± 148 μm	Nr. of neurons	8,381
Distance of L5st Su8W synapses		Active fraction	0.35
Proximal	209 ± 86 μm	Convergence	0.5
Distal	1,199 ± 167 μm	Nr. of spikes	1
		Avg. spike time	150 ms
		Spike time dispersion	50 ms

mapping approach by structural overlap yielded that this particular cell at this particular location (at the lateral border of its PW barrel column) may receive a total of 460 synapses from the VPM, i.e., roughly 2 % of its synapses thus originate in the thalamus. Interestingly, only 245 of the VPM synapses originate from the respective D2 barreloid. Nearly the same amount of synapses (215) originates from the eight surrounding barreloids (i.e., C1–3, D1, D3, E1–E3). This estimate of the structural thalamocortical connectivity may in part explain why L5tt neurons usually respond well to deflections of the PW and Su8W, while responses of VPM neurons are largely confined to a single whisker (Fig. 8.2).

Figure 8.3b shows one possible configuration how the VPM synapses may be distributed along the dendrites of the L5tt neuron. This allows quantifying the subcellular distribution in terms of path length distance to the soma (Table 8.1). Both the distribution of PW and Su8W synapses display a bimodal innervation pattern, in agreement with measurements of functional synapses in vitro (Petreanu et al. 2009). Approximately 79 % of the VPM synapses are located along the basal and apical oblique dendrites and are thus referred to as proximal contacts. The remaining 21 % are mainly located along the apical trunk in L4, with a few synapses located on the apical tuft.

The innervation pattern of L5st synapses is very different (Fig. 8.3c). First, the total number of L5st synapses that this particular L5tt cell may receive is 3,368 (i.e., ~16 % of all synapses). The majority (2,055) originates from neurons located in the eight surrounding barrel columns. Second, in contrast to VPM synapses, most L5st synapses, around 60 %, are located within the apical tuft dendrite, almost none are located along the apical trunk, and the remaining 40 % are proximal contacts, located along the basal and apical oblique dendrites. Third, interestingly the subcellular distribution, as quantified by the path length distance to the soma, for PW and Su8W L5st synapses differs. While the absolute numbers of distal PW and Su8W synapses from L5st neurons are similar (i.e., 878 vs. 1,171), the numbers of proximal synapses located along the basal and apical oblique dendrites deviate strongly, by more than a factor of two (i.e., 435 vs. 884) (Table 8.1).

In summary, the present approach of recording from single neurons in vivo, combined with reconstructing the complete 3D dendrite/axon morphologies with respect to standardized anatomical landmarks, allows embedding neurons into an average anatomically realistic model of the vibrissal cortex. Consequently, each recorded neuron is described not only by its functional properties in response to various sensory stimuli but also by a detailed set of anatomical parameters, such as (1) 3D location, (2) dendrite morphology, (3) axon projection pattern, (4) morphological cell type, (5) numbers, and (6) subcellular distribution of cell type- and location-specific innervating synapses. Thus, our approach allows interpreting functional recordings within the context of the underlying neural network structure, which will decrease the ambiguity of functional measurements and allow for in silico experiments that investigate potential mechanistic principles of sensory information processing and ultimately of behavior.

8.2.4 *Simulation of Sensory-Evoked Signal Flow in Rat Vibrissal Cortex*

The present approach yields the necessary information to perform in silico experiments at single-neuron and network levels within the context of the anatomically realistic structure of the underlying circuits. For example at the network level, we previously performed Monte Carlo (MC) simulations to investigate the activation of an ensemble of L4 spiny stellate (L4ss) neurons by thalamocortical VPM input after passive whisker deflection (Lang et al. 2011). Because the synaptic connectivity within our network models is given as a connection probability with 50 μm resolution and synapses are placed randomly on dendrites within each voxel, we investigated the influence of the exact synapse locations (i.e., anatomical connectivity) on the network response.

Further, the convergence between the thalamus and the cortex is smaller than 1, for example 0.43 between the VPM and L4ss neurons (Bruno and Sakmann 2006). Thus we investigated the influence of the exact locations of active synapses (i.e., functional connectivity) on the network response. Finally, the input from the VPM is distributed in time (Brecht and Sakmann 2002), i.e., 65 % of the VPM neurons elicit one action potential (AP) 10.0 ± 1.7 ms after deflection of the related PW. Thus we investigated the influence of input synchrony on the network response.

Performing a sensitivity analysis by using MC simulations, we found that the network response is largely determined by the functional connectivity pattern and the synchrony of the VPM input, as was suggested by previous functional measurements in vivo (Bruno and Sakmann 2006). Changes in anatomical connectivity had almost no influence on the network response. Furthermore, we found that the spiking probability of L4ss neurons in all MC simulation trails decreased from the BC towards the barrel column borders. Such a location-specific spiking activity has indeed been observed in L2/3 neurons, one of the major target populations of L4ss neurons (Feldmeyer et al. 2002), using two-photon calcium imaging during passive whisker deflection experiments (Kerr et al. 2007).

Apart from investigating mechanistic principles of sensory-evoked information processing at the network level, our approach further allows performing single-neuron simulations within the context of the entire network. This may help elucidating mechanistic principles at the single-neuron level by comparing the simulation results directly with the respective functional recordings of sensory-evoked activity in the very same neuron. In the following we illustrate one possible in silico experiment that may help to determine a cellular mechanism for object localization during active whisking.

Based on the results that L5tt neurons are reliably activated by passive touch, while L5st neurons carry phase information on whisker location during periods of active whisker movement, we suggested the following hypothesis (Oberlaender et al. 2011): after passive touch, L5tt neurons receive input from the VPM primarily within their basal and apical oblique dendrites, resulting in single AP activity. During active whisker movement, phase information about whisker location is relayed by L5st neurons to large portions of L2/3. L5tt neurons display abundant apical tuft dendrites in L2/3. Thus, innervation of L5tt apical tuft dendrites by L5st neurons may result in phase locking of the apical tuft dendrite. When locating an object during active whisking periods (active touch), proximal input from the VPM may coincide with phase input at the apical tuft dendrites, resulting in an amplification of the inputs and thus burst spiking output from L5tt neurons (Schaefer et al. 2003). The transition from single AP to bursting output may render one potential mechanism to encode for object locations. Because such a mechanism is crucially influenced by the neuron's location, dendrite morphology, and input distribution, simulations of this scenario require a quantitative and anatomically realistic mapping of potential structural and functional inputs.

Figure 8.3d–g illustrates our first simulation experiments using a full-compartmental model of the in vivo-characterized and completely reconstructed neuron shown in Figs. 8.1 and 8.3a–c, embedded within the anatomically realistic model of the vibrissal cortex. As a starting point, no active conductance distributions were included. Membrane resistance was set to 12,000 Ωcm^2 with reversal potential −70 mV, membrane capacitance to 0.75 $\mu F/cm^2$, and axial resistance to 150 Ωcm. The surface area increase due to spines was accounted for by scaling the membrane capacitance and membrane conductance with a factor of 1.6 (Holmes and Rall 1992). Synapses were modeled as a two-state AMPA conductance with time constants 1.885/ms (transmitter binding rate) and 0.89/ms (transmitter unbinding rate) and a peak conductance value of 0.6 nS (Destexhe et al. 1994). The parameters of each synapse were tuned to elicit somatic EPSPs that match in shape and amplitude with previously reported measurements in vivo (Bruno and Sakmann 2006). Using this simplified model, we investigated the potential spread of depolarization caused by VPM input after passive whisker touch and during L5st input during periods of active whisking.

During passive touch simulations, the input was derived from Brecht and Sakmann (2002) and Bruno and Sakmann (2006) and was as follows: (1) 65 % of the VPM synapses originating from the related PW barreloid elicit one spike, 10.0 ± 1.7 ms after the stimulus; (2) 5.8 % of the VPM synapses originating from the Su8W barreloids elicit one spike, 15.0 ± 1.0 ms after the stimulus; (3) the functional convergence for both PW and Su8W VPM synapses was assumed to be 0.43. Consequently, in case of the present L5tt neuron, between approximately 8 and 16 ms after the stimulus, 145 of 245 PW and 8 of 215 Su8W synapses were active (Fig. 8.3d). Figure 8.3e shows the resulting depolarization 12 ms (left panel) and 17 ms (right panel) after the stimulus. Without active conductances, the depolarization by VPM input after passive touch remains largely restricted to the basal and apical oblique dendrites and does not spread into the apical tuft.

During active whisking simulations, the input was derived from de Kock and Sakmann (2009) and Oberlaender et al. (2011) and was as follows: 35 % of the L5st neurons increased spiking activity during whisking, resulting in spike rates of typically 5Hz. One whisking cycle (back and forth movement) comprises about 200 ms, resulting in approximately one spike per cycle. Each L5st neuron responds to a different whisker location, i.e., represents a different whisking phase. Simulating one whisking cycle, we thus activated 35 % of the PW (444 of 1,313) and Su8W (736 of 2,055) L5st synapses and distributed their activation across the cycle (Fig. 8.3f). Phase preference of L5st neurons to whisker protraction was simulated by drawing spike times from a normal distribution with its mean at 150 ms and a standard deviation of 50 ms (modulo 200 ms, the duration of one whisking cycle). Figure 8.3g shows the resulting depolarization 125 ms (left panel) and 175 ms (right panel) after the onset of the whisker movement. In contrast to the VPM activation after passive touch and in the absence of active conductances, the depolarization by L5st input remains largely confined to the apical tuft dendrite, leaving the proximal dendrites (i.e., basal and apical oblique) largely at rest. Thus, even though the total number of active L5st synapses is about eight times higher than the number of active VPM synapses, the largely asynchronous L5st input is most likely insufficient to elicit additional spiking of this L5tt neuron, which is in agreement with previous functional measurements of L5tt neurons during active whisking (de Kock and Sakmann 2009).

In summary, these simulation experiments demonstrate that a typical L5tt neuron may have two structurally and functionally distinct domains, the proximal and apical tuft dendrites, respectively. The two compartments receive very different synaptic input and may thus process and encode different aspects of whisker-evoked sensory stimuli. Depending on the spatial and temporal superposition of the two inputs, the output of an L5tt neuron may change, encoding information of the environment, such as the location of an object.

8.3 Conclusion

In this chapter we reviewed a novel approach to elucidate mechanistic principles underlying sensory perception and behavior by simulating whisker-evoked activity within an anatomically realistic model of rat vibrissal cortex. Combining complete 3D reconstructions of functionally characterized neurons in vivo with (1) a standard model of the vibrissal cortex that allows integrating anatomical data obtained from many animals into a common reference frame with 50 μm precision; (2) a realistic number and 3D distribution of excitatory neuron somata, obtained from counting all neurons in three vibrissal cortices; (3) objective clustering to identify morphological cell types; and (4) synapse mapping by structural overlap between neuron- and cell type-specific spine and bouton distributions, our approach allows for a detailed functional and anatomical characterization of individual neurons. Specifically, for each neuron we obtain spontaneous and whisker-evoked activity patterns, the 3D location within the vibrissal cortex, dendrite morphology, axonal projections, morphological cell type, and subcellular synaptic innervation patterns. This rich set of parameters does not only allow correlating structural and functional data but gives rise to simulations at the single-neuron and network levels. Furthermore, the present example of one specific L5tt neurons that is activated by input from VPM neurons after passive whisker touch and by input from L5st neurons during active whisking illustrates that functional measurements need to be interpreted within the context of the underlying network structure. Depending on the behavioral state, sensory stimulus, and timing between individual stimuli, a L5tt neuron will receive very different synaptic input patterns. In conclusion, the 3D location and morphology of a neuron, as well as the specifics of sensory stimuli, dramatically influence a neuron's synaptic input pattern and thus its computations and output. The present approach is one way to investigate this interplay between structural and functional properties and may ultimately result in network models that allow monitoring sensory-evoked streams of excitation from the periphery throughout the brain.

Acknowledgments We thank Bert Sakmann for advice and initiation of the collaborative project; Randy Bruno for labeling thalamocortical axon morphologies; and Hanno-Sebastian Meyer for immunohistochemistry staining of neuron somata.

Author contributions: Conceived and designed the project: MO. Performed the experiments: CPJdK. Performed the simulations: RE. Analyzed the data: RE MO. Contributed network assembly and analysis tools: VJD. Wrote the paper: MO.

Funding: Funding was provided by the Bernstein Center for Computational Neuroscience, Tuebingen (funded by the German Federal Ministry of Education and Research (BMBF; FKZ: 01GQ1002)) (MO and RE); by the Max Planck Institute for Biological Cybernetics, Tuebingen (MO and RE); by the Max Planck Florida Institute for Neuroscience, Jupiter (MO); by the Werner Reichardt Center for Integrative Neuroscience, Tuebingen (MO); by the Studienstiftung des Deutschen Volkes (RE); by the Center for Neurogenomics and Cognitive Research—CNCR (CPJdK); by the Zuse Institute Berlin (VJD); and by the Max Planck Institute of Neurobiology, Martinsried (VJD). The funders had no role in study design, data collection and analysis, decision to publish, or preparation of the manuscript.

Competing interests: The authors have declared that no competing interests exist.

References

Alloway KD (2008) Information processing streams in rodent barrel cortex: the differential functions of barrel and septal circuits. Cereb Cortex 18(5):979–989

Ankerst M, Breunig M, Kriegel HP, Sander J (1999) OPTICS: ordering points to identify the clustering structure. In: Delis A, Faloutsous C, Ghandeharizadeh S (eds) ACM SIGMOD'99 International. Conference on Management of Data, Philadelphia, 1999. ACM Press New York, NY, pp 49–60

Armstrong-James M, Fox K (1987) Spatiotemporal convergence and divergence in the rat S1 "barrel" cortex. J Comp Neurol 263(2):265–281

Aronoff R, Matyas F, Mateo C, Ciron C, Schneider B, Petersen CC (2010) Long-range connectivity of mouse primary somatosensory barrel cortex. Eu J Neurosci 31(12):2221–2233

Barth AL, Poulet JF (2012) Experimental evidence for sparse firing in the neocortex. Trends Neurosci 35(6):345–355

Binzegger T, Douglas RJ, Martin KA (2004) A quantitative map of the circuit of cat primary visual cortex. J Neurosci 24(39):8441–8453

Bock DD, Lee WC, Kerlin AM, Andermann ML, Hood G, Wetzel AW, Yurgenson S, Soucy ER, Kim HS, Reid RC (2011) Network anatomy and in vivo physiology of visual cortical neurons. Nature 471(7337):177–182

Brecht M, Sakmann B (2002) Whisker maps of neuronal subclasses of the rat ventral posterior medial thalamus, identified by whole-cell voltage recording and morphological reconstruction. J Physiol 538(Pt 2):495–515
Brecht M, Roth A, Sakmann B (2003) Dynamic receptive fields of reconstructed pyramidal cells in layers 3 and 2 of rat somatosensory barrel cortex. J Physiol 553(Pt 1):243–265
Briggman KL, Helmstaedter M, Denk W (2011) Wiring specificity in the direction-selectivity circuit of the retina. Nature 471(7337):183–188
Bruno RM, Sakmann B (2006) Cortex is driven by weak but synchronously active thalamocortical synapses. Science 312(5780):1622–1627
Celikel T, Sakmann B (2007) Sensory integration across space and in time for decision making in the somatosensory system of rodents. Proc Natl Acad Sci USA 104(4):1395–1400
de Kock CP, Sakmann B (2009) Spiking in primary somatosensory cortex during natural whisking in awake head-restrained rats is cell-type specific. Proc Natl Acad Sci USA 106(38): 16446–16450
de Kock CP, Bruno RM, Spors H, Sakmann B (2007) Layer and cell type specific suprathreshold stimulus representation in primary somatosensory cortex. J Physiol 581(1):139–154
Denk W, Horstmann H (2004) Serial block-face scanning electron microscopy to reconstruct three-dimensional tissue nanostructure. PLoS Biol 2(11):e329
Dercksen VJ, Weber B, Guenther D, Oberlaender M, Prohaska S, Hege HC (2009) Automatic alignment of stacks of filament data. IEEE Int Symp on Biomedical Imaging: From Nano to Macro (ISBI) 971–974
Dercksen VJ, Egger R, Hege HC, Oberlaender M (2012a) Synaptic connectivity in anatomically realistic neural networks: modeling and visual analysis. In: Eurographics workshop on visual computing for biology and medicine
Dercksen VJ, Oberlaender M, Sakmann B, Hege HC (2012b) Interactive visualization: a key pre-requisite for reconstruction of anatomically realistic neural networks. In: Proceedings of the 2009 workshop on visualization in medicine and life sciences (VMLS 09)
Destexhe A, Mainen ZF, Sejnowski TJ (1994) Synthesis of models for excitable membranes, synaptic transmission and neuromodulation using a common kinetic formalism. J Comp Neurosci 1(3):195–230
Egger R, Narayanan RT, Helmstaedter M, de Kock CP, Oberlaender M (2012) 3D reconstruction and standardization of the rat vibrissal cortex for precise registration of single neuron moprhology. PLoS Comput Biol 8(12):1–18
Feldmeyer D, Lubke J, Silver RA, Sakmann B (2002) Synaptic connections between layer 4 spiny neurone-layer 2/3 pyramidal cell pairs in juvenile rat barrel cortex: physiology and anatomy of interlaminar signalling within a cortical column. J Physiol 538(Pt 3):803–822
Grewe BF, Helmchen F (2009) Optical probing of neuronal ensemble activity. Curr Opin Neurobiol 19(5):520–529
Groh A, de Kock CP, Wimmer VC, Sakmann B, Kuner T (2008) Driver or coincidence detector: modal switch of a corticothalamic giant synapse controlled by spontaneous activity and short-term depression. J Neurosci 28(39):9652–9663
Groh A, Meyer HS, Schmidt EF, Heintz N, Sakmann B, Krieger P (2010) Cell-type specific properties of pyramidal neurons in neocortex underlying a layout that is modifiable depending on the cortical area. Cereb Cortex 20(4):826–836
Hay E, Hill S, Schurmann F, Markram H, Segev I (2011) Models of neocortical layer 5b pyramidal cells capturing a wide range of dendritic and perisomatic active properties. PLoS Comput Biol 7(7):e1002107
Helmstaedter M, Briggman KL, Denk W (2008) 3D structural imaging of the brain with photons and electrons. Curr Opin Neurobiol 18(6):633–641
Helmstaedter M, Sakmann B, Feldmeyer D (2009) The relation between dendritic geometry, electrical excitability, and axonal projections of L2/3 interneurons in rat barrel cortex. Cereb Cortex 19(4):938–950
Hill DN, Curtis JC, Moore JD, Kleinfeld D (2011) Primary motor cortex reports efferent control of vibrissa motion on multiple timescales. Neuron 72(2):344–356

Holmes WR, Rall W (1992) Estimating the electrotonic structure of neurons with compartmental models. J Neurophysiol 68(4):1438–1452
Horikawa K, Armstrong WE (1988) A versatile means of intracellular labeling: injection of biocytin and its detection with avidin conjugates. J Neurosci Methods 25(1):1–11
Hubel DH, Wiesel TN (1959) Receptive fields of single neurones in the cat's striate cortex. J Physiol 148:574–591
Hutson KA, Masterton RB (1986) The sensory contribution of a single vibrissa's cortical barrel. J Neurophysiol 56(4):1196–1223
Kerr JN, de Kock CP, Greenberg DS, Bruno RM, Sakmann B, Helmchen F (2007) Spatial organization of neuronal population responses in layer 2/3 of rat barrel cortex. J Neurosci 27(48): 13316–13328
Kozloski J, Sfyrakis K, Hill S, Schurmann F, Peck C, Markram H (2008) Identifying, tabulating, and analyzing contacts between branched neuron morphologies. IBM J Res Dev 52(1–2):43–55
Land PW, Buffer SA Jr, Yaskosky JD (1995) Barreloids in adult rat thalamus: three-dimensional architecture and relationship to somatosensory cortical barrels. J Comp Neurol 355(4):573–588
Lang S, Dercksen VJ, Sakmann B, Oberlaender M (2011) Simulation of signal flow in 3D reconstructions of an anatomically realistic neural network in rat vibrissal cortex. Neural Netw 24(9):998–1011
Margrie TW, Brecht M, Sakmann B (2002) In vivo, low-resistance, whole-cell recordings from neurons in the anaesthetized and awake mammalian brain. Pflugers Arch 444(4):491–498
Martinez LM, Wang Q, Reid RC, Pillai C, Alonso JM, Sommer FT, Hirsch JA (2005) Receptive field structure varies with layer in the primary visual cortex. Nat Neurosci 8(3):372–379
Meyer HS, Wimmer VC, Hemberger M, Bruno RM, de Kock CP, Frick A, Sakmann B, Helmstaedter M (2010a) Cell type-specific thalamic innervation in a column of rat vibrissal cortex. Cereb Cortex 20(10):2287–2303
Meyer HS, Wimmer VC, Oberlaender M, de Kock CP, Sakmann B, Helmstaedter M (2010b) Number and laminar distribution of neurons in a thalamocortical projection column of rat vibrissal cortex. Cereb Cortex 20(10):2277–2286
Meyer HS, Schwarz D, Wimmer VC, Schmitt AC, Kerr JN, Sakmann B, Helmstaedter M (2011) Inhibitory interneurons in a cortical column form hot zones of inhibition in layers 2 and 5A. Proc Natl Acad Sci USA 108(40):16807–16812
Meyer HS, Egger R, Guest JM, Foerster R, Reissl S, Oberlaender M (in press) The cellular organization of cortical barrel columns is whisker-specific. Proc Natl Acad Sci USA
Mishchenko Y, Hu T, Spacek J, Mendenhall J, Harris KM, Chklovskii DB (2010) Ultrastructural analysis of hippocampal neuropil from the connectomics perspective. Neuron 67(6): 1009–1020
Mountcastle VB (1957) Modality and topographic properties of single neurons of cat's somatic sensory cortex. J Neurophysiol 20(4):408–434
Oberlaender M, Bruno RM, Sakmann B, Broser PJ (2007) Transmitted light brightfield mosaic microscopy for three-dimensional tracing of single neuron morphology. J Biomed Opt 12(6):064029
Oberlaender M, Broser PJ, Sakmann B, Hippler S (2009a) Shack-Hartmann wave front measurements in cortical tissue for deconvolution of large three-dimensional mosaic transmitted light brightfield micrographs. J Microsc 233(2):275–289
Oberlaender M, Dercksen VJ, Egger R, Gensel M, Sakmann B, Hege HC (2009b) Automated three-dimensional detection and counting of neuron somata. J Neurosci Methods 180(1): 147–160
Oberlaender M, Boudewijns ZS, Kleele T, Mansvelder HD, Sakmann B, de Kock CP (2011) Three-dimensional axon morphologies of individual layer 5 neurons indicate cell type-specific intracortical pathways for whisker motion and touch. Proc Natl Acad Sci USA 108(10):4188–4193
Oberlaender M, de Kock CP, Bruno RM, Ramirez A, Meyer HS, Dercksen VJ, Helmstaedter M, Sakmann B (2012a) Cell type-specific three-dimensional structure of thalamocortical circuits in a column of rat vibrissal cortex. Cereb Cortex 22(10):2375–2391

Oberlaender M, Ramirez A, Bruno RM (2012b) Sensory experience restructures thalamocortical axons during adulthood. Neuron 74(4):648–655

O'Connor DH, Clack NG, Huber D, Komiyama T, Myers EW, Svoboda K (2010a) Vibrissa-based object localization in head-fixed mice. J Neurosci 30(5):1947–1967

O'Connor DH, Peron SP, Huber D, Svoboda K (2010b) Neural activity in barrel cortex underlying vibrissa-based object localization in mice. Neuron 67(6):1048–1061

Petreanu L, Mao T, Sternson SM, Svoboda K (2009) The subcellular organization of neocortical excitatory connections. Nature 457(7233):1142–1145

Pinault D (1996) A novel single-cell staining procedure performed in vivo under electrophysiological control: morpho-functional features of juxtacellularly labeled thalamic cells and other central neurons with biocytin or neurobiotin. J Neurosci Methods 65(2):113–136

Poulet JF, Petersen CC (2008) Internal brain state regulates membrane potential synchrony in barrel cortex of behaving mice. Nature 454(7206):881–885

Sakata S, Harris KD (2009) Laminar structure of spontaneous and sensory-evoked population activity in auditory cortex. Neuron 64(3):404–418

Sarid L, Bruno R, Sakmann B, Segev I, Feldmeyer D (2007) Modeling a layer 4-to-layer 2/3 module of a single column in rat neocortex: interweaving in vitro and in vivo experimental observations. Proc Natl Acad Sci USA 104(41):16353–16358

Schaefer AT, Larkum ME, Sakmann B, Roth A (2003) Coincidence detection in pyramidal neurons is tuned by their dendritic branching pattern. J Neurophysiol 89(6):3143–3154

Simons DJ, Carvell GE, Hershey AE, Bryant DP (1992) Responses of barrel cortex neurons in awake rats and effects of urethane anesthesia. Exp Brain Res 91(2):259–272

Staiger JF, Flagmeyer I, Schubert D, Zilles K, Kotter R, Luhmann HJ (2004) Functional diversity of layer IV spiny neurons in rat somatosensory cortex: quantitative morphology of electrophysiologically characterized and biocytin labeled cells. Cereb Cortex 14(6):690–701

Svoboda K, Denk W, Kleinfeld D, Tank DW (1997) In vivo dendritic calcium dynamics in neocortical pyramidal neurons. Nature 385(6612):161–165

Wallace DJ, Sakmann B (2008) Plasticity of representational maps in somatosensory cortex observed by in vivo voltage-sensitive dye imaging. Cereb Cortex 18(6):1361–1373

Welker C (1976) Receptive fields of barrels in the somatosensory neocortex of the rat. J Comp Neurol 166(2):173–189

White EL (1979) Thalamocortical synaptic relations: a review with emphasis on the projections of specific thalamic nuclei to the primary sensory areas of the neocortex. Brain Res 180(3): 275–311

Woolsey TA, Van der Loos H (1970) The structural organization of layer IV in the somatosensory region (SI) of mouse cerebral cortex. The description of a cortical field composed of discrete cytoarchitectonic units. Brain Res 17(2):205–242

Yu C, Derdikman D, Haidarliu S, Ahissar E (2006) Parallel thalamic pathways for whisking and touch signals in the rat. PLoS Biol 4(5):e124

Chapter 9
Optimized Dendritic Morphologies for Noisy Inputs

Klaus M. Stiefel and Benjamin Torben-Nielsen

Abstract Dendrites are the cellular protrusions of neurons receiving the majority of synaptic inputs. We investigated the structure–function relationship of the dendrites of model neurons optimized for input order detection of stochastic inputs. For this purpose, we used an inverse method based on a genetic algorithm. In this method, via iterative test and selection steps, the genetic algorithm finds a dendritic structure as good as possible for a user-selected neural computation. In a previous study, we generated model neurons optimized for reacting strongly to two groups of synaptic inputs occurring in one but not the reverse temporal order. In the current study, we added both temporal noise (synapse activation times) and spatial noise (synapse placement) to this computational task. We observed that the model neurons which were exposed to a more noisy input generally had smaller dendritic trees. We explain this finding by the fact that for input–order detection, sampling from more varied responses is advantageous and that positive outliers in a population are selected for. We conclude with a general discussion of signal integration in neurons, dendrites, and noise.

K.M. Stiefel (✉)
The MARCS Institute, University of Western Sydney, Sydney, Australia
e-mail: K.Stiefel@uws.edu.au

B. Torben-Nielsen
Department of Neurobiology, Hebrew University of Jerusalem, Edmond J. Safra Campus, Givat Ram, 91904 Jerusalem, Israel

Blue Brain Project, Swiss Federal Institute of Technology in Lausanne, Quartier d'Innovation, Batiment J, 1015 Lausanne, Switzerland
e-mail: btorbennielsen@gmail.com

H. Cuntz et al. (eds.), *The Computing Dendrite: From Structure to Function*, Springer Series in Computational Neuroscience 11, DOI 10.1007/978-1-4614-8094-5_9,

9.1 Introduction

Most neurons in the brains of animals possess dendrites, elongated and branched protrusions which are receiving synaptic contacts from other neurons. The shapes of these dendrites are highly varied between different animals, different brain areas, and different neuronal types in one brain area. The branching patterns of these dendrites vary between highly elaborate trees many 100 μm long with several separate regions to short, simple extensions with no more than one degree of branching. Both the amount and the exact pattern of dendritic branches vary widely, with some neurons' dendrites entirely confined to a 2-dimensional plane while others sample a volume of brain tissue and with significantly different degree of sampling of the area the dendrites pass through (Segev and London 2000; Stuart et al. 2007).

Besides these morphological differences, dendrites are also equipped with vastly different sets of ionic conductances. Ion channels conducting Na^+, K^+, Ca^{2+}, and K^+ ions are distributed spatially heterogeneously over the dendrites of neurons (Migliore and Shepherd 2002). The exact sets of conductances, their densities, and spatial distribution further add to the diversity of dendrites. The different conductance distributions can make dendritic trees essentially passive or dampened or excitable and allow no, weak, or completely regenerative action potential back-propagation into the dendrites (Stuart et al. 1997). The difference between different neurons' dendrites is well demonstrated when looking at mammalian cortical layer V pyramidal neurons and cerebellar Purkinje neurons, two of the most thoroughly studied neuronal cell types. To illustrate how dendrites are diverse on several levels of biological organization (ion channels, excitability, morphology, synaptic connections), we use the occasion of this book chapter to briefly review the dendrites of these two neuron types. With this, we hope to illustrate to the reader the complexities found in the dendrites of neurons.

In layer V pyramidal neurons, the dendrites are divided into several subgroups. Close to the soma, in layer V of the cortex, are the basal dendrites. Typically between two and five thin (<2 μm) basal dendrites grow horizontally from the soma, remaining in layer V, and branch several times. Also growing from the soma is the main apical trunk that grows vertically towards the pia. In the region immediately above the soma, the oblique dendrites branch off from the apical dendrite. From there, the apical dendrite projects further upwards to cortical layer I, where it branches out into the apical tuft. The dendrites of a cortical layer V pyramidal neuron are covered in spines, tiny (several microns long) protrusions, which are the target of excitatory synapses from a variety of cortical and subcortical sources. Inhibitory synapses target both spines and dendritic stems. There are several types of spines, and their density is about 0.2/μm. Cortical layer V pyramidal neurons fire fast Na^+ action potentials, which are initiated in the somato-proximal parts of the axon and back-propagate into the axon and dendrites. Due to a moderate density of Na^+ conductance in the dendrites, the action potentials are not completely regenerated during this back-propagation. They lose amplitude and widen by the time they reach the distal parts of the dendrites (Hay et al. 2013). Layer V pyramidal dendrites also show very interesting Ca^{2+} electrogenesis. In the basal dendrites, NMDA-type glutamate receptors can cause Ca^{2+} spikes (Major et al. 2008). In the distal dendrites, voltage-gated Ca^{2+} conductances can cause such Ca^{2+} spikes (Schiller et al. 2000). Back-propagating action potentials can evoke Ca^{2+} spikes in the distal apical dendrites when they coincide with synaptic depolarizations at these locations. These Ca^{2+} spikes can then propagate to the soma and axon, where they evoke more action potentials in a ping-pong-like manner of interaction between the distal dendrites and the soma/axon (Larkum et al. 1999). A diverse set of K^+ and mixed cation conductances, several of them increasing in density with path distance from the soma (Stuart et al. 1997), hold this dendritic excitability in check and alter the integration of synaptic inputs.

The dendrites of cerebellar Purkinje neurons are organized in one large, 2-dimensional fan. Emerging from the soma is a thick primary dendrite which then branches out evenly into a large array of higher order dendrites densely covering the plane of the Purkinje neuron. The dendrites of Purkinje neurons are densely covered by spines (>3/μm). The spines are the target of excitatory synapses from the parallel fibers, one of the two main groups of inputs to Purkinje neurons. The other main inputs are the climbing fibers. These inputs synapse on the main dendrite and via multiple synaptic connections cause a massive Ca^{2+} spike-mediated depolarization of the whole Purkinje neuron, a complex spike. In such a complex spike, several Na^+ action potentials ride on a Ca^{2+} spike envelope. In contrast, the simple spikes fired by Purkinje neurons are single Na^+ action potentials. They do not back-propagate into the dendritic trees due to the low Na^+ conductance density of the Purkinje neuron dendrites and the large current sink the multiple dendritic branches constitute. Besides Na^+ and Ca^{2+} conductances, the dendrites of Purkinje neurons contain a diverse set of K^+ and mixed cation conductances which regulate spiking and excitability (Martina et al. 2003).

The function of the dendrites of central neurons is not completely clear. While a general understanding of the function of most brain regions exists, the exact nature of the input–output transformations of individual neurons in vivo is not known with certainty. Hence, we know that primary visual cortical neurons fire in response to the presence of a properly oriented stimulus in their receptive field (DeAngelis et al. 1993; Henry et al. 1974). But we do not know how exactly the individual neurons integrate all their synaptic inputs to arrive at this specific response. This issue is aggravated by the fact that in an area like the cortex the synaptic input to any neuron is many synapses removed from any sensory input or motor output. Furthermore, activity reaching cortical pyramidal neurons has often traveled trough recursive cortical connections: the representational meaning of input spike trains is simply not clear. Due to the lack of understanding of the coding function (what aspect of the external world does neuronal/dendritic activity code for?), discussion has often centered on the underlying computational function (what is the signal transformation performed by dendrites). One proposal is that dendrites are merely collecting synaptic inputs in the regions they are traversing. According to this view, it is the function of different branching patterns to sample inputs from different regions (Cuntz et al. 2008; Cuntz 2012). Another proposal is that dendrites are involved in purposely modifying the incoming synaptic signals. Dendrites have been shown to boost synaptic inputs, normalize its waveform, and locally integrate a subset of inputs before passing the result on to the soma (Berger et al. 2003; Polsky et al. 2004). Both the passive electrotonic properties of dendritic trees as well as their ionic conductances contribute to the modification of synaptic potentials. While we agree that dendrites serve both input sampling and input modification, the current study is only concerned with the latter.

So what are the signals dendrites have to process in the functioning brain? One class of signals are obviously specific sensory information, retrieved memory contents, and motor programs. But, noise and background fluctuations are also ubiquitously present in the brain. Ion channel noise, spontaneous synaptic release, and background network activity are some examples. Some neurobiological processes which generate membrane potential fluctuations most likely do not represent relevant information, such as the stochastic opening and closing of ion channels caused by thermal protein structure fluctuations (ion channel flicker, Hille 2001). However, even though the ion channel flicker is not thought to represent information about the external environment, this stochastic process is still important in setting the operation regimes for neurons (Schneidman et al. 1998; Cannon et al. 1998).

Other forms of background activity, such as the synaptic background activity prevalent in several brain regions, do likely carry information (Destexhe and Paré 1999—but see Waters and Helmchen 2006 for a differing view). Since this information is probably contextual and not directly correlated to a stimulus presented by an experimenter, the exact nature of this information is difficult to elucidate, and the distinction between noise and contextual fluctuations is difficult to make. Furthermore, the statistical properties of these background fluctuations (amplitude, power spectrum) could be of relevance, while the precise instance of the signal could be irrelevant. In any case, neurons are not existing in a quiescent resting state but are exposed to ongoing noise and background fluctuations (used interchangeably here) from diverse sources. Background activity in the cortex is significant and state dependent. During wakefulness, rapid-eye-movement sleep, and ketamine anesthesia, cortical neurons experience a high-conductance state. A constant barrage of partially synchronized excitatory and inhibitory synaptic potentials causes significant (standard deviation >2 mV) membrane potential fluctuations which can be approximated as self-correlated noise (Destexhe et al. 2001). Background activity in the cerebellum is equally important, with constant (>1 Hz) synaptic input from 10^6 parallel fibers depolarizing the Purkinje neurons (Rapp et al. 1992). Mechanistically, this background activity depends more on the single cells than on networks of neurons. This is the case since Purkinje neurons fire tonically in the absence of synaptic input.

Naturally, signal processing and transmission work best in low-noise environments. The nervous system is no such low-noise environment, and the background noise could interfere with nervous system function. Hence, likely processes evolved which compensate for the noise or even take advantage of it. But little is known how dendrites cope with the brain's background activity. A number of studies by Destexhe and colleagues (Hô and Destexhe 2000; Destexhe and Paré 1999; Destexhe et al. 2002) have shown how the cortical synaptic background activity influences the membrane properties of cortical pyramidal neurons. They found that neurons

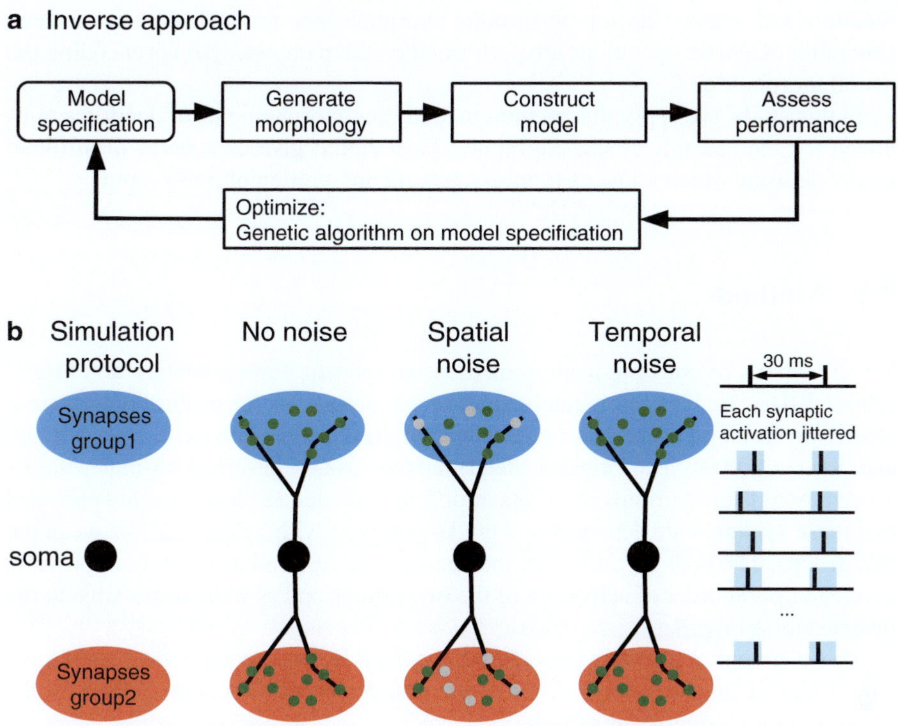

Fig. 9.1 Inverse approach and the simulation protocol. (**a**) Schematic of the inverse approach. A pool of parametrized model specifications is translated to a model neuron. Then, the performance of each model at the task at hand is assessed. Based on this performance, a genetic algorithm selects and slightly mutates the model specification. The algorithm iteratively increases the performance of the model neurons at the predefined computation. (**b**) Simulation protocol used to assess the model's performance. *Left*: Initial setup defining the soma position and the locations where synapses can be added. In the noiseless case, synapses are inserted when a dendrite is inside the "target location" (*green dots*). In case of spatial noise, a stochastic distribution defines whether a synapse is actually added (*green dots*) or not (*gray dot*). In case of temporal noise, a uniform distribution jitters the exact activation time of each synapse. (*Panel a* reproduced from Torben-Nielsen and Stiefel (2010a, b))

became depolarized and had shortened passive integration time constants and action potential back-propagation was enhanced (Hô and Destexhe 2000). The effect of cerebellar background activity is equally believed to be a depolarization of neurons and a shortening of passive time constants (Jaeger and Bower 1994; Rapp et al. 1992). While these results are intriguing, general principles of how dendrites cope with noisy inputs are not understood.

We have recently developed an inverse method for relating dendritic function to structure (Stiefel and Sejnowski 2007; Torben-Nielsen and Stiefel 2009, 2010a, b). This method is based on applying a genetic algorithm to morphologies generated by a morphogenetic algorithm (Fig. 9.1). The genetic algorithm tests for a user-selected function and selects the top-performing morphologies for inclusion in the next generation. Genetic operations are performed on the parameter strings encoding the neural morphologies.

In this study, we apply this method to study the morphologies of dendrites optimized for increasingly stochastic inputs. This should give us a series of artificial model neurons which allow us to investigate the integration of noisy inputs.

9.2 Methods

For the details of our optimization method we refer to Torben-Nielsen and Stiefel (2009, 2010a, b). The computational tasks we optimized for in this study were a stochastic variation of the input-order detection task investigated in Torben-Nielsen and Stiefel (2009). The original, deterministic task is designed so that neurons receive two distinct groups of inputs at different predefined locations in space and that these synapses are activated at a predefined time with a delay (Δt) between the two groups. The goal we define for the neuron is to respond as strongly as possible to one temporal order of activation of the two groups and as weak as possible to the inverse temporal order.

This differential response (=performance, *P*) is defined as the ratio between the amplitudes of the compound EPSP, in the non-preferred direction, divided by the response in the preferred direction:

$$\mathrm{p} = \frac{\mathrm{A}_{null}}{\mathrm{A}_{pref}},$$

with *A* being the EPSP amplitude at the soma.

The soma is located at the center, and two synaptic target zones are defined in opposing directions at 200 μm away from the soma. The synaptic zones are spherical and have a radius of 50 μm.

In this work we introduce both spatial and temporal noise (but never both simultaneously) as sketched in Fig. 9.1b. In the deterministic task, the model neuron receives one synapse per 5 μm of dendrite that grows through the synaptic zones. In the spatial noise case, the synapses are added with some level of stochasticity: not every 5 μm a synapse is necessarily inserted. Higher spatial noise levels mean that less and less synapses get inserted into the model neuron. Thus, the actual location is not modified, but the probability (*P*) that a synapse will be inserted at that location is. Because we want to exclude neurons that are “optimized” for one noise level, we assess the performance of each model neuron at five noise levels: once for five noise levels between 0 and 25 % of noise and once with five noise levels between 0 and 50 % noise. In the deterministic case, all synapses belonging to a group are activated at either time t or at $t+\Delta t$ ($\Delta t=30$ ms, because 30 ms proved to illicit strongest responses in passive model neurons (Torben-Nielsen and Stiefel 2009). Temporal noise assigns to each synapse a uniformly distributed activation time on $[T-\varepsilon, T+\varepsilon]$ where T stands for the appropriate time (either $T=t$ or $T=t+\Delta t$) and ε is the magnitude of the temporal jitter, which is here set to 5 ms (Fig. 9.1b). All optimization parameters and electrical defaults are described in Tables 1 and 2 from Torben-Nielsen and Stiefel (2009).

In contrast to the original study, where the optimization goal was a linear combination of multiple objectives, we used a multi-objective optimization algorithm. The two objectives are (1) the performance in the task as outlined above and (2) minimization of the total length. The latter objective is inspired by the parsimony principle generally found in nature. In multi-objective optimization, selection of good performing solutions is based on the ranking "performs equally," "performs worse," or "performs better." More specifically, in theory, the algorithm finds a set of solutions which "perform equally"; this set is called the Pareto front. All individuals on the Pareto front are equally good solutions, which means that increasing the performance on one of the objectives will result in a decrease in the other objectives. Thus, while there can be a large variation in the solutions, they are said to be not worse than one another as the objectives are often conflicting. In the simulations performed here, we used an initial population size of 250 individuals and ran the optimization for 5,000 generations. To avoid overfitting a particular instance generated by one random seed, we tested each model with five random seeds and used the average outcome to measure the performance. We have no way of knowing if the genetic algorithm converged to an optimum and how close the optimization products are to a true optimum. Nevertheless, we argue that they are reasonably close because (1) in previous work the solution found by the genetic algorithm was essentially at the analytically computed optimum (Torben-Nielsen and Stiefel 2009) and (2) when distinct runs (starting at different locations in the fitness landscape due to different starting random seeds) converge to a similar output, we can interpret that phenomenon as the algorithm having found a global optimum.

9.3 Results

The optimization runs yielded model neurons that performed well as input-order detectors. By adopting a multi-objective optimization strategy, a set of model neurons is obtained after the optimization. These models are performing "not worse" than one another and reside on the Pareto front: the location that represents their performance on both optimization objectives. Smaller values are better. A relevant part of the set of solutions (input-order detection ratio $P<0.8$ and total length $<2{,}100$ µm) is shown in Fig. 9.2 (left). Clearly, a trade-off can be seen between both objectives. We consider better models to be aligned with the dotted arrow as these models score well on both objectives (rather than standing out in one of the objectives while neglecting the other objective). For fairness model comparison, we compared the different noise-affected models when they perform input-order detection at $P=0.7$. One important observation is that the noiseless models

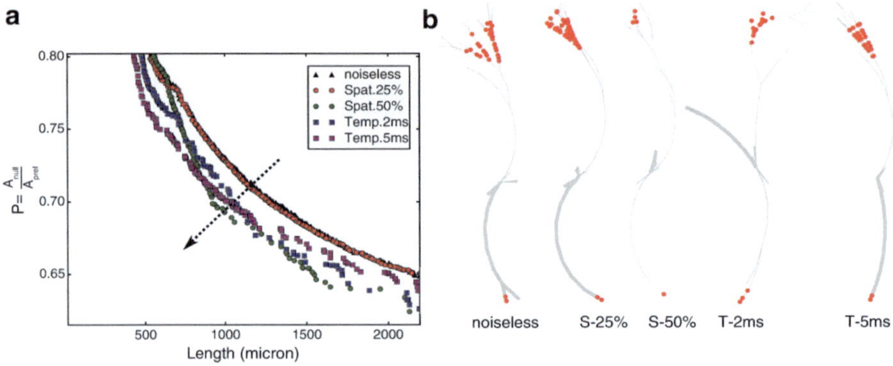

Fig. 9.2 Optimization results. (**a**) Relevant portion of the optimized models displayed in terms of their performance on both objectives. The *dotted arrow* indicates the axis along which we consider models "better" as they perform well on both objectives simultaneously. The noiseless models perform worse than model subjected to noise. Models achieving the same performance in input-order detection (*y*-axis) are smaller when subjected to noise during optimization. Higher noise levels (50 % spatial noise or 5 ms jitter) result in smaller neurons with less synapses. (**b**) Representative model morphologies after optimization under different noise levels: spatial noise of 25 % (S-25 %) or 50 % (S-50 %) and temporal noise with 2 (T-2 ms) and 5 (T-5 ms) ms jitter. The models are chosen to have similar input-order detection performance $P=0.7$. *Red dots* indicate the synapses. In case of spatial noise, the synapses at the highest noise level are shown

perform worse than the models in which spatial or temporal noise is introduced (Fig. 9.2a, black triangles). Additionally, the models subjected to higher noise levels (50 % spatial noise or uniform jitter of 5 ms) perform best. The model neuron performing input-order detection nearest to $P=0.7$ is plotted in Fig. 9.2b. In case of the spatial noise, we plotted the synapses at the highest noise levels (either 25 or 50 %). A second important observation is that at the same performance of input-order detection (say at $P=0.7$), the model neurons subject to noise are smaller compared to the noiseless model neurons. Consequently, the model neurons subject to higher noise levels also contain less synapses.

In order to further elucidate how smaller dendritic trees would be beneficial for performing input-order detection on noisy inputs, we used a simplified ball-and-stick model neuron. This model neuron is composed of only a soma and two unbranched dendrites, bearing groups of four or less synapses each; electrical properties were the same as in the morphologically elaborate models.

We tested how the ball-and-stick model integrated stochastic and deterministic synaptic inputs. For this, we tested these ball-and-stick models for input-order detection over a range of time intervals (Δt) in both temporal orders between the two groups of synapses. We then calculated the ratio between the maximum compound EPSP amplitudes in the preferred and non-preferred direction as a function of Δt (Fig. 9.3). In the deterministic case, the inputs arrived exactly Δt apart, whereas in the stochastic case (as in the optimization runs), they were jittered around that time.

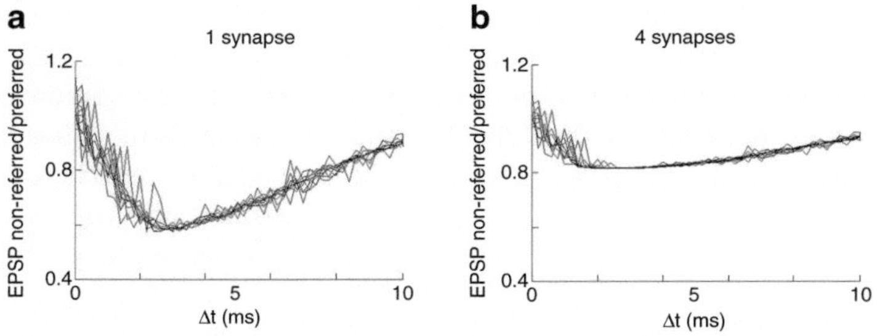

Fig. 9.3 Simulations of stochastic input-order detection by a ball-and-stick model. (**a**) Simulations with one synapse per dendrite. (**b**) Four synapses per dendrite. Input-order detection performance (ratio between the maximum EPSP amplitudes in the preferred and non-preferred directions, deterministic simulations in *bold*). The *thick lines* represent the average, the *thin lines* the individual simulation runs. Please note that the higher performance for a smaller number of synapses, due to the smaller EPSPs, is corrected for in the optimizations, where larger EPSPs were part of the optimization goal

While we took measures in the optimization procedure against unrealistically small compound EPSPs (that lead to trivial solutions), we employed no such measures for EPSP size in the ball-and-stick model simulations. In these neurons, smaller EPSPs performed better due to a more linear addition of small voltage deletions. The relevant comparison here, however, is between models with the same number of unitary synapses at different noise levels.

We found that on average, all models performed the same with and without temporal input noise. Independently of the number of synapses, the input-order detection performance did not increase or decrease with increasing stochasticity. What did change, however, was the variance around the mean performance, which was higher (better), the fewer synapses a neuron contained. The maximum input-order detection performance increased with fewer synapses with respect to the case in which all synapses were activated simultaneously (Fig. 9.3).

Our interpretation of the prevalence of smaller neurons with fewer synapses in the optimizations hence is as follows:

1. The mean input-order detection performance of neurons (independently of the number of synapses) does not change with increased noise.
2. In contrast, the variance of the input-order detection performance is higher in neurons with few synapses, and positive outliers (very high performance) are more likely.
3. When such cases of very high performance occur, the corresponding neurons are always rewarded by the genetic algorithm and selected for inclusion in the next generation.
4. Over time, this enriches the population in neurons with fewer synapses.

The genetic algorithm acts on a population of neurons, in which it enriches high-performing neurons. This is also the case if the high performances are not consistent but just more likely, as in the case of neurons with fewer synapses.

The biological interpretation of this finding is that a *population* of neurons with fewer synapses per neuron can perform better than a population in which all input is relayed to all neurons. This originates from the fact that some neurons with less input will perform better than the average neuron that receives more but stochastic inputs. Such populations will be more likely to have at least one neuron reacting very strongly. In neurons with more synapses, the stochasticity will average out, and the response of the different neurons will be more similar to each other. A population with such many-synapsed neurons will be more likely to miss positive outliers. Therefore, we hypothesize that any network which aims to detect surprisingly large events in heterogeneous input streams should be composed of neurons with few synapses.

9.4 Discussion

Our findings show a surprising decrease in dendritic size when optimizing neural morphologies for optimal response to increasingly noisy inputs. A standard way to overcome the stochasticity to individual elements of a system is to use a large number of them. In this way, the fluctuations of the individual elements will cancel out. However, this was not the solution found by the genetic algorithm we employed to find optimized dendritic trees for input-order detection on stochastic inputs. Surprisingly, the genetic algorithm found smaller neurons receiving less synapses with increasing stochasticity of the input streams.

This finding suggested to us that for input-order detection it is *not* crucial to cancel out noise. In fact, variation between different instances of stimulus presentations might be beneficial as is getting different opinions on one question. Simulations with a simplified ball-and-stick model demonstrated this mechanism and supported this proposal. Neurons with fewer synapses will be more likely to pick up positive outliers when performing input-order detection. This is because that the selection algorithm picked neurons from a population which, by chance, picked up a positive outlier. This selection mechanism favored small neurons, which will also respond well to positive outliers in a population of neurons in a brain area or ganglion. However, the population of neurons in the genetic algorithm is not meant as a model for a population of neurons in such a brain area or ganglion.

The computational task we investigated here, stochastic input-order detection, is an interesting and time-critical task most likely performed by a number of neural systems (Stiefel et al. 2012). However, neurons will certainly perform a large number of other computational functions as well, some of which will not be enhanced by noise. Additionally, while we did study the influence of active conductance in the deterministic input-order detection task (Torben-Nielsen and Stiefel 2009), we have not investigated the role of active conductances in the stochastic version nor did we investigate spike initiation in the soma/axon, which could further influence neuronal morphologies.

We speculate that in contrast, neurons which are mainly concerned with carrying out computational functions harmed or at least not benefited by noise will be large. The cortical layer V pyramidal and cerebellar Purkinje neurons described in the introduction are unlikely to execute such noise-benefited computational functions on a whole-cell level. Smaller neurons, such as cerebellar granule cells or cortical inhibitory interneurons, are more likely to execute computational functions aided by noise. Large neurons could still compute such functions on a smaller scale of neural organization, such as the level of individual spines (Branco et al. 2010; Stiefel et al. 2012).

This study also demonstrates the usefulness of our inverse method for mapping computational functions on dendritic morphologies. Human intuition often fails at complex issues, such as relating a neural morphology to its response to noise. In essence, our approach finds the approximate inverse solution to the stochastic nonlinear partial differential equation describing the multi-compartmental model of a neuron. This is no trivial task, and surprises about function–structure relationships are inevitable. We hope to employ this approach in the future to probe a variety of computational functions of neurons.

Acknowledgments We thank Vandana Padala Reddy, Andre van Schaik, Jonathan Tapson, and Terry Sejnowski for comments on the manuscript and helpful discussion.

References

Berger T, Senn W, Lüscher H-R (2003) Hyperpolarization-activated current Ih disconnects somatic and dendritic spike initiation zones in layer V pyramidal neurons. J Neurophysiol 90:2428–2437

Branco T, Clark BA, Häusser M (2010) Dendritic discrimination of temporal input sequences in cortical neurons. Science 329:1671–1675

Cannon R, Turner D, Pyapali G, Wheal H (1998) An on- line archive of reconstructed hippocampal neurons. J Neuroscience Methods 84(1–2):49–54

Cuntz H, Forstner F, Haag J, Borst A (2008) The morphological identity of insect dendrites. PLoS Comput Biol 4:e1000251

Cuntz H (2012) The dendritic density field of a cortical pyramidal cell. Front Neuroanat. 6:2

DeAngelis GC, Ohzawa I, Freeman RD (1993) Spatiotemporal organization of simple-cell receptive fields in the cat's striate cortex. I. General characteristics and postnatal development. J Neurophysiol 69:1091–1117

Destexhe A, Paré D (1999) Impact of network activity on the integrative properties of neocortical pyramidal neurons in vivo. J Neurophysiol 81:1531–1547

Destexhe A, Rudolph M, Fellous J-M, Sejnowski TJ (2001) Fluctuating synaptic conductances recreate in vivo-like activity in neocortical neurons. Neuroscience 107:13–24

Destexhe A, Rudolph M, Pare D (2002) The high-conductance state of neocortical neurons in vivo. Nat Rev Neurosci 4:739–751

Hay E, Schürmann F, Markram H, Segev I (2013) Preserving axosomatic spiking features despite diverse dendritic morphology. J Neurophysiol 109(12):2972–2981

Henry GH, Dreher B, Bishop PO (1974) Orientation specificity of cells in cat striate cortex. J Neurophysiol 37:1394–1409

Hille B (2001) Ion channels of excitable membranes, 3rd edn. Sinauer Associates, Sunderland, MA, USA, pp 2001–2007

Hô N, Destexhe A (2000) Synaptic background activity enhances the responsiveness of neocortical pyramidal neurons. J Neurophysiol 84:1488–1496
Jaeger D, Bower JM (1994) Prolonged responses in rat cerebellar Purkinje cells following activation of the granule cell layer: an intracellular in vitro and in vivo investigation. Exp Brain Res 100:200–214
Larkum ME, Zhu JJ, Sakmann B (1999) A new cellular mechanism for coupling inputs arriving at different cortical layers. Nature 398:338–341
Major G, Polsky A, Denk W, Schiller J, Tank DW (2008) Spatiotemporally graded NMDA spike/plateau potentials in basal dendrites of neocortical pyramidal neurons. J Neurophysiol 99:2584–2601
Martina M, Yao GL, Bean BP (2003) Properties and functional role of voltage-dependent potassium channels in dendrites of rat cerebellar purkinje neurons. J Neurosci 23:5698–5707
Migliore M, Shepherd GM (2002) Emerging rules for the distributions of active dendritic conductances. Nat Rev Neurosci 3:362–370
Polsky A, Mel BW, Schiller J (2004) Computational subunits in thin dendrites of pyramidal cells. Nat Neurosci 7:621–627
Rapp M, Yarom Y, Segev I (1992) The impact of parallel fiber background activity on the cable properties of cerebellar Purkinje cells. Neural Comput 4:518–533
Schiller J, Major G, Koester HJ, Schiller Y (2000) NMDA spikes in basal dendrites of cortical pyramidal neurons. Nature 404:285–289
Schneidman E, Freedman B, Segev I (1998) Ion channel stochasticity may be critical in determining the reliability and precision of spike timing. Neural Comput 10:1679–1703
Segev I, London M (2000) Untangling dendrites with quantitative models. Science 290(744):750
Stiefel KM, Sejnowski TJ (2007) Mapping function onto neuronal morphology. J Neurophysiol 98:513–526
Stiefel KM, Tapson J, van Schaik A (2012) Temporal order detection and coding in nervous systems. Neural Comput 25(2):510–531
Stuart G, Spruston N, Häusser M (eds) (2007) Dendrites, 2nd edn. Oxford University Press, USA
Stuart G, Spruston N, Sakmann B, Häusser M (1997) Action potential initiation and backpropagation in neurons of the mammalian CNS. Trends Neurosci 20:125–131
Torben-Nielsen B, Stiefel KM (2009) Systematic mapping between dendritic function and structure. Network 20:69–105
Torben-Nielsen B, Stiefel KM (2010a) An inverse approach for elucidating dendritic function. Front Comput Neurosci 4
Torben-Nielsen B, Stiefel KM (2010b) Wide-field motion integration in fly VS cells: insights from an inverse approach. PLoS Comput Biol 6:e1000932
Waters J, Helmchen F (2006) Background synaptic activity is sparse in neocortex. J Neurosci 26:8267–8277

Part II
Dendritic Computation

Chapter 10
Introduction to Dendritic Computation

Michiel W.H. Remme and Benjamin Torben-Nielsen

Abstract Dendrites receive the far majority of synaptic inputs to a neuron. The spatial distribution of inputs across the dendrites can be exploited by neurons to increase their computational repertoire. The role of dendrites in neural computation is the theme of the second part of this book to which this chapter forms the introduction. We review the various mechanisms that dendritic neurons can implement to introduce selectivity to spatiotemporal input patterns or to alter firing patterns, and briefly introduce the theoretical methods that are used to study this.

10.1 Introduction

The focus in the second part of this book is on the function of dendrites in the context of neural computation. Specifically, what is the role of dendrite morphology and membrane properties in the processing of synaptic input and its transformation into neural output?

The exploration of the role of dendrites in neural input integration was pioneered by Wilfrid Rall. This started in the 1950s with experimental work by Eccles and others that suggested surprisingly brief membrane time constants for certain cat spinal motoneurons. Those time constant estimates relied on the assumption that

M.W.H. Remme (✉)
Institute for Theoretical Biology, Humboldt-Universität zu Berlin, Invalidenstrasse 43, 10115 Berlin, Germany
e-mail: michiel.remme@hu-berlin.de

B. Torben-Nielsen
Department of Neurobiology, Hebrew University of Jerusalem, Edmond J. Safra Campus, Givat Ram, 91904 Jerusalem, Israel

Blue Brain Project, Swiss Federal Institute of Technology in Lausanne, Quartier d'Innovation, Batiment J, 1015 Lausanne, Switzerland
e-mail: btorbennielsen@gmail.com

H. Cuntz et al. (eds.), *The Computing Dendrite: From Structure to Function*, Springer Series in Computational Neuroscience 11, DOI 10.1007/978-1-4614-8094-5_10,

motoneurons could be described as point neurons and, therefore, that voltage transients followed exponential time courses. Rall realized that the current flow from the soma into the dendrites should shape the time course and elegantly demonstrated in a note to Science that the voltage response in a dendritic neuron follows a significantly faster mathematical function (Rall 1957). This work was followed by several years of controversy about the role of dendrites in neural function, but Rall's view on their relevance was eventually accepted. In the meantime Rall went on to develop the compartmental modeling method and the equivalent cylinder concept, and to analyze the response of dendritic neurons to different spatiotemporal synaptic input patterns, laying the foundation of the study of dendritic function (see Segev et al. 1995).

Many reviews have been written that summarize and interpret the five decades of work on dendritic computation that started with Rall's explorations (for example, Magee 2000; Segev and London 2000; Häusser and Mel 2003; Williams and Stuart 2003; Gulledge et al. 2005; London and Häusser 2005; Stuart et al. 2008; Johnston and Narayanan 2008; Larkum and Nevian 2008; Spruston 2008; Branco and Häusser 2010). In this chapter, we introduce the key concepts that all this work has yielded and show how the various chapters in this second part of the book fit into the overall scheme.

10.2 Dendritic Mechanisms that Compensate for Cable Filtering or that Exploit the Spatial Dimension

A key question that comes up when discussing the role of dendrites is whether they are a "bug or a feature" to the neuron (Häusser and Mel 2003). For particular types of neurons, a principal function of dendrites can be only the collection of input from presynaptic cells and the cell should essentially function as a point neuron, meaning that it should produce the same output in response to a specific temporal input pattern, irrespective of how the inputs are distributed across the dendrites. In that case, the neuron has to deal with the problems introduced by the dendrites themselves ("the bug"), that is, compensate for signal attenuation and temporal distortion of synaptic inputs that result from passive dendritic filtering (Fig. 10.1a).

A dendritic neuron can employ various mechanisms to make it function similar to a point neuron. As a first example, distal synaptic inputs to cortical pyramidal cells can be electrotonically so remote that passively spreading EPSPs would hardly affect the somatic voltage. The presence of regenerative currents (e.g., voltage-dependent sodium or calcium currents) throughout the dendrites can boost those distal inputs as they propagate towards the soma, such that they can have a stronger effect on somatic output, potentially resulting in an equal "vote" for all synaptic inputs ("dendritic democracy," Häusser 2001). This boosting of the amplitude of distal inputs can be a subthreshold process when passive signal attenuation is not too strong (Fig. 10.1b; Schwindt and Crill 1995; Remme and Rinzel 2011) or can involve the generation of dendritic spikes. The chapter by Destexhe and

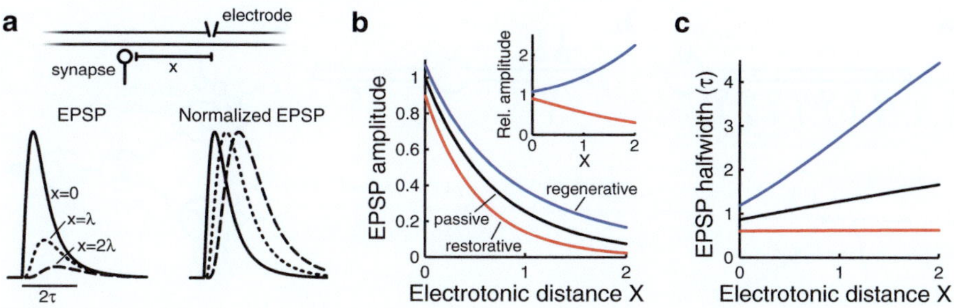

Fig. 10.1 Dendritic filtering of synaptic inputs. (**a**) A synaptic input current is injected in a passive infinite cable and recorded at 0, 1, and 2 space constants distance from the synapse. (**b**) The peak EPSP amplitude attenuates as it spreads along the cable. A regenerative membrane current (*blue*; e.g., a persistent sodium current) boosts the EPSP amplitude compared to the passive cable (*black*). A restorative current (*red*; e.g., most potassium currents) causes stronger attenuation. *Inset* shows EPSP amplitude relative to that of the passive response. (**c**) The width of the EPSP at half-maximal amplitude increases as the EPSP spreads in a passive cable. A regenerative current broadens the EPSP further, while a restorative current can maintain a constant halfwidth. Adapted from Remme and Rinzel (2011)

Rudolph-Lilith (Chap. 11) in this book reviews dendritic input integration of cortical pyramidal cells under in vivo conditions, including how these cells can achieve synaptic democracy through dendritic spikes.

Another example can be found in neurons in the mammalian auditory brainstem. These cells typically process synaptic input on an extremely fast timescale, since the input reaching those cells fluctuates up to the timescale of sound-wave carrier frequencies (<1–2 kHz). The auditory brainstem cells are typically quite compact electrotonically, such that input attenuation does not pose a major problem; however, EPSPs do broaden significantly when the membrane is passive. One specialization that certain auditory brainstem cells show is the presence of low-threshold voltage-gated potassium ion channels in their dendrites (Mathews et al. 2010). These currents counter the passive broadening of EPSPs and preserve the very high temporal resolution (Fig. 10.1c; Remme and Rinzel 2011).

The above types of dendritic membrane mechanisms can be central to the functioning of certain neurons, but it is important to realize that they do not by themselves enhance the computational power compared to that of a point neuron. However, it is also possible that dendrites are a "feature" for certain types of neurons, expanding the computational repertoire of those neurons. It is imaginable that dendrites originally evolved simply to allow for high interconnectivity of neurons, but that nature has found ways to make use of the spatial separation of synaptic inputs over the dendritic extent. Many studies indeed demonstrate a role for dendrites in enhancing computational power. We next give an overview of the mechanisms that dendritic neurons can utilize to increase their computational power with links to the relevant chapters in this book.

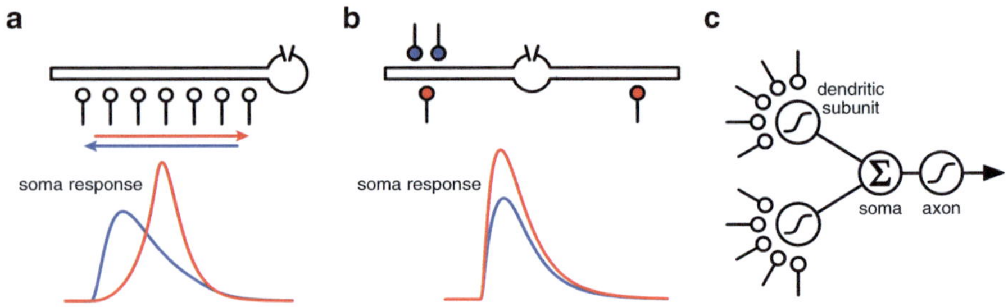

Fig. 10.2 Exploiting spatial separation of synaptic inputs to increase a neuron's computational power. (**a**) Dendritic filtering introduces propagation delays, which results in selectivity to spatio-temporal input patterns. (**b**) Synaptic conductance input leads to sublinear summation of colocalized inputs, while synapses on different branches can summate linearly. (**c**) Segregation of the dendritic tree into subunits with local spiking nonlinearities turn a single neuron into a network. The schematic depicts that input to dendritic subunits is passed through an effectively sigmoidal nonlinearity; the subunit outputs are summed at the soma and then passed through a global nonlinearity at the axon

10.2.1 Dendritic Filtering: Selectivity to Spatiotemporal Patterns

How can neurons exploit the spatial dimension to enhance computational power? This does, in fact, not even require nonlinear mechanisms: the linear dendritic filtering of input currents that was discussed above, can already endow the neuron with a selectivity to spatiotemporal input patterns that cannot be achieved by a point neuron model. Rall already demonstrated that activation of synaptic inputs in a sequence along a dendritic cable towards the soma leads to a larger-amplitude composite EPSP at the soma, than when the sequence is directed away from the soma (Fig. 10.2a; Rall 1964). This direction-selectivity relies on the delays that result from passive dendritic filtering and is key to several studies in this book: the chapters by Grimes and Diamond (Chap. 12) and by Smith and Taylor (Chap. 13) describe direction-selective interneurons and ganglion cells in the retina; Stiefel and Torben-Nielsen (Chap. 9) discuss how passive dendritic morphologies can be optimized to implement input-order detection. Dendritic signal propagation delays also play a key role in the chapter by McGinley (Chap. 14), who discusses how octopus cells in the auditory brainstem can make use of those delays to compensate for the delays between sound frequency channels that emerge in the cochlea.

To generate significant propagation delays and thereby produce the selectivity to spatiotemporal patterns, dendrites need to be electrotonically fairly long (though subthreshold voltage-dependent membrane currents can modulate those delays; Koch 1984; Remme and Rinzel 2011). The chapter by Branco (Chap. 15) in this book shows that selectivity to spatiotemporal input patterns is also achieved by more compact dendrites with a mechanism that relies on an impedance gradient along dendritic branches and the nonlinear activation of NMDA receptors.

Finally, passive dendritic filtering also plays a key role in the chapter by Cuntz, Haag, and Borst (Chap. 16) who discuss optic flow processing in the fly. They show that dendro-dendritic electric coupling between cells leads to a spatial blurring—resulting from dendritic filtering—of the signals being passed from one cell to the other. This blurred signal can then, in fact, be used to sharpen the original signal.

10.2.2 Synaptic Conductance Input: Local Sublinear Summation

The above mechanisms that rely on dendritic filtering already operate when EPSPs are described by input current transients (i.e., the system is linear). In reality, chemical synapses affect the postsynaptic cell through local conductance changes, which introduces a nonlinearity because of the voltage-dependence of the driving force of a synaptic current: the closer the voltage is to the reversal potential of the synaptic current, the smaller the current will be. Hence, when considering simultaneous activation of multiple conductance inputs in a passive point-neuron model, one observes sublinear summation of synaptic inputs. In a dendritic tree, such sublinear summation is found when inputs are colocalized; however, this nonlinear interaction decreases with distance between the inputs (Rall 1967; Koch et al. 1990). Because of this, a more linear summation of synaptic inputs can be achieved in a neuron with a dendritic tree than in a point neuron, by a distribution of the synapses over the different dendritic branches. Moreover, this mechanism also introduces pattern selectivity, even if all inputs are at the same distance from the soma, depending on which inputs are colocalized and which are dispersed over the tree.

The saturation nonlinearity plays a central role in the work reviewed by Peron (Chap. 17) on visual interneurons in locusts. There, the saturating nonlinearity serves to output the logarithm of the input variable that is signaled to these neurons through strong excitatory synapses. The inhibitory input to these cells signals different information and is segregated to another part of the dendritic tree, such that there is no interference between the local computations.

A study on dendritic computation that is worth mentioning here is the work by Agmon-Snir et al. (1998). They demonstrated that segregation of two excitatory synaptic pathways over the dendrites of a bipolar neuron can form the basis of a coincidence detector that not only has increased sensitivity but is also very robust in the sense that a somatic voltage threshold can only be reached when the two pathways are coactive, even when the number of inputs that are active from either pathway is variable, whereas a single pathway (to one dendrite) can never reach threshold, due to local saturation of the synaptic driving force (Fig. 10.2b). Such robustness cannot be achieved in a point neuron model.

The local effects of synaptic conductance transients also allow for targeted inhibition of excitatory synapses in dendritic neurons. In a point neuron, an inhibitory input will have the same effect on all excitatory inputs. Such global inhibition can also be achieved in a dendritic neuron by having the inhibitory inputs impinge on the soma. However, an inhibitory synapse on a dendritic branch will particularly inhibit the colocalized excitatory inputs (see for example Jack et al. 1975; Koch et al. 1983). A single inhibitory cell typically contacts specific domains of a postsynaptic neuron with multiple (10–20) synapses rather than with a single synapse. Gidon (Chap. 18) reveals that this has various nonintuitive consequences, specifically when focusing on how inhibitory input affects local and global dendritic excitability rather than how it affects the subthreshold propagation of EPSPs to the soma.

Finally, the large amount of synaptic input activity that many types of neurons experience under in vivo conditions also affects their integrative properties. The chapter by Destexhe and Rudolph-Lilith (Chap. 11) details how background input decreases both the input impedance and membrane time constant, leads to stronger spatial segregation between all inputs, and introduces a stochastic operating mode.

10.2.3 Voltage-Dependent Membrane Conductances: Resonance, Oscillations, Dendritic Spikes, Firing Patterns

The two principal ingredients to expand a neuron's computational power described above—dendritic filtering and synaptic conductance input—only required passive dendrites. However, many types of neurons display voltage-dependent membrane conductances in their dendrites (for review see for example Magee 2008). Such nonlinearities open up many possibilities for dendrites to further expand the computational repertoire of a neuron.

Membrane currents can be nonuniformly distributed throughout the neuron (Nusser 2009). This can of course already be the case for passive membrane currents and is, in fact, a key finding in the work presented by Mathiä, Bartos, and Vida (Chap. 19) in their chapter on input integration by inhibitory basket cells in the hippocampus. A prominent voltage-dependent conductance in the dendrites of many types of neurons is the hyperpolarization-activated mixed cation current I_h. In cortical pyramidal neurons this current has a strongly polarized distribution, with most of the current present in the distal dendrites. Zhuchkova, Remme, and Schreiber (Chap. 20) show that, because of this nonuniform distribution, synaptic inputs are differentially filtered depending on where they arrive in the neuron. They focus in particular on the ability of the h-current to create a band-pass filter, i.e., produce subthreshold membrane-potential resonances (Hutcheon and Yarom 2000).

Interactions between voltage-dependent conductances can also underlie the generation of intrinsic subthreshold oscillations, which have been demonstrated in various cell types (see for example Alonso and Llinas 1989). Such nonlinear dynamics lie at the heart of hypotheses aiming to explain the generation of place fields in hippocampus (O'Keefe and Recce 1993) and grid fields in entorhinal cortex (Burgess et al. 2007). These hypotheses require that the oscillation-generating membrane segments are segregated from the soma and that the dendritic oscillators integrate synaptic input independently from the rest of the neuron. Remme, Lengyel, and Gutkin (Chap. 21) analyze the biophysical requirements for such a setup and show that subthreshold oscillators tend to synchronize very quickly because of the ongoing intracellular interactions between them. Hence, instead of supporting multiple independent oscillators, the dendritic tree collapses to a single oscillator.

One influential theory of dendritic function—building on strong local dendritic nonlinearities—hypothesizes that the dendritic tree consists of multiple subunits, such as single dendritic branches (Branco and Häusser 2010), that independently process input and convey the signal to the soma through an amplitude boosting nonlinearity (NMDA, calcium, or sodium spikes). The first studies that put this idea forward have focused on instantaneous processing of continuous variables (firing rates). Each dendritic subunit sums its input and passes it through a local sigmoidal nonlinearity; the outputs of these subunits are summated at the soma and passed through a global sigmoidal nonlinearity (Fig. 10.2c; Poirazi et al. 2003; Polsky et al. 2004). Formulated in this way, a single neuron represents a two-layer feedforward neural network, and is, in theory, able to approximate any (instantaneous) input–output transformation. This framework is analyzed by Cazé, Humphries, and Gutkin (Chap. 22), who consider how the computational capacity of a neuron is increased when it has (a finite number of) independent dendritic subunits with either saturating (through synaptic conductances, see above) or spike-type nonlinearities. Furthermore, Smith and Taylor (Chap. 13) describe retinal ganglion cells that consist of partially isolated subunits that signal to the soma via dendritic spikes. Grimes and Diamond (Chap. 12) discuss retinal interneurons that consist of many independent dendritic subunits, each, in fact, providing output to postsynaptic neurons through local dendritic release of neurotransmitters.

Finally, voltage-dependent conductances distributed throughout dendritic neurons can also qualitatively alter the membrane-potential dynamics. Classic studies by Pinsky and Rinzel (1994) and Mainen and Sejnowski (1996) showed that neural morphology can determine whether a cell operates in a tonic-firing or in a burst-firing mode. The chapter by van Ooyen and van Elburg (Chap. 23) in this book further details how the size and branching pattern of a dendritic tree shapes a neuron's firing dynamics. An even more detailed perspective on this theme is presented in the chapter by O'Donnell and Nolan (Chap. 24). They study the consequences of stochastic gating of individual ion channels in neuron models with detailed morphologies. Using specially developed software, they focus on how stochastic gating impacts membrane-potential dynamics and ultimately dendritic computation.

Clearly, there are many other ways in which voltage-dependent conductances can shape membrane-potential dynamics and neural computation that we have not reviewed here. We have given an overview of the key mechanisms with links to chapters in this book and want to refer the reader to the aforementioned reviews for further information.

10.3 Synaptic and Structural Plasticity

As we have seen above, there is a range of dendritic mechanisms that can increase a neuron's computational power. However, to actually exploit those mechanisms, it is crucial that the synaptic inputs are properly distributed across the dendritic tree (Larkum and Nevian 2008). Also, for a neuron to learn to respond appropriately to its input, and to store information about its history of input and output, a variety of ongoing plasticity mechanisms is required. These synaptic, homeostatic, and structural plasticity processes in dendrites are reviewed in the chapter by Kastellakis and Poirazi (Chap. 25). An example of storage and retrieval of input–output patterns by a single neuron is described in the chapter by de Sousa, Maex, Adams, Davey, and Steuber (Chap. 26). They focus on learning by cerebellar Purkinje cells and use a morphologically realistic conductance-based model to examine the capacity to learn and recognize input patterns through long-term depression of the parallel fiber inputs to these cells.

10.4 Theoretical Approaches to Study Dendritic Function

We will next briefly introduce the main theoretical methods that are currently in use to study dendritic input integration. The backbone of this is cable theory, which was already well established in the first half of the twentieth century to describe passive signal flow in neurites (see for example Hodgkin and Rushton 1946). However, it was Rall who pioneered the extensive use and development of cable theory to analyze dendritic function (Rall 1959, 1960). In cable theory a dendritic branch is represented by a one-dimensional partial differential equation describing how the voltage profile along the branch evolves over time. The partial differential equation, with appropriate boundary conditions, can be solved analytically with various methods (see for example Tuckwell 1988). In this way one can obtain explicit expressions for, for example, the steady state voltage profile along a dendrite in response to a step current, the dendrite's frequency-dependent input filtering, and the impulse response function or Green's function of the branch with which one can compute the response to an arbitrary stimulus (see also Jack et al. 1975). Cable theory can be used to describe a complex, branched dendritic morphology and even multiple coupled cells. Timofeeva and Coombes (Chap. 27) illustrate this in their chapter in which they compute—using the "sum-over-trips" method—the impulse response function for two neurons that are coupled through a dendritic gap junction.

Cable theory is not restricted to describe passive neurons but can also be used to analyze neurons with voltage-dependent membrane conductances. This relies on linearization of such currents around a holding membrane potential by which a "quasi-active" cable description is obtained that can reproduce subthreshold dendritic input filtering of the full nonlinear model (Koch 1984; Remme 2013). Such a linearization approach is used by Zhuchkova, Remme, and Schreiber (Chap. 20) to describe a single active compartment that is coupled to a passive cable.

The chapter by Remme, Lengyel, and Gutkin (Chap. 21) uses the quasi-active cable approximation to analyze the dynamics of an active cable that connects two dendritic oscillators.

Of course, there are limits to the complexity of neurons and voltage dynamics that can be conveniently described and/or analyzed using cable theory. The most common approach is then to make use of compartmental modeling, a method that was also pioneered by Rall (1964). Compartmental modeling relies on spatial discretization of the cable equation, resulting in a set of connected compartments: the partial differential equation becomes a family of ordinary differential equations. Each compartment can include any type of nonlinear membrane mechanism. To solve the compartmental model one relies on numerical simulation of the system. For this one can use software packages such as NEURON (Hines and Carnevale 1997) and GENESIS (Bower and Beeman 1998), which are designed to conveniently construct and efficiently simulate compartmental models. Most work described in the second part of this book makes use of compartmental modeling.

A central challenge in compartmental modeling is how to fit the many model parameters. Druckman (Chap. 28) discusses the various key issues in this problem and the methods that are available to constrain the parameters of single neuron models. A further potential issue is that the number of equations that need to be solved numerically can become so large that it is no longer feasible to simulate the system, especially when intending to simulate networks of detailed model neurons. One can, of course, try to reduce the number of compartments of the model; however, this comes (at least) at the cost of losing spatial accuracy. Hedrick and Cox (Chap. 29) discuss alternative strategies rooted in linear algebra, in which the spatial distribution of synaptic inputs within the dendrites are retained, while the computational load to simulate the neuron models is strongly reduced.

10.5 Future Perspectives

Recent innovations in experimental techniques provide us with an unprecedented wealth of information about dendritic function. Optogenetics allow for precise manipulation of input and output activity of neurons (Fenno et al. 2011), and advances in calcium and voltage imaging make it possible to record dendritic responses at high spatial and temporal resolution, all under in vivo conditions (Akemann et al. 2010; Grienberger and Konnerth 2012). Putting all these findings together in a coherent framework and increasing the understanding of the immensely complex nervous system gives a central role to thorough analytical and computational approaches such as the ones outlined in this book. At the same time, there are limits to what can be done experimentally, and computational work can also be used there as a spotlight to direct experimental work. Given the intricate relation between experiments, model validation and model predictions, experimental techniques en par with computational approaches will yield the largest advances in understanding in dendritic computation.

A further recent development relying on both experimental and theoretical neuroscience is the aim of several research consortia to build extremely detailed models of cortical columns or even of the entire brain (e.g., the Blue Brain Project, Markram 2006). The computational demands to simulate such models might be prohibitively large. Therefore, researchers have taken up the challenge to reduce the computational complexity of morphological neuron models while preserving their integrative properties resulting from the spatial distribution of synapses and nonuniform membrane properties. Some of these new approaches are described in this book (Chap. 29) and we expect development in this field will continue expanding.

The central question remains: what do neurons compute? As reported in this book and elsewhere, many computations are attributed to dendrites. However, what dendrites *can* compute is not necessarily identical to what they actually *do* compute in vivo. For example, the findings from many studies depend on very specific spatiotemporal input activation patterns and/or neglect the various sources of noise. This issue poses the most important challenge: to find out what a neuron really computes rather than what a neuron can theoretically compute. What are the in vivo constraints put on the system and how do they affect what can be computed? Single neuron computations are part of a system of parallel and serial input–output transformations. How are the computations distributed and combined? Cuntz and colleagues (Chap. 16) point at this concern and demonstrate that in the fly Lobula Plate, optic flow signal processing is a collaborative computation rather than a computation performed by a single neuron. The issue comes down to what computations are too hard (in practice) to be computed by a single neuron and require network interactions. Indeed, many challenges lie ahead and we believe that the work described in this second part of the book lead in the right direction and pave the way for important future discoveries.

References

Agmon-Snir H, Carr CE, Rinzel J (1998) The role of dendrites in auditory coincidence detection. Nature 393:268–272

Akemann W, Mutoh H, Perron A et al (2010) Imaging brain electric signals with genetically targeted voltage-sensitive fluorescent proteins. Nat Methods 7:643–649

Alonso AA, Llinas RR (1989) Subthreshold Na^+-dependent theta-like rhythmicity in stellate cells of entorhinal cortex layer II. Nature 342:175–177

Bower JM, Beeman D (1998) The book of GENESIS: exploring realistic neural models with the GEneral NEural SImulation System. Springer, New York, NY

Branco T, Häusser M (2010) The single dendritic branch as a fundamental functional unit in the nervous system. Curr Opin Neurobiol 20:494–502

Burgess N, Barry C, O'Keefe J (2007) An oscillatory interference model of grid cell firing. Hippocampus 17:801–812

Fenno L, Yizhar O, Deisseroth K (2011) The development and application of optogenetics. Annu Rev Neurosci 34:389–412

Grienberger C, Konnerth A (2012) Imaging calcium in neurons. Neuron 73:862–885

Gulledge AT, Kampa BM, Stuart GJ (2005) Synaptic integration in dendritic trees. J Neurobiol 64:75–90
Häusser M (2001) Synaptic function: dendritic democracy. Curr Biol 11:R10–R12
Häusser M, Mel B (2003) Dendrites: bug or feature? Curr Opin Neurobiol 13:372–383
Hines ML, Carnevale NT (1997) The NEURON simulation environment. Neural Comput 9:1179–1209
Hodgkin AL, Rushton WA (1946) The electrical constants of a crustacean nerve fibre. Proc R Soc Med 134:444–479
Hutcheon B, Yarom Y (2000) Resonance, oscillation and the intrinsic frequency preferences of neurons. Trends Neurosci 23:216–222
Jack JJB, Noble D, Tsien RW (1975) Electric current flow in excitable cells. Oxford University Press, Oxford
Johnston D, Narayanan R (2008) Active dendrites: colorful wings of the mysterious butterflies. Trends Neurosci 31:309–316
Koch C, Poggio T, Torre V (1983) Nonlinear interactions in a dendritic tree: localization, timing, and role in information processing. Proc Natl Acad Sci 80:2799–2802
Koch C (1984) Cable theory in neurons with active, linearized membranes. Biol Cybern 50:15–33
Koch C, Douglas R, Wehmeier U (1990) Visibility of synaptically induced conductance changes: theory and simulations of anatomically characterized cortical pyramidal cells. J Neurosci 10:1728–1744
Larkum ME, Nevian T (2008) Synaptic clustering by dendritic signalling mechanisms. Curr Opin Neurobiol 18:321–331
London M, Häusser M (2005) Dendritic computation. Annu Rev Neurosci 28:503–532
Magee JC (2000) Dendritic integration of excitatory synaptic input. Nat Rev Neurosci 1:181–190
Magee JC (2008) Dendritic voltage-gated ion channels. In: Stuart GJ, Spruston N, Häusser M (eds) Dendrites, 2nd edn. Oxford University Press, Oxford, pp 225–250
Mainen ZF, Sejnowski TJ (1996) Influence of dendritic structure on firing pattern in model neocortical neurons. Nature 382:363–366
Markram H (2006) The blue brain project. Nat Rev Neurosci 7:153–160
Mathews PJ, Jercog PE, Rinzel J et al (2010) Control of submillisecond synaptic timing in binaural coincidence detectors by Kv1 channels. Nat Neurosci 13:601–609
Nusser Z (2009) Variability in the subcellular distribution of ion channels increases neuronal diversity. Trends Neurosci 32:267–274
O'Keefe J, Recce ML (1993) Phase relationship between hippocampal place units and the EEG theta rhythm. Hippocampus 3:317–330
Pinsky PF, Rinzel J (1994) Intrinsic and network rhythmogenesis in a reduced Traub model for CA3 neurons. J Comput Neurosci 1:39–60
Poirazi P, Brannon T, Mel BW (2003) Pyramidal neuron as two-layer neural network. Neuron 37:989–999
Polsky A, Mel BW, Schiller J (2004) Computational subunits in thin dendrites of pyramidal cells. Nat Neurosci 7:621–627
Rall W (1957) Membrane time constant of motoneurons. Science 126:454
Rall W (1959) Branching dendritic trees and motoneuron membrane resistivity. Exp Neurol 1:491–527
Rall W (1960) Membrane potential transients and membrane time constant of motoneurons. Exp Neurol 2:503–532
Rall W (1964) Theoretical significance of dendritic trees for neuronal input-output relations. In: Reis R (ed) Neural theory and modeling. Stanford University Press, Stanford, CA, pp 73–97
Rall W (1967) Distinguishing theoretical synaptic potentials computed for different soma-dendritic distributions of synaptic input. J Neurophysiol 30:1138–1168
Remme MWH, Rinzel J (2011) Role of active dendritic conductances in subthreshold input integration. J Comput Neurosci 31:13–30

Remme MWH (2013) Quasi-active approximation of nonlinear dendritic cables. In: Jaeger D, Jung R (eds) Encyclopedia of computational neuroscience: SpringerReference. Springer-Verlag, Berlin, Heidelberg

Schwindt PC, Crill WE (1995) Amplification of synaptic current by persistent sodium conductance in apical dendrite of neocortical neurons. J Neurophysiol 74:2220–2224

Segev I, London M (2000) Untangling dendrites with quantitative models. Science 290:744–750

Segev I, Rinzel J, Shepherd GM (1995) The theoretical foundation of dendritic function: selected papers of Wilfrid Rall with commentaries. MIT Press, Cambridge, MA

Spruston N (2008) Pyramidal neurons: dendritic structure and synaptic integration. Nat Rev Neurosci 9:206–221

Stuart GJ, Spruston N, Häusser M (2008) Dendrites, 2nd edn. Oxford University Press, Oxford

Tuckwell HC (1988) Introduction to theoretical neurobiology. Cambridge University Press, Cambridge

Williams SR, Stuart GJ (2003) Role of dendritic synapse location in the control of action potential output. Trends Neurosci 26:147–154

Chapter 11
Noisy Dendrites: Models of Dendritic Integration In Vivo

Alain Destexhe and Michelle Rudolph-Lilith

Abstract While dendritic processing has been well characterized in vitro, there is little experimental data and models available about the integrative properties of dendrites in vivo. Here, we review existing computational models to infer the dendritic processing of neocortical pyramidal neurons in vivo. We start by summarizing experimental measurements of the "high-conductance states" of cortical neurons in vivo. Next, we show models predicting that, in such states, the responsiveness of cortical neurons should be greatly enhanced, in particular due to the presence of high-amplitude fluctuations ("synaptic noise"). We infer that in dendrites this effect should be particularly strong, leading to the spontaneous activation of dendritic spikes. The presence of noise in dendrites also enhances spike propagation. We show that opposite distance dependencies of spike initiation and propagation result in roughly location-independent synaptic efficacies. In addition, in high-conductance states, dendrites display sharper temporal processing capabilities. Thus, we conclude that noisy active dendrites behave more "democratically," and that dendrites should have enhanced processing capabilities in vivo.

11.1 Introduction

Activated states of the brain are characterized by intense, irregular, and desynchronized neuronal activity. In awake animals, neurons in different cortical structures display high spontaneous firing rates, from 5 to 20 Hz in cats (Evarts, 1964; Steriade and McCarley, 1990). Moreover, the cerebral cortex shows a very dense connectivity, with each pyramidal neuron receiving between 5,000 and 60,000

A. Destexhe (✉) • M. Rudolph-Lilith
Unit for Neuroscience, Information and Complexity (UNIC), UPR-3293, CNRS, Gif-sur-Yvette, France
e-mail: Destexhe@iaf.cnrs-gif.fr; destexhe@unic.cnrs-gif.fr

H. Cuntz et al. (eds.), *The Computing Dendrite: From Structure to Function*, Springer Series in Computational Neuroscience 11, DOI 10.1007/978-1-4614-8094-5_11,

synaptic contacts, most of which originate from the cortex itself (DeFelipe and Fariñas, 1992; Braitenberg and Schüz, 1998). As a consequence, many synaptic inputs are simultaneously activated onto cortical neurons in vivo. Indeed, intracellular recordings in awake animals reveal that cortical neurons are subjected to an intense synaptic bombardment and, as a result, are depolarized and have a low input resistance (Matsumura et al., 1988; Baranyi et al., 1993; Steriade et al., 2001) compared to brain slices kept in vitro. This activity is also responsible for a considerable amount of subthreshold fluctuations, called "synaptic noise." Together these properties characterize what is called the "high-conductance state" of cortical neurons. How such high-conductance and high-noise conditions affect the integrative properties of neurons remains an intense subject of research (reviewed in Destexhe et al., 2003; Destexhe and Contreras, 2006; Destexhe, 2007; Destexhe and Rudolph-Lilith, 2012). Besides few exceptions (Helmchen et al., 1999; Branco et al., 2010; Branco and Häusser, 2011; Lavzin et al., 2012), little is known about the integrative properties of dendrites in vivo.

In this chapter, we start by overviewing experimental measurements of high-conductance states in cortical neurons in vivo. We next review computational models to explore the impact of such states on integrative properties, and what sort of computational advantages may be conferred by synaptic noise in dendrites.

11.2 Characterization of High-Conductance States In Vivo

11.2.1 The Synaptic Noise in Neocortical Neurons In Vivo

In awake animals, the cerebral cortex (and more generally the entire brain) displays an "activated" state, with distinct characteristics compared to other states like slow-wave sleep or anesthesia. These characteristics include a low-amplitude "desynchronized" electroencephalogram (EEG), a depolarized V_m, and irregular firing activity (Fig. 11.1a, Awake). During slow-wave sleep, the EEG and V_m activity follow low-frequency rhythms (Fig. 11.1a, Slow-Wave Sleep). The most prominent rhythm consists of slow-wave complexes in the EEG, which are paralleled with up/down-state dynamics in the V_m. During the up-state (Fig. 11.1a, gray bars), the V_m is depolarized and the activity is similar to wakefulness; during the down-state, all cortical neurons are hyperpolarized and do not fire. Several anesthetics, such as urethane or ketamine–xylazine, induce EEG and V_m dynamics very similar to slow-wave sleep. For instance, ketamine–xylazine anesthesia generates an up/down-state pattern very similar to sleep (Destexhe et al., 2003). For recent reviews on EEG and V_m dynamics during activated and sleep states, see Steriade (2001, 2003), Steriade and McCarley (1990), and McCormick and Bal (1997).

One of the main interests in dealing with up/down-state patterns is that there is good evidence that the up-states follow dynamics very similar to that of activated states of the brain (for a recent review, see Destexhe et al., 2007). Indeed, at the level of EEG and intracellular activities, the dynamics seen during up-states are

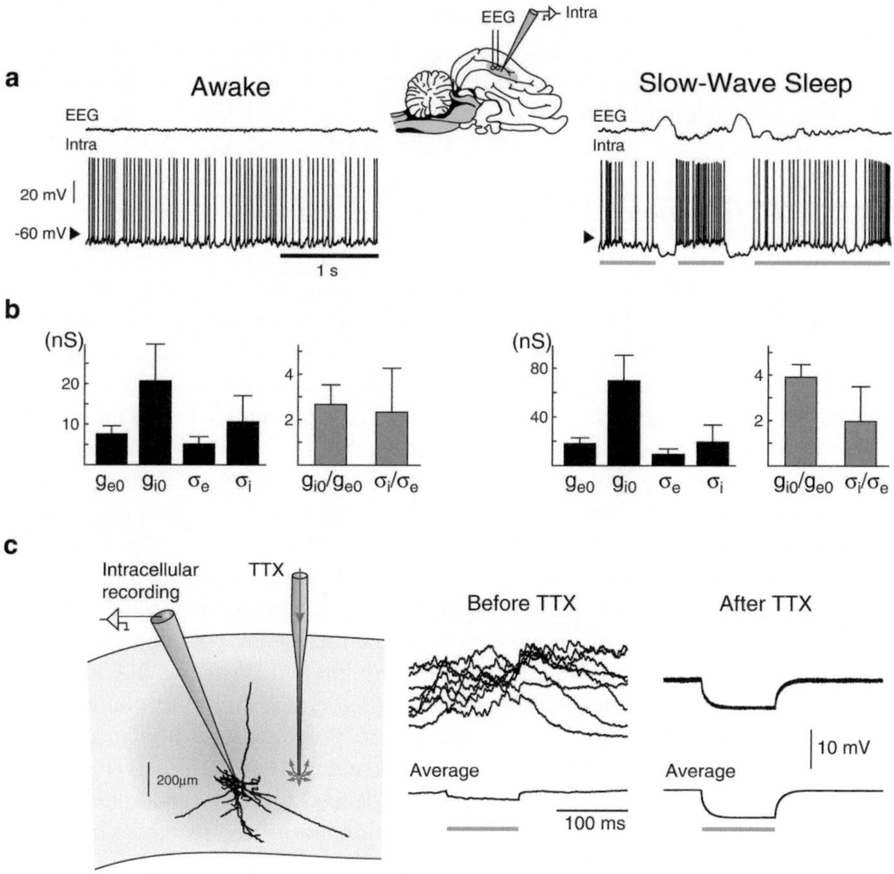

Fig. 11.1 High-conductance states in neocortical neurons in vivo. (**a**) Intracellular recordings and electroencephalogram (EEG) in the association cortex of cats (area 5–7, see scheme), during the waking state (*left*) and during slow-wave sleep (*right*). The activity during sleep consists of up- and down-states (up-states indicated by *gray bars*; modified from Steriade et al., 2001). (**b**) Conductance measurements during wake and sleep states. Intracellular recordings such as in (**a**) were used to estimate the mean excitatory and inhibitory conductances (g_{e0}, g_{i0}) and their standard deviations (σ_e, σ_i), respectively (modified from Rudolph et al., 2007). (**c**) Quantification of the conductance state of the membrane in active states compared to quiescent states by microperfusion of TTX to cortex (see scheme). Hyperpolarizing current pulses were injected during up-states (Before TTX), and later after total suppression of network activity (After TTX), in the same neurons. The overall conductance was about five times lower after TTX for this particular cell (modified from Paré et al., 1998)

almost undistinguishable from that during wakefulness. An illustrative example of this similarity is that electrical stimulation of the brain stem (pedonculopontine tegmentum, or PPT) can transform the up/down-state dynamics into the typical desynchronized EEG of activated states, which appears as a "prolonged" up-state (see Steriade et al., 1993; Rudolph et al., 2005). Thus, it seems that the up-states constitute a relatively good approximation of the network state during activated states. It is important to stress that these states are close, but not identical, as shown for example by conductance measurements (reviewed in Destexhe et al., 2007; Destexhe and Rudolph-Lilith, 2012).

11.2.2 Conductance Measurements

The total excitatory and inhibitory conductance can be estimated from V_m measurements in different ways. Injection of constant current during up/down-states induces a marked change of the V_m during down-states, while the up-state seems much less sensitive to the injected current (Paré et al., 1998; Destexhe et al., 2003). In other words, the up-state has a much larger conductance compared to the down-state. These measurements constitute a first indication that the up-states correspond to a high-conductance state. The same results were obtained with purely subthreshold activity, suggesting that they are not due to the conductances of action potentials.

A second, more direct measurement was obtained by using the "VmD method" (Rudolph et al., 2004; Piwkowska et al., 2008), which estimates the total conductances and their variances by fitting experimental V_m distributions to the Gaussian approximation of an analytical expression for the membrane potential distribution of an effective stochastic membrane model (Rudolph and Destexhe, 2003c, 2005). This approach leads to estimates of the mean excitatory and inhibitory conductances (g_{e0}, g_{i0}) and their standard deviations (σ_e, σ_i), respectively, and was successfully applied to intracellular recordings from awake and naturally sleeping cats (Fig. 11.1b; Rudolph et al., 2007). These measurements evidenced that the membrane in these states is indeed in a high-conductance state, with inhibitory conductances in general several-fold larger than excitatory conductances.

This result was consistent with another type of direct measurement, in which intracellular recordings were compared during up-states and after suppression of network activity using TTX (Fig. 11.1c; Paré et al., 1998). Because TTX blocks all sodium channels, it effectively suppresses all action potential-dependent activity and reveals the resting V_m of the neuron. Input resistance (R_{in}) measurements showed that, taking the up-states of ketamine–xylazine anesthesia as reference, these active states have about five times more synaptic conductance compared to the resting V_m of the cell (Fig. 11.1c; Paré et al., 1998; Destexhe and Paré, 1999). These results are not affected by the V_m level or by spiking activity as identical results are obtained at hyperpolarized and subthreshold levels. Furthermore, R_{in} measurements correspond to the linear portion of the *I*–*V* curve, suggesting little or no contamination by intrinsic voltage-dependent currents (Destexhe and Paré, 1999; see also discussion in Monier et al., 2008).

Similar measurements have also been obtained during active states in vivo in other studies, by comparing up- and down-states under various anesthetics such as ketamine–xylazine or urethane, in different species. These estimates are very variable, ranging from up to several-fold smaller R_{in} in up-states (Contreras et al., 1996;

Paré et al., 1998; Petersen et al., 2003; Leger et al., 2005), to nearly identical R_{in} between up- and down-states or even larger R_{in} in up-states (Metherate and Ashe, 1993; Zou et al., 2005; Waters and Helmchen, 2006). It was argued that the latter, rather paradoxical observation, only found in rats so far, is the result of the presence of potassium currents in down-states (Zou et al., 2005), or voltage-dependent rectification (Waters and Helmchen, 2006). Consistent with the latter, blocking K^+ currents using cesium-filled electrodes has negligible effects on the up-state, but abolishes the hyperpolarization during the down-states (Timofeev et al., 2001). Moreover, in cats, the R_{in} of the down-state differs from that of the resting V_m (after TTX) by about twofold (Paré et al., 1998). It is, thus, clear that at least the down-state is very different from the true resting V_m of the neuron. Finally, conductance measurements in awake and naturally sleeping animals have revealed a wide diversity between cells in cat cortex (Rudolph et al., 2007), ranging from large synaptic conductances, much larger than the resting conductance, to synaptic conductances smaller or equal to the resting conductance. However, on average, the synaptic conductance was estimated as about three times the resting conductance, where the inhibitory conductance is about twice the excitatory conductance (Rudolph et al., 2007). Strong inhibitory conductances were also found in artificially evoked active states using PPT stimulation (Rudolph et al., 2005), as shown in Fig. 11.1b.

In conclusion, the data reviewed here indicate that in brain activity states with desynchronized EEG, neocortical neurons display a high-conductance state characterized by the following features: (1) a large membrane conductance, which corresponds to a threefold to fivefold decrease in input resistance; (2) an average membrane potential (around −65 to −60 mV), which is significantly depolarized compared to the natural resting V_m (−70 to −80 mV); and (3) large amplitude membrane potential fluctuations (σ_V of 2–6 mV), which are at least tenfold larger than those seen in the absence of network activity. In addition, the data indicate that these characteristics are attributable mostly to network activity, and that inhibitory conductances account for most of the large membrane conductance. The consequences of this strong synaptic bombardment are considered in the next section.

11.3 Computational Consequences of High-Conductance States

11.3.1 Models of High-Conductance States

Since several decades, theoretical studies have been designed to understand the impact of noise on the integrative properties of neurons. The notion of high-conductance state itself, as well as the fact that neurons could integrate differently in such states, was first proposed by modeling studies. By integrating the sustained synaptic conductance arising from network activity into models, Barrett (1975) for motoneurons, and later Holmes and Woody (1989) for pyramidal cells, predicted that synaptic activity could have a profound impact on dendritic integration. This theme was then investigated using biophysically and morphologically more precise models in cortex (Bernander et al., 1991; Destexhe and Paré, 1999) and cerebellum (Rapp et al., 1992; De Schutter and Bower, 1994). Such models have predicted a number of computational consequences of background activity and high-conductance states characterizing the in vivo state in neurons which will be briefly summarized in the next section.

In addition to morphologically precise computational models, a large number of theoretical studies have also designed simplified and mathematically treatable models to study the effect of noise on neurons. Synaptic activity is commonly modeled by a source of current noise in the neuron (Levitan et al., 1968; Tuckwell, 1988), which leads to a description of the membrane potential in terms of a stochastic process. However, this type of model is too simple to account for the effect of conductances, in particular high-conductance states. For that reason, more recently, another approach was followed which modeled background activity by fluctuating conductances instead of fluctuating currents (Destexhe et al., 2001). In this case, the synaptic conductances are stochastic processes, which, in turn, give rise to a stochastic V_m dynamics. The advantage of this representation is that the high-conductance state of the membrane can be directly reproduced and modulated, for instance, through the independent control of the conductance mean and the variance. The realism and computational simplicity of these models also enable injection into real neurons in order to recreate high-conductance states artificially using the dynamic-clamp technique (reviewed in Destexhe and Bal, 2009).

Another advantage is that such models are simple enough to allow analytical treatment. Various mathematical studies of the firing dynamics of neurons with conductance-based inputs were performed (see for example Burkitt et al., 2003; Moreno-Bote and Parga, 2005; Muller et al., 2007) and have consequences on network dynamics with conductance-based inputs (Meffin et al., 2004; see also Shelley et al., 2002). The VmD method mentioned above is also a direct consequence of this mathematical tractability (Rudolph et al., 2004).

11.3.2 Impact of High-Conductance States on Integrative Properties

Computational models have predicted several interesting computational consequences of high-conductance states and synaptic noise (Rudolph and Destexhe, 2003a; reviewed in Destexhe et al., 2003; Destexhe and Rudolph-Lilith, 2012).

11.3.2.1 Probabilistic Responses

A first consequence of the presence of synaptic noise is that the V_m behaves stochastically and, therefore, neuronal responses in high-conductance states are highly

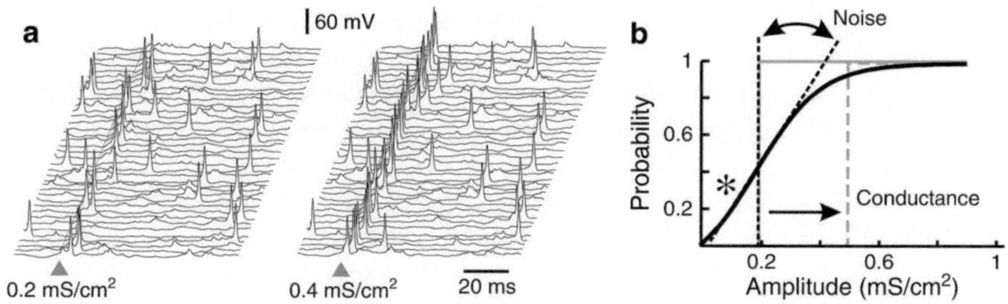

Fig. 11.2 Increased responsiveness in the presence of synaptic noise. (**a**) In the presence of synaptic noise, the response to additional inputs is highly variable, and spikes are evoked at different latencies at each trial (40 trials are shown for two different input amplitudes; modified from Hô and Destexhe, 2000). (**b**) Modulation of the response curve. The response curve (probability of evoking spikes as a function of input amplitude) is all-or-none in quiescent conditions (*gray*). Decomposing synaptic activity in "conductance" and "noise" components shows that conductance shifts the curve rightward (*gray dashed*), while the noise changes its slope (gain modulation; *black*). The response of the neuron is a combination of these effects, showing enhanced responsiveness (*star*) for otherwise subthreshold input regime

variable (Fig. 11.2a). In such states, it is necessary to use repeated trials for any given stimulus, and the appropriate measure of the response is to compute the probability of emitting spikes. The use of such probabilistic measures are well known in in vivo electrophysiology, where routinely "post-stimulus time histograms" (PSTH) from data are calculated. Here, integrating the response (total "output" of the neuron) after a stimulus yields the total probability that a spike is emitted in response to the given stimulus (Fig. 11.2b).

11.3.2.2 Noise-Induced Enhanced Responsiveness

An important consequence of high-conductance states is not only the transformation of neurons into probabilistic devices, but also the profound impact on their response properties. The response curve (or transfer function), which is obtained by representing the total response probability (integrated over time after stimulus) against stimulus amplitude, is all-or-none for a neuron not subjected to stochastic synaptic activity. This behavior reflects the presence of a fixed spiking threshold (Fig. 11.2b, gray), in which case the emitted spike can only tell whether the stimulus is larger than the threshold. In the presence of synaptic noise, however, the response curve is qualitatively different. It is no longer step-like, but spans a whole range of input amplitudes (Fig. 11.2b, black). In this case, the probability of spiking is indicative of the whole range of input amplitude. Specifically, for small-amplitude inputs (those in the physiological range), which are normally subthreshold, the neuron's response probability is enhanced (Fig. 11.2b, star). This enhanced responsiveness is a very robust feature of neurons in the presence of synaptic background activity

(Hô and Destexhe, 2000; Shu et al., 2003). Moreover, the shape of the response function can be altered by changes in the statistics of the background activity. For that reason, this phenomenon has been coined "gain modulation" (Chance et al., 2002), reflecting the fact that the slope of the response curve is modulated by synaptic noise.

Using computational models to independently control the total amount of conductance, and the amount of fluctuations, it was possible to determine their respective role. The conductance alone shifts the response curve (Fig. 11.2b, rightward arrow), while the noise component alone modulates the slope (gain) of the response curve (Hô and Destexhe, 2000; Chance et al., 2002; Shu et al., 2003; Mitchell and Silver, 2003; Prescott and De Koninck, 2003). It is important to note that the type of modulation by noise will depend strongly on the intrinsic properties of the neurons. For that reason, also an inverse gain modulation can be observed (Fellous et al., 2003) and may be explained by potassium conductances (Higgs et al., 2006). Similarly, the dual response (burst vs. single-spike) of thalamic relay neurons is also strongly affected by the presence of synaptic noise, and the two modes may no longer be distinguishable (Wolfart et al., 2005).

It is important to note that the phenomenon of enhanced responsiveness is similar to stochastic resonance phenomena, which have been thoroughly studied by physicists (reviewed in Gammaitoni et al., 1998; Wiesenfeld and Moss, 1995). Stochastic resonance is a noise-induced enhancement of the signal-to-noise ratio in nonlinear systems. It typically presents itself as a peak in the signal-to-noise ratio when the latter is considered as a function of the noise amplitude. Thus, the system appears to "resonate" or to respond optimally for an intermediate but non-vanishing amount of noise. While neurons can also show such behavior when subjected to noise (Levin and Miller, 1996; Stacey and Durand, 2000), the situation is more complex than for classical stochastic resonance phenomena, because in neurons the noise sources are synaptic conductances, and these conductances lead to an additional shunting effect of the cellular membrane (see details in Rudolph and Destexhe, 2001b). As will be shown below, such a modulation of neuronal responsiveness by conductance noise is the basis for the explanation of other neuronal response properties.

11.3.2.3 Noisy Dendrites: Equalization of Synaptic Efficacies

Focusing on the dendrites of neocortical neurons, the presence of massive synaptic bombardment may fundamentally change dendritic integration properties, as illustrated in Fig. 11.3. An inherent property of neuronal dendrites and other electrical cable structures is the attenuation of voltage. This is true in particular for pyramidal neurons: synaptic inputs can experience strong attenuation in the neuron at rest (Fig. 11.3a, left). If the high-conductance state of the membrane is integrated as a static conductance component through an increase in the leak conductance of the membrane, the attenuation is much more severe (Fig. 11.3a, middle): already

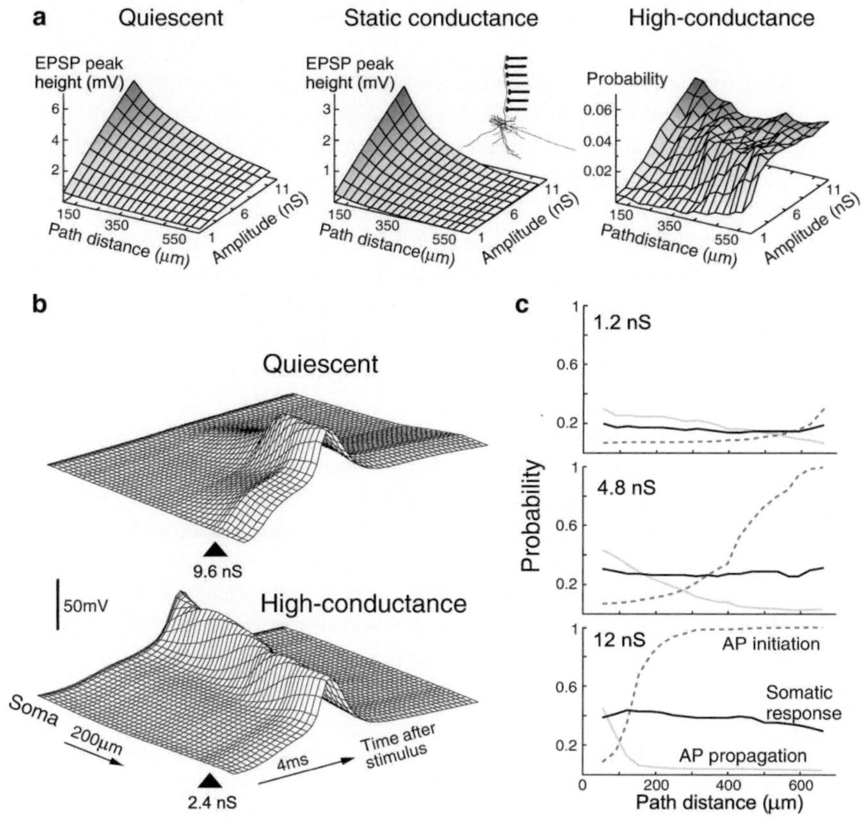

Fig. 11.3 Location independence of synaptic inputs in high-conductance states. (**a**) Somatic response amplitudes for excitatory synaptic inputs located at different positions in dendrites (*inset*). In a quiescent neuron (*left*), distal inputs are moderately attenuated, as predicted by cable theory. With a static high conductance (*middle*), this attenuation is stronger. By including both conductance and noise (*right*), the efficacy of synaptic inputs (defined as the total probability that a synaptic stimulus evokes a somatic spike) was weakly dependent on the dendritic location of the synapse. (**b**) Illustration of forward propagating spike in a quiescent neuron (*top*), and in the presence of background activity (*bottom*). In quiescent conditions, the propagation of the spike is fragile and might be terminated before reaching the soma. In contrast, under in vivo-like conditions, dendritic spikes were found to reliably propagate up to the soma, even for small stimulus amplitudes. (**c**) Comparison of the probability of evoking a dendritic spike (AP initiation) and the probability that an evoked spike translated into a somatic-axonal spike (AP propagation). Both were represented as a function of the location of the stimulus for three stimulus amplitudes. In all cases, the probability of stimulus-evoked somatic spikes, which was obtained by multiplying the probability for AP initiation and propagation, was found to be nearly location independent (modified from Rudolph and Destexhe, 2003a; see this paper for details about the model)

synaptic inputs at distances of a few hundred microns from the soma are almost totally attenuated, and resulting post-synaptic potentials (PSPs) at the soma are undetectable, a phenomenon which is perfectly predictable by cable theory. Remarkably, however, if the full high-conductance state is simulated, the spiking probability shows a surprisingly low dependence on the location of inputs in dendrites (Fig. 11.3a, right).

For the explanation of this equalization of synaptic efficacy, the dendritic excitability, combined with noise, plays a critical role. In quiescent conditions, synaptic inputs arising in distal dendrites can elicit a local dendritic spike, but such a spike is hard to evoke and typically does not propagate well across the dendritic structure (Fig. 11.3b, top). With synaptic noise, the V_m activity is highly variable, but as was outlined above, the presence of noise can boost the effect of small inputs. Indeed, numerical simulations show that there is a small probability that evoked spikes propagate all the way to the soma (Rudolph and Destexhe, 2003a; Fig. 11.3b, bottom). The probability that a local dendritic spike propagates to the soma is therefore nonzero, although it would be zero in a quiescent neuron, similar to the enhanced responsiveness of Fig. 11.2b.

The situation is, however, more complex here because the probability to evoke and propagate spikes depends as well on the position in the dendrite. The probability of evoking a somatic spike increases with distance (Fig. 11.3c, dotted curves) and, thus, is higher for more distal inputs, because distal branches have a higher local input resistance caused by a smaller dendritic radius. Conversely, the probability that an evoked dendritic spikes propagates towards the soma is inversely proportional to distance (Fig. 11.3c, light gray): it is high for positions close to the soma, but decreases with distance as there is a higher chance that the evoked AP fails to propagate.

The probability for a dendritic input to evoke a somatic spike is given by multiplying these two probabilities. Because of their inverse distance dependence, the product of these two probabilities is necessarily less dependent on location (Fig. 11.3c, black). Remarkably, although different intensities of synaptic bombardment give different profiles for the distance dependence of the probabilities of evoking and propagating spikes, their product is in all cases almost independent on distance (Fig. 11.3c). Thus, according to this "stochastic integrative mode" (Rudolph and Destexhe, 2003a), the neuron could solve one long-standing problem, namely how to equally integrate inputs situated at different locations in extended dendritic trees. This equalization mechanism depends on both intrinsic properties (dendritic excitability) and the presence of synaptic noise. It is also seen for different dendritic morphologies (Fig. 11.4; Rudolph and Destexhe, 2003a).

11.3.2.4 Sharper Temporal Processing

A major consequence of the presence of synaptic bombardment is that it will greatly affect temporal processing. The large conductance is necessarily associated with a reduced membrane time constant, which is visible in the faster response to injected current (Fig. 11.1c, averaged traces). As proposed more than 30 years ago (Barrett, 1975), this reduction in the membrane time constant should favor finer temporal discrimination (Holmes and Woody, 1989; Bernander et al., 1991; Destexhe and Paré, 1999). In excitable dendrites, small membrane time constants

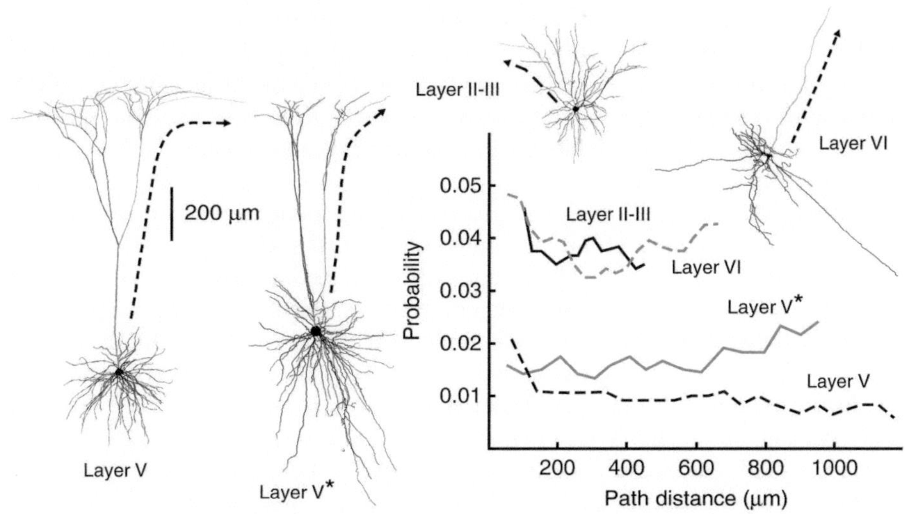

Fig. 11.4 Location independence for different dendritic morphologies. The probability of somatic response to AMPA stimulation (12 nS amplitude) is indicated for different dendritic sites (corresponding branches are indicated by *dashed arrows*; equivalent electrophysiological parameters and procedures as in Fig. 11.3c) for four different cells (one layer II–III, two layer V, and one layer VI), based on cellular reconstructions from cat cortex (Douglas et al. 1991; Contreras et al. 1997; modified from Rudolph and Destexhe, 2003a; see this paper for details about the model)

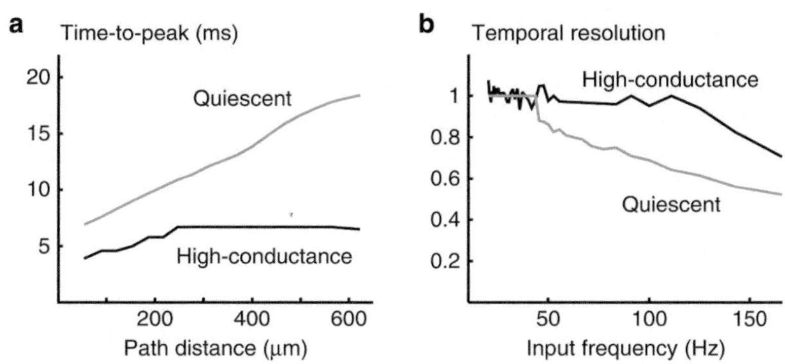

Fig. 11.5 Increased temporal resolution with synaptic noise. (**a**) Location-dependence of the timing of EPSPs. A similar stimulation paradigm as in Fig. 11.3a was used, except that the time-to-peak of averaged synaptic responses is represented against distance to soma. In quiescent conditions (*gray*), the time-to-peak of EPSPs increased approximately linearly with distance from soma. This location dependence was markedly reduced in the high-conductance state (*black*). (**b**) Enhanced temporal resolution of synaptic inputs during high-conductance state. Excitatory synaptic inputs were delivered at different frequencies. The temporal resolution was quantified by computing the ratio between the stimulus interval and the mode of the corresponding interspike interval histogram. In quiescent conditions (*gray*), the model could respond reliably up to a stimulus frequency of about 50 Hz. In high-conductance states (*black*), the frequency following was reliable up to much higher frequencies (beyond 100 Hz). (**a**) modified from Rudolph and Destexhe, 2003a; (**b**) modified from Destexhe et al., 2003; see these papers for details about the model

also promote fast-propagating action potentials, resulting in a reduced location-dependence of EPSP timing (Fig. 11.5a; Rudolph and Destexhe, 2003a). Here, synaptic noise seems to set the dendrite into a fast-conducting mode, in which the timing of inputs shows reduced dependence on their distance from the soma. This remarkable property is likely to facilitate the association of synaptic inputs arising at distant locations from each other.

The mechanisms underlying this fast-conducting mode depend on several factors. First, the reduction of the membrane time constant due to the high conductance, which can be modeled with an equivalent static conductance, shows a partial increase in the conduction speed of dendritic spikes. Second, varying the presence of sodium conductances also led to intermediate effects on the timing due to the presence of dendritic spikes. Combining both of these effects suggests that this fast conducting mode is due to fast propagation of dendritic spikes in a membrane of fast time constant (see details in Rudolph and Destexhe, 2003a).

Neurons in high-conductance states also display a superior ability to distinguish and process high-frequency inputs, when compared to low-conductance states. This is illustrated in Fig. 11.5b, which shows the temporal resolution of a neuron represented against the input frequency. In quiescent or low-conductance states, neurons can follow inputs (i.e., produce a spike) up to a maximal frequency which is typically around 40–50 Hz (Fig. 11.5b, gray). With synaptic noise, the neuron can lock its response to larger frequencies (up to more than 100 Hz in the example of Fig. 11.5b, black). This property is attributable to the smaller time constant associated with high membrane conductances.

Modeling studies have explored other computational advantages of synaptic noise on temporal processing. If both excitatory and inhibitory conductances are large during high-conductance states, slight variations of either excitation or inhibition can be very effective in modifying spiking probability. As a consequence, neurons can reliably detect faint changes in temporal correlation of their synaptic inputs (Halliday, 1999; Salinas and Sejnowski, 2000; Rudolph and Destexhe, 2001a). This type of response is interesting, because changes in correlation do not change the average conductance nor the average V_m, but they uniquely appear as changes of the level of fluctuations (variances) of the conductances and of the V_m. In this case, neurons respond to a signal which is not carried by the mean activity of conductances, which, thus, constitutes an example of a paradigm which cannot be modeled by rate-based models.

Finally, high-conductance states also impact on the operating mode of cortical neurons. Neurons can operate either as coincidence detectors or as temporal integrators, which determine whether the cortex encodes information by the precise timing of spikes, or by average firing rates. Modeling studies monitored the spike output of neurons submitted to a full spectrum of multisynaptic input patterns, from highly coincident to temporally dispersed (Maršálek et al., 1997; Kisley and Gerstein, 1999). It was found that, in general, the spike output jitter tends to be lower than the input jitter, indicating that neurons tend to synchronize the responses and reduce their temporal dispersion. However, we found that this conclusion is different when simulating high-conductance states. In this case, the temporal dispersion is nearly identical between input and output (Rudolph and Destexhe, 2003b). This suggests that in high-conductance states, both operating modes can be used robustly and in parallel.

11.4 Conclusions

In this chapter, we have reviewed models of the high-conductance states in cortical neurons, and their predicted consequences on dendritic integration. We have first overviewed experimental results (Sect. 11.2), showing that cortical neurons in vivo are subject to a considerable amount of synaptic noise, and experience what is called a high-conductance state. In Sect. 11.3, we have reviewed models of synaptic noise which have predicted a number of computational consequences of synaptic noise and high-conductance states on neuronal processing.

It is important to note that models are not limited to the sole task of predicting consequences on integrative properties. Simple models, such as the point-conductance model of synaptic noise (Destexhe et al., 2001), can be used to add artificial synaptic noise in neurons in vitro using the dynamic-clamp technique. This mixed modeling-experimental technique is of primary importance, because it allows one to directly test the predictions of the models in real neurons. It can also be used to realize experiments that would not be possible in vivo, such as controlling the amount of synaptic noise, or controlling independently the amount of mean conductances and their fluctuations. Another important application of the point-conductance model is that it can form the basis for methods to analyze experimental data. Such methods include the estimation of synaptic conductances (the VmD method; Rudolph et al., 2004), the estimation of spike-triggered average conductances (Pospischil et al., 2007), or the estimation of synaptic parameters from the power spectrum of the V_m (Destexhe and Rudolph, 2004). These quantities are evaluated from the sole knowledge of the V_m activity, which makes it applicable to standard intracellular recording conditions. In the past years, each of these methods was tested using computational models, as well as in real neurons using the dynamic-clamp technique (reviewed in Piwkowska et al., 2008).

One main consequence of high-conductance states is that the presence of synaptic noise drastically affects the responsiveness of the neurons by changing their response curve or transfer function. The finding that the gain and responsiveness of neurons are enhanced by synaptic noise was first reported from modeling studies (Hô and Destexhe, 2000), and then investigated experimentally using dynamic-clamp injection of in vivo-like synaptic noise (Destexhe et al., 2001; Chance et al., 2002; Fellous et al., 2003; Prescott and De Koninck, 2003; Shu et al., 2003; Wolfart et al., 2005; Higgs et al., 2006; Piwkowska et al., 2008), thus confirming some of the predictions formulated by models. A fascinating possible consequence is that the enhanced responsiveness due to synaptic noise could be used as an attentional mechanism (Hô and Destexhe, 2000; Shu et al., 2003). By modulating the amount of synaptic noise, it should be possible to switch entire networks from unresponsive to responsive states, an intriguing possibility which should be investigated by designing appropriate experiments and models.

Given these drastic effects of noise on neurons, the fact that synaptic noise has also profound impact on dendritic processing comes to no surprise. Here, the presence of noise affects not only the initiation and propagation of dendritic spikes, but its consequences are very different in different parts of the dendritic tree. Remarkably, it was found that there is a compensation between two opposite distance dependence, leading to a roughly location-independent effect of synaptic inputs on somatic spiking (Fig. 11.3c). Thus, the probabilistic dendrites seem to behave more "democratically," giving to each synaptic input a roughly equal vote on somatic spiking.

This remarkable property still awaits to be tested experimentally, but such a test is not easy. It would require to modulate the conductance state of the whole extent of the dendrites, and control synaptic inputs of similar weights at different positions. These two conditions constitute clear experimental challenges. However, recently it was shown that it is possible to maintain irregular states of activity (up-down states) in cortical slices (Sanchez-Vives and McCormick, 2000; Silberberg et al., 2004), which would be a possible way towards performing such experiments. From a mathematical point of view, the almost perfect compensation observed for very different distance-dependent profiles (Fig. 11.3c) suggests a simple theoretical explanation for the observed effects. Specifically, it should be possible to reproduce the present observations using simplified models, endowed, for example, with multicompartment tapering dendrites and integrate-and-fire mechanism with noise. This, as well, constitutes a possible direction for future theoretical studies.

Note that the location independence was only studied for isolated excitatory synaptic inputs, and should still be investigated for more complex input combinations, such as multiple excitations and combined excitatory and inhibitory inputs. In such cases, we expect that the proximity of the different inputs on the dendrite will play an important role and may reveal local aspects of dendritic computations, as analyzed previously (Mel, 1994).

Models also showed that synaptic noise enhances the temporal resolution of dendrites, both by setting a fast propagating mode and by enhancing the ability of the neuron to follow high-frequency inputs (Fig. 11.5). Interestingly, the latter type of temporal processing was the first consequence put forward by modeling studies (Barrett, 1975; Holmes and Woody, 1989; Bernander et al., 1991; Rudolph and Destexhe, 2003a,b). Neurons in high-conductance states necessarily have a faster membrane which allows sharper temporal processing. Remarkably, the fast-propagating mode would make the neuron more suitable to act as a coincidence detector. Surprisingly, few experimental studies have investigated these temporal aspects, which constitute other interesting directions for future investigations.

Finally, we did not review here dendrites with calcium currents, although this constitutes an interesting subject to investigate in the presence of synaptic noise as well. It was shown previously that synaptic noise has strong effect on bursting mediated by T-type calcium currents in thalamic relay neurons (Wolfart et al., 2005).

This study revealed that the classic duality of "tonic" and "burst" modes of firing in thalamic relay cells no longer holds in high-conductance states, because, with synaptic noise, bursts and single-spikes participate to all responses. Interestingly, it was observed that intrinsically bursting neurons are much less frequent in vivo compared to in vitro recordings (reviewed in Steriade et al., 1993). Noise modulation of the calcium current in dendrites, as in thalamic neurons, as well as the high synaptic conductances, which are likely to compete with dendritic calcium conductances, may explain why cortical cells are more reticent to burst under in vivo conditions.

In conclusion, modeling studies predict that noisy membranes integrate according to different rules. This concept also applies to dendrites and may be responsible for properties such as equalization of synaptic efficacies and finer temporal processing. Thus, like single neurons, the presence of synaptic noise may also be beneficial to dendrites.

Acknowledgements The experimental data reviewed here were obtained by Drs. D. Paré and I. Timofeev. This research was supported by the Centre National de la Recherche Scientifique (CNRS, France), Agence Nationale de la Recherche (ANR), Future and Emerging Technologies (FET, European Union; FACETS and BrainScales projects) and the Human Frontier Science Program (HFSP). Additional information is available at http://cns.iaf.cnrs-gif.fr.

References

Baranyi A, Szente MB, Woody CD (1993) Electrophysiological characterization of different types of neurons recorded in vivo in the motor cortex of the cat. II. Membrane parameters, action potentials, current-induced voltage responses and electrotonic structures. J Neurophysiol 69:1865–1879

Barrett JN (1975) Motoneuron dendrites: role in synaptic integration. Fed Proc 34:1398–1407

Bernander O, Douglas RJ, Martin KA, Koch C (1991) Synaptic background activity influences spatiotemporal integration in single pyramidal cells. Proc Natl Acad Sci USA 88:11569–11573

Braitenberg V, Schüz A (1998) Cortex: statistics and geometry of neuronal connectivity, 2nd edn. Springer, Berlin

Branco T, Häusser M (2011) Synaptic integration gradients in single cortical pyramidal cell dendrites. Neuron 69:885–892

Branco T, Clark BA, Häusser M (2010) Dendritic discrimination of temporal input sequences in cortical neurons. Science 329:1671–1675

Burkitt AN, Meffin H, Grayden DB (2003) Study of neuronal gain in a conductance-based leaky integrate-and-fire neuron model with balanced excitatory and inhibitory synaptic input. Biol Cybern 89:119–125

Chance FS, Abbott LF, Reyes AD (2002) Gain modulation from background synaptic input. Neuron 35:773–782

Contreras D, Destexhe A, Steriade M (1997) Intracellular and computational characterization of the intracortical inhibitory control of synchronized thalamic inputs in vivo. J Neurophysiol 78:335–350

Contreras D, Timofeev I, Steriade M (1996) Mechanisms of long lasting hyperpolarizations underlying slow sleep oscillations in cat corticothalamic networks. J Physiol 494:251–264

DeFelipe J, Fariñas I (1992) The pyramidal neuron of the cerebral cortex: morphological and chemical characteristics of the synaptic inputs. Prog Neurobiol 39:563–607

De Schutter E, Bower JM (1994) Simulated responses of cerebellar Purkinje cells are independent of the dendritic location of granule cell synaptic inputs. Proc Natl Acad Sci USA 91:4736–4740
Destexhe A (2007) High-conductance state. Scholarpedia 2(11):1341
Destexhe A, Bal T (eds.) (2009) The dynamic-clamp: from principles to applications. Springer, New York
Destexhe A, Contreras D (2006) Neuronal computations with stochastic network states. Science 314:85–90
Destexhe A, Paré D (1999) Impact of network activity on the integrative properties of neocortical pyramidal neurons in vivo. J Neurophysiol 81:1531–1547
Destexhe A, Rudolph M (2004) Extracting information from the power spectrum of synaptic noise. J Comput Neurosci 17:327–345
Destexhe A, Rudolph-Lilith M (2012) Neuronal noise. Springer, New York
Destexhe A, Rudolph M, Fellous J-M, Sejnowski TJ (2001) Fluctuating synaptic conductances recreate in-vivo-like activity in neocortical neurons. Neuroscience 107:13–24
Destexhe A, Rudolph M, Paré D (2003) The high-conductance state of neocortical neurons in vivo. Nat Rev Neurosci 4:739–751
Destexhe A, Hughes SW, Rudolph M, Crunelli V (2007) Are corticothalamic 'up' states fragments of wakefulness? Trends Neurosci 30:334–342
Douglas RJ, Martin KA, Whitteridge D (1991) An intracellular analysis of the visual responses of neurones in cat visual cortex. J Physiol 440:659–696
Evarts EV (1964) Temporal patterns of discharge of pyramidal tract neurons during sleep and waking in the monkey. J Neurophysiol 27:152–171
Fellous JM, Rudolph M, Destexhe A, Sejnowski TJ (2003) Synaptic background noise controls the input/output characteristics of single cells in an in vitro model of in vivo activity. Neuroscience 122:811–829
Gammaitoni L, Hanggi P, Jung P, Marchesoni F (1998) Stochastic resonance. Rev Mod Phys 70:223–287
Halliday DM (1999) Weak, stochastic temporal correlation of large-scale synaptic inputs is a major determinant of neuronal bandwidth. Neural Comput 12:693–707
Helmchen F, Svoboda K, Denk W and Tank DW (1999) In vivo dendritic calcium dynamics in deep-layer cortical pyramidal neurons. Nat Neurosci 2:989–996
Higgs MH, Slee SJ, Spain WJ (2006) Diversity of gain modulation by noise in neocortical neurons: regulation by the slow after-hyperpolarization conductance. J Neurosci 26:8787–8799
Hô N, Destexhe A (2000) Synaptic background activity enhances the responsiveness of neocortical pyramidal neurons. J Neurophysiol 84:1488–1496
Holmes WR, Woody CD (1989) Effects of uniform and non-uniform synaptic "activation-distributions" on the cable properties of modeled cortical pyramidal neurons. Brain Res 505:12–22
Kisley MA, Gerstein GL (1999) The continuum of operating modes for a passive model neuron. Neural Comput 11:1139–1154
Lavzin M, Rapoport S, Polsky A, Garion L, Schiller J (2012) Nonlinear dendritic processing determines angular tuning of barrel cortex neurons in vivo. Nature 490:397–401
Leger J-F, Stern EA, Aertsen A, Heck D (2005) Synaptic integration in rat frontal cortex shaped by network activity. J Neurophysiol 93:281–293
Levin JE, Miller JP (1996) Broadband neural coding in the cricket sensory system enhanced by stochastic resonance. Nature 380:165–168
Levitan H, Segundo JP, Moore GP, Perkel DH (1968) Statistical analysis of membrane potential fluctuations. Biophys J 8:1256–1274
Maršálek P, Koch C, Maunsell J (1997) On the relationship between synaptic input and spike output jitter in individual neurons. Proc Natl Acad Sci USA 94:735–740
Matsumura M, Cope T, Fetz EE (1988) Sustained excitatory synaptic input to motor cortex neurons in awake animals revealed by intracellular recording of membrane potentials. Exp Brain Res 70:463–469

McCormick DA, Bal T (1997) Sleep and arousal: thalamocortical mechanisms. Annu Rev Neurosci 20:185–215
Meffin H, Burkitt AN, Grayden DB (2004) An analytical model for the "large, fluctuating synaptic conductance state" typical of neocortical neurons in vivo. J Comput Neurosci 16:159–175
Mel BW (1994) Information processing in dendritic trees. Neural Comput 6:1031–1085
Metherate R, Ashe JH (1993) Ionic flux contributions to neocortical slow waves and nucleus basalis-mediated activation: whole-cell recordings in vivo. J Neurosci 13:5312–5323
Mitchell SJ, Silver RA (2003) Shunting inhibition modulates neuronal gain during synaptic excitation. Neuron 38:433–445
Monier C, Fournier J, Frégnac Y (2008) In vitro and in vivo measures of evoked excitatory and inhibitory conductance dynamics in sensory cortices. J Neurosci Meth 169:323–365
Moreno-Bote R, Parga N (2005) Membrane potential and response properties of populations of cortical neurons in the high conductance state. Phys Rev Lett 94:088103
Muller E, Buesing L, Schemmel J, Meier K (2007) Spike-frequency adapting neural ensembles: beyond mean adaptation and renewal theories. Neural Comput 19:2958–3010
Paré D, Shink E, Gaudreau H, Destexhe A, Lang EJ (1998) Impact of spontaneous synaptic activity on the resting properties of cat neocortical neurons in vivo. J Neurophysiol 79:1450–1460
Petersen CC, Hahn TT, Mehta M, Grinvald A, Sakmann B (2003) Interaction of sensory responses with spontaneous depolarization in layer 2/3 barrel cortex. Proc Natl Acad Sci USA 100:13638–13643
Piwkowska Z, Pospischil M, Brette R, Sliwa J, Rudolph-Lilith M, Bal T, Destexhe A (2008) Characterizing synaptic conductance fluctuations in cortical neurons and their influence on spike generation. J Neurosci Meth 169:302–322
Pospischil M, Piwkowska Z, Rudolph M, Bal T, Destexhe A (2007) Calculating event-triggered average synaptic conductances from the membrane potential. J Neurophysiol 97:2544–2552
Prescott SA, De Koninck Y (2003) Gain control of firing rate by shunting inhibition: roles of synaptic noise and dendritic saturation. Proc Natl Acad Sci USA 100:2076–2081
Rapp M, Yarom Y, Segev I (1992) The impact of parallel fiber background activity on the cable properties of cerebellar Purkinje cells. Neural Comput 4:518–533
Rudolph M, Destexhe A (2001a) Correlation detection and resonance in neural systems with distributed noise sources. Phys Rev Lett 86:3662–3665
Rudolph M, Destexhe A (2001b) Do neocortical pyramidal neurons display stochastic resonance? J Comput Neurosci 11:19–42
Rudolph M, Destexhe A (2003a) A fast-conducting, stochastic integrative mode for neocortical neurons in vivo. J Neurosci 23:2466–2476
Rudolph M, Destexhe A (2003b) Tuning neocortical pyramidal neurons between integrators and coincidence detectors. J Comput Neurosci 14:239–251
Rudolph M, Destexhe A (2003c) Characterization of subthreshold voltage fluctuations in neuronal membranes. Neural Comput 15:2577–2618
Rudolph M, Destexhe A (2005) An extended analytic expression for the membrane potential distribution of conductance-based synaptic noise. Neural Comput 17:2301–2315
Rudolph M, Piwkowska Z, Badoual M, Bal T, Destexhe A (2004) A method to estimate synaptic conductances from membrane potential fluctuations. J Neurophysiol 91:2884–2896
Rudolph M, Pelletier J-G, Paré D, Destexhe A (2005) Characterization of synaptic conductances and integrative properties during electrically-induced EEG-activated states in neocortical neurons in vivo. J Neurophysiol 94:2805–2821
Rudolph M, Pospischil M, Timofeev I, Destexhe A (2007) Inhibition controls action potential generation in awake and sleeping cat cortex. J Neurosci 27:5280–5290
Salinas E, Sejnowski TJ (2000) Impact of correlated synaptic input on output firing rate and variability in simple neuronal models. J Neurosci 20:6193–6209
Sanchez-Vives MV, McCormick DA (2000) Cellular and network mechanisms of rhythmic recurrent activity in neocortex. Nat Neurosci 10:1027–1034

Shelley M, McLaughlin D, Shapley R, Wielaard J (2002) States of high conductance in a large-scale model of the visual cortex. J Comput Neurosci 13:93–109
Shu Y, Hasenstaub A, Badoual M, Bal T, McCormick DA (2003) Barrages of synaptic activity control the gain and sensitivity of cortical neurons. J Neurosci 23:10388–10401
Silberberg G, Wu C, Markram H (2004) Synaptic dynamics control the timing of neuronal excitation in the activated neocortical microcircuit. J Physiol 556:19–27
Stacey WC, Durand DM (2000) Stochastic resonance improves signal detection in hippocampal CA1 neurons. J Neurophysiol 83:1394–1402
Steriade M (2001) Impact of network activities on neuronal properties in corticothalamic systems. J Neurophysiol 86:1–39
Steriade M (2003) Neuronal substrates of sleep and epilepsy. Cambridge University Press, Cambridge
Steriade M, McCarley RW (1990) Brainstem control of wakefulness and sleep. Plenum Press, New York
Steriade M, Amzica F, Nuñez A (1993) Cholinergic and noradrenergic modulation of the slow (∼0.3 Hz) oscillation in neocortical cells. J Neurophysiol 70:1384–1400
Steriade M, Timofeev I, Grenier F (2001) Natural waking and sleep states: a view from inside neocortical neurons. J Neurophysiol 85:1969–1985
Timofeev I, Grenier F, Steriade M (2001) Disfacilitation and active inhibition in the neocortex during the natural sleep-wake cycle: an intracellular study. Proc Natl Acad Sci USA 98:1924–1929
Tuckwell HC (1988) Introduction to theoretical neurobiology. Cambridge University Press, Cambridge
Waters J, Helmchen F (2006) Background synaptic activity is sparse in neocortex. J Neurosci 26:8267–8277
Wiesenfeld K, Moss F (1995) Stochastic resonance and the benefits of noise: from ice ages to crayfish and SQUIDs. Nature 373:33–36
Wolfart J, Debay D, Le Masson G, Destexhe A, Bal T (2005) Synaptic background activity controls spike transfer from thalamus to cortex. Nat Neurosci 8:1760–1767
Zou Q, Rudolph M, Roy N, Sanchez-Vives M, Contreras D, Destexhe A (2005) Reconstructing synaptic background activity from conductance measurements in vivo. Neurocomputing 65:673–678

Chapter 12
Distributed Parallel Processing in Retinal Amacrine Cells

Jeffrey S. Diamond and William N. Grimes

Abstract In most central neurons, dendritic arbors integrate and shape synaptic input before it reaches the soma and axon initial segment, where action potentials are generated to relay the processed information down the axon to distant synaptic outputs. These input and output regions are typically clearly segregated into separate areas of the cell. One fascinating exception to this rule occurs in the retina, where amacrine interneurons typically receive synaptic inputs and make synaptic outputs within the same dendritic arbor. This morphological multiplexing, more typically observed in invertebrate neurons, adds a whole new dimension to dendritic processing, one that has recently been studied in greater detail, both at the level of synaptic mechanism and in the context of visual processing. Here we review recent work examining dendritic input/output signaling in two different amacrine cell subtypes that play distinct roles in visual processing in the retina.

12.1 Introduction

Amacrine cells, which constitute the most diverse cell class in the vertebrate retina (~40 different subtypes; MacNeil and Masland 1998), provide feedback and feedforward inhibition to bipolar cells and ganglion cells, respectively, to help sculpt early neural representations of the visual world. Although amacrine cells typically exhibit distinct, occasionally spectacular dendritic arbors, most subtypes lack an

J.S. Diamond (✉)
Synaptic Physiology Section, NINDS Intramural Research Program, 35 Convent Drive, Building 45, Room 3C-1000, Bethesda, MD 20892-3701, USA
e-mail: diamond@ninds.nih.gov

W.N. Grimes
Department of Physiology and Biophysics, Howard Hughes Medical Institute, University of Washington, 1705 N.E. Pacific Street, Seattle, WA 98195, USA

H. Cuntz et al. (eds.), *The Computing Dendrite: From Structure to Function*, Springer Series in Computational Neuroscience 11, DOI 10.1007/978-1-4614-8094-5_12,

axon, leading a bemused Ramon y Cajal to name them from the Greek, *a-makrós-inos*: "without long fiber." Despite the general lack of obvious output processes, amacrine cells collectively release a pharmacopeia of different neurotransmitters, including GABA, glycine, acetylcholine, dopamine, and various peptide transmitters, perhaps conjuring an alternative Latin/Greek etymology (ama-crine: "love to secrete"). In most cases, amacrine cells dispatch their synaptic output from the same dendrites onto which they receive their synaptic input, with no obvious route through the soma. This feature appears to have confused Cajal: amacrine cells are typically unaccompanied by the feathered arrows that he used to indicate the direction of information flow in his prescient drawings.

This chapter will focus on two of the best-studied amacrine cells: Starburst amacrine cells (SACs) and A17 amacrine cells. Although they have different morphologies and contribute to very different aspects of network function and visual processing, they both share a common, distinctive feature: their function within the retinal network is thought to rely on compartmentalized input/output signaling within small regions of their dendrites. In the case of SACs, functionally isolated dendritic branches that are precisely connected to ganglion cell dendrites mediate feedforward inhibitory signaling that underlies directional selectivity; in A17s, compartmentalized signaling enables single neurons to provide independent feedback to hundreds of bipolar cells in parallel. In both cases, distinct morphological and biophysical properties combine to optimize local dendritic processing of visual information.

12.2 SACs Compute Direction Selectivity Through Specialized Connectivity and Physiology

One of the most intriguing models for neural computation is direction-selective encoding of motion in the visual system (Fig. 12.1a; Demb 2007). Direction-selective neurons—cells that respond preferentially to motion in a particular direction (Fig. 12.1b)—are observed in both cortex (Hubel and Wiesel 1959), LGN

Fig. 12.1 (continued) across the receptive field in different directions. Responses to the onset of light (ON responses) are shown in *closed symbols*; OFF responses are shown in *open symbols*. From Oesch et al. 2005. (**c**) Original asymmetric feedforward inhibitory mechanism proposed by Barlow and Levick (1965). The asymmetric inhibitory signal causes excitatory input to be vetoed by inhibition only when the stimulus travels in the "null" direction. (**d**) typical SAC morphology (*red*) amidst a mosaic of SAC somata, from mouse retina. From Keeley et al. (2007). (**e**) SAC (*black skeleton*) with output varicosities (*black dots*). Connected DSGC dendritic trees are indicated by *dashed ellipses*, and synapses are color-coded according to the preferred direction of the postsynaptic DSGC. (**f**) Output synapses made by 24 different SACs (relative to each SAC soma, *centered* at the origin), color-coded according to DSGC preferred direction, show that SAC dendrites connect to DSCGs in an antiparallel fashion. From Briggman et al. (2011)

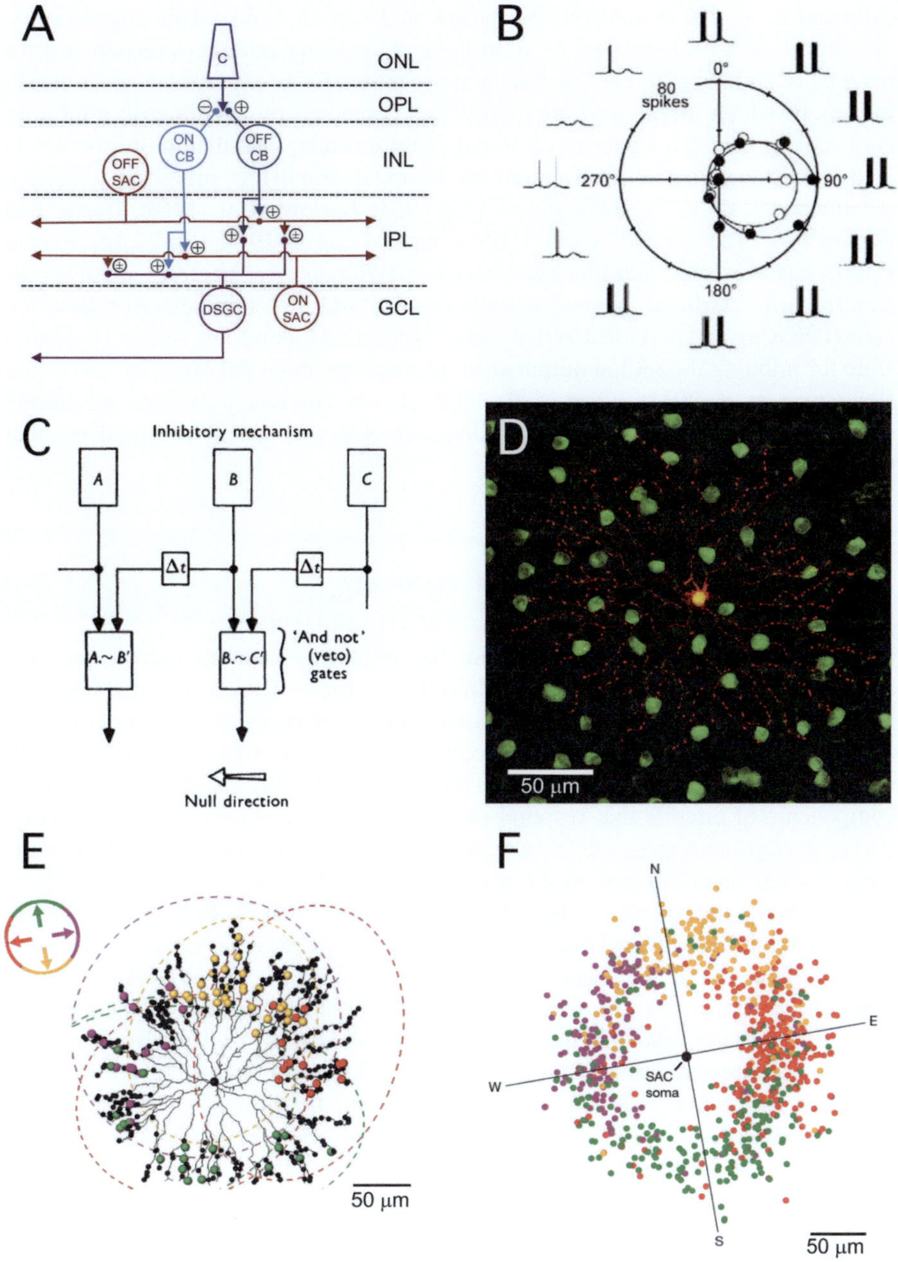

Fig. 12.1 Starburst amacrine cells (SACs) help compute directional selectivity. (**a**) retina schematic showing direction selectivity circuitry. ON and OFF SACs receive glutamatergic input from bipolar cells and make excitatory (cholinergic) and inhibitory (GABAergic) connections to direction-selective ganglion cells (DSGC). Labels for panel **a**: *C* cone, *CB* cone bipolar cell, *ONL* outer nuclear layer, *OPL* outer plexiform layer, *INL* inner nuclear layer, *IPL* inner plexiform layer, *GCL* ganglion cell layer. (**b**) spike responses in an ON-OFF DSGC elicited by bars of light moving

(Marshel et al. 2012), and retina (Barlow and Levick 1965). This suggests that direction-selective signals are used throughout the visual pathways responsible for perception and must, in part, originate in the retina. Early experiments gave rise to models in which directional selectivity was generated by asymmetric inhibitory connectivity between amacrine cells and ganglion cells, such that inhibition would lead excitation in the "null" direction and lag excitation in the "preferred" direction (Maturana et al. 1960; Barlow and Hill 1963; Barlow et al. 1964; Oyster and Barlow 1967; Barlow and Levick 1965; Torre and Poggio 1978; Fig. 12.1c). Recent studies have revealed that this hypothesized asymmetric connectivity involves the dendrites of "starburst" amacrine cells (SACs) and direction-selective ganglion cells (DSGCs), and that localized signaling within SAC dendrites is largely responsible for imbuing the retinal output with information about the direction of motion (Briggman et al. 2011; Euler et al. 2002). In the following sections we briefly outline the major conclusions regarding the SAC-dependent mechanisms that underlie this retinal computation.

12.2.1 SAC Morphology and Physiology

SACs are easily distinguished in mammalian retinas by their exclusive expression of choline acetyltransferase (ChAT; Schmidt et al. 1987). They form a dense mosaic across the retina: each point in retinal space is covered by the dendrites of as many as 70 SACs (Vaney 1984). The SAC cell class comprises two apparently complimentary subtypes: SACs with their cell bodies located in the ganglion cell layer are components of circuits that respond preferentially to brighter moving objects on darker backgrounds (i.e., ON SACs), whereas SACs with their cell bodies in the inner nuclear layer are part of the retinal circuitry that responds preferentially to darker moving objects on brighter backgrounds (OFF SACs). Both SAC subtypes release the inhibitory neurotransmitter, GABA, and the excitatory neurotransmitter, acetylcholine, onto DSGC dendrites (O'Malley et al. 1992), but physiology experiments have shown that the feedforward GABAergic component of the DSGC's input has the strongest directional-selectivity and is most critical to the retina's encoding of directed motion (Fried et al. 2005; Kittila and Massey 1997).

True to their name, individual SACs exhibit a radially symmetric, burst-like morphology (Fig. 12.1d): several primary dendritic trunks plunge from the soma through one-third of the inner plexiform layer (IPL), turn radially (outward) and, within a narrow stratum of the IPL, bifurcate in a stereotypical manner, with branch points increasing in frequency with distance from the soma. ON SAC dendrites occupy a narrow layer of the ON region; OFF SACs occupy a corresponding layer in the OFF region (Fig. 12.1a). Microscopy studies have shown that SACs receive direct glutamatergic input from bipolar cells throughout their dendritic arbors, whereas their GABAergic (and cholinergic) output synapses are expressed only within the more heavily branched distal dendrites (Fig. 12.1e; Brecha et al. 1988; Famiglietti 1991; O'Malley and Masland 1989; Vaney and Young 1988). This asymmetric distribution of input and output synapses along individual dendrites suggests that centrifugally (center → periphery)-propagating signals should trigger synaptic release from distal tips whereas centripetally (periphery → center)-propagating signals would have no proximal output synapses to activate (Tukker et al. 2004; Vaney and Young 1988).

In 2002, Thomas Euler and colleagues directly demonstrated two critical features of SAC physiology (Euler et al. 2002). By monitoring dendritic calcium signals (via 2-photon microscopy) while presenting centripetal and centrifugal motion stimuli to the retinal preparation, they showed (1) that calcium signals in the distal dendrites of SACs are preferentially sensitive to centrifugal motion and (2) that SAC dendrites operate independently. They hypothesized that each SAC comprises multiple, independent computational units (each maximally sensitive to a slightly different direction) and that the retina's sensitivity to directed motion originates within these units.

Additional morphological and biophysical mechanisms appear to augment the directed flow of information and isolation of activity within individual dendrites. Morphologically, even the most proximal part of each radiating dendritic unit is a thin segment that may restrict electrical and biochemical signals to and from the more distal parts of the dendrite (Fig. 12.1d). These proximal dendritic segments connect directly to the soma, not to each other, thereby forming an impedance mismatch with the larger somatic compartments. Consequently, the soma acts as a current sink, effectively isolating dendrites from each other (Goldstein and Rall 1974; Tukker et al. 2004). Distinctive biophysical membrane properties of SAC dendrites, such as TTX-insensitive sodium channels (Oesch and Taylor 2010) and gradients in both synaptic activity (Hausselt et al. 2007) and chloride transport (Gavrikov et al. 2003), may also enhance their preferential response to centrifugal (center → peripheral) stimuli (Taylor and Smith 2012).

12.2.2 Wiring Asymmetries Confer Direction Selectivity onto Motion-Sensitive Signals

Although individual SAC dendrites respond preferentially to light bars moving in a particular direction, recordings from SAC somata show relatively weak direction selectivity (Euler et al. 2002). Moreover, most DSGCs exhibit relatively symmetric dendritic arbors (e.g., Briggman et al. 2011, but see Trenholm et al. 2011). How, then, do two classes of morphologically symmetric neurons give rise to directional selectivity in the four cardinal directions (dorsal, ventral, medial, and lateral; Oyster and Barlow 1967)? The answer requires that SAC dendrites operate independently and that the GABAergic input to a particular DSGC comes primarily from SAC dendrites that are aligned antiparallel to the DSGCs preferred direction.

As mentioned above, direct measurements of intracellular calcium signals within the dendrites of SACs have shown that each radially connected dendritic unit functions independently in response to visual stimuli (Euler et al. 2002), but a detailed assessment of wiring asymmetry remained lacking. This critical aspect of the direction-selective circuitry was recently elucidated using a combination of microscopy techniques (Briggman et al. 2011; Denk and Horstmann 2004). First, ganglion cells were indiscriminately loaded with a calcium-sensitive fluorophore and then imaged (using 2-photon microscopy) while delivering moving stimuli to the retina, enabling a subset of DSGCs and their directional preferences to be identified. The tissue was then fixed and prepared for serial block-face electron microscopy, a new technique that enables high-resolution three-dimensional reconstruction of large volumes ($>10^6$ μm^3) of tissue (Briggman and Denk 2006). From these anatomical data, Briggman et al. (2011) reconstructed the dendritic arbors of several DSGCs from which they recorded directional light responses, as well as many of the SACs making synaptic connections onto those DSGCs. They found that the dendrites of each SAC provided direct GABAergic input to all four DSGC subtypes (Fig. 12.1e) and that the preferred direction of SAC dendrites (i.e., proximal to distal) was indeed antiparallel to the preferred direction of each contacted DSGC (Fig. 12.1f; Fried et al. 2002; Wei et al. 2011).

12.2.3 Remaining Questions

Despite the exciting progress described above, major questions regarding direction selectivity and SAC physiology remain unanswered. For example, what computational role do cholinergic SAC synapses play? Direction-selectivity is not abolished in the absence of cholinergic signaling (Kittila and Massey 1997), suggesting that cholinergic transmission is not required for directional processing. Acetylcholine release exhibits a different threshold compared to GABA release (Lee et al. 2011), suggesting that it may modulate the gain of direction-selective signals, or operate over particular spatial frequencies (Grzywacz et al. 1998). Another mystery concerns the role of the dendrodendritic GABAergic synapses that SACs make with each other. Dendrodendritic inhibition between the tips of antiparallel SAC dendrites could suppress GABA release onto "preferentially responding" DSGCs when a moving bar crosses over the dendritic intersection between a pair of reciprocally connected SACs (Lee and Zhou 2006). Finally, do ON and OFF SACs operate identically, or do they exhibit functional differences that are critical to the computation of directional motion? The answer to this question likely will require exploration of SAC physiology across a broad range of light stimulus conditions and may involve synaptic interactions between the ON and OFF layers of the IPL (Rivlin-Etzion et al. 2012).

12.3 The A17 Paradox: A Large Neuron Containing Hundreds of Local Processors

Most dendritic computations require signal integration and/or propagation. An interesting counterexample is the retinal A17 amacrine cell, in which morphological and biophysical specializations create hundreds of highly compartmentalized, functionally independent computational units within their dendrites (Ellias and Stevens 1980; Grimes et al. 2010). Consequently, A17 amacrine cells appear to forsake typical dendritic integration (i.e., collecting and combining synaptic inputs to form a single output signal that is distributed via the axon) in favor of distributed, parallel processing.

12.3.1 Morphology and Physiology of the A17 Amacrine Cell

A17 amacrine cells were originally identified by their distinct morphological and biochemical properties: Dozens of thin (<150 nm diameter), non-branching dendrites emanate from a soma at the border of the INL and IPL and are studded with small (~1 μm) varicosities located at roughly 20 μm intervals (Zhang et al. 2002; Fig. 12.2a). In addition, A17s take up indoleamines and can be identified via fluorescent staining for serotonin (Negishi et al. 1981; Sandell and Masland 1986; Dong and Hare 2003; Fig. 12.2b). Each of the tiny compartments (varicosities) contains the essential machinery to receive excitatory (glutamatergic) synaptic input from one rod bipolar cell (RBC) and provide reciprocal inhibitory (GABAergic) synaptic output to that same RBC, forming a complete feedback microcircuit within a volume of ~1 μm^3 (Fig. 12.2c).

The extent of functional interactions between these dendritic varicosities has been debated. On one hand, active conductances (e.g., Na_v channels) in A17 dendrites could boost postsynaptic signals and enhance signal propagation, thereby ensuring that feedback varicosities in the same A17 amacrine exert a cooperative ("global") circuit influence (Bloomfield and Xin 2000; Zhang et al. 2002). Because the dendritic fields of A17s overlap extensively, this particular cell physiology could give rise to classic "center-surround inhibition" in response to local stimulation (Bloomfield and Xin 2000). On the other hand, studies combining serial electron microscopy and computational modeling have suggested that passive membrane properties and dendritic morphology alone might be sufficient to restrict locally generated signals to individual varicosities, possibly leading to their independent operation (Ellias and Stevens 1980; Grimes et al. 2010).

More recently, electrophysiological recordings in rat retinal slices demonstrated that A17s express very little Na_v conductance and are incapable of firing action potentials (Grimes et al. 2010). These biophysical and electrotonic properties, when combined into a compartmental model, suggested that significant voltage attenuation ought to occur between varicosities (Grimes et al. 2010; Fig. 12.2f). To test experimentally whether each varicosity functions independently, electrophysiological recordings from A17 somata were made concomitantly with 2-photon imaging of synaptically evoked calcium signals in individual varicosities (Grimes et al. 2010; Fig. 12.2e, f). Recordings in voltage-clamp mode confirmed that varicosities could be stimulated independently (Fig. 12.2e). In current clamp mode (i.e., permitting dendritic depolarization and active membrane contributions), synaptic stimulation of one varicosity elicited calcium influx in a neighbor only if it was significantly closer than the typical distance between varicosities (~20 μm; Grimes et al. 2010;

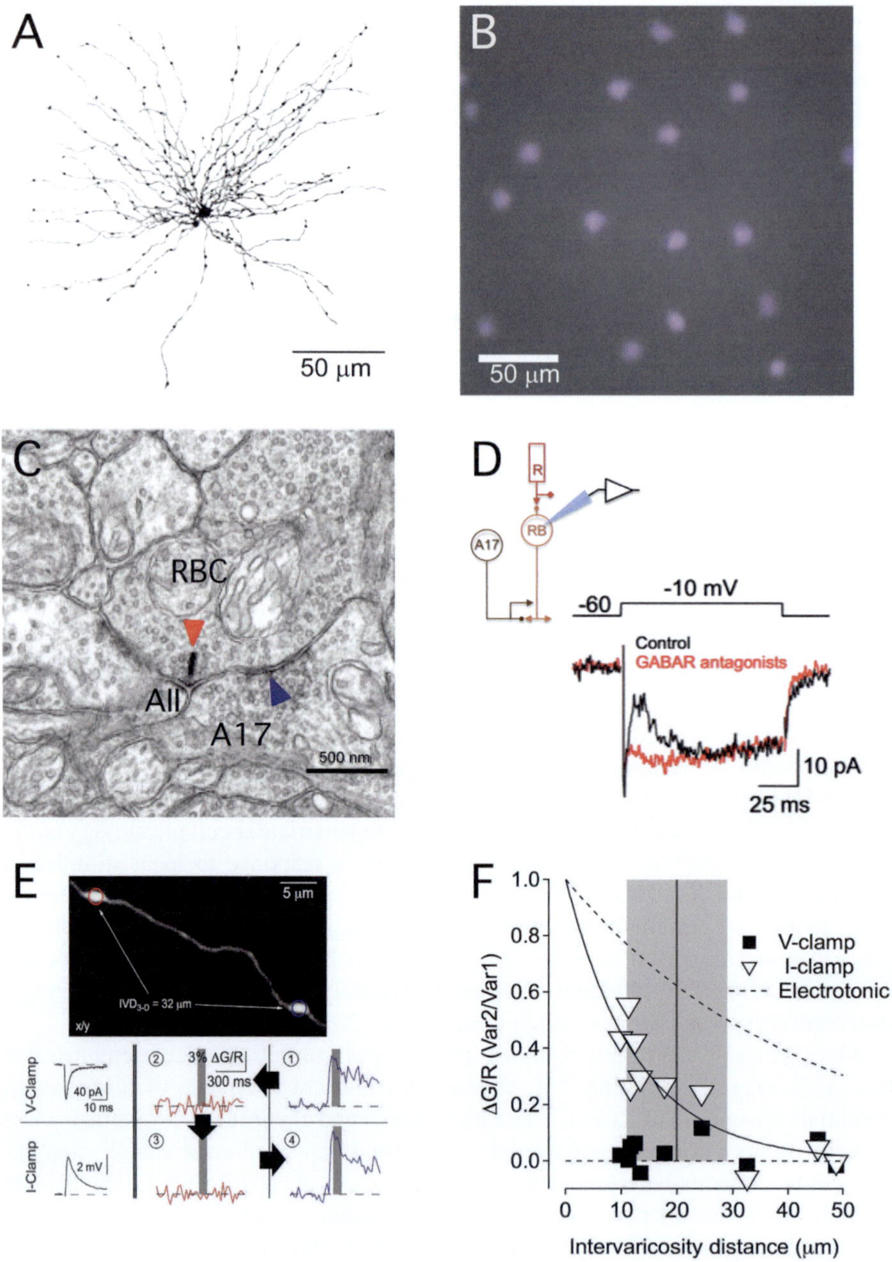

Fig. 12.2 (**a**) Typical A17 morphology, from rat retina. (**b**) A17 somata in flat-mount rabbit retina labeled with 5,7-dihydroxytryptamine. From Dong and Hare (2003). (**c**) Electron microscopic section from rat retina showing a rod bipolar cell (RBC) terminal apposed to postsynaptic processes from an AII amacrine cell and an A17 amacrine cell. Note the synaptic ribbon (*red arrow*) in the RBC terminal and the reciprocal feedback synapse (*blue arrow*) in the A17 process. (**d**) Reciprocal

Zhang et al. 2002; Fig. 12.2f). These results suggested that the >100 varicosities on each A17 typically operate independently, especially under low-light (scotopic) conditions when light-evoked signals are sparse in both space and time. These interneurons therefore provide an opportunity to study synaptic and cellular mechanisms that govern the function of tiny microcircuits in the context of sensory processing.

Perhaps not surprisingly, synaptic input and output appear to be tightly coupled within these small A17 varicosities. Under certain conditions, GABA release from A17s is triggered primarily by calcium influx through calcium-permeable AMPA receptors, indicating an intimate relationship between the postsynaptic glutamate receptors and presynaptic GABA release machinery within A17 varicosities (Chavez et al. 2006). These results suggest a mechanism by which input–output interactions may be exclusively local and relatively unaffected by the local dendritic membrane potential.

A17s express voltage-gated calcium (Ca_v) conductances (Menger and Wassle 2000) in both their somas and synaptic varicosities (Grimes et al. 2009), but Ca_v activation in varicosities is limited by large conductance, calcium-activated potassium (BK) channels that suppress postsynaptic depolarization. Accordingly, when BK channels are blocked, Ca_vs are activated and enhance GABA release (Grimes et al. 2009).

More recent ultrastructural data indicates that most feedback varicosities contain one excitatory input synapse and two reciprocal inhibitory synapses. These feedback synapses are not distributed randomly across the varicosity membrane; rather they are typically aligned to one side of the excitatory input synapse. The closest ("proximal") synapse is typically located within 200 nm of the center of the excitatory ribbon synapse, while the "distal" synapse is typically 500–800 nm away (Grimes et al. unpublished observations). One possibility is that calcium influx through postsynaptic calcium-permeable glutamate receptors triggers GABA release at the proximal synapse (Chavez et al. 2006) but, due to calcium buffering in the varicosity, fails to trigger release at the distal synapse. The distal release site, instead, may require membrane depolarization to activate calcium influx through Ca_v channels, a process likely regulated by BK channels. BK channels themselves are highly regulated (Hou et al. 2009), and those expressed in A17 amacrine cells exhibit particularly rapid inactivation (Grimes et al. 2009). BK channels in A17 varicosities may, therefore, underlie activity-dependent computations.

Fig. 12.2 (continued) inhibitory feedback recorded in a RBC from a rat retinal slice. Inward Ca_v current and outward IPSC were evoked by depolarizing the RBC membrane from −60 to −10 mV. From Chavez et al. (2006). (**e**) Synaptically activated Ca^{2+} responses in an A17 dendrite. The RBC above varicosity 1 (*blue circle*) was electrically stimulated to elicit a Ca^{2+} signal (*blue traces*). Stimulation failed to elicit a Ca^{2+} signal in varicosity 2 (*red circle*, *red traces*), even when the A17 membrane potential was allowed to move freely (current clamp). From Grimes et al. (2010). (**f**) Ratio of responses (varicosity 2:varicosity 1) from ten experiments like that shown in (**e**). *Gray region* indicates spacing (mean ± SD, $n = 47$) of synaptic varicosities along A17 dendrites. *Dashed line* indicates predicted electrotonic decay of membrane depolarizations, given the measured active and passive properties of A17 dendrites. From Grimes et al. (2010)

12.3.2 A17 Amacrine Cells Influence Visual Processing Under Dim Lighting Conditions

From a visual processing standpoint, A17 amacrine cells receive excitatory synaptic input from, and provide reciprocal inhibitory feedback to second order neurons (i.e., RBCs) in the retinal pathway/circuitry responsible for night vision. Local processing by A17 varicosities must influence temporal aspects of signals and/or noise traversing this specialized pathway. The retinal circuitry responsible for night vision has an extremely high gain in darkness, sufficient to detect and transmit signals arising from the absorption of single photons (Field et al. 2005). In dimly lit environments, conditions for which this circuitry is optimized, signals are sparse in both space and time. A high level of convergence within this circuitry confers the retina's output neurons (i.e., ganglion cells) with high sensitivity. As mean light levels approach absolute darkness, however, neural noise converging onto downstream neurons must be tightly controlled so that spatially and temporally infrequent single-photon signals are perceivable under these conditions.

Indeed, our recent experiments indicate that A17-mediated reciprocal feedback inhibition *enhances* the sensitivity of AII amacrine cells to extremely dim flashes delivered in absolute darkness (Grimes et al. unpublished observations). Interestingly, this result appears to be due to the fact that feedback inhibition, by regulating tonic release from RBCs in absolute darkness, maintains a larger pool of releasable vesicles, thereby increasing synaptic gain in darkness.

As signal/flash strength increases the discrimination of signals on a noisy background becomes a trivial task. Under these conditions A17 seems to play a role in shaping signal kinetics and dynamic range. In vivo experiments using electroretinograms from rabbits have shown that reciprocal inhibition from A17 amacrine cells shapes the time course of the retinal's response to a range of flash strengths from darkness (Dong and Hare 2003). A recent in vivo report on mice has suggested that BK channels in A17 amacrine cells limit the dynamic range of RBC signals (Tanimoto et al. 2012), a result that is in direct agreement with the effects of BK channels antagonists on RBC dynamic range in vitro (Grimes et al. 2009).

12.3.3 Remaining Questions

As the anatomy and biophysical mechanisms underlying A17-mediated dendritic computations become better understood, we pause to consider the remaining questions surrounding A17 amacrine cells. Do A17 amacrine cells contribute to visual processing at higher light levels? And, if so, do these contributions require independent processing? What cells, other than RBCs, provide input to A17 amacrine cells and how do these inputs influence A17 function (e.g., A17 receive inhibition on their descending dendrites)? Finally, given the markedly atypical dendritic integration in A17s (Grimes et al. 2010), what parameters dictate the optimal size of A17 dendritic arbors (Chklovskii 2004)?

12.4 Conclusions

Despite the wealth of interesting information gleaned from studying both SACs and A17 amacrine cells, the physiology and circuit functions of the majority of amacrine cell types remain elusive. Estimates indicate that the retinas of mammalian species contain 30 or more different amacrine cell types, so an abundance of interesting information remains to be discovered. What is clear so far is that the specific details of each amacrine cell type's connectivity, morphology, and physiology critically underlie their computational roles in the surrounding circuitry. For example, colocalization of input and output synapses within individual dendrites of SACs and A17s is a morphological specialization that affords the capacity for independent parallel processing. This particular synaptic organization has been observed more often in invertebrate nervous systems (e.g., the terminal abdominal ganglion in crayfish; Takahata et al. 2000), although dendrodendritic synapses also have been reported in the mammalian suprachiasmatic nucleus (Guldner and Wolff 1974) and granule cells in mammalian and reptile olfactory bulb (Isaacson and Strowbridge 1998; Jahr and Nicoll 1980). In a cell class with tremendous morphological diversity, some amacrine cells (e.g., AII amacrine cells and polyaxonal amacrine cells) exhibit greater separation between inputs and outputs. Correspondingly, at least one amacrine cell type with more conventional morphology has been shown to be involved in a computation that requires signal integration over a significant area of visual space (Baccus et al. 2008).

Enhanced access to other amacrine cells types, with new cell type-specific markers (Siegert et al. 2009) and other genetic manipulations, will enable us to address in many more amacrine cells the same questions that have been posed in SACs and A17s. Hopefully, this will help us to understand how various forms of compartmentalized dendritic processing contribute to specific processing tasks of the retinal circuitry. In particular, it will be interesting to learn whether each amacrine cell type performs functions required by a particular RGC, or whether they each process information that is used by a number of RGC subtypes. And how does dendritic processing and amacrine cell function change from the night, when photons are scarce and rods mediate visual signaling, and the day, when cone photoreceptors provide color information to be processed? There is much to be done but the field (thanks, no doubt, to amacrine cells) appears to be headed in the right direction.

Acknowledgement This work was supported by the NINDS Intramural Research Program (NS003039 to JSD).

References

Baccus SA, Olveczky BP, Manu M, Meister M (2008) A retinal circuit that computes object motion. J Neurosci 28(27):6807–6817

Barlow HB, Hill RM (1963) Selective sensitivity to direction of movement in ganglion cells of the rabbit retina. Science 139(3553):412–414
Barlow HB, Hill RM, Levick WR (1964) Retinal ganglion cells responding selectively to direction and speed of image motion in the rabbit. J Physiol 173:377–407
Barlow HB, Levick WR (1965) The mechanism of directionally selective units in rabbit's retina. J Physiol 178(3):477–504
Bloomfield SA, Xin D (2000) Surround inhibition of mammalian AII amacrine cells is generated in the proximal retina. J Physiol 523(Pt 3):771–783
Brecha N, Johnson D, Peichl L, Wassle H (1988) Cholinergic amacrine cells of the rabbit retina contain glutamate decarboxylase and gamma-aminobutyrate immunoreactivity. Proc Natl Acad Sci 85(16):6187–6191
Briggman KL, Denk W (2006) Towards neural circuit reconstruction with volume electron microscopy techniques. Curr Opin Neurobiol 16(5):562–570
Briggman KL, Helmstaedter M, Denk W (2011) Wiring specificity in the direction-selectivity circuit of the retina. Nature 471(7337):183–188
Chavez AE, Singer JH, Diamond JS (2006) Fast neurotransmitter release triggered by Ca influx through AMPA-type glutamate receptors. Nature 443(7112):705–708
Chklovskii DB (2004) Synaptic connectivity and neuronal morphology: two sides of the same coin. Neuron 43(5):609–617
Demb JB (2007) Cellular mechanisms for direction selectivity in the retina. Neuron 55(2): 179–186
Denk W, Horstmann H (2004) Serial block-face scanning electron microscopy to reconstruct three-dimensional tissue nanostructure. PLoS Biol 2(11):e329
Dong CJ, Hare WA (2003) Temporal modulation of scotopic visual signals by A17 amacrine cells in mammalian retina in vivo. J Neurophysiol 89(4):2159–2166
Ellias SA, Stevens JK (1980) The dendritic varicosity: a mechanism for electrically isolating the dendrites of cat retinal amacrine cells? Brain Res 196(2):365–372
Euler T, Detwiler PB, Denk W (2002) Directionally selective calcium signals in dendrites of starburst amacrine cells. Nature 418(6900):845–852
Famiglietti EV (1991) Synaptic organization of starburst amacrine cells in rabbit retina: analysis of serial thin sections by electron microscopy and graphic reconstruction. J Comp Neurol 309(1):40–70
Field GD, Sampath AP, Rieke F (2005) Retinal processing near absolute threshold: from behavior to mechanism. Annu Rev Physiol 67:491–514
Fried SI, Munch TA, Werblin FS (2002) Mechanisms and circuitry underlying directional selectivity in the retina. Nature 420(6914):411–414
Fried SI, Munch TA, Werblin FS (2005) Directional selectivity is formed at multiple levels by laterally offset inhibition in the rabbit retina. Neuron 46(1):117–127
Gavrikov KE, Dmitriev AV, Keyser KT, Mangel SC (2003) Cation–chloride cotransporters mediate neural computation in the retina. Proc Natl Acad Sci 100(26):16047–16052
Goldstein SS, Rall W (1974) Changes of action potential shape and velocity for changing core conductor geometry. Biophys J 14(10):731–757
Grimes WN, Li W, Chavez AE, Diamond JS (2009) BK channels modulate pre- and postsynaptic signaling at reciprocal synapses in retina. Nat Neurosci 12(5):585–592
Grimes WN, Zhang J, Graydon CW, Kachar B, Diamond JS (2010) Retinal parallel processors: more than 100 independent microcircuits operate within a single interneuron. Neuron 65(6):873–885
Grimes WN, Zhang J, Tian H, Graydon CW, Hoon M, Rieke F, Diamond JS (unpublished observations) Complex inhibitory feedback microcircuitry regulates retinal signaling near visual threshold.
Grzywacz NM, Merwine DK, Amthor FR (1998) Complementary roles of two excitatory pathways in retinal directional selectivity. Vis Neurosci 15(6):1119–1127
Guldner FH, Wolff JR (1974) Dendro-dendritic synapses in the suprachiasmatic nucleus of the rat hypothalamus. J Neurocytol 3(2):245–250

Hausselt SE, Euler T, Detwiler PB, Denk W (2007) A dendrite-autonomous mechanism for direction selectivity in retinal starburst amacrine cells. PLoS Biol 5(7):e185
Hou S, Heinemann SH, Hoshi T (2009) Modulation of BKCa channel gating by endogenous signaling molecules. Physiology (Bethesda) 24:26–35
Hubel DH, Wiesel TN (1959) Receptive fields of single neurones in the cat's striate cortex. J Physiol 148:574–591
Isaacson JS, Strowbridge BW (1998) Olfactory reciprocal synapses: dendritic signaling in the CNS. Neuron 20(4):749–761
Jahr CE, Nicoll RA (1980) Dendrodendritic inhibition: demonstration with intracellular recording. Science 207(4438):1473–1475
Keeley PW, Whitney IE, Raven MA, Reese BE (2007) Dendritic spread and functional coverage of starburst amacrine cells. J Comp Neurol 505(5):539–546
Kittila CA, Massey SC (1997) Pharmacology of directionally selective ganglion cells in the rabbit retina. J Neurophysiol 77(2):675–689
Lee S, Kim K, Zhou ZJ (2011) Role of ACh-GABA cotransmission in detecting image motion and motion direction. Neuron 68(6):1159–1172
Lee S, Zhou ZJ (2006) The synaptic mechanism of direction selectivity in distal processes of starburst amacrine cells. Neuron 51(6):787–799
MacNeil MA, Masland RH (1998) Extreme diversity among amacrine cells: implications for function. Neuron 20(5):971–982
Marshel JH, Kaye AF, Nauhaus I, Callaway EM (2012) Anterior-posterior direction opponency in the superficial mouse lateral geniculate nucleus. Neuron 76:713–720
Maturana HR, Lettvin JY, McCulloch WS, Pitts WH (1960) Anatomy and physiology of vision in the frog (Rana pipiens). J Gen Physiol 43(6):129–175
Menger N, Wassle H (2000) Morphological and physiological properties of the A17 amacrine cell of the rat retina. Vis Neurosci 17(5):769–780
Negishi K, Kato S, Teranishi T (1981) Indoleamine-accumulating cells and dopaminergic cells are distributed similarly in carp retina. Neurosci Lett 25(1):1–5
O'Malley DM, Masland RH (1989) Co-release of acetylcholine and gamma-aminobutyric acid by a retinal neuron. Proc Natl Acad Sci 86(9):3414–3418
O'Malley DM, Sandell JH, Masland RH (1992) Co-release of acetylcholine and GABA by the starburst amacrine cells. J Neurosci 12(4):1394–1408
Oesch N, Euler T, Taylor WR (2005) Direction-selective dendritic action potentials in rabbit retina. Neuron 47(5):739–750
Oesch NW, Taylor WR (2010) Tetrodotoxin-resistant sodium channels contribute to directional responses in starburst amacrine cells. PLoS One 5(8):e12447
Oyster CW, Barlow HB (1967) Direction-selective units in rabbit retina: distribution of preferred directions. Science 155(3764):841–842
Rivlin-Etzion M, Wei W, Feller MB (2012) Visual stimulation reverses the directional preference of direction-selective retinal ganglion cells. Neuron 76(3):518–525
Sandell JH, Masland RH (1986) A system of indoleamine-accumulating neurons in the rabbit retina. J Neurosci 6(11):3331–3347
Schmidt M, Humphrey MF, Wassle H (1987) Action and localization of acetylcholine in the cat retina. J Neurophysiol 58(5):997–1015
Siegert S, Scherf BG, Del Punta K, Didkovsky N, Heintz N, Roska B (2009) Genetic address book for retinal cell types. Nat Neurosci 12(9):1197–1204
Takahata M, Takashima A, Hikosaka R (2000) Information processing by nonspiking interneurons: passive and active properties of dendritic membrane determine synaptic integration. Biosystems 58(1–3):143–149
Tanimoto N, Sothilingam V, Euler T, Ruth P, Seeliger MW, Schubert T (2012) BK channels mediate pathway-specific modulation of visual signals in the in vivo mouse retina. J Neurosci 32(14):4861–4866
Taylor WR, Smith RG (2012) The role of starburst amacrine cells in visual signal processing. Vis Neurosci 29(1):73–81

Torre V, Poggio T (1978) A synaptic mechanism possibly underlying directional selectivity to motion. Proc R Soc Lond B 202:409–416

Trenholm S, Johnson K, Li X, Smith RG, Awatramani GB (2011) Parallel mechanisms encode direction in the retina. Neuron 71(4):683–694

Tukker JJ, Taylor WR, Smith RG (2004) Direction selectivity in a model of the starburst amacrine cell. Vis Neurosci 21(4):611–625

Vaney DI (1984) 'Coronate' amacrine cells in the rabbit retina have the 'starburst' dendritic morphology. Proc R Soc Lond B Biol Sci 220(1221):501–508

Vaney DI, Young HM (1988) GABA-like immunoreactivity in cholinergic amacrine cells of the rabbit retina. Brain Res 438(1–2):369–373

Wei W, Hamby AM, Zhou K, Feller MB (2011) Development of asymmetric inhibition underlying direction selectivity in the retina. Nature 469(7330):402–406

Zhang J, Li W, Trexler EB, Massey SC (2002) Confocal analysis of reciprocal feedback at rod bipolar terminals in the rabbit retina. J Neurosci 22(24):10871–10882

Chapter 13
Dendritic Computation of Direction in Retinal Neurons

Robert G. Smith and W. Rowland Taylor

Abstract The retina utilizes a variety of dendritic mechanisms to compute direction from image motion. The computation is accomplished by starburst amacrine cells (SBACs) which are GABAergic neurons presynaptic to direction-selective ganglion cells (DSGCs). SBACs are symmetric neurons with several branched dendrites radiating out from the soma. Larger EPSPs are produced in the dendritic tips of SBACs as a stimulus sequentially activates inputs from the base of each dendrite outwards. The directional difference in EPSP amplitude is further amplified near the dendritic tips by voltage-gated channels to produce directional release of GABA. Reciprocal inhibition between adjacent SBACs may also amplify directional release. Directional signals in the independent SBAC branches are preserved because each dendrite makes selective contacts only with DSGCs of the appropriate preferred-direction. Directional signals are further enhanced within the dendritic arbor of the DSGC, which essentially comprises an array of distinct dendritic compartments. Each of these dendritic compartments locally sum excitatory and inhibitory inputs, amplifies them with voltage-gated channels, and generates spikes that propagate to the axon via the soma. Overall, the computation of direction in the retina is performed by several local dendritic mechanisms both presynaptic and postsynaptic, with the result that directional responses are robust over a broad range of stimuli.

R.G. Smith (✉)
Department of Neuroscience, University of Pennsylvania, Philadelphia, PA 19104, USA
e-mail: rob@retina.anatomy.upenn.edu

W.R. Taylor
Department of Ophthalmology, School of Medicine, Portland, OR 97239, USA

H. Cuntz et al. (eds.), *The Computing Dendrite: From Structure to Function*,
Springer Series in Computational Neuroscience 11, DOI 10.1007/978-1-4614-8094-5_13,

13.1 Introduction

Since their discovery over 40 years ago, direction-selective ganglion cells (DSGCs) in the vertebrate retina have captured the imagination of neuroscientists as a model for understanding how neural computations are implemented (Barlow and Levick 1965). Many details about the underlying retinal circuitry have been determined (for a recent review, see Vaney et al. 2012), and it is apparent that diverse mechanisms contribute, both at the synaptic level, and within dendrites. In this chapter we outline presynaptic and postsynaptic dendritic mechanisms that contribute to generating directional responses.

Retinal ganglion cells (GCs) represent the output from the retina, and receive glutamatergic excitatory inputs from bipolar cells, and GABAergic and glycinergic inhibitory inputs from amacrine cells. Mammalian retinas contain about 20 different types of GC each with characteristic receptive field properties (Rodieck 1998; Masland 2012). The DSGCs respond with a vigorous burst of action potentials to stimuli in a "preferred" direction, and weakly to the same stimulus moving in the opposite "null" direction. The response amplitude approximates a sinusoidal dependence on motion angle, suggesting that the cell computes the motion vector component aligned along the preferred-null axis (Sivyer et al. 2010). The earliest study indicated that the computation did not require activation across the full receptive field, but that motion within smaller regions, termed "subunits," was sufficient to generate directional responses (Barlow and Levick 1965). A subunit inhibits the DSGC when it sees a stimulus moving in the null direction. Since the directional preferences of all subunits are aligned, the combined activity of the subunits prevents the DSGC from responding during null motion. Spatially asymmetric inputs from the GABAergic starburst amacrine cell (SBAC) to the DSGCs were hypothesized to form the basis for the null inhibition underlying these directional responses (Vaney et al. 1989), a prediction that has been borne out in the last decade (Vaney et al. 2012). The subunits are assumed to be generated by the convergent input of many SBACs that have dendritic overlap with the DSGC. The spatial asymmetry of synaptic connections between SBACs and DSGCs arises due to selective contacts between the two cells that preserve the directional signals generated in the SBACs.

For a response to be asymmetric for direction of motion, a spatiotemporal asymmetry is required. An early example of such an asymmetry that has been posited in recent models of dendritic integration (e.g., Branco et al. 2010; McGinley et al. 2012) was outlined by Rall (1964), who showed in a purely passive model of a neuron that a sequence of synaptic inputs progressing from its dendritic tips to the soma gave a larger signal at the soma than when the sequence started at the soma and progressed to the dendritic tips. Such directional signals depend critically upon the asymmetric location of the integration point. For example, for a location that is symmetric with respect to the dendritic structure, such as the DSGC soma, the Rall-effect will produce symmetric, nondirectional responses. In the following sections we will show that the integration points in both the DSGCs and the presynaptic SBACs are located asymmetrically, thus allowing utilization of the Rall-effect.

However, even with such directional asymmetries in the structure of the circuitry, an additional problem remains. A purely linear system can only modify the amplitude and phase of an input signal. Thus directional signals can be viewed, in the frequency domain, as resulting from constructive or destructive interference. Such a system is not robust, in the sense that there will be stimuli with frequency and phase characteristics that produce "anomalous" responses. For example, the directional response of a linear neural model to a spatially repeating stimulus such as a sine-wave grating can produce nondirectional or even reversed-directional, responses (Egelhaaf et al. 1989; Tukker et al. 2004). Thus robust biological direction-selective systems contain nonlinearities to enhance their directionality. For example, cortical direction-selective cells, which lack input from retinal DSGCs, are thought to generate their direction-selective properties by linearly summing input from "non-lagged" and spatially offset "lagged" cells in the lateral geniculate nucleus (LGN) (Adelson and Bergen 1985; McLean et al. 1994). But the lagged LGN cells generate their lagged properties by integrating excitation and inhibition with a nonlinear spike threshold (Vigeland et al. 2013).

We will outline current thinking about the retinal mechanisms for presynaptic and postsynaptic nonlinearities that greatly enhance the functional utility of the "Rall effect." Much of our discussion will consider the mechanisms that generate directional responses in the most widely studied DSGCs, the so-called On-Off DSGCs of rabbit retina. The anatomical structure of these cells provides no indication as to the preferred direction (Yang and Masland 1994), and for comparison we will briefly consider the mechanisms proposed to generate directional responses within mouse DSGCs that show a strong asymmetry in dendritic structure that is aligned with the preferred-null axis.

13.2 Dendritic Processing Presynaptic to the DSGCs

In this section we focus on presynaptic mechanisms that generate directional GABAergic inhibition from the SBAC to the DSGC that is strongest in the DSGC's null direction. In Sect. 13.3 we consider three further mechanisms involving postsynaptic integration of synaptic inputs to the DSGC that contribute to generating directional responses.

At the first retinal stage, photoreceptors transduce light into electrical signals that are transmitted synaptically to second order neurons, called bipolar cells. The responses of photoreceptors and bipolar cells show no dependence on the direction of stimulus motion. Bipolar cells provide excitatory inputs to the starburst amacrine cells and the DSGC, the two key neuronal elements for generating directional signals in the retina. An early finding was that directional responses in DSGCs are dependent upon GABAergic transmission (Ariel and Daw 1982; Caldwell et al. 1978). The critical GABAergic transmission originates from starburst amacrine cells (SBAC), because when these cells are selectively ablated in mouse or rabbit, the DS responses of the retina are abolished (Amthor et al. 2002; Yoshida et al. 2001). This striking neuron has a radially symmetrical dendritic tree, 250–500 μm in diameter with a unique morphology, in which successive bifurcations generate profuse branching near the dendritic tips (Fig. 13.1) (Famiglietti 1983). In recordings from the soma, the SBAC appears not to be DS for bars or gratings (Taylor and Wässle 1995; Peters and Masland 1996), which is to be expected at the soma of a radially symmetric cell that integrates responses from all directions with equal weight (Tukker et al. 2004). However, the soma is not a functionally relevant integration point, since these cells have no axon and do not generate action potentials (Taylor and Wässle 1995; Peters and Masland 1996; Zhou and Fain 1996; Cohen 2001). Rather, the output is generated from synaptic boutons located in the outer third of the dendritic arbor, and these local integration points receive summed activity from excitatory inputs arrayed along each dendrite (Famiglietti 1991). The asymmetric location of these output synapses within the dendritic arbor results in a preferred axis of stimulation from the soma along the dendrite to the bouton. When each bouton is considered as an integration point, the Rall-effect is predicted to produce an "intrinsic" DS signal at that bouton (Tukker et al. 2004). The summation of excitatory inputs along the dendrite generates a larger signal in a dendritic bouton when a stimulus proceeds outwards from soma (centrifugal), because the EPSPs generated near the distant soma are relatively delayed and thus superimpose with EPSPs subsequently generated at the bouton (Tukker et al. 2004). In the opposite direction (centripetal), the EPSP at the bouton rises and falls before the EPSPs from the distant soma arrive, resulting in a smaller peak depolarization.

The morphological parameters that maximize this dendrite intrinsic DS signal were examined in detail for SBACs (Tukker et al. 2004). Computational modeling showed that the precise dendritic structure is not critical. For example, simplified models of the SBAC, in which the fine details of the branching pattern were removed, produced strong intrinsic DS signals at the dendritic tips (Tukker et al. 2004). However, the degree of isolation from the soma is critical (Velte and Miller 1997; Miller and Bloomfield 1983). For models with thick dendrites, which are nearly isopotential, or for those with thin dendrites, which are electrically isolated from the soma, the directional difference between centrifugal and centripetal signals is reduced when compared with dendrites with an intermediate thickness that provide a partial isolation from the soma. Partial electrical isolation on the order of one space constant produces optimal DS at the dendritic tips (Tukker et al. 2004), because the directional difference depends on the superposition of the "global" depolarization, generated across the entire dendritic arbor, with the local depolarization generated at a given bouton. This superposition is asymmetric for stimulus direction because each bouton is located asymmetrically relative to the global signal. Since the SBACs are radially symmetric, boutons on opposite sides will have opposite intrinsic preferred directions. Thus, the boutons in each dendritic branch will have an intrinsic preferred direction roughly aligned along a centrifugal vector connecting the soma with the bouton.

Calcium imaging experiments have demonstrated, in agreement with the intrinsic preferred direction, that centrifugal motion stimuli generate stronger calcium

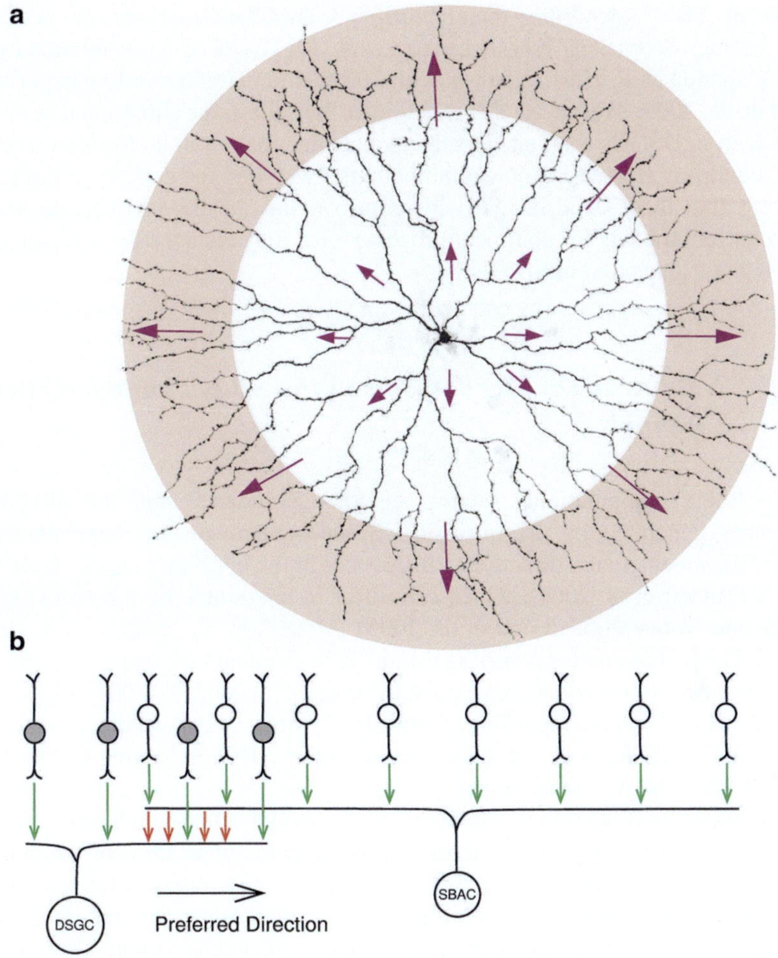

Fig. 13.1 (**a**) Morphology of starburst amacrine cell (SBAC) and directional preference of the dendritic tips. The SBAC has a distinctive symmetric morphology, in which a few primary dendrites emanate from the soma and bifurcate several times to generate a bushy appearance at the dendritic tips (Famiglietti 1983). Each dendrite integrates excitatory postsynaptic potentials (EPSPs) from bipolar cells. The amplitude of the summed EPSP is larger for outward motion (soma to dendrite, *purple arrows*), and generates GABA release from the outer third of the dendritic tree (*gray*) with a directional preference aligned with the orientation of the dendrite relative to the soma. This directional-difference in EPSP amplitude represents an "intrinsic" mechanism that arises from the morphology and spatiotemporal activation of the excitatory inputs. Image of starburst amacrine, David Vaney, unpublished personal communication. (**b**) Connectivity of the DSGC pathway. Arrays of bipolar cells make glutamatergic synapses with the SBAC and the DSGC (*green arrows*). The SBAC makes GABAergic contacts (*red arrows*) onto DSGCs from the outer third of its dendritic arbor. Each SBAC dendrite makes selective inhibitory contacts only to DSGCs with the opposite preferred direction. The DSGC contains two dendritic arborization layers, each with input from bipolar cells and SBACs; only one layer is shown. Note that the bipolar cells differ, since SBACs receive tonic excitatory input but DSGCs do not (see text)

signals in SBAC dendrites than centripetal stimulation (Euler et al. 2002). Electron-microscopic analysis (Briggman et al. 2011), cell-specific ablation experiments (Yoshida et al. 2001; Amthor et al. 2002), and paired recording experiments (Fried et al. 2005; Wei et al. 2011), indicate that the large directional inhibitory inputs seen in DSGCs originate in selective connections from multiple overlapping SBAC dendrites aligned along a common axis. However, the genesis of directional signals within these dendrites remains controversial. In the next section we will review current models for nonlinear processes that may enhance the intrinsic directional signals generated in SBACs.

13.2.1 Nonlinear DS Mechanisms in the SBAC: Voltage-Gated Channels

Passive linear computational models of SBACs indicate that the directional-differences generated by the intrinsic DS mechanism are not large enough to produce the much greater directional differences in the inhibitory inputs observed in DSGCs (Tukker et al. 2004), which can differ in magnitude by threefold or more (Taylor and Vaney 2002; Fried et al. 2005). Moreover, the models show that an intrinsic DS preference for centrifugal motion in the dendritic tips is accompanied by a preference for centripetal motion at the soma (Tukker et al. 2004; also see Rall 1964); however, voltage recordings from SBAC somas show a preference for centrifugal motion (Euler et al. 2002; Oesch and Taylor 2010), indicating that the linear model of DS in SBACs is incomplete.

An obvious modification to the simple linear model was the addition of voltage-gated calcium channels to the dendritic boutons in the outer third of the dendritic arbor, since such channels are required to support neurotransmitter release, and might also produce nonlinear amplification of the small intrinsic directional signals in SBACs. Indeed, addition of voltage-gated calcium channels to an otherwise passive model resulted in large directional signals, commensurate with those inferred from measuring the inhibition in DSGCs or the calcium transients in the SBACs (Tukker et al. 2004). Other investigations indicate that voltage-gated sodium channels, although not dense enough to generate spikes in SBACs (Taylor and Wässle 1995; Zhou and Fain 1996; Cohen 2001), might also contribute to the DS nonlinearity (Oesch and Taylor 2010). Moreover, the addition of voltage-gated channels to the dendrites correctly predicts the *centrifugal* preference for motion that is observed for somatic recordings. The voltage-gated channels nonlinearly amplify the intrinsic directional difference in the amplitude of EPSPs at the dendritic tip, producing larger EPSPs for centrifugal motion (Fig. 13.2) (Trenholm et al. 2011). The amplified directional EPSPs at the dendritic tips propagate back to the soma and override the centripetal preference for motion that is expected for the linear Rall effect. The boosting effects of voltage-gated channels will presumably depend on their biophysical characteristics. The activation range, and steady-state voltage-dependent

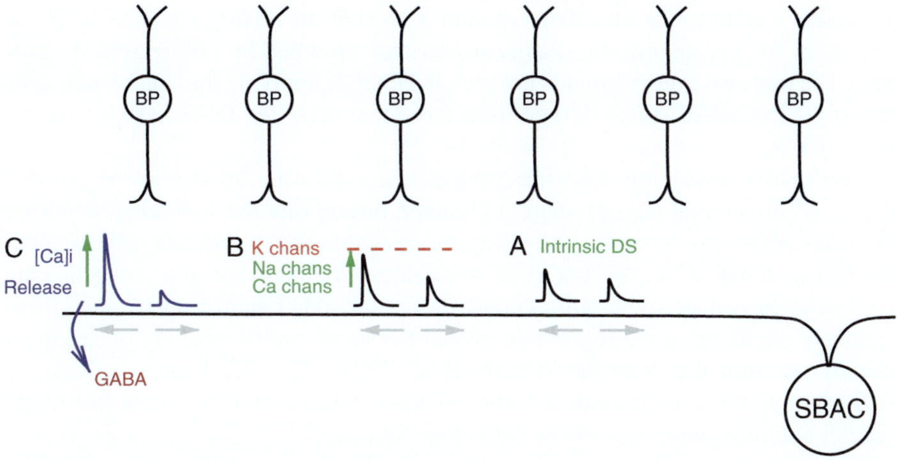

Fig. 13.2 Biophysical mechanisms amplifying DS at dendritic tips of the SBAC. An array of bipolar cells (BP) provides visually mapped excitatory input over the entire dendritic arbor. (**a**) The directional preference of each dendrite is defined by the direction it points relative to the soma. EPSPs vary in amplitude (*black traces*) in response to stimuli moving towards and away from the soma (*gray arrows*). Due to the cable properties of the dendrite that partially isolate the tip from the soma, stimuli that sequentially activate bipolar cell inputs from soma to tip produce a larger EPSP at the tip than stimuli moving from tip to soma. This gives an "intrinsic DS" centrifugal preference to the dendritic tip, since the larger centrifugal EPSP can generate more local transmitter release (Tukker et al. 2004). (**b**) The directional difference in EPSP amplitude from (**a**) is enhanced by voltage-gated channels (Na, Ca, *green arrow*) which nonlinearly amplify larger EPSPs, thus increasing the directional difference (Oesch and Taylor 2010). Kv channels in the soma and proximal dendrites limit depolarization to −20 mV (*red dashed line*) and may function to isolate dendritic signals (Ozaita et al. 2004). (**c**) The directional difference in the dendritic tips is likely further enhanced by the synaptic release mechanism (*blue arrow*), for example a nonlinear dependence on calcium influx ([Ca]i, *blue traces*) (Hausselt et al. 2007; Lee et al. 2010). Thus the directional synaptic release at the dendritic tips is dependent on the extended spatial activation of the synaptic inputs, coupled with nonlinear amplification. Other mechanisms, such as reciprocal inhibition from neighboring SBACs, and a gradient in the membrane potential or inhibitory reversal potential from dendritic tip to soma, may also contribute

inactivation range of the calcium and sodium channels must be such that enough channels are available for activation at typical resting potentials of SBACs, and that the channels are activated sufficiently by the light-evoked EPSPs during motion stimulation. While these parameters have not been exhaustively investigated, there is indirect evidence that the resting potential of SBACs is set dynamically. Under steady background illumination (resting state), SBACs receive a constant tonic excitatory input that is large enough to depolarize the cell by several millivolts (Taylor and Wässle 1995; Peters and Masland 1996). Hausselt et al. (2007) proposed that this tonic excitation generates a voltage gradient within the dendrites of the SBACs that is essential component of the mechanism for directional voltage responses in SBACs. The model included activation of voltage-gated calcium channels and was found to operate only within a relatively narrow voltage-range.

This adds weight to the idea that dynamic control of the resting potential might be important for optimizing the nonlinear boosting provided by voltage-gated channels. The tonic excitatory input is not seen in DSGCs, implying that the bipolar cells that drive the SBAC differ in type from those that drive the DSGC (Helmstaedter et al. 2013).

The models involving dendritic mechanisms within the SBACs as the primary source of directional signals share a common notion that the individual dendritic branches represent separate processing mechanisms that essentially act independently (see Chap. 12 in this book). In this context, the presence of a fast-activating Kv3 delayed rectifier potassium channel takes on added significance, since these channels are found in the highest density in the soma, with a gradient of declining density out into the dendrites (Ozaita et al. 2004). This K channel activates at −20 mV and thus may provide a barrier to depolarization near the soma that might restrict nonlinear amplification to individual dendrites.

13.2.2 Nonlinear DS Mechanisms in the SBAC: Reciprocal Inhibition

Network interactions between SBACs have also been proposed to contribute to generating the directional inhibitory input to the On-Off DSGCs. SBACs are highly branched and overlap extensively, with each point in the retina "covered" by the dendritic arbors of 30–70 individual cells (coverage factor calculated as [dendritic arbor area (μm^2) × cell density ($n/\mu m^2$)]; Tauchi and Masland 1985; Famiglietti 1991, 1992; Vaney 1984). These overlapping SBAC dendrites run together in tight fascicles along with the DSGC dendrites (Tauchi and Masland 1985; Famiglietti 1992; Vaney and Pow 2000; Dong et al. 2004; Briggman et al. 2011), an arrangement that allows for numerous contacts between SBACs and DSGCs, but also between adjacent SBACs. Indeed, recordings from pairs of SBACs provide evidence for inhibitory and reciprocal synaptic connections between adjacent cells (Lee and Zhou 2006). Such an arrangement raises the possibility for positive feedback within the network that could enhance directional signaling (Lee and Zhou 2006; Taylor and Smith 2012). For example, when the dendritic tip of one SBAC is depolarized by centrifugal motion, it will increase GABA release and hyperpolarize the adjacent dendrite of a neighboring SBAC, which in turn will reduce the reciprocal inhibition from that dendrite. The reduced reciprocal inhibition will enhance the depolarization and inhibitory output of the original SBAC, and so on. This scheme represents positive feedback, which could enhance the DS signals in SBAC dendritic tips. Such a model provides an alternative explanation for the function of tonic excitation to SBACs, because if a reciprocal negative feedback mechanism is to produce positive feedback it requires tonic excitation to drive both cells to tonically inhibit each other (Taylor and Wässle 1995; Lee and Zhou 2006; Taylor and Smith 2012).

Such a positive feedback mechanism implies that as a stimulus enhances the inhibitory output of one SBAC dendrite, it will suppress the output of the reciprocally connected SBAC dendrite which has an opposite preferred direction. Since the preferred directions of SBAC dendrites are aligned with the orientation of each dendritic branch, the positive feedback model further requires that, despite the extensive coverage factor of 30–70 for SBACs, reciprocal connections are limited to opposing dendrites. In a cell such as the SBAC that makes synaptic output only in the outer third of dendritic arbor, this can only occur if the somas of reciprocally connected SBACs lie outside the dendritic extent of the partner cell. Contrary to this expectation, reciprocal inhibitory connections are seen between more closely spaced SBACs (Lee and Zhou 2006). Moreover, even with sufficient separation of somas, contacts between orthogonally oriented dendrites of adjacent SBACs would be counterproductive as they would inhibit the responses of the preferred direction dendrite regardless of stimulus direction (Taylor and Smith 2012). It is not known whether reciprocal contacts between SBACs display the selectively required for the positive feedback model.

An alternative mechanism has been proposed for enhancing DS and appears to include both intrinsic DS mechanisms and enhancement by network inhibition between SBACs. In this model, a chloride gradient along each dendrite causes GABA receptors to be depolarizing near the SBAC soma and hyperpolarizing near the dendritic tips (Gavrikov et al. 2006; Enciso et al. 2010). A model including these mechanisms can approximately duplicate somatic recordings from the SBAC. Apparently this mechanism, by preventing GABAergic inputs near the soma from generating inhibition, may enhance DS in the dendritic tips.

While the involvement of reciprocal network connections between SBACs represents an attractive model for enhancing DS signals, the evidence for such mechanisms remains equivocal. Models of the SBAC can develop DS at the dendritic tips without any inhibition (Tukker et al. 2004), and moreover, recordings from SBACs show DS signals with GABAergic inhibition blocked (Euler et al. 2002; Hausselt et al. 2007). Finally recordings from SBACs, under conditions that produce strong DS signals in DSGCs, fail to elicit strong inhibitory inputs in SBACs (Oesch and Taylor 2010), implying that inhibition between SBACs is not required to generate DS responses within DSGCs. However, such negative results are hardly compelling, and further experimental evidence is required to establish whether reciprocal inhibition plays an active role in generating directional signals.

13.2.3 Asymmetric Synaptic Connectivity from SBACs

Due to the large coverage factor of SBACs (30–70), at each point in the retina, the co-stratifying DSGCs have access to SBAC dendrites with all possible preferred directions, yet On-Off DSGCs display only four, orthogonally oriented, preferred directions. This observation suggested that the four populations of DSGCs make selective contacts with appropriately oriented SBAC dendritic branches (Taylor and Vaney 2003). Recent large-scale EM reconstructions have confirmed this prediction, by demonstrating the angular specificity of the synaptic connections made from SBACs onto DSGCs. SBAC dendrites pointing in the DSGC's null direction make more synaptic contacts than SBAC dendrites pointing in the preferred direction (Briggman et al. 2011). Although not definitively identified, the synaptic connections were assumed to be GABAergic, which is consistent with a physiological analysis demonstrating functionally asymmetric GABAergic connections (Wei et al. 2011). However, the SBAC is unusual among central neurons in that it releases an inhibitory neurotransmitter (GABA) and an excitatory neurotransmitter (acetylcholine, ACh) (O'Malley et al. 1992). Paired recordings between SBACs and DSGCs indicate that the calcium sensitivity and threshold for release are different for the two transmitters, with the release of acetylcholine requiring stronger depolarization and larger calcium influx (Lee et al. 2010). Since transmitter release from SBACs is assumed to be driven by the directional calcium signals in the SBAC dendritic tips (Euler et al. 2002) the release of both GABA and ACh is expected to have the same directional preference, i.e., the release of ACh should also be larger in the null direction. Contrary to this expectation, all reports to date have indicated that excitation is larger in the preferred direction (Chiao and Masland 2002; Lee et al. 2010). Possible resolutions to this apparent contradiction include the hypothesis that the release of acetylcholine from SBAC dendrites may be less directional than that of GABA, perhaps due to differences in the calcium-sensitivity of release (Lee et al. 2010), or that the cholinergic synapses from a SBAC dendrite may connect to different DSGCs than its GABAergic synapses. An alternative explanation is that the directional excitation observed in somatic recordings of DSGCs results from space-clamp errors within the DSGC that are rendered directional by the large directional difference in inhibition (Poleg-Polsky and Diamond 2011). If true, then the cholinergic synapses from SBAC to DSGC might be made symmetrically from all directions. Another possibility is that inhibitory feedback from the SBAC to bipolar cells could produce a directional bias in the release of glutamate from individual axon terminals or axonal branches of the bipolar cells. Further research will be required to resolve this issue.

13.3 Postsynaptic Dendritic Processing in DSGCs

In the foregoing we have focused on presynaptic mechanisms that generate directional GABAergic inhibition from the SBAC to the DSGC that is strongest in the DSGC's null direction. In this section we consider three further mechanisms involving postsynaptic integration of synaptic inputs in the DSGC that contribute to generating directional responses.

13.3.1 Spatial Offset

Each SBAC dendrite collects excitatory inputs along its entire length, but only makes inhibitory outputs from the outer third of the dendrite. Thus, relative to a stimulus edge moving through the receptive field in the null direction, the inhibition received by the DSGC is spatially offset ahead of the stimulus, roughly by the dendritic radius of the SBAC (see Fig. 13.1b). This spatial offset will be evident as a difference in the temporal delay of the inhibition for motion in opposite directions. By contrast, such temporal delays for excitatory inputs to DSGCs from bipolar cells, which have relatively small dendritic and axonal extents, will tend to be small. Thus in the preferred direction, where inhibition lags behind the stimulus edge, excitation will drive the DSGC, but in the null direction, prior activation of the relatively sustained inhibition will countermand the excitation and suppress spiking (Fig. 13.3). This mechanism relies entirely on postsynaptic integration of the inhibitory and excitatory inputs, and could in principle operate in the absence of presynaptic mechanisms. That is, the postsynaptic DS mechanism depends only on the temporal offset of inhibition, not on any directional modulation of its amplitude (Taylor and Vaney 2002; Schachter et al. 2010).

13.3.2 Dendritic Intrinsic Directional Signals and Active Dendrites

The "intrinsic DS" outlined for the SBACs above applies to any dendritic structure, and thus potentially affects postsynaptic integration in the dendrites of DSGCs. In computational analyses of reconstructed DSGC dendrites, we have shown that synaptic integration at each point in the dendritic arbor is asymmetric, with the EPSP amplitude at a specific locus being slightly larger for motion stimuli directed centrifugally from the center of the dendritic arbor, with this difference becoming greater for a more distal locus than a more proximal one. Mapping the local preferred-directions across the dendritic arbor results in a pattern of vectors radiating out from the center, and becoming progressively larger at more distal locations (Fig. 13.3e, f) (Schachter et al. 2010). Similar to the effect of dendritic morphology described above for the SBACs (Fig. 13.1), the strength of the local DS vectors relies on partial electrotonic isolation of each dendritic locus from the remainder of the cell, i.e., the electrotonic space-constant needs to be roughly the same size as the dendritic extent.

The consequences of intrinsic DS within DSGC dendrites are less obvious than for SBACs, since the morphology of a DSGC bears no relation to its preferred direction (Huberman et al. 2009; Yang and Masland 1994; Vaney et al. 2012). For example, on the Null side of the DSGC (the side where a null stimulus first reaches the cell), the intrinsic DS at each dendritic locus is aligned with the DS of the cell, but on the Preferred side, it is opposed (Fig. 13.3). Consequently, if DSGCs were to sum inputs across the entire dendritic arbor before generating spikes, such asymmetries would tend to cancel out, and thus become functionally irrelevant. However, in their seminal description of DSGCs, Barlow and Levick (1965) recognized that directional spike responses could be obtained for stimuli that traversed only a faction of the dendritic arbor, indicating that synaptic integration and directional spike initiation could occur within subregions of the dendritic arbor, which they termed "subunits," repeated many times across a single dendritic arbor. Subsequently it was shown that spikes can be initiated within the dendrites of

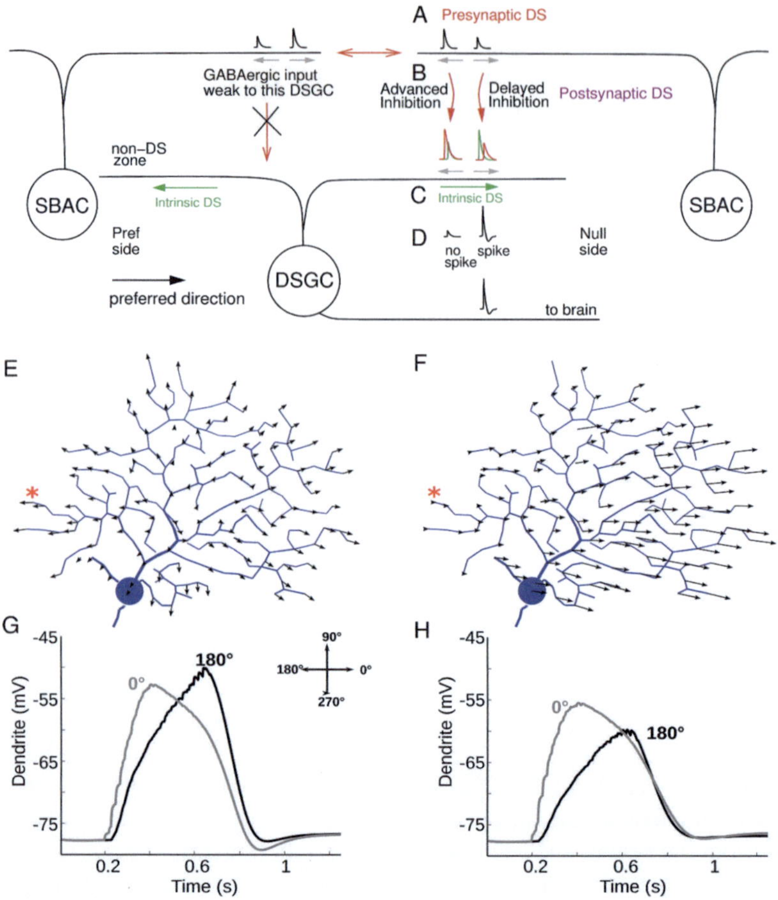

Fig. 13.3 *Top*, Mechanisms in the direction-selective ganglion cell (DSGC) and its presynaptic circuit that contribute to its directional preference. (**a**) Presynaptic DS in EPSPs (*black traces*) originates in the SBAC (see Fig. 13.2), which inhibits (*red*) the DSGC in the null direction. Dendrites from many SBACs contact the DSGC over its entire dendritic arbor. The SBAC dendrites that make synaptic contact with the DSGC all point in the direction opposite to its preferred direction (e.g., the SBAC shown on *right*). Neighboring SBAC dendrites that point in the DSGC's preferred direction do not connect to it (*red arrow* with *black cross*) but are interconnected into the DSGC's presynaptic circuit by reciprocal GABAergic synapses (*red double arrow*), which may further enhance DS via positive feedback (see main text). (**b**) The excitatory conductance (*green trace*) from bipolar cells overlaps the inhibitory conductance (*red trace*) from the SBAC. The spatial extent of the SBAC dendrite temporally advances the SBAC inhibition in the null direction and delays the inhibition in the Pref direction. (**c**) Intrinsic DS in the DSGC dendrites (*green arrow*, see Fig. 13.2) amplifies responses to motion away from the soma, which enhances the directionality of the DSGC on the Null side, but opposes directionality of the DSGC on the Preferred (Pref) side, likely contributing to a "non-DS zone." (**d**) Nonlinear spike initiation within the subunits further enhances the directional difference between preferred and null direction PSPs (Schachter et al. 2010). Schematic view shows only one dendritic arborization layer of the bistratified DSGC. A stimulus moving in the preferred direction (*black arrow*) goes from the Preferred (Pref) side of the cell to the Null side. *Bottom* (**e–h**), Interaction between intrinsic DS and presynaptic DS (taken from Fig. 10 of

DSGCs, and that the dendritic spikes propagate to the soma where they elicit axo-somatic spikes (Oesch et al. 2005). The implied spike initiation zones within the dendrites presumably correspond to the subunits of Barlow and Levick (1965). If synaptic integration within the dendrites can control local spike initiation, then it seems plausible that the intrinsic DS within the local subunits might also contribute to generating directional responses. As noted above, the intrinsic DS is aligned with the preferred direction on the Null side but opposes DS on the Preferred side, and therefore, DS signals should be weaker on the Preferred side. This prediction is in accord with the presence of a "non-DS" zone on the Preferred side of DSGCs where DS is weak or absent (Barlow and Levick 1965), which has been confirmed subsequently in physiological and modeling studies (He et al. 1999; Schachter et al. 2010; Trenholm et al. 2011).

Our modeling studies further demonstrated that due to the high current densities required to support dendritic spiking, realistic levels of inhibitory synaptic input are insufficient to suppress active dendritic spikes even when interposed between the spike initiation point and the soma (Schachter et al. 2010). Similar observations have been made for cortical neurons (Williams 2004), and led us to propose that DS inhibitory inputs most likely control the local PSP amplitude, and thus can prevent spike initiation. This immunity of dendritic spikes to on-the-path inhibitory inputs further supports the notion that dendritic processing within computational subunits can amplify DS signals while remaining relatively unaffected by processing in other dendritic compartments (Fig. 13.4).

13.3.3 DS in a Morphologically Asymmetric Cell

As noted above, the morphology of On-Off DSGCs is not predictive of the preferred direction of a particular neuron (Huberman et al. 2009; Yang and Masland 1994), and, the activity of $GABA_A$ receptor mediated inhibition is absolutely essential for generating DS responses. However, a genetically labeled On-Off DSGC found in mouse, the so-called HB9 cell, differs in both these respects; it has an asymmetrical dendritic arbor that is aligned with the preferred direction, and can generate DS without inhibition (Trenholm et al. 2011). The authors proposed that in the absence

Fig. 13.3 (continued) Schachter et al. 2010). (**e**) Model of a DSGC digitized from a labeled cell, without presynaptic DS or inhibition, illustrates the relative strength of the intrinsic DS mechanism across the dendritic arbor. *Arrows* represent the strength and direction of the preferred DS vector in responses to moving bars at each dendritic location. (**f**) Same model as (**e**) but including presynaptic DS, with excitation and inhibition whose amplitude varied according to the direction of the stimulus bars. (**g**, **h**) Responses to bars moving at 0 and 180° (from the *right* and *left*, respectively, see Fig. 13.4b), recorded in the model from the left side of the dendritic arbor, at the *red star* in (**e**) and (**f**), respectively. The effect of the presynaptic DS in (**f**) and (**h**) is to sum with the intrinsic DS of each dendrite, biasing it towards the preferred direction of the presynaptic DS

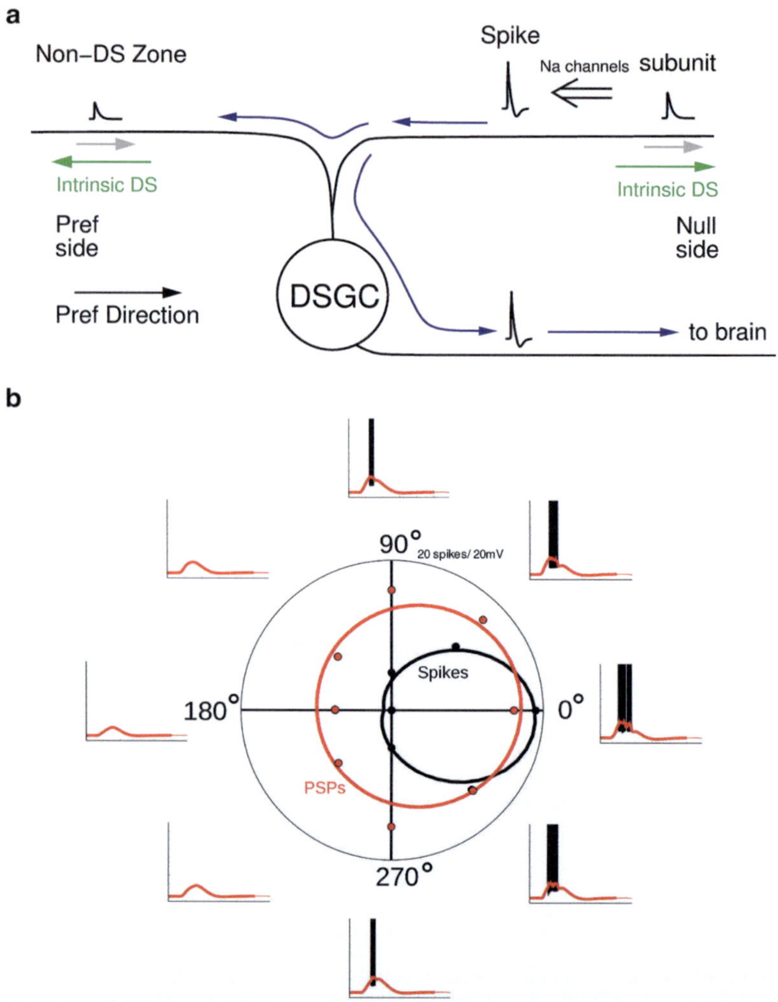

Fig. 13.4 (**a**) The DSGC comprises partially isolated subunits that amplify DS from its presynaptic circuit. Each dendritic tip is partially isolated from the soma and represents a computational subunit that integrates its synaptic and dendritic inputs. For a stimulus moving in the preferred direction (*gray arrows*), EPSPs on the Preferred side (*smaller black trace* at *left*) are attenuated by the dendrite's intrinsic DS (*green arrows*) that opposes the presynaptic and postsynaptic DS mechanisms, thus contributing to the non-DS zone. When the excitation and inhibition summed with the intrinsic DS generates a large EPSP within the subunit (*larger black trace* at *right*), this activates Na channels which initiate a spike that propagates throughout the cell (*long blue arrows*). This allows the cell to respond to local motion within small subunits. The spike back-propagates into all the dendrites, which resets the other subunits and prevents them from generating spikes. Thus, the dendritic computational subunits of the DSGC function in "winner-take-all" mode, in which the subunit with the strongest input signal drives the cell's spikes (Schachter et al. 2010). (**b**) Dendritic spikes initiated in subunits amplify DS from EPSPs (taken from Fig. 12a of Schachter et al. 2010). Polar plots of the number of spikes (*black*) and peak PSP amplitude (*red*) as a function of eight stimulus directions, fit to von Mises functions (circular Gaussians). The surrounding traces show the spikes (*black*), and the PSPs (*red*) at each direction

of inhibition, the morphological intrinsic DS mechanism generates DS signals that are amplified by voltage-gated channels in the dendrites (Tukker et al. 2004; Trenholm et al. 2011). They noted that the lack of dendrites on the Preferred side is functionally beneficial, because the intrinsic DS mechanism for Preferred side dendrites, if they existed, would countermand the cell's preferred direction (Schachter et al. 2010). This point was further supported by experiments showing that in symmetric DSGCs, in the absence of inhibition, the preferred directions on opposite sides of the dendritic arbor were opposite, and pointed away from the center of the dendritic field, as expected for DS signals generated by the intrinsic DS mechanism (Schachter et al. 2010; Trenholm et al. 2011).

A second type of On-Off DSGC with strongly asymmetric dendritic arbors has been documented in the mouse, the so-called JAM-B cell (Kim et al. 2008). Similar to the HB9 cells, JAM-B cells displayed preferred directions tightly aligned with the axis of asymmetry of the dendrites, and pointing in the same direction. Thus the intrinsic DS mechanism might also be important for generating directional responses in these cells, but unfortunately the effect of inhibitory blockers was not tested, and so the relative roles of inhibition and other possible DS mechanisms remains unknown.

13.3.4 How Much Does Each of the Mechanisms Contribute to DS Signals?

To generate directional signals requires an asymmetric spatiotemporal correlation coupled with a nonlinearity (Barlow and Levick 1965). In the foregoing we have argued that within the DSGC circuitry, nonlinearities occur within local dendritic compartments on the presynaptic and postsynaptic sides, and that on both sides the nonlinearity appears to result from the activity of voltage-gated channels that effectively amplify local directional-differences in PSP amplitudes. Moreover, on both sides, directional signals arise in part from intrinsic asymmetries arising naturally from the structure of dendrites (the "Rall effect"). The distributed nature of the computation, built up presynaptically and postsynaptically, makes teleological sense, because one can imagine that directionally asymmetric features at any point within a circuit might be incrementally selected to improve overall performance. Perhaps the asymmetries represented by the Rall-effect were the seeds that provided the selective pressure for the eventual evolution of these complex circuits.

Although the relative contributions of the various DS mechanisms to the overall computation remain unclear, it seems likely that the distributed nature of the computation may have implications for the temporal properties of the system. For example, in its simplest form, a system for computing direction contains two sensors and a delay line (Barlow and Levick 1965). Such a system will exhibit narrow speed tuning that can be calculated from the time-constant of the delay and the separation of the sensors. However, DS signals in DSGCs operate over a broad range of stimulus speeds (Grzywacz and Amthor 2007; He and Levick 2000). The processes underpinning this broad speed tuning have not been identified, but the availability of different synaptic and biophysical mechanisms, at different levels in the circuit, could support broad speed tuning by providing multiple time-constants for signal integration. An example of different temporal tuning by DS mechanisms is evident in the HB9 DSGCs, in which the intrinsic DS mechanism generates directional responses in the absence of the inhibitory mechanisms, but only at low stimulus velocities (Trenholm et al. 2011). A second example is provided by measurements from the JAM-B DSGCs, in which Off and On responses are more directional at high and low velocities respectively (Kim et al. 2008).

Acknowledgments Thanks to David Vaney for the image of the starburst amacrine cell in Fig. 13.1a. This study was supported by NEI grant EY022070.

References

Adelson EH, Bergen JR (1985) Spatiotemporal energy models for the perception of motion. J Opt Soc Am A 2:284–299

Amthor FR, Keyser KT, Dmitrieva NA (2002) Effects of the destruction of starburst-cholinergic amacrine cells by the toxin AF64A on rabbit retinal directional selectivity. Vis Neurosci 19:495–509

Ariel M, Daw NW (1982) Pharmacological analysis of directionally sensitive rabbit retinal ganglion cells. J Physiol 324:161–185

Barlow HB, Levick WR (1965) The mechanism of directionally selective units in rabbit's retina. J Physiol 178:477–504

Branco T, Clark BA, Häusser M (2010) Dendritic discrimination of temporal input sequences in cortical neurons. Science 329:1671–1675

Briggman KL, Helmstaedter M, Denk W (2011) Wiring specificity in the direction-selectivity circuit of the retina. Nature 471:183–188

Caldwell JH, Daw NW, Wyatt HJ (1978) Effects of picrotoxin and strychnine on rabbit retinal ganglion cells: lateral interactions for cells with more complex receptive fields. J Physiol 276:277–298

Chiao CC, Masland RH (2002) Starburst cells nondirectionally facilitate the responses of direction-selective retinal ganglion cells. J Neurosci 22:10509–10513

Cohen ED (2001) Voltage-gated calcium and sodium currents of starburst amacrine cells in the rabbit retina. Vis Neurosci 18:799–809

Dong W, Sun W, Zhang Y, Chen X, He S (2004) Dendritic relationship between starburst amacrine cells and direction-selective ganglion cells in the rabbit retina. J Physiol 556:11–17

Egelhaaf M, Borst A, Reichardt W (1989) Computational structure of a biological motion-detection system as revealed by local detector analysis in the fly's nervous system. J Opt Soc Am A 6:1070–1087

Euler T, Detwiler PB, Denk W (2002) Directionally selective calcium signals in dendrites of starburst amacrine cells. Nature 418:845–852

Enciso GA, Rempe M, Dmitriev AV, Gavrikov KE, Terman D, Mangel SC (2010) A model of direction selectivity in the starburst amacrine cell network. J Comput Neurosci 28:567–578

Famiglietti EV Jr (1983) 'Starburst' amacrine cells and cholinergic neurons: mirror-symmetric on and off amacrine cells of rabbit retina. Brain Res 261:138–144

Famiglietti EV (1991) Synaptic organization of starburst amacrine cells in rabbit retina: analysis of serial thin sections by electron microscopy and graphic reconstruction. J Comp Neurol 309:40–70

Famiglietti EV (1992) Dendritic co-stratification of ON and ON-OFF directionally selective ganglion cells with starburst amacrine cells in rabbit retina. J Comp Neurol 324:322–335
Fried SI, Münch TA, Werblin FS (2005) Directional selectivity is formed at multiple levels by laterally offset inhibition in the rabbit retina. Neuron 46:117–127
Gavrikov KE, Nilson JE, Dmitriev AV, Zucker CL, Mangel SC (2006) Dendritic compartmentalization of chloride cotransporters underlies directional responses of starburst amacrine cells in retina. Proc Natl Acad Sci USA 103:18793–18798
Grzywacz NM, Amthor FR (2007) Robust directional computation in on-off directionally selective ganglion cells of rabbit retina. Vis Neurosci 24:647–661
Hausselt SE, Euler T, Detwiler PB, Denk W (2007) A dendrite-autonomous mechanism for direction selectivity in retinal starburst amacrine cells. PLoS Biol 5:e185
He S, Jin ZF, Masland RH (1999) The nondiscriminating zone of directionally selective retinal ganglion cells: comparison with dendritic structure and implications for mechanism. J Neurosci 19:8049–8056
He S, Levick WR (2000) Spatial-temporal response characteristics of the ON-OFF direction selective ganglion cells in the rabbit retina. Neurosci Lett 285:25–28
Helmstaedter M, Briggman KL, Turaga SC, Jain V, Seung HS, Denk W (2013) Connectomic reconstruction of the inner plexiform layer in the mouse retina. Nature 500:168–174
Huberman AD, Wei W, Elstrott J, Stafford BK, Feller MB, Barres BA (2009) Genetic identification of an On-Off direction-selective retinal ganglion cell subtype reveals a layer-specific subcortical map of posterior motion. Neuron 62:327–334
Kim IJ, Zhang Y, Yamagata M, Meister M, Sanes JR (2008) Molecular identification of a retinal cell type that responds to upward motion. Nature 452:478–482
Lee S, Kim K, Zhou ZJ (2010) Role of ACh-GABA cotransmission in detecting image motion and motion direction. Neuron 68:1159–1172
Lee S, Zhou ZJ (2006) The synaptic mechanism of direction selectivity in distal processes of starburst amacrine cells. Neuron 51:787–799
Masland RH (2012) The neuronal organization of the retina. Neuron 76:266–280
McGinley MJ, Liberman MC, Bal R, Oertel D (2012) Generating synchrony from the asynchronous: compensation for cochlear traveling wave delays by the dendrites of individual brainstem neurons. J Neurosci 32:9301–9311
McLean J, Raab S, Palmer LA (1994) Contribution of linear mechanisms to the specification of local motion by simple cells in areas 17 and 18 of the cat. Vis Neurosci 11:271–294
Miller RF, Bloomfield SA (1983) Electroanatomy of a unique amacrine cell in the rabbit retina. Proc Natl Acad Sci USA 80:3069–3073
Oesch N, Euler T, Taylor WR (2005) Direction-selective dendritic action potentials in rabbit retina. Neuron 47:739–750
Oesch NW, Taylor WR (2010) Tetrodotoxin-resistant sodium channels contribute to directional responses in starburst amacrine cells. PLoS One 5(8):e12447
O'Malley DM, Sandell JH, Masland RH (1992) Co-release of acetylcholine and GABA by the starburst amacrine cells. J Neurosci 12:1394–1408
Ozaita A, Petit-Jacques J, Völgyi B, Ho CS, Joho RH, Bloomfield SA, Rudy B (2004) A unique role for Kv3 voltage-gated potassium channels in starburst amacrine cell signaling in mouse retina. J Neurosci 24:7335–7343
Peters BN, Masland RH (1996) Responses to light of starburst amacrine cells. J Neurophysiol 75:469–480
Poleg-Polsky A, Diamond JS (2011) Imperfect space clamp permits electrotonic interactions between inhibitory and excitatory synaptic conductances, distorting voltage clamp recordings. PLoS One 6(4):e19463
Rall W (1964) Theoretical significance of dendritic trees for neuronal input-output relations. In: Reis RF (ed) Neural theory and modeling. Stanford University Press, Stanford, CA, pp 72–97
Rodieck RW (1998) The first steps in seeing. Sinauer Associates, Sunderland, MA. ISBN 9780878937578

Schachter MJ, Oesch N, Smith RG, Taylor WR (2010) Dendritic spikes amplify the synaptic signal to enhance detection of motion in a simulation of the direction-selective ganglion cell. PLoS Comput Biol 6(8):e1000899

Sivyer B, van Wyk M, Vaney DI, Taylor WR (2010) Synaptic inputs and timing underlying the velocity tuning of direction-selective ganglion cells in rabbit retina. J Physiol 588:3243–3253

Tauchi M, Masland RH (1985) Local order among the dendrites of an amacrine cell population. J Neurosci 5:2494–2501

Taylor WR, Vaney DI (2002) Diverse synaptic mechanisms generate direction selectivity in the rabbit retina. J Neurosci 22:7712–7720

Taylor WR, Vaney DI (2003) New directions in retinal research. Trends Neurosci 26:379–385

Taylor WR, Wässle H (1995) Receptive field properties of starburst cholinergic amacrine cells in the rabbit retina. Eur J Neurosci 7:2308–2321

Taylor WR, Smith RG (2012) The role of starburst amacrine cells in visual signal processing. Vis Neurosci 29:73–81

Trenholm S, Johnson K, Li X, Smith RG, Awatramani GB (2011) Parallel mechanisms encode direction in the retina. Neuron 71:683–694

Tukker JJ, Taylor WR, Smith RG (2004) Direction selectivity in a model of the starburst amacrine cell. Vis Neurosci 21:611–625

Vaney DI (1984) 'Coronate' amacrine cells in the rabbit retina have the 'starburst' dendritic morphology. Proc R Soc Lond B Biol Sci 220:501–508

Vaney DI, Collin SP, Young HM (1989) Dendritic relationships between cholinergic amacrine cells and direction-selective retinal ganglion cells. In: Osborne NN, Weiler R (eds) Neurobiology of the inner retina. Springer, Berlin, pp 157–168

Vaney DI, Pow DV (2000) The dendritic architecture of the cholinergic plexus in the rabbit retina: selective labeling by glycine accumulation in the presence of sarcosine. J Comp Neurol 421:1–13

Vaney DI, Sivyer B, Taylor WR (2012) Direction selectivity in the retina: symmetry and asymmetry in structure and function. Nat Rev Neurosci 13:194–208

Velte TJ, Miller RF (1997) Spiking and nonspiking models of starburst amacrine cells in the rabbit retina. Vis Neurosci 14:1073–1088

Vigeland LE, Contreras D, Palmer LA (2013) Synaptic mechanisms of temporal diversity in the lateral geniculate nucleus of the thalamus. J Neurosci 33:1887–1896

Williams SR (2004) Spatial compartmentalization and functional impact of conductance in pyramidal neurons. Nat Neurosci 7:961–967

Wei W, Hamby AM, Zhou K, Feller MB (2011) Development of asymmetric inhibition underlying direction selectivity in the retina. Nature 469:402–406

Yang G, Masland RH (1994) Receptive fields and dendritic structure of directionally selective retinal ganglion cells. J Neurosci 14:5267–5280

Yoshida K, Watanabe D, Ishikane H, Tachibana M, Pastan I, Nakanishi S (2001) A key role of starburst amacrine cells in originating retinal directional selectivity and optokinetic eye movement. Neuron 30:771–780

Zhou ZJ, Fain GL (1996) Starburst amacrine cells change from spiking to nonspiking neurons during retinal development. Proc Natl Acad Sci USA 93:8057–8062

Chapter 14
Rapid Integration Across Tonotopy by Individual Auditory Brainstem Octopus Cells

Matthew J. McGinley

Abstract Broadband transient sounds, such as clicks, are transduced in a traveling wave in the cochlea that spreads from base to apex. This traveling wave causes delays in the activation of auditory nerve fibers that vary systematically as a function of the tonotopy in the ventral cochlear nucleus (VCN) in the brainstem, activating high-frequency fibers first. Octopus cells in the mammalian VCN consistently spread their dendrites across the tonotopic axis so that the tips receive input from fibers tuned to the highest frequencies. As a result, broadband transient sounds produce a somatopetal (soma-directed) sweep of activation in octopus cells' dendrites. Low-voltage-activated potassium channels (g_{KL}) in the dendrites and soma sharpen the sensitivity to sweep duration. Branch points in octopus cells' dendrites show significant impedance mismatch, resulting in violation of Rall's "3/2 power law" and shaping of sweep sensitivity. Thus, the morphology, connectivity, and membrane biophysics of octopus cells allow them to compensate for the cochlear traveling wave delay and respond to clicks with exquisite temporal precision. In the context of the time–frequency (Gabor) uncertainty principle, octopus cells can be seen to solve a general problem of encoding frequency-dispersed but temporally restricted patterns using somatopetal sweep sensitivity. Compensation for longer delays in low-frequency hearing animals, implications for downstream processing, and relationship to other systems are discussed.

M.J. McGinley, Ph.D. (✉)
Neurobiology Department, School of Medicine, Yale University,
333 Cedar Street, New Haven, CT 06510, USA
e-mail: matthew.mcginley@yale.edu

H. Cuntz et al. (eds.), *The Computing Dendrite: From Structure to Function*, Springer Series in Computational Neuroscience 11, DOI 10.1007/978-1-4614-8094-5_14,

14.1 Temporally Precise Coding: The Problem of Broadband Transients

Speed and temporal precision are at the heart of auditory processing. Auditory nerve fibers (ANFs) in mice can detect sound pressure fluctuations with periods as short as 12 μs (Taberner and Liberman 2005), and humans can detect differences of as little as 10 μs in the time of arrival of a sound between the ears (Mills 1958). By contrast, sensory systems that rely on G-protein-coupled receptors, such as vision and olfaction, process information ~10,000 times more slowly (Hille 1994). The phenomenal speed of mammalian auditory systems is made possible by the precise transduction of sound pressure waves in air into a traveling wave down the basilar membrane in the cochlea (Fig. 14.1a). This traveling wave performs a spectral decomposition into a spatial frequency map along a logarithmically spaced frequency axis (Dallos 1992). The cochlea can thus be thought of as a log-spaced parallel bank of band-pass filters (Yang et al. 1992) with exquisite temporal and frequency resolution that approaches the physical limits of sound (Weisburn and Parks 1992).

Temporal precision in the processing of broadband sounds presents a special problem to auditory systems. In the filter bank model of the cochlea, for each frequency channel (i.e., each hair cell or ANF) there is an inherent trade-off between its spectral and temporal resolution, encapsulated by the Gabor uncertainty principle (Gabor 1946): $\Delta t \times \Delta f \geq 1/\pi$; Δt is the uncertainty in time, and Δf is the uncertainty in frequency, originally inspired by the Heisenberg uncertainty principle (Heisenberg 1927). Simply put, it takes time to determine the frequency of an oscillation with precision. As a result of this trade-off, the uncertainty in, for example, the time of onset of each spectral component of a broadband sound is inversely related to the uncertainty in the frequency of that component. Since the uncertainty in frequency is related to the frequency itself (approximately proportional to the log of the frequency) (Taberner and Liberman 2005), the temporal uncertainty varies systematically with the frequency of the sound component.

The cochlear traveling wave can thus be viewed as a particular solution to the Gabor uncertainty problem. This idea can be tested by analyzing the timing of firing of the first action potential by an ANF in response to a click sound (Fig. 14.1b). The latency from tone onset to first action potential consists of a constant delay, T_c, reflecting the acoustic system, axonal propagation, etc., and a group delay that varies with the sound frequency, $T(f)$, corresponding to the traveling wave delay. The total delay is the sum of the two: $T_t = T_c + T(f)$. From the Gabor uncertainty perspective, $T(f)$, the frequency-dependent delay, is equal to the uncertainty in the time of the signal, Δt. Equating these terms makes the assertion that the group delay at each frequency reflects an accumulation of evidence overcoming the temporal uncertainty, after which time an action potential codes the occurrence of sound energy in that frequency channel. Next, we assume that the uncertainty in frequency, Δf, is proportional to the log of the frequency, reflecting the log-based spacing of the cochlear

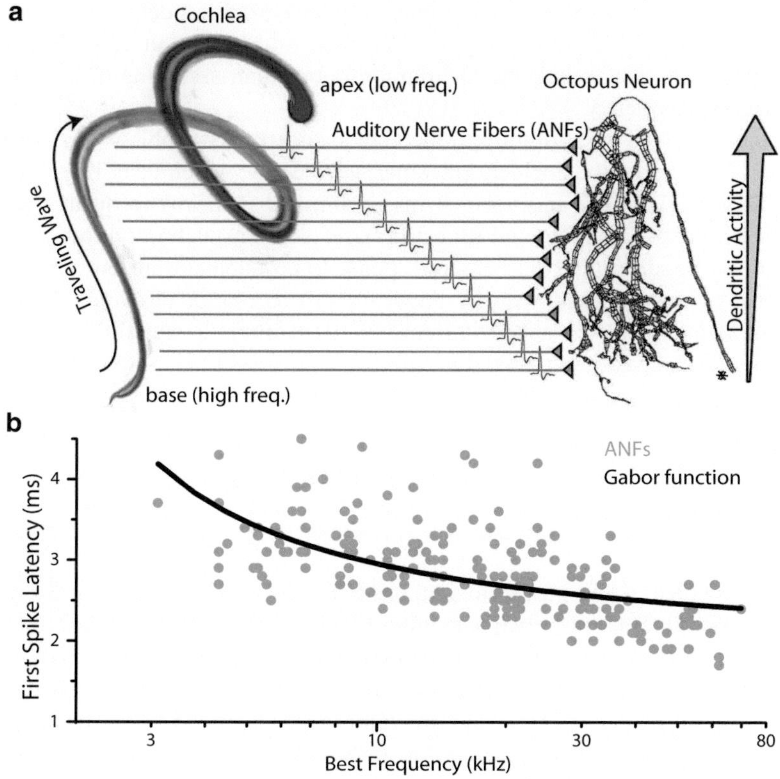

Fig. 14.1 The cochlear traveling wave delay results in a somatopetal wave of activation down the dendrites of octopus cells. (**a**) A schematic representation of the traveling wave spreading down the basilar membrane of the cochlea of a mouse from base to apex (*left*) resulting in a delay line in the activation of auditory nerve fibers (*middle*). This delay profile results in a somatopetal wave of dendritic activity in the dendrites of octopus cells in responses to broadband transient sounds. The *asterisk* indicates the axon of the reconstructed octopus cell. (**b**) The traveling wave delay in mice is apparent in first-spike latencies in responses recorded in vivo to clicks. The *black curve* is the result of a least-squares regression fit of the Gabor uncertainty function ($T_c = 1.8 \pm 0.1$ ms and $\alpha = 0.36 \pm 0.04$). Parts of panels (**a**) and (**b**) are reproduced with permission from Fig. 14.1 of ref. McGinley et al. (2012)

filter bank. Substituting the uncertainty relationship into the equation for total delay yields $T_t = T_c + \frac{1}{(\pi \times \alpha \times \ln(f))}$, where α is the proportionality constant.

A fit to the first spike latency data with a delay function derived from the Gabor uncertainty relationship is in reasonably good agreement with the data (Fig. 14.1b). In particular, it captures the tendency of traveling wave delays to be proportionately larger at lower frequencies, corresponding to the slowing of the traveling wave as it approaches the apex of the cochlea. This slowing effect is prominent in animals with low-frequency hearing (Ruggero 1992). A better prediction would be generated by incorporating the experimentally measured frequency dependence of sharpness of tuning of ANFs (e.g., ref. Taberner and Liberman 2005, Fig. 14.3b) into the function for frequency uncertainty, rather than assuming a linear dependence on log frequency.

Whether viewed as a solution to the Gabor uncertainty problem or simply as a temporal dispersion, the cochlear traveling wave delay, which lasts 1.6 ms in mice (McGinley et al. 2012) and up to 8 ms in animals with low-frequency hearing, including humans (Ruggero and Rich 1987), results in asynchronous activity in the population of auditory nerve fibers in response to broadband transient sounds. This asynchrony must be dealt with by downstream neurons or circuits if broadband transient sounds are to be encoded precisely, for example to signal the snapping of a branch under the foot of a predator.

14.2 Biophysical Specializations of Octopus Cells: Built for Speed

From the cochlea, ANFs project their axons to the cochlear nucleus where they synapse on several classes of principal neurons that in turn carry acoustic information to the next stage of the auditory pathway (Fig. 14.1a). One of the principal classes, octopus cells, is in all respects built for speed. They have extraordinarily brief membrane time constants (~300 μs) and low input resistances (~5 MΩ) (Golding et al. 1995, 1999), making them perhaps the fastest neurons in the mammalian brain (Spruston et al. 2007). Their fast membranes result from a high density of low-voltage-activated potassium conductance (g_{KL}) and hyperpolarization-activated, cyclic nucleotide-gated, channels in their soma and dendrites (Oertel et al. 2008), which oppose each other in a balance at the resting potential (Bal and Oertel 2000, 2001; Cao and Oertel 2011). Octopus cells also have brief synaptic currents, mediated through rapid calcium-permeable AMPA receptors (Gardner et al. 1999, 2001; Raman and Trussell 1992). Octopus cells express voltage-gated calcium channels (Golding et al. 1999; Bal and Oertel 2007), but their physiological role is not known. There is an atypical developmental profile of cholinergic receptors in octopus cells, which may influence their intracellular calcium dynamics (Morley and Happe 2000).

The high density of g_{KL} channels in octopus cells results in an unusual action potential generation mechanism. If an octopus cell is depolarized slowly, g_{KL} channels open, shunt the sodium current, and prevent the generation of an action potential (Fig. 14.2a). If instead an octopus cell is depolarized quickly, g_{KL} channels do not have time to open, creating a window for sodium channel activation to cause an action potential (Fig. 14.2a). Thus, octopus cells have a threshold in rate of depolarization or dV/dt (Fig. 14.2b, left) (Ferragamo and Oertel 2002; Cai et al.

1997). This d*V*/d*t* threshold is a feature of other auditory neurons with high g_{KL}, including bushy cells (McGinley and Oertel 2006) and neurons in the medial superior olive (Golding and Oertel 2012; Cai et al. 2000), which have intrinsic membrane properties that are very similar to octopus cells. The d*V*/d*t* threshold creates an integration window or time during which somatic depolarization must occur to result in an action potential (McGinley and Oertel 2006), which results in extremely stringent coincidence detection (Oertel et al. 2009; Golding and Oertel 2012; König et al. 1996). The integration window in octopus cells is ~1 ms (Fig. 14.2b, right) (McGinley and Oertel 2006).

As a result of the d*V*/d*t* threshold, excitatory postsynaptic potentials (EPSPs) in octopus cells must rise quickly in order to elicit an action potential. EPSPs must also result in sufficient depolarization to activate sodium channels, and thus octopus cells also have a voltage threshold. The relationship between the rate-of-depolarization threshold and the voltage threshold for an action potential has not been explored. Determining the function describing this relationship, $V_{\text{thresh}}(\mathrm{d}V/\mathrm{d}t)$, would provide a more complete picture of the input–output relationship of neurons with a d*V*/d*t* threshold. Furthermore, knowing this relationship would motivate a natural definition of the efficacy of a composite EPSP waveform:

$$\varepsilon_{syn} = \begin{cases} A_{syn} / (V_{thresh}\left(\mathrm{d}V_{syn}\right) - V_{rest}) & if\,\mathrm{d}V_{syn} \geq \mathrm{d}V / \mathrm{d}t_{thresh} \\ 0 & if\,\mathrm{d}V_{syn} < \mathrm{d}V / \mathrm{d}t_{thresh} \end{cases}$$

where A_{syn} is the EPSP amplitude and $\mathrm{d}V_{\text{syn}}$ is the rate of depolarization during the rising phase of the EPSP. Visual inspection of the voltage response to current ramps suggests that the voltage threshold does not change dramatically with rate of depolarization (Fig. 14.2a). Therefore, the amplitude of an EPSP is a reasonable first approximation to its efficacy when the rate of rise is above d*V*/d*t* threshold. Steady depolarization—which may arise from ongoing synaptic activity or modulatory influences—does not affect the d*V*/d*t* threshold, at least in bushy cells (McGinley and Oertel 2006). However, it may affect the voltage threshold (Levy and Kipke 1998), also warranting further investigation.

Octopus cells put their extraordinary speed to use when processing sounds. They respond rapidly and with great temporal precision to onset transients in broadband sounds such as clicks or in loud narrowband sounds. In cats, they respond with sub-millisecond temporal precision to clicks (Godfrey et al. 1975), having a jitter that is less than 200 μs during a 500 Hz click train (Oertel et al. 2000), and can fire at every cycle of a loud 800 Hz tone (Rhode and Smith 1986). They encode rapid temporal components of vocalizations by strongly phase locking to the fundamental frequency in the presence of formants (Rhode 1998) and to the signal envelope of rapid amplitude modulations (Rhode 1994), which are prominent in speech and music.

They may contribute to pitch representation, which occurs explicitly in primary auditory cortex by unexplained mechanisms (Rhode 1995; Bendor and Wang 2005). Thus, octopus cells respond with temporal precision to broadband transients in sounds in a variety of contexts, despite the cochlear traveling wave delay. The sound responses and cellular and synaptic biophysics of octopus cells have been reviewed in detail elsewhere (Oertel et al. 2000, 2009; Golding and Oertel 2012).

14.3 Morphology and Connectivity of Octopus Cells: Positioned to Integrate Broadband Sounds

Octopus cells were named on the basis of their striking morphology; they have several thick primary dendrites that emanate and branch from only one side of the cell body (Fig. 14.2c) (Osen 1969). Octopus cells also have thick daughter branches throughout their dendritic trees. This can be seen when comparing the diameter of parent and daughter branches to the so-called 3/2 power law of Rall: $d_p^{3/2} = d_{d,1}^{3/2} + d_{d,2}^{3/2}$, where d_p is the diameter of the parent dendrite, and d_d "1" and "2" are the diameters of the daughter dendrites at a branch point (Rall 1964). If a branch point obeys this power law, the impedance at the branch will be matched. Octopus dendritic branch points consistently violate the power law, with the right side of the equation being larger than the left for nearly all branches due to large daughter branches (see ref. McGinley et al. 2012, Fig. 8d). This could serve several functions for octopus cell dendritic integration, including increased distal dendritic surface area, rapid transmission of dendritic signals, or delays due to impedance mismatch.

The tendency of thick daughter dendrites to increase distal dendritic surface area can be seen in a simple box representation of octopus cell morphology; substantial surface area is distributed across many branch orders (Fig. 14.2d). Since the surface area of a cylinder is linearly proportional to its diameter, branches obeying the 3/2

Fig. 14.2 (continued) each action potential is indicated with an *asterisk* symbol. (**b**) The d*V*/d*t* threshold and integration window vary between several principal cell types of the VCN. Octopus cells have higher d*V*/d*t* thresholds and shorter integration windows than bushy cells. Stellate cells exhibit classical "integrator" characteristics. They do not have a d*V*/d*t* threshold, so their integration window is not limited by the action potential generation mechanism. (**c**) Photomicrograph of a biocytin-labeled octopus cell against a Nissel stain background (*left*) and 3D digital reconstruction of the same neuron (*right*). *D* dorsal; *M* medial. (**d**) A *box* representation of the dendritic morphology of octopus cells. Each *box* corresponds to a branch order. The *x*-axis is length, and the *y*-axis is total surface area. Each *box* is the average across cells of the average across branches within each order for each cell. The *oval* corresponds to the soma; the *x*-axis is the diameter, and the *y*-axis is the surface area of a sphere with equivalent surface area to a reconstruction of the soma. Panel (**b**) is reproduced with permission from Fig. 5 of ref. McGinley and Oertel 2006 and panels (**c**) and (**d**) from Fig. 8 of ref. McGinley et al. (2012)

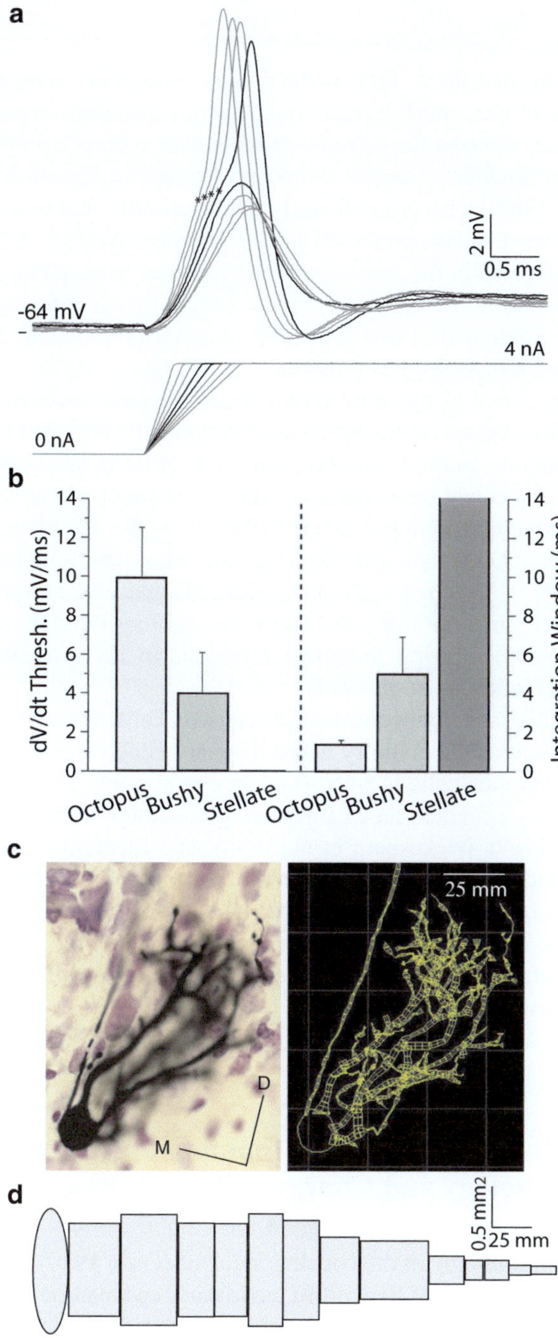

Fig. 14.2 The unique biophysics and morphology of octopus cells. (**a**) The membrane voltage of an octopus cell (*top*) during application of linear current ramps of varying durations (*bottom*). *Black traces* show the membrane voltage just below and above the rate-of-depolarization (d*V*/d*t*) threshold. Longer duration subthreshold ramps do not trigger an action potential (not shown; see refs. Morley and Happe 2000; Cai et al. 1997). The visually determined approximate voltage threshold for

power law would distribute less surface area into their daughter dendrites. Thick daughter dendrites could increase the speed of signal transmission by reducing the axial resistance, whereas the impedance mismatch at branch points would delay propagation from smaller to larger diameter dendrites (Agmon-Snir and Segev 1993; Lanzinger 1987). The original study demonstrating that octopus cells compensate traveling wave delays predicted a delay compensation of ~0.5 ms, matching in vivo delays measured in the same species (McGinley et al. 2005, 2012). A more recent study that did not incorporate detailed morphological information predicted compensation for about half of that delay (~0.28 ms) (Spencer et al. 2012), suggesting that the precise morphology is important.

The dominant source of synaptic input to octopus cells is excitation from the auditory nerve (Oertel et al. 2000). ANFs are tonotopically organized in the cochlear nucleus of all mammals studied, including mice (Osen 1970; Wickesberg and Oertel 1988). Octopus cells orient their dendrites across the tonotopic axis, so ANFs that encode higher frequencies synapse closer to the tips of the dendrites, and those that encode lower frequencies synapse closer to the soma (Fig. 14.1a, see also ref. Golding et al. 1995). Octopus cells show some diversity in morphology (Rhode et al. 1983) and, in mice but not cats, excite one another through axon collaterals (Golding et al. 1995), serving unknown functions in the processing of sounds (Godfrey et al. 1975; Rhode et al. 1983).

The role of inhibitory connections onto octopus cells is uncertain but may be minimal. Stimulation of the auditory nerve does not elicit GABAergic or glycinergic inhibition in octopus cells in mice (Golding et al. 1995). Consistent with this observation, there is a marked lack of glycinergic staining in the octopus cell area (Wenthold et al. 1987; Wickesberg et al. 1994), and tuberculoventral glycinergic interneurons exclude their projection from the octopus cell area (Wickesberg et al. 1991). However, GABAergic terminals exist on octopus cell dendrites (Marie et al. 1989), and glycine may be expressed in terminals in the octopus cell area of guinea pigs (Kolston et al. 1992), though glycine expression is sensitive to experimental conditions (Wickesberg et al. 1994).

14.4 Rapid Integration Across Tonotopy: A Marriage of Morphology and Connectivity with Biophysics

The relationship of tonotopy in the cochlear nucleus (Osen 1970) with the dendritic morphology of octopus cells (Osen 1969) led to early speculation about their role in sound processing. It was suggested that octopus cells' dendritic orientation might determine the sensitivity to frequency-modulated (FM) sweeps of different rates or direction (Morest et al. 1973). This idea was later related to Rall's developing view of spatiotemporal integration in dendrites (Szentagothai and Arbib 1974). The discovery in mice that octopus cells orient their dendrites in a consistent orientation and direction with respect to the tonotopic organization led to the hypothesis that octopus cells compensate the traveling wave delay in their dendrites (Golding et al. 1995; Wickesberg and Oertel 1988).

This hypothesis was directly tested in mice by combining in vivo measurements of the traveling wave delay, in vitro recordings from octopus cells, morphological reconstruction, and passive and active compartmental modeling (McGinley et al. 2012). The traveling wave delay was measured in mice and found to be ~1.6 ms across the full hearing range (see Fig. 14.1b). Because octopus cells' dendrites span about 1/3 of the tonotopy in the VCN (Golding et al. 1995), during a broadband transient sound each neuron will be exposed to an ~0.5-ms sweep of dendritic activation spreading somatopetally (towards the soma) (see Fig. 14.1a). In separate experiments, whole-cell recordings in the current-clamp configuration were made from octopus cells in acute slices of the VCN from mice. Families of current steps were applied to determine the passive and active response of the membrane (Fig. 14.3a, b, black traces). After fixation, morphological reconstructions were made of recorded neurons (Fig. 14.2c, left) and imported into the software NEURON (Hines and Carnevale 1997) for computation modeling (Fig. 14.2c, right).

Passive compartmental models of octopus cells were built from current-clamp recordings and morphological reconstructions, using simulated annealing (Kirkpatrick and Vecchi 1983) to optimize fits of the models to the data (Fig. 14.3a). Active models were constructed using a model of the low-voltage-activated potassium conductance (g_{KL}) derived from medial superior olive neurons (Khurana et al. 2011), which have very similar g_{KL} channels to octopus cell neurons (Golding and Oertel 2012; Bal and Oertel 2001; Scott et al. 2005; Mathews et al. 2010), and a model of hyperpolarization-activated, cyclic nucleotide-gated channels derived from octopus cells (Bal and Oertel 2000; McGinley et al. 2012). Model parameters were adjusted for each model to result in a close match to the current-clamp recordings from the modeled neuron (Fig. 14.3b). Because g_{KL}- and hyperpolarization-activated channels are known to reside in the dendrites of octopus cells and medial superior olivary neurons (Scott et al. 2005; Mathews et al. 2010; Oertel et al. 2008), these channels were included in the dendrites in active models. The kinetics of excitatory synaptic conductances were set to match previous recordings from octopus cells (Gardner et al. 1999).

To test the sensitivity of octopus cells to somatopetal sweeps of synaptic activation arising from the traveling wave delay, the waveform of the somatic EPSP was determined in passive and active models in response to somatopetal sweeps of excitation with different durations. A single-compartment (point) model with identical input resistance and membrane time constant to each passive model was compared in order to evaluate the role of dendrites in shaping the somatic EPSP. Comparison was also made to a passive model with the location of arrival of each input randomized in order to evaluate the role of tonotopy and the resulting somatopetal wave of activation in EPSP shape. A final model was generated with the magnitude of the synaptic conductance at each dendritic location adjusted so that the somatic EPSP in response to individual synaptic activation was approximately constant. Such

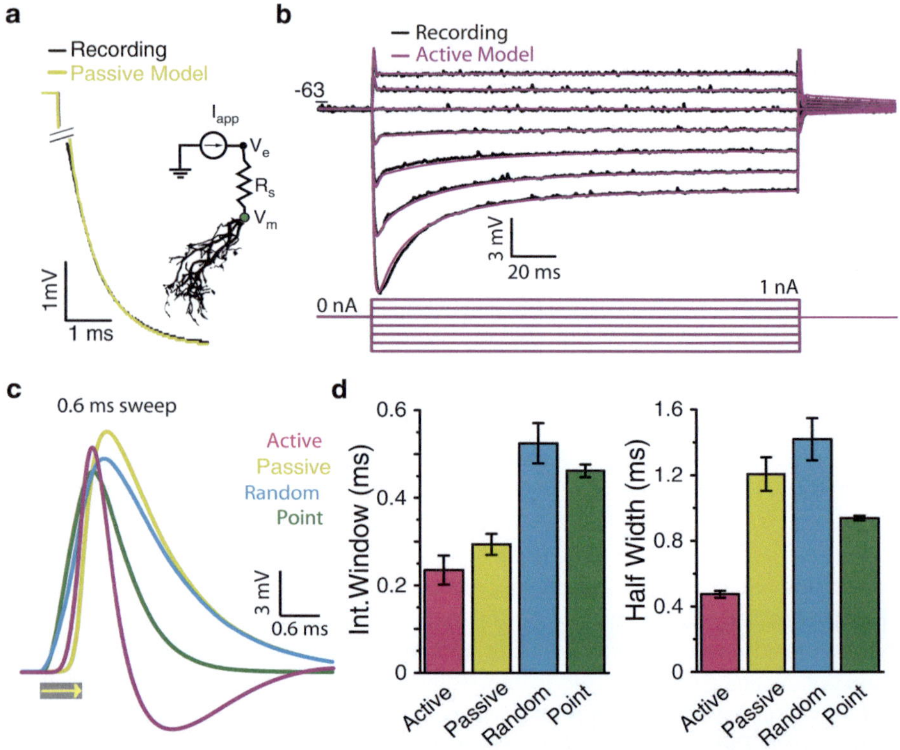

Fig. 14.3 Sensitivity to somatopetal sweeps of synaptic activation in biophysically and morphologically realistic models of octopus cells. (**a**) An example recording corrected for I_h (*black curve*) and the superimposed best-fit passive simulation (*yellow curve*). The series resistance drop, indicated by the *break*, was removed. The *inset* shows a diagram of the circuit used in passive simulations: I_{app} applied current; V_e electrode voltage; R_s series resistance; V_m membrane voltage. (**b**) Response of an octopus cell (*top black traces*) to a family of long current steps (*bottom traces*). Superimposed *magenta* traces show the responses to current steps of the same magnitude in an active compartmental model of the same octopus cell. (**c**) Somatic EPSP in response to a sweep of 0.6-ms duration from a typical passive model (*yellow*), active model (*magenta*), active model with spatially randomized inputs (*blue*), and a passive point model (*green*). (**d**) A comparison of the shape parameters between models for a sweep duration of 0.6 ms. *Color* scheme same as in panel (**c**). Statistical comparisons are in Table 3 of McGinley et al. (McGinley et al. 2012). Panels (**a**) and (**b**) are reproduced with permission from Fig. 14.2, panel (**c**) from Fig. 14.3, and panel (**d**) from Fig. 14.4 of ref. McGinley et al. (2012)

"input normalization" was discovered and thought to be important for reducing the location dependence of synaptic efficacy in hippocampal pyramidal neurons (Magee and Cook 2000). Surprisingly, input normalization had no effect on the dependence of somatic EPSP shape on somatopetal sweep duration (ref. McGinley et al. 2012, Fig. 14.3).

In active and passive models, somatopetal sweeps of activation lasting 0.6 ms—approximately equal to the duration expected in each octopus cell from the traveling wave delay—resulted in somatic EPSPs that were tall and sharp, and sharpest in active models (Fig. 14.3c, magenta and yellow curves). Responses in point models or passive models with randomized synaptic locations had smaller and more slowly rising EPSPs (Fig. 14.3c, blue and green curves). Active and passive models had short integration windows (the duration of the rising phase of the EPSP; ~0.2 and 0.3 ms, respectively), whereas randomized or point models had long integration windows (~0.5 ms; Fig. 14.3d, left). Thus, octopus cells are effective at compensating traveling wave delays in part due to their morphology and fast membranes.

14.5 The Role of Low-Voltage-Activated Potassium Channels

Although both active and passive models responded to sweeps matching the traveling wave delay with comparable rising rates and integration windows, half-widths were dramatically shorter in active than in passive models (0.5 vs. 1.2 ms; Fig. 14.3d, right). This reduction in half-width resulted from activation of g_{KL} channels during the EPSP, consistent with experimental observations in octopus cells and medial superior olivary neurons (Golding et al. 1999; Ferragamo and Oertel 2002; Svirskis et al. 2002; Scott et al. 2007). The large effect of g_{KL} channels on the falling phase of EPSPs is similar to the shortening of the falling phase of the action potential by potassium currents in classical and contemporary observations (Hodgkin et al. 1952; Bean 2007). The shortening of the falling phase of the EPSP by g_{KL} channels rapidly repolarizes EPSPs, which is important when they occur at high rates in the face of ongoing activity (Svirskis et al. 2004; Scott et al. 2007).

Activation of g_{KL} channels is also important in shaping the *rising* phase of the EPSP during somatopetal sweeps of activation in octopus cells. This can be seen in calculations of the sweep duration that results in the fastest rising and largest EPSP in passive versus active models. In active models, EPSPs were largest for sweeps lasting ~0.5 ms, with little variance between models (Fig. 14.4a, magenta symbol) in good agreement with the measured traveling wave delay of 0.53 ms per octopus cell (McGinley et al. 2012). Too slow of a sweep allows g_{KL} to activate and thereby reduce the EPSP amplitude. In passive models, however, the sweep duration with maximum EPSP amplitude varied between models from 0.8 to 1.6 ms (Fig. 14.4a, black symbol). Similarly, the rate of rise (dV/dt) of the EPSP was significantly larger in active than in passive models for physiological sweep durations; active models had a dV/dt of 50–70 mV/ms compared to 40–48 mV/ms in passive models for sweep durations of 0.4–0.6 ms (see ref. McGinley et al. 2012, Fig. 14.3b). The half-widths of EPSPs in passive and active models were similar to amplitudes in their dependence on sweep duration (Fig. 14.4a), further evidence of the role of g_{KL} in octopus cells' precise responses in the face of the cochlear traveling wave.

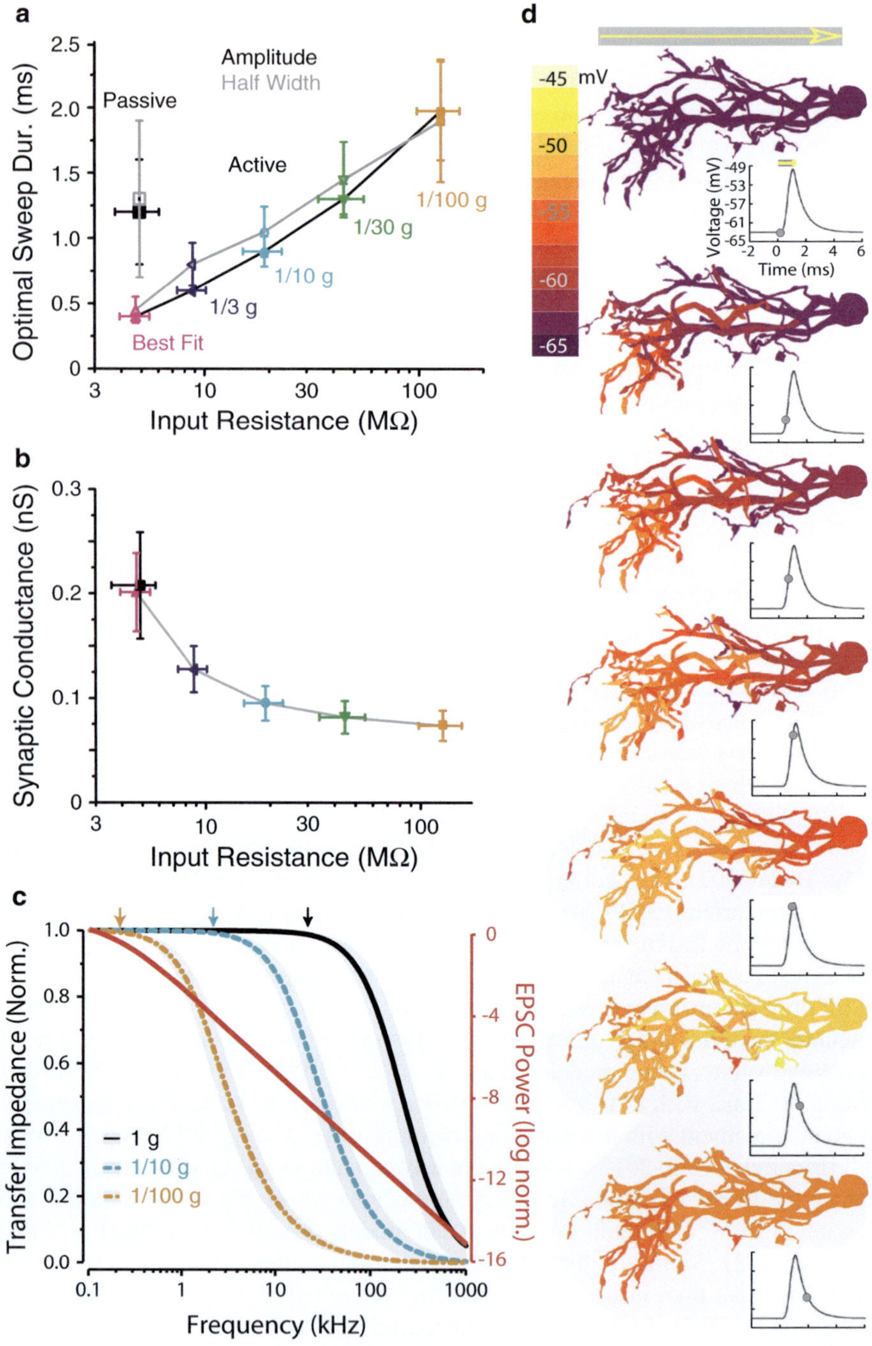

Fig. 14.4 Determinants of rapid signal transmission of octopus cells' dendrites. **(a)** The sweep duration that produces EPSPs with the largest amplitude or the narrowest half-width—the optimal sweep duration—is plotted against the input resistance for best-fit passive models (*black* and *grey squares*) and best-fit active models (*magenta triangles*) or active models with scaled membrane conductance (*other colored symbols*). *Black lines* and *filled symbols* are for amplitude; *grey lines*

14.6 Compensating Longer Delays in Low-Frequency-Hearing Animals

Octopus cells in mice have specialized dendritic morphology, connectivity, and biophysics tailored to compensate cochlear traveling wave delays and encode broadband transient sounds. However, several important aspects of octopus cell physiology remain unexplored, particularly relating to species with lower frequency hearing and thus longer traveling wave delays (Ruggero 1992; Ruggero and Rich 1987). Current findings motivate several possible mechanisms for octopus cells to compensate longer delays, including larger size, lower overall membrane conductance, lower dendritic g_{KL} conductance, or differences in the span of best frequencies in ANFs. Octopus cells in humans are large (Adams 1986). Octopus cells in cats and dogs have lower membrane conductance than in mice, but measurements have been made only in relatively immature animals (Bal and Baydas 2009; Bal et al. 2009).

The effect of overall membrane conductance on dendritic delay compensation was studied in active compartmental models of octopus cells (McGinley et al. 2012). Decreasing active and passive ion channel concentrations by 3-, 10-, 30-, or 100-fold changed the dendritic sweep duration that resulted in the largest and sharpest EPSPs over a range from 0.5 to 2.0 ms (Fig. 14.4a). Surprisingly, somatic input resistance changed only 25-fold, and the necessary synaptic conductance to elicit a large EPSP changed only 3-fold with the 100-fold change in ion channel concentration (Fig. 14.4b), demonstrating the importance of cytoplasmic resistance in dendritic integration. Bigger animals could use larger diameter dendrites to overcome cytoplasmic resistivity. Channels for g_{KL} have an ~3-fold lower density in the dendrites of MSO neurons (Mathews et al. 2010). Octopus cells might use a similar conductance gradient to compensate longer delays. However, changing the dendritic-to-somatic leak conductance over a 10,000-fold range had little effect on dendritic delay compensation in passive octopus cell models (see Fig. 7 of ref. McGinley et al. 2012). The role of conductance gradients of active channels in dendritic delay compensation should be studied further.

Fig. 14.4 (continued) and *open symbols* are for half-width. Conductance magnitudes for active models indicated below with colors matching the symbols. (**b**) The total synaptic conductance is plotted against the input resistance, for the same models as in panel A. To facilitate comparisons, the synaptic conductance for each cell in each model was adjusted so that the EPSP amplitude for zero sweep duration was 15 mV. (**c**) The transfer impedance in best-fit passive models (*black*) and passive models with 10-fold (*light blue*) or 100-fold (*orange*) reduced membrane conductance superimposed on the EPSC power spectrum plotted against frequency (*red*). *Arrows* indicate maximum frequency up to which each model has relatively constant transfer impedance. *Grey bands* indicate ±1 SD, calculated across models ($n=4$). (**d**) Color maps of the spatial profile of the voltage for a sweep duration of 0.6 ms in an example simulated passive octopus cell. The time corresponding to each color map is indicated on the simulated EPSP traces (*insets*) with a *green dot*. Panels (**a**) and (**b**) are reproduced with permission from Fig. 5 and panel (**d**) from Fig. 8 of ref. McGinley et al. (2012)

There is a trade-off between rapid membrane responsiveness and slow dendritic integration that limits the length of delays that octopus cells can compensate in their dendrites and still fire rapidly. Bushy cells in mice and octopus cells in immature cats and dogs have lower g_{KL} conductances resulting in lower d*V*/d*t* thresholds and thus less stringent coincident detection (McGinley and Oertel 2006; Bal and Baydas 2009; Bal et al. 2009). Octopus cells' low membrane resistance allows them to transmit fast synaptic signals down their dendrites. This can be seen by calculating the frequency dependence of the transfer impedance between the tips of the dendrites and the soma, using the ztransfer tool in NEURON (Hines and Carnevale 1997). Normalized transfer impedance values can be calculated by dividing by the value at 0.1 kHz in that model. In best-fit passive models of octopus cells, the normalized transfer impedance is flat up to ~20 kHz (Fig. 14.4c, black curve and arrow).

Thus, octopus cells' dendrites in mice transmit the frequency content in their excitatory postsynaptic currents (EPSCs) (Gardner et al. 1999). This can be seen in the power spectrum of a double-exponential fit to EPSC waveforms from the synaptic conductance, g_{syn} measured in octopus cells, which can be derived, and is

$$\mathrm{P}(f) = \frac{g_{\max}^2 \times \left(w_f - w_r\right)^2}{\eta^2 \times \left(w_r^2 + f^2\right) \times \left(w_f^2 + f^2\right)},$$

where f is the frequency in kHz, $w = 1/\tau$, "*r*" and "*d*" refer to the rising and falling time constants of the double exponential, and η is a normalization factor, following ref. McGinley et al. (2012). Defining normalized EPSG power as log(P(*f*)/*P*(0.1)), where *P*(0.1) is the power at 0.1 kHz, illustrates that the EPSC power spectrum in octopus cells is significant out to very high frequencies as a result of the extremely fast rise of the EPSC (Fig. 14.4c, red curve) (Gardner et al. 1999). When the conductance values in active models are divided by 10 (Fig. 14.4c, blue curve) or 100 (Fig. 14.4c, orange curve) while keeping all other parameters the same, they no longer reliably transfer the high-frequency components of EPSCs. For 10-fold reduction, the transfer impedance was flat up to ~2 kHz (Fig. 14.4c, blue arrow), and with 100-fold it was flat up to ~0.2 kHz (Fig. 14.4c, orange arrow). As a result of octopus cells' effective transfer of fast signals down their dendrites, a rapid somatopetal sweep of excitation results in progressive activation of the dendritic tree and rapidly rising EPSPs (Fig. 14.4d). The role of active conductances, particularly gKL, and conductance gradients in shaping the dendrite-to-soma transfer impedance warrants further exploration.

14.7 Downstream Processing and Comparison to Other Systems

Octopus cells are perhaps the fastest dendritic integrators in the brain (Golding and Oertel 2012; McGinley et al. 2012; McGinley and Oertel 2006; Oertel et al. 2000, 2009; Spruston et al. 2007) and encode sounds with extraordinary temporal precision (Godfrey et al. 1975; Oertel et al. 2000; Rhode 1994, 1998; Rhode and Smith 1986). Yet, our understanding of how this exquisite spectro-temporal processing is utilized by downstream circuits is based on indirect evidence and speculation. Octopus cells project predominantly to the contralateral ventral subdivision of the ventral nucleus of the lateral lemniscus (VNLLv) with exceptionally thick axons (Adams 1997; Smith et al. 2005; Schofield and Cant 1997) where they form large endbulb endings onto somas (Adams 1997; Vater and Feng 1990). This configuration results in rapid and powerful excitation of their targets (Schofield and Cant 1997; Vater and Feng 1990; Willard and Ryugo 1983), which are similar morphologically—and perhaps physiologically—to the temporally precise large spherical bushy cells in the VCN (Adams 1997; Wu 1999; McGinley and Oertel 2006; Oertel et al. 2009). VNLLv spherical bushy cells are inhibitory neurons (Saint Marie et al. 1997; Moore and Moore 1987; Winer et al. 1995) and project predominantly non-topographically to the ipsilateral inferior colliculus (IC) (Adams 1979; Brunso-Bechtold et al. 1981; Roth et al. 1978; Whitley and Henkel 1984; Loftus et al. 2010).

Consistent with anatomical predictions, neurons in the VNLL synchronize to rapid sound modulations (Zhang and Kelly 2006). Some neurons in VNLL respond to clicks with large-amplitude, short-duration, onset inhibition, probably further sharpening the output to IC (Nayagam et al. 2005). There is substantial physiological evidence of rapid and broadly tuned inhibition in the IC. Short latency inhibition is seen in a large fraction of unit responses in the IC of cats and rabbits to a wide range of sounds, particularly when unanesthetized (Kuwada et al. 1989; Carney and Yin 1989; Fitzpatrick et al. 1995, 1999; Bock et al. 1972). Pharmacological evidence supports that short latency inhibition is a direct GABAergic or glycinergic input to the IC (Casseday et al. 2000). Rapid and/or side-band hyperpolarization of the membrane potential plays a key role in response area of many intracellularly recorded IC neurons (Kuwada et al. 1997). Much of this rapid and broadband input could be driven by the octopus cell-to-VNLL pathway. However, broadly tuned inputs to the central nucleus of IC, including from the DNLL and elsewhere in the VNLL, are also inhibitory, possibly because broadband excitation would degrade the tonotopic organization (Oertel and Wickesberg 2002).

The temporally precise disynaptic inhibitory pathway from octopus cells, through the VNLL, to the IC may be preserved from birds and reptiles, through rodents, bats, and cats, to humans (Adams 1979, 1997; Carr and Code 2000), and yet its function is unknown. It has been hypothesized that the function relates to pattern recognition in natural sounds (Oertel and Wickesberg 2002). Alternatively, the pathway could serve to sharpen tonotopy in the IC by providing lateral inhibition (Blakemore et al. 1970), though it is not clear why extraordinary speed would be necessary for this purpose (Amari 1977). Instead, the disynaptic inhibitory pathway could aid in sound source separation or stream segregation (Bregman 1994) by "wiping the slate" of activity in IC neurons at the initiation of sounds, providing a signal for "binding" immediately subsequent activity into a percept (Treisman 1996; Shadlen and Movshon 1999). This binding mechanism could contribute to the precedence effect (Langner 1992; Litovsky et al. 1999) and echo suppression (Yin 1994), because the first instance of the sound would usually contain the sharpest onsets and thus engage the binding mechanism most strongly. The disynaptic pathway likely contributes to transforming the rapid temporal code in the auditory periphery into a rate code or a slower temporal code in the IC suitable for processing by thalamocortical circuits (Eggermont 2001; Joris et al. 2004).

Octopus cells compensate the cochlear traveling wave delay using sensitivity to somatopetal sweeps of excitation, with passive cable properties playing a significant role. Thus, octopus cells may be the clearest example to date of a computation performed using Rall's prediction in 1964 of dendritic sensitivity to sweeps (Rall 1964). Some amacrine and ganglion cells in the retina use sensitivity to dendritic sweeps to compute object motion (Euler et al. 2002; Vaney et al. 2012). However, motion occurs over a wide range of temporal and spatial scales, and thus multiple complex circuit, synaptic, and intrinsic mechanisms play a role in this computation (Briggman et al. 2011; Vaney et al. 2012; Chap. 13 in this book). It has been demonstrated that cortical neurons are sensitive to sweeps of excitation, possibly to aid in detection of temporal input sequences (Branco et al. 2010; Chap. 15 in this book). However, sweep sensitivity in cortical neurons occurs on a much slower time scale than in octopus cells, supported by impedance gradients and other nonlinearities in cortical dendrites (Branco et al. 2010). Given the primitive state of our understanding of the precise distribution of ion channels in dendrites and the in vivo patterns of dendritic activation in most brain areas (Briggman et al. 2011; Bock et al. 2011; Johnston and Narayanan 2008), dendritic sweep sensitivity—predicted by Rall and exploited elegantly by octopus cells—will be the subject of study for decades to come.

Acknowledgments I thank Donata Oertel and John Rinzel for critical feedback on the manuscript and Doug Oliver, Charlie Liberman, and Nace Golding for valuable input.

References

Adams JC (1979) Ascending projections to the inferior colliculus. J Comp Neurol 183(3):519–538

Adams JC (1986) Neuronal morphology in the human cochlear nucleus. Arch Otolaryngol Head Neck Surg 112(12):1253

Adams J (1997) Projections from octopus cells of the posteroventral cochlear nucleus to the ventral nucleus of the lateral lemniscus in cat and human. Aud Neurosci 3(4):335–350

Agmon-Snir H, Segev I (1993) Signal delay and input synchronization in passive dendritic structures. J Neurophysiol 70(5):2066–2085

Amari S (1977) Dynamics of pattern formation in lateral-inhibition type neural fields. Biol Cybern 27(2):77–87

Bal R, Baydas G (2009) Electrophysiological properties of octopus neurons of the cat cochlear nucleus: an in vitro study. J Assoc Res Otolaryngol 10(2):281–293

Bal R, Oertel D (2000) Hyperpolarization-activated, mixed-cation current (I h) in octopus cells of the mammalian cochlear nucleus. J Neurophysiol 84(2):806–817

Bal R, Oertel D (2001) Potassium currents in octopus cells of the mammalian cochlear nucleus. J Neurophysiol 86(5):2299–2311

Bal R, Oertel D (2007) Voltage-activated calcium currents in octopus cells of the mouse cochlear nucleus. J Assoc Res Otolaryngol 8(4):509–521
Bal R, Baydas G, Naziroglu M (2009) Electrophysiological properties of ventral cochlear nucleus neurons of the dog. Hear Res 256(1):93–103
Bean BP (2007) The action potential in mammalian central neurons. Nat Rev Neurosci 8(6):451–465
Bendor D, Wang X (2005) The neuronal representation of pitch in primate auditory cortex. Nature 436(7054):1161–1165
Blakemore C, Carpenter RH, Georgeson MA (1970) Lateral inhibition between orientation detectors in the human visual system. Nature 228:37–39
Bock G, Webster W, Aitkin L (1972) Discharge patterns of single units in inferior colliculus of the alert cat. J Neurophysiol 35:265–277
Bock DD, Lee WCA, Kerlin AM, Andermann ML, Hood G, Wetzel AW, Yurgenson S, Soucy ER, Kim HS, Reid RC (2011) Network anatomy and in vivo physiology of visual cortical neurons. Nature 471(7337):177–182
Branco T, Clark B, Häusser M (2010) Dendritic discrimination of temporal input sequences in cortical neurons. Science 329(5999):1671–1675
Bregman AS (1994) Auditory scene analysis: the perceptual organization of sound. MIT Press, Cambridge
Briggman KL, Helmstaedter M, Denk W (2011) Wiring specificity in the direction-selectivity circuit of the retina. Nature 471(7337):183–188
Brunso-Bechtold J, Thompson G, Masterton R (1981) HRP study of the organization of auditory afferents ascending to central nucleus of inferior colliculus in cat. J Comp Neurol 197(4):705–722
Cai Y, Walsh EJ, McGee JA (1997) Mechanisms of onset responses in octopus cells of the cochlear nucleus: implications of a model. J Neurophysiol 78(2):872–883
Cai Y, McGee JA, Walsh EJ (2000) Contributions of ion conductances to the onset responses of octopus cells in the ventral cochlear nucleus: simulation results. J Neurophysiol 83(1):301–314
Cao XJ, Oertel D (2011) The magnitudes of hyperpolarization-activated and low-voltage-activated potassium currents co-vary in neurons of the ventral cochlear nucleus. J Neurophysiol 106(2):630–640
Carney LH, Yin T (1989) Responses of low-frequency cells in the inferior colliculus to interaural time differences of clicks: excitatory and inhibitory components. J Neurophysiol 62(1):144–161
Carr CE, Code RA (2000) The central auditory system of reptiles and birds. Comparative hearing: birds and reptiles. Springer, New York, pp 197–248
Casseday JH, Ehrlich D, Covey E (2000) Neural measurement of sound duration: control by excitatory–inhibitory interactions in the inferior colliculus. J Neurophysiol 84(3):1475–1487
Dallos P (1992) The active cochlea. J Neurosci 12(12):4575–4585
Eggermont JJ (2001) Between sound and perception: reviewing the search for a neural code. Hear Res 157(1):1–42
Euler T, Detwiler PB, Denk W (2002) Directionally selective calcium signals in dendrites of starburst amacrine cells. Nature 418(6900):845–852
Ferragamo MJ, Oertel D (2002) Octopus cells of the mammalian ventral cochlear nucleus sense the rate of depolarization. J Neurophysiol 87(5):2262–2270
Fitzpatrick DC, Kuwada S, Batra R, Trahiotis C (1995) Neural responses to simple simulated echoes in the auditory brain stem of the unanesthetized rabbit. J Neurophysiol 74(6):2469–2486
Fitzpatrick DC, Kuwada S, Kim D, Parham K, Batra R (1999) Responses of neurons to click-pairs as simulated echoes: auditory nerve to auditory cortex. J Acoust Soc Am 106:3460
Gabor M (1946) Theory of Communication. J Inst Elec Eng 93:429–457

Gardner SM, Trussell LO, Oertel D (1999) Time course and permeation of synaptic AMPA receptors in cochlear nuclear neurons correlate with input. J Neurosci 19(20):8721–8729
Gardner SM, Trussell LO, Oertel D (2001) Correlation of AMPA receptor subunit composition with synaptic input in the mammalian cochlear nuclei. J Neurosci 21(18):7428–7437
Godfrey DA, Kiang N, Norris BE (1975) Single unit activity in the posteroventral cochlear nucleus of the cat. J Comp Neurol 162(2):247–268
Golding NL, Oertel D (2012) Synaptic integration in dendrites: exceptional need for speed. J Physiol 590:5563–5569
Golding NL, Robertson D, Oertel D (1995) Recordings from slices indicate that octopus cells of the cochlear nucleus detect coincident firing of auditory nerve fibers with temporal precision. J Neurosci 15(4):3138–3153
Golding NL, Ferragamo MJ, Oertel D (1999) Role of intrinsic conductances underlying responses to transients in octopus cells of the cochlear nucleus. J Neurosci 19(8):2897–2905
Heisenberg W (1927) Über den anschaulichen inhalt der quantentheoretischen kinematik und mechanik. Zeitschrift Für Physik 43(3–4):172–198
Hille B (1994) Modulation of ion-channel function by G-protein-coupled receptors. Trends Neurosci 17(12):531–536
Hines ML, Carnevale NT (1997) The NEURON simulation environment. Neural Comput 9(6):1179–1209
Hodgkin AL, Huxley AF, Katz B (1952) Measurement of current–voltage relations in the membrane of the giant axon of Loligo. J Physiol 116(4):424
Johnston D, Narayanan R (2008) Active dendrites: colorful wings of the mysterious butterflies. Trends Neurosci 31(6):309–316
Joris P, Schreiner C, Rees A (2004) Neural processing of amplitude-modulated sounds. Physiol Rev 84(2):541–577
Khurana S, Remme MWH, Rinzel J, Golding NL (2011) Dynamic interaction of Ih and IK-LVA during trains of synaptic potentials in principal neurons of the medial superior olive. J Neurosci 31(24):8936–8947
Kirkpatrick S, Vecchi M (1983) Optimization by simmulated annealing. Science 220(4598):671–680
Kolston J, Osen K, Hackney C, Ottersen O, Storm-Mathisen J (1992) An atlas of glycine-and GABA-like immunoreactivity and colocalization in the cochlear nuclear complex of the guinea pig. Anat Embryol 186(5):443–465
König P, Engel AK, Singer W (1996) Integrator or coincidence detector? The role of the cortical neuron revisited. Trends Neurosci 19(4):130
Kuwada S, Batra R, Stanford TR (1989) Monaural and binaural response properties of neurons in the inferior colliculus of the rabbit: effects of sodium pentobarbital. J Neurophysiol 61(2):269–282
Kuwada S, Batra R, Yin TCT, Oliver DL, Haberly LB, Stanford TR (1997) Intracellular recordings in response to monaural and binaural stimulation of neurons in the inferior colliculus of the cat. J Neurosci 17(19):7565–7581
Langner G (1992) Periodicity coding in the auditory system. Hear Res 60(2):115–142
Lanzinger DJ (1987) Group delay caused by impedance mismatch. In ARFTG Conference Digest-Spring, 29th. IEEE 11:247–264
Levy KL, Kipke DR (1998) Mechanisms of the cochlear nucleus octopus cell's onset response: synaptic effectiveness and threshold. J Acoust Soc Am 103:1940
Litovsky RY, Colburn HS, Yost WA, Guzman SJ (1999) The precedence effect. J Acoust Soc Am 106(4):1633–1654
Loftus WC, Bishop DC, Oliver DL (2010) Differential patterns of inputs create functional zones in central nucleus of inferior colliculus. J Neurosci 30(40):13396–13408
Magee JC, Cook EP (2000) Somatic EPSP amplitude is independent of synapse location in hippocampal pyramidal neurons. Nat Neurosci 3:895–903
Marie RLS, Morest DK, Brandon CJ (1989) The form and distribution of GABAergic synapses on the principal cell types of the ventral cochlear nucleus of the cat. Hear Res 42(1):97–112

Mathews PJ, Jercog PE, Rinzel J, Scott LL, Golding NL (2010) Control of submillisecond synaptic timing in binaural coincidence detectors by Kv1 channels. Nat Neurosci 13(5):601–609
McGinley MJ, Oertel D (2006) Rate thresholds determine the precision of temporal integration in principal cells of the ventral cochlear nucleus. Hear Res 216:52–63
McGinley MJ, Lazimy YM, Bal R, Oertel D (2005) Measurement and modeling of cable properties and optimal input delay profiles in octopus cells of the ventral cochlear nucleus. Ass Res Otolaryngol 28
McGinley MJ, Liberman MC, Bal R, Oertel D (2012) Generating synchrony from the asynchronous: compensation for cochlear traveling wave delays by the dendrites of individual brainstem neurons. J Neurosci 32(27):9301–9311
Mills AW (1958) On the minimum audible angle. J Acoust Soc Am 30(4):237–246
Moore JK, Moore RY (1987) Glutamic acid decarboxylase-like immunoreactivity in brainstem auditory nuclei of the rat. J Comp Neurol 260(2):157–174
Morest DK, Kiang NYS, Kane EC, Guinan Jr J, Godfrey D (1973) Stimulus coding at caudal levels of the cat's auditory nervous system. II. Patterns of synaptic organization (Moller AR and Boston P., eds.) New York Academic, 479–504
Morley BJ, Happe HK (2000) Cholinergic receptors: dual roles in transduction and plasticity. Hear Res 147(1):104–112
Nayagam DAX, Clarey JC, Paolini AG (2005) Powerful, onset inhibition in the ventral nucleus of the lateral lemniscus. J Neurophysiol 94(2):1651–1654
Oertel D, Wickesberg RE (2002) Ascending pathways through ventral nuclei of the lateral lemniscus and their possible role in pattern recognition in natural sounds. Integrative functions in the mammalian auditory pathway. Springer Verlag, New York, pp 207–237
Oertel D, Bal R, Gardner SM, Smith PH, Joris PX (2000) Detection of synchrony in the activity of auditory nerve fibers by octopus cells of the mammalian cochlear nucleus. Proc Natl Acad Sci USA 97(22):11773–11779
Oertel D, Shatadal S, Cao X-J (2008) In the ventral cochlear nucleus Kv1. 1 and subunits of HCN1 are colocalized at surfaces of neurons that have low-voltage-activated and hyperpolarization-activated conductances. Neurosci 154(1):77–86
Oertel D, McGinley MJ, Cao XJ (2009) Temporal processing in the auditory pathway. Encyclopedia Neurosci 9:909–919
Osen KK (1969) Cytoarchitecture of the cochlear nuclei in the cat. J Comp Neurol 136(4):453–483
Osen K (1970) Course and termination of the primary afferents in the cochlear nuclei of the cat. An experimental anatomical study. Arch Ital Biol 108(1):21
Rall W (1964) Theoretical significance of dendritic trees for neuronal input–output relations. In: Reiss RF (ed) Neural theory and modeling. Stanford Univ. Press, Stanford, CA
Raman IM, Trussell LO (1992) The kinetics of the response to glutamate and kainate in neurons of the avian cochlear nucleus. Neuron 9(1):173–186
Rhode WS (1994) Temporal coding of 200 % amplitude modulated signals in the ventral cochlear nucleus of cat. Hear Res 77(1):43–68
Rhode WS (1995) Interspike intervals as a correlate of periodicity pitch in cat cochlear nucleus. J Acoust Soc Am 97(4):2414
Rhode WS (1998) Neural encoding of single-formant stimuli in the ventral cochlear nucleus of the chinchilla. Hear Res 117(1–2):39
Rhode WS, Smith PH (1986) Encoding timing and intensity in the ventral cochlear nucleus of the cat. J Neurophysiol 56(2):261–286
Rhode W, Oertel D, Smith P (1983) Physiological response properties of cells labeled intracellularly with horseradish peroxidase in cat ventral cochlear nucleus. J Comp Neurol 213(4):448–463
Roth G, Aitkin L, Andersen R, Merzenich M (1978) Some features of the spatial organization of the central nucleus of the inferior colliculus of the cat. J Comp Neurol 182(4):661–680
Ruggero MA (1992) Responses to sound of the basilar membrane of the mammalian cochlea. Curr Opin Neurobiol 2(4):449–456

Ruggero MA, Rich NC (1987) Timing of spike initiation in cochlear afferents: dependence on site of innervation. J Neurophysiol 58(2):379–403
Saint Marie RL, Shneiderman A, Stanforth DA (1997) Patterns of γ-aminobutyric acid and glycine immunoreactivities reflect structural and functional differences of the cat lateral lemniscal nuclei. J Comp Neurol 389(2):264–276
Schofield BR, Cant NB (1997) Ventral nucleus of the lateral lemniscus in guinea pigs: cytoarchitecture and inputs from the cochlear nucleus. J Comp Neurol 379(3):363–385
Scott LL, Mathews PJ, Golding NL (2005) Posthearing developmental refinement of temporal processing in principal neurons of the medial superior olive. J Neurosci 25(35):7887–7895
Scott LL, Hage TA, Golding NL (2007) Weak action potential backpropagation is associated with high-frequency axonal firing capability in principal neurons of the gerbil medial superior olive. J Physiol 583(2):647–661
Shadlen MN, Movshon JA (1999) Synchrony unbound: review a critical evaluation of the temporal binding hypothesis. Neuron 24:67–77
Smith PH, Massie A, Joris PX (2005) Acoustic stria: anatomy of physiologically characterized cells and their axonal projection patterns. J Comp Neurol 482(4):349–371
Spencer MJ, Grayden DB, Bruce IC, Meffin H, Burkitt AN (2012) An investigation of dendritic delay in octopus cells of the mammalian cochlear nucleus. Front Comput Neurosci 6:83
Spruston N, Stuart G, Häusser M (2007) Dendritic integration. Dendrites. eds. Oxford University Press. 231–271
Svirskis G, Kotak V, Sanes DH, Rinzel J (2002) Enhancement of signal-to-noise ratio and phase locking for small inputs by a low-threshold outward current in auditory neurons. J Neurosci 22(24):11019–11025
Svirskis G, Kotak V, Sanes DH, Rinzel J (2004) Sodium along with low-threshold potassium currents enhance coincidence detection of subthreshold noisy signals in MSO neurons. J Neurophysiol 91(6):2465–2473
Szentagothai J, Arbib MA (1974) Conceptual models of neural organization. Neurosci Res Program Bull 12:305–510
Taberner AM, Liberman MC (2005) Response properties of single auditory nerve fibers in the mouse. J Neurophysiol 93(1):557–569
Treisman A (1996) The binding problem. Curr Opin Neurobiol 6(2):171–178
Vaney DI, Sivyer B, Taylor WR (2012) Direction selectivity in the retina: symmetry and asymmetry in structure and function. Nat Rev Neurosci 13:194–208
Vater M, Feng AS (1990) Functional organization of ascending and descending connections of the cochlear nucleus of horseshoe bats. J Comp Neurol 292(3):373–395
Weisburn BA, Parks, TW (1992) Cochlear filters: resolution and the uncertainty principle. In: Digital signal processing workshop, 1992. The (pp. 1–6). IEEE
Wenthold R, Huie D, Altschuler R, Reeks K (1987) Glycine immunoreactivity localized in the cochlear nucleus and superior olivary complex. Neurosci 22(3):897–912
Whitley JM, Henkel CK (1984) Topographical organization of the inferior collicular projection and other connections of the ventral nucleus of the lateral lemniscus in the cat. J Comp Neurol 229(2):257–270
Wickesberg RE, Oertel D (1988) Tonotopic projection from the dorsal to the anteroventral cochlear nucleus of mice. J Comp Neurol 268(3):389–399
Wickesberg RE, Whitlon D, Oertel D (1991) Tuberculoventral neurons project to the multipolar cell area but not to the octopus cell area of the posteroventral cochlear nucleus. J Comp Neurol 313(3):457–468
Wickesberg RE, Whitlon D, Oertel D (1994) In vitro modulation of somatic glycine-like immunoreactivity in presumed glycinergic neurons. J Comp Neurol 339(3):311–327
Willard FH, Ryugo DK (1983) Anatomy of the central auditory system. The auditory psychobiology of the mouse. Springfield: CC Thomas. p 201–304
Winer JA, Larue DT, Pollak GD (1995) GABA and glycine in the central auditory system of the mustache bat: structural substrates for inhibitory neuronal organization. J Comp Neurol 355(3):317–353

Wu SH (1999) Synaptic excitation in the dorsal nucleus of the lateral lemniscus. Prog Neurobiol 57(3):357–375

Yang X, Wang K, Shamma SA (1992) Auditory representations of acoustic signals. IEEE Transactions on information. Theory 38(2):824–839

Yin T (1994) Physiological correlates of the precedence effect and summing localization in the inferior colliculus of the cat. J Neurosci 14(9):5170–5186

Zhang H, Kelly JB (2006) Responses of neurons in the rat's ventral nucleus of the lateral lemniscus to amplitude-modulated tones. J Neurophysiol 96(6):2905–2914

Chapter 15
Computing Temporal Sequence with Dendrites

Tiago Branco

Abstract This chapter describes recent experimental results showing that dendrites of cortical pyramidal neurons can compute the temporal sequence of synaptic input. Electrophysiological recordings combined with two-photon glutamate uncaging, calcium imaging, and compartmental modeling have shown that single cortical dendrites have a gradient of nonlinear synaptic integration, which relies on dendritic impedance gradients and nonlinear synaptic NMDA receptor activation. This gradient confers high sensitivity to the temporal input sequence, allowing single dendrites and individual neurons to implement a fundamental cortical computation.

15.1 Introduction

The detection and discrimination of temporal input sequences is a fundamental computation implemented by the brain. It underlies basic elements of animal and human behavior, such as communication via vocalizations, where specific sequences of sounds have a social meaning, or the sensory perception of directionality, where the order of activation of sensory receptors carries information about where the stimulus is coming from or going to (Meister et al. 1995; deCharms and Merzenich 1996; Wehr and Laurent 1996; Johansson and Birznieks 2004). While the early stages of sensory systems have specific mechanisms to implement the detection of temporal sequences (Barlow and Levick 1965; Borg-Graham and Grzywacz 1992; Euler et al. 2002; Johansson and Birznieks 2004; Hausselt et al. 2007), this sensitivity to the temporal order of the stimulus is carried to high-order sensory brain areas (Simons 1978; deCharms and Merzenich 1996; Livingstone 1998; Zhang et al. 2003) and

T. Branco (✉)
MRC Laboratory of Molecular Biology, Francis Crick Avenue,
Cambridge Biomedical Campus, Cambridge, CB2 0QH, UK
e-mail: tbranco@mrc-lmb.cam.ac.uk

H. Cuntz et al. (eds.), *The Computing Dendrite: From Structure to Function*,
Springer Series in Computational Neuroscience 11, DOI 10.1007/978-1-4614-8094-5_15,

ultimately should be transmitted to motor areas to generate outputs that are conditional to particular sensory stimulus sequences. The transmission of temporal selectivity can in principle be implemented in several ways. For example, a network of simple excitatory and inhibitory neurons can selectively amplify particular sequences of activation of an input layer and generate varying levels of output depending on the activation sequence (Borg-Graham and Grzywacz 1992). An alternative means for temporal discrimination is to implement it at the level of the neuron itself, instead of relying on network-based mechanisms. This requires more complex individual neurons that have mechanisms that decode temporal sequences but has the advantage that it requires less neurons to implement, which could be an advantage for small neuronal networks.

The first work to point out that neurons could respond selectively to sequences of synaptic inputs was the theoretical work of Wilfrid Rall in 1964 (Rall 1964). Pioneering the application of mathematical and cable analysis to the study of propagation of electrical impulses in dendrites, Rall showed that because dendrites act as a delay line, activation of synapses along a dendrite in different directions should produce different responses at the soma. Starting from the tip of the dendrites (IN sequence), if the interval between activating each synapse is such that the sequence of activation is equal to the speed of propagation, the peak of all synaptic potentials will arrive at the same time at the soma and summate to produce a large potential. Activation in the inverse direction (OUT sequence) can never produce peaks that align temporally at the soma and will tend to produce a plateau response (Fig. 15.1a). Dendrites thus seem to be ideally placed to implement the computation of temporal order detection. In addition, they are highly nonlinear devices because of the presence of multiple voltage-gated conductances (Schiller et al. 1997; Golding and Spruston 1998; Magee 2000; Gulledge et al. 2005; London and Häusser 2005; Johnston and Narayanan 2008) and because synaptic input is a conductance and not a pure current source. These mechanisms could in principle also be exploited to implement temporal sequence detection, and it has been shown that auditory neurons rely on sublinear summation of inputs in the same dendrite to compute coincidence detection (Agmon-Snir et al. 1998) and that starburst amacrine cells in the retina use a dendritic mechanism that contributes to direction selectivity of retinal ganglion cells (Euler et al. 2002; Hausselt et al. 2007). Despite Rall's prediction being more than 50 years old, until recently it was not technically feasible to directly test the sensitivity of dendrites to the sequence of excitatory synaptic input. This chapter reviews and summarizes the experimental and modeling work previously published in Branco et al. (2010), showing that cortical dendrites can compute temporal input sequences.

15.2 Technical Challenges and Solutions

The main challenge in testing the response of dendrites to sequences of synaptic input is that it requires exceptionally high spatial and temporal precision of synaptic stimulation. Synapses are about 1 μm wide and act on a millisecond time scale, so a technique for activating single synapses in specific orders has to act on the same scales.

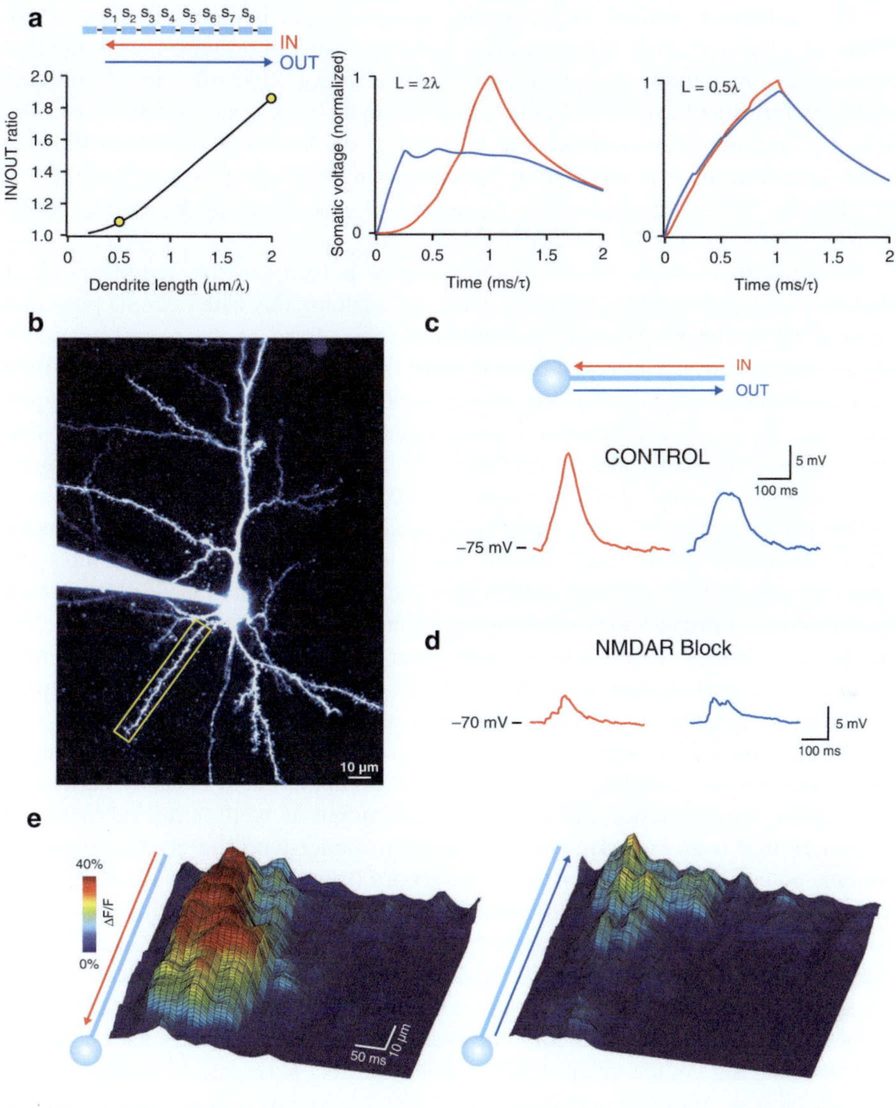

Fig. 15.1 (**a**) The original Rall model was recreated as faithfully as possible, and the cable length progressively shortened (*left panel*). The original model used a cable 2 length constants long, resulting in prominent direction selectivity (*middle panel*). Dendritic branches in layer 2/3 pyramidal neurons and most neurons in the brain have a length of 0.5 λ or less where direction selectivity is negligible (*right panel*). (**b**) Two-photon image of a layer 2/3 pyramidal cell used in experiments. (**c**) Two-photon glutamate uncaging on a single dendrite (*yellow box* in (**b**)) shows highly direction selective responses. The difference between the two directions requires activation of NMDA receptors (**d**). (**e**) Simultaneous calcium imaging reveals pronounced differences in the amount and spatial profile of calcium transients for the two different sequences of activation

The traditional method for activating synaptic input is electrical stimulation. While this has very high temporal precision (the electrical impulse for stimulation is less than 1 ms), the spatial precision is hard to control. Most often the stimulation activates a group of axons, and unless they run in an orderly and parallel direction, it is hard to control the number and location of the activated synapses. While in some preparations it is possible to activate a single synapse, testing a sequence of synapses would require careful placement of several electrodes next to target synapses, which would be extremely challenging.

An alternative means of activating synapses is to apply the neurotransmitter directly to the postsynaptic receptors. One way of doing this with extreme precision is to apply to the preparation a chemically caged form of the neurotransmitter (Ellis-Davies 2007). The caged form is inert, but the cage can be removed by light of a specific wavelength, allowing the neurotransmitter to bind to the postsynaptic receptors. By using a diffraction-limited laser spot it is possible to uncage neurotransmitter with high lateral resolution (<0.5 μm) and activate single synapses (Matsuzaki et al. 2001; DiGregorio et al. 2007). The laser light intensity is controlled using an electro- or acousto-optic modulator, and sub-millisecond flashes can be achieved. While single-photon lasers are more efficient at releasing most cages, single-photon excitation suffers from lack of resolution in the Z plane, which is a problem for preparations where synapses are several tens of micrometers below the surface. In these cases, a two-photon laser can be used to achieve excellent Z precision (~1 μm) (Matsuzaki et al. 2001). In the work described in this chapter, excitatory synapses were activated using two-photon glutamate uncaging.

For measuring the electrical response of a neuron to synaptic activation, the most convenient site of recording is the soma, as this compartment is easily accessible and reports the generation of action potential output as well as the subthreshold responses that lead to it. However, in order to understand the electrogenesis of somatic potentials it is informative to also record from the dendrites where the activated synapses are located. While multiple and simultaneous dendritic recordings have been obtained from thin dendrites (Larkum et al. 2007; Nevian et al. 2007), these recordings remain extremely challenging, and unless multiple electrodes on the same dendrite are used, they report the electrical activity visible only at the site of recording. An alternative is to combine electrical somatic recordings with dendritic calcium imaging. While calcium entry is not a direct surrogate of voltage, it can provide valuable information about the local voltage dynamics. In this work, a second two-photon laser was used to perform calcium imaging in several line-scan configurations, in combination with whole-cell somatic recordings.

15.3 Sequence Detection in Single Dendritic Branches

To test the sensitivity of single dendrites to the order of activation of a defined set of synapses, spatiotemporal input patterns were delivered using two-photon glutamate uncaging in spines of layer 2/3 pyramidal neurons (Branco et al. 2010). The first pattern tested was the one explored by Rall in his simulations: an ordered sequence of synaptic activation in opposite directions. This systematically produced electrical responses at the soma that were larger when activation started from the tip, just as Rall had predicted (Fig. 15.1b, c). The differences in the size of the two electrical responses were large enough that they had a pronounced effect in the probability of generating action potentials, demonstrating that the output of cortical neurons can represent the direction of synaptic input sequences onto a single dendritic branch using a local computation mechanism. While these experimental results are in accordance to Rall's simulations, simultaneous calcium imaging of the stimulated dendrite showed that there were dramatic differences in the amount of calcium between the two patterns (Fig. 15.1e). This suggested that local dendritic electrogenesis was contributing to the differences between the two patterns, which was not expected from Rall's theory, because in this case the difference results from summation of the peaks at the soma. Furthermore, there was also a noticeable difference in the integral of the two responses, which again was not expected from Rall's prediction, as the differences should be for the peak of the response but the charge should be the same in both directions. The contribution of dendritic electrogenesis was confirmed by blocking NMDA receptors, which abolished the differences between the two sequences (Fig. 15.1d).

At first sight, the lack of directional selectivity in the presence of NMDA-receptor blockers seems to contradict Rall's prediction. However, passive direction selectivity requires a cable long enough to generate significant filtering and delays to the EPSP peak. Basal dendrites of layer 2/3 cells are around 0.5 length constants long (Larkman et al. 1992), which is not long enough to introduce significant peak delays. Indeed, Rall's simulations were done with a cable which was 2 length constants long, and repeating these simulations for a shorter cable produces virtually no differences in the peak EPSP (Fig. 15.1a).

To understand the biophysical basis of the NMDAR-dependent direction sensitivity, a detailed compartmental model was constructed, based on a reconstructed layer 2/3 pyramidal cell. The initial model included excitatory synapses with NMDA receptors but it was otherwise passive, and it was able to accurately reproduce the experimentally observed sensitivity to the direction of input sequences, which was markedly reduced by removing NMDA receptors from the model (Fig. 15.2a, b). Analysis of the simulations showed that the mechanism for direction sensitivity in single dendrites can be explained by the interaction between the voltage-dependent Mg^{++} block of NMDAR channels (Mayer et al. 1984; Nowak et al. 1984) with the spatial gradient in local input impedance along the dendrite. When distal synapses are activated first, the local depolarization is large due to the high local input impedance, and this produces substantial relief of Mg^{++} block in the NMDAR channel. Subsequent activation of more proximal synapses in turn generates significantly greater NMDAR current, as the distally generated EPSP propagates inwards, leading to a regenerative wave of depolarization spreading towards the soma. If the order of activation is reversed, the small local depolarization produced at the proximal synapses is less efficient in relieving the Mg^{++} block, and therefore substantial regenerative activation of NMDAR conductance is only achieved at the distal synapses.

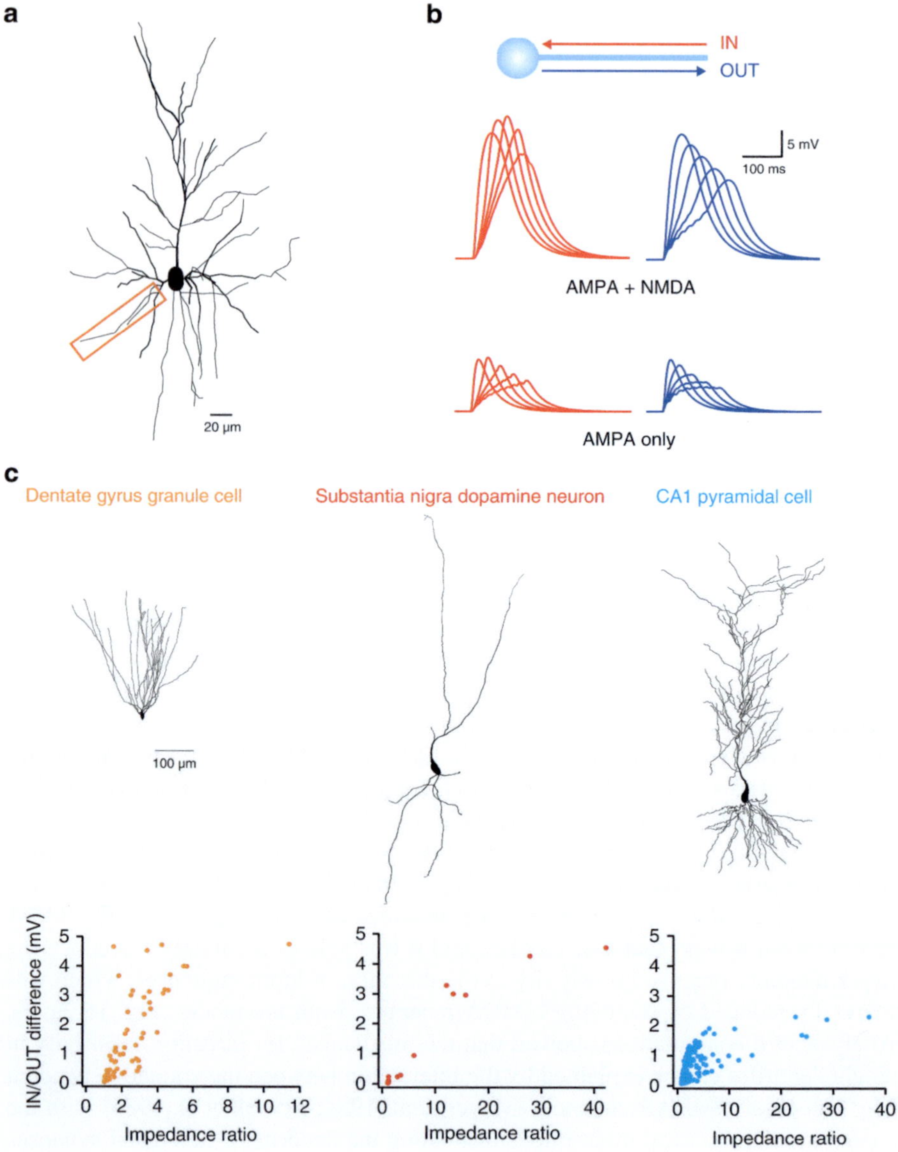

Fig. 15.2 (**a**) A cell used in one of the experiments was reconstructed and its morphology used for the simulations. (**b**) A passive model with AMPA and NMDA receptors reproduces the robust direction selectivity in short dendrites, in an NMDAR-dependent manner. Different traces for each direction show progressively slower speeds of activation. (**c**) Direction selectivity was tested for different cell morphologies (each datapoint in the plots corresponds to one dendritic branch). While IN/OUT differences increase with the impedance ratio of the dendrite, longer dendrites generate more pronounced differences (impedance ratio is the impedance at the tip over the impedance at the base of the branch)

Thus, more NMDAR conductance is recruited when activating synapses towards the soma than in the opposite direction, and the spatial profile of the experimentally recorded calcium transients is fully consistent with this explanation and matched the NMDAR conductance profile along the dendrite in the model.

Given the simple requirements for the generation of NMDAR-dependent direction sensitivity, the model was used to show that dendrites of a variety of cell types exhibit sufficient impedance gradients to exploit this mechanism (Fig. 15.2c). Different cell types have IN/OUT differences of different magnitude, which is due to two factors: first, the absolute value of the local impedance, which determines the overall amount of local depolarization and NMDA conductance and thus the amplitude of the IN/OUT difference, and second, the length of the dendritic branch, which sets the maximal inter-synaptic distance. This distance is important because the magnitude of the IN/OUT difference depends not only on the facilitation of NMDAR opening by previously activated synapses but also on the local driving force. In shorter dendrites, like those of CA1 pyramidal cells, synaptic depolarization spreads to the whole branch with little attenuation, temporarily reducing the driving force available to subsequent synaptic activation. In such dendrites, the benefit of activating distal synapses first is therefore partially attenuated by the smaller driving force available to the proximal synapses, thus counteracting the enhanced NMDAR opening and yielding smaller direction sensitivity when compared to the longer dendrites of granule cells of the dentate gyrus or substantia nigra dopamine neurons.

Given the prominent presence of different types of active dendritic conductances in pyramidal cells, models of increasing complexity were used to explore the impact of different types of voltage-gated conductances on direction sensitivity. These simulations showed that in general, increasing dendritic excitability increases the sensitivity to the direction of input sequences, which can also be achieved by depolarizing the membrane potential. This increase in sensitivity has however a limit, given that if the dendrite is too excitable, the less favorable OUT sequence can become as efficient as the IN sequence in recruiting NMDAR conductance, and both sequences of activation will generate maximal NMDAR recruitment and produce very similar responses. While the distribution of voltage-gated conductances is not known for basal dendrites of cortical neurons, other dendrites such as the apical trunk of hippocampal CA1 pyramidal cells and cortical L5 cells have a gradient of Ih channels, which will impact on the impedance gradient and can in principle counteract the sequence discrimination mechanism described here. However, the predominant effect of experimentally blocking Ih channels in L5 neurons is hyperpolarization and reduction of NMDAR recruitment (Branco and Häusser 2011), indicating that the presence of Ih can be favorable to NMDAR-dependent dendritic mechanisms. In general dendritic gradients of voltage-gated channels (Lörincz et al., 2002; Magee 1999; Mathews et al. 2010; Williams and Stuart 2000) or synaptic properties (Katz et al. 2009; Magee and Cook 2000) will be superimposed on, and may modify, the impedance gradient and the dendritic integration mechanisms that depend on it.

While testing and analyzing the activation of the two perfectly ordered IN and OUT sequences was essential for determining the underlying mechanisms of sequence sensitivity, it seems unlikely that such sequences will occur often in vivo, given the requirement for very precise subcellular connectivity and the stochastic nature of synaptic transmission. Therefore, the model was used to test the discrimination of random temporal patterns of input and showed that different sequences produced a wide range of somatic electrical responses, in an NMDAR-dependent manner. This happens because synapses at different locations along the dendrite differentially influence each other, depending on their relative timing. The use of the model had the great advantage that more than 400,000 sequences could be tested, which showed that not all sequences are different from each other. Instead, groups of sequences with similar temporal structures produce similar electrical responses, and the overall likelihood of discriminating any two sequences (for a >1 mV EPSP peak difference) is around 40 %. Using the information from the model, a subset of temporal sequences was selected and tested experimentally, confirming that single dendrites discriminate multiple temporal input patterns (Fig. 15.3a, b). This particular result is a great example of the benefits of the close interplay between modeling and experiments, as experimentally it would have not been feasible to test a large number of different sequences, and without the model it would have been impossible to predict which sequences should produce significantly different responses.

15.4 Sequence Discrimination with Multi-Dendrite Patterns

Voltage-dependent synaptic interactions are maximal if synapses are on the same dendrite, and thus single dendrite spatial patterns should exhibit maximal sensitivity to temporal sequences using an NMDAR-dependent mechanism. However, neurons in the cortex are connected by synapses that spread over multiple dendritic branches (Watts and Thomson 2005; Helmstaedter et al. 2007), and while it is possible that activation of specific subsets of input neurons results in activation of a large number of synapses in the same branch, it seems likely that in most cases the complement of activated synapses will be spread over multiple dendritic branches of the same neuron. Given that voltage can spread between dendritic branches (Zador et al. 1995), the extent depending on the tree geometry and biophysical properties, it is possible that the sequence discrimination mechanism described here for single dendrites might be exploited when synapses are distributed over the dendritic tree. This was tested both experimentally and using modeling, which revealed that robust sequence discrimination was possible in these conditions, in an NMDAR-dependent manner (Fig. 15.3c). Exploration of a wide range of spatial input distributions using the model showed that while the reliability of sequence discrimination scaled with the average number of synapses per dendritic branch, discrimination was still possible with very small number of synapses per branch. To confirm that sequence discriminability with patterns of synapses on different dendrites relies on voltage interactions between different branches, we performed experiments where the

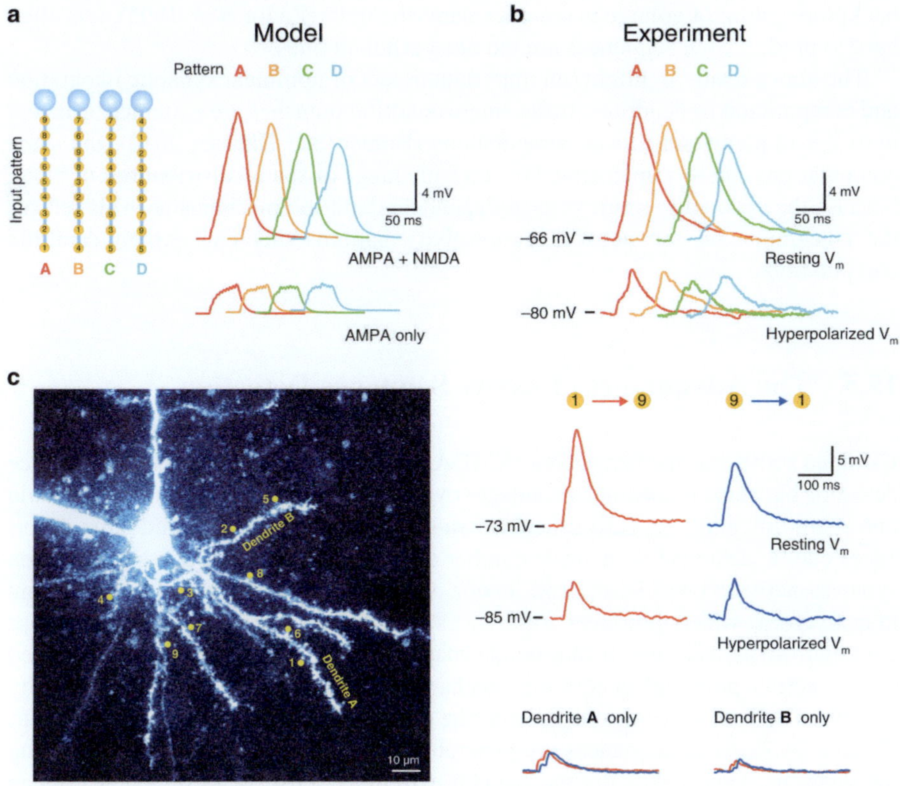

Fig. 15.3 (**a**) Four example temporal sequences tested in the model and corresponding somatic responses. All responses are different from each other in an NMDAR-dependent manner. (**b**) The same four sequences tested experimentally in a single dendrite show responses that are equally distinguishable. (**c**) Spatial patterns distributed across multiple dendrites also show sensitivity to the temporal sequence of activation. Isolated activation of each of the two dendrites that have more than one synapse shows small negligible differences for the two temporal sequences, suggesting that the overall difference results from interactions between all dendritic branches

somatic EPSP was recorded in response to the combined activation of all branches as well as to the separate activation of individual branches that had more than one synapse. The assumption that the differences between the two tested sequences with distributed input resulted entirely from the differences generated within a single dendrite was tested by comparing the sum of the differences from each branch with the recorded difference in response to multi-branch activation. The measured difference was almost ten times larger than expected, directly confirming that interactions between different dendrites underlie sequence discrimination with distributed input (Fig. 15.3c). Importantly, there was no relationship between sequence discrimination and the net directionality of the pattern. The sequences that generate the largest responses are the ones that maximize voltage transfer between synapses activated closely in time, and given the complex nature of forward and backpropagation of voltage in complex dendritic trees (Zador et al. 1995), it is often hard to predict what sequences are the most efficient ones.

The above result highlights an important aspect of nonlinear synaptic integration and computation in dendrites: while single dendritic branches are extremely efficient units for implementing local computations (Branco and Häusser 2010), the same computations can be implemented over a wide range of spatial distributions of input, because they can still recruit voltage-dependent dendritic mechanisms. This relaxes the constraints on the network connectivity patterns that can exploit dendritic computations.

15.5 The Advantages of Active Sequence Detection

Using an active mechanism such as NMDAR activation for implementing sequence detection has several potential advantages over a purely passive mechanism. The main one is amplification: an active mechanism generates more charge, and large differences can be achieved with small number of synapses and short dendrites. For ten synapses activated on a basal dendrite of a cortical neuron this amplification is more than fivefold, which produces differences that are robust against synaptic noise coming from activation of background synapses and that can be reliably thresholded by the action potential generation mechanism. The passive sequence sensitivity described by Rall depends crucially on the precise timing of synaptic activation, as differences between sequences are generated because of propagation delays along the dendrite. This makes the mechanism extremely sensitive to temporal jitter in synaptic activation (Faisal and Laughlin 2007) and could entirely abolish differences between sequences. On the other hand, an NMDAR-dependent mechanism requires less temporal precision, because the selective amplification of favorable sequences depends on the overlap between depolarization coming from activation of other synapses and glutamate binding to the NMDA receptor, which lasts for several milliseconds (Lester and Jahr 1992). Another potential advantage of using NMDA receptors is that the degree of sequence sensitivity can be quickly tuned by changing the membrane potential as well as by changing the amount of NMDA receptors, which can be done in a dendrite- or a synapse-specific manner, thus allowing the same neuron to have dendrites or groups of synapses with different ability for discriminating temporal sequences.

It is important to note that while NMDA receptors are the main channel underlying the mechanism described here, in principle similar results could be obtained with other types of voltage-gated channels, such as voltage-gated calcium channels. However, NMDA receptors have the advantage of being recruited in a more graded manner when compared to voltage-gated calcium or sodium channels (Major et al. 2008; Branco and Häusser 2011), which allows for a wide analog range of response sizes to which sequences can be mapped to, whereas sodium or calcium voltage-gated channels will tend to produce a binary discrimination of sequences, generating either a large or a small response.

15.6 Conclusion

The experimental and theoretical results described in this chapter show that single neurons can discriminate temporal sequences of synaptic input using a dendritic mechanism. This is a complex computation that can in principle be exploited by neural networks for implementing sequence detection tasks that are important for behavior. Like for other dendritic computations that have been described (Mel 1993; Schiller et al. 2000; Poirazi et al. 2003; London and Häusser 2005; Losonczy and Magee 2006; Branco and Häusser 2011), successful implementation depends on the subcellular organization of networks and could prove especially relevant for circuits with layered input such as the hippocampus, where for example this mechanism could be used by dentate gyrus granule cells to directly detect the sequence of entorhinal cortex activation. Given that sequence detection can be implemented across multiple dendrites, a general bias in the distribution of inputs, which has been described for several cortical areas (Petreanu et al. 2009; Richardson et al. 2009; Little and Carter 2012), could be sufficient to differentially process the order of activation of different classes of input to single neurons. A final answer requires detailed knowledge of the entire spatiotemporal pattern of inputs onto single neurons during behavior. While this is a formidable task, technical advances in electrical recording and imaging methods are rapidly making this an achievable goal, and of the amazing library of mechanisms that are available to neurons, we will soon know which are the ones used to implement computations in real life.

References

Agmon-Snir H, Carr CE, Rinzel J (1998) The role of dendrites in auditory coincidence detection. Nature 393(6682):268–272

Barlow HB, Levick WR (1965) The mechanism of directionally selective units in rabbit's retina. J Physiol 178(3):477–504

Borg-Graham LJ, Grzywacz NM (1992) A model of the directional selectivity circuit in retina: transformations by neurons singly and in concert. In: McKenna T, Davis J, Zornetzer ZF (eds) Single neuron computation. Academic, San Diego, pp 347–376

Branco T, Häusser M (2010) The single dendritic branch as a fundamental functional unit in the nervous system. Curr Opin Neurobiol 20(4):494–502

Branco T, Häusser M (2011) Synaptic integration gradients in single cortical pyramidal cell dendrites. Neuron 69(5):885–892

Branco T, Clark BA, Häusser M (2010) Dendritic discrimination of temporal input sequences in cortical neurons. Science 329(5999):1671–1675

deCharms RC, Merzenich MM (1996) Primary cortical representation of sounds by the coordination of action-potential timing. Nature 381(6583):610–613

DiGregorio DA, Rothman JS, Nielsen TA, Silver RA (2007) Desensitization properties of AMPA receptors at the cerebellar mossy fiber granule cell synapse. J Neurosci 27(31):8344–8357

Ellis-Davies GC (2007) Caged compounds: photorelease technology for control of cellular chemistry and physiology. Nat Methods 4(8):619–628

Euler T, Detwiler PB, Denk W (2002) Directionally selective calcium signals in dendrites of starburst amacrine cells. Nature 418(6900):845–852

Faisal AA, Laughlin SB (2007) Stochastic simulations on the reliability of action potential propagation in thin axons. PLoS Comput Biol 3(5):e79

Golding NL, Spruston N (1998) Dendritic sodium spikes are variable triggers of axonal action potentials in hippocampal CA1 pyramidal neurons. Neuron 21(5):1189–1200

Gulledge AT, Kampa BM, Stuart GJ (2005) Synaptic integration in dendritic trees. J Neurobiol 64(1):75–90

Hausselt SE, Euler T, Detwiler PB, Denk W (2007) A dendrite-autonomous mechanism for direction selectivity in retinal starburst amacrine cells. PLoS Biol 5(7):e185

Helmstaedter M, de Kock CP, Feldmeyer D, Bruno RM, Sakmann B (2007) Reconstruction of an average cortical column in silico. Brain Res Rev 55(2):193–203

Johansson RS, Birznieks I (2004) First spikes in ensembles of human tactile afferents code complex spatial fingertip events. Nat Neurosci 7(2):170–177

Johnston D, Narayanan R (2008) Active dendrites: colorful wings of the mysterious butterflies. Trends Neurosci 31(6):309–316

Katz Y, Menon V, Nicholson DA, Geinisman Y, Kath WL, Spruston N (2009) Synapse distribution suggests a two-stage model of dendritic integration in CA1 pyramidal neurons. Neuron 63(2):171–177

Larkman AU, Major G, Stratford KJ, Jack JJ (1992) Dendritic morphology of pyramidal neurones of the visual cortex of the rat. IV: electrical geometry. J Comp Neurol 323(2):137–152

Larkum ME, Waters J, Sakmann B, Helmchen F (2007) Dendritic spikes in apical dendrites of neocortical layer 2/3 pyramidal neurons. J Neurosci 27(34):8999–9008

Lester RA, Jahr CE (1992) NMDA channel behavior depends on agonist affinity. J Neurosci 12(2):635–643

Little JP, Carter AG (2012) Subcellular synaptic connectivity of layer 2 pyramidal neurons in the medial prefrontal cortex. J Neurosci 32(37):12808–12819

Livingstone MS (1998) Mechanisms of direction selectivity in macaque V1. Neuron 20(3):509–526

London M, Häusser M (2005) Dendritic computation. Annu Rev Neurosci 28:503–532

Lörincz A, Notomi T, Tamás G, Shigemoto R, Nusser Z (2002) Polarized and compartment-dependent distribution of HCN1 in pyramidal cell dendrites. Nat Neurosci 5(11):1185–1193

Losonczy A, Magee JC (2006) Integrative properties of radial oblique dendrites in hippocampal CA1 pyramidal neurons. Neuron 50(2):291–307

Magee JC (1999) Dendritic lh normalizes temporal summation in hippocampal CA1 neurons. Nat Neurosci 2(6):508–514

Magee JC (2000) Dendritic integration of excitatory synaptic input. Nat Rev Neurosci 1(3):181–190

Magee JC, Cook EP (2000) Somatic EPSP amplitude is independent of synapse location in hippocampal pyramidal neurons. Nat Neurosci 3(9):895–903

Major G, Polsky A, Denk W, Schiller J, Tank DW (2008) Spatiotemporally graded NMDA spike/plateau potentials in basal dendrites of neocortical pyramidal neurons. J Neurophysiol 99(5):2584–2601

Mathews PJ, Jercog PE, Rinzel J, Scott LL, Golding NL (2010) Control of submillisecond synaptic timing in binaural coincidence detectors by K(v)1 channels. Nat Neurosci 13(5):601–609

Matsuzaki M, Ellis-Davies GC, Nemoto T, Miyashita Y, Iino M, Kasai H (2001) Dendritic spine geometry is critical for AMPA receptor expression in hippocampal CA1 pyramidal neurons. Nat Neurosci 4(11):1086–1092

Mayer ML, Westbrook GL, Guthrie PB (1984) Voltage-dependent block by Mg2+ of NMDA responses in spinal cord neurones. Nature 309(5965):261–263

Meister M, Lagnado L, Baylor DA (1995) Concerted signaling by retinal ganglion cells. Science 270(5239):1207–1210

Mel BW (1993) Synaptic integration in an excitable dendritic tree. J Neurophysiol 70(3):1086–1101

Nevian T, Larkum ME, Polsky A, Schiller J (2007) Properties of basal dendrites of layer 5 pyramidal neurons: a direct patch-clamp recording study. Nat Neurosci 10(2):206–214

Nowak L, Bregestovski P, Ascher P, Herbet A, Prochiantz A (1984) Magnesium gates glutamate-activated channels in mouse central neurones. Nature 307(5950):462–465

Petreanu L, Mao T, Sternson SM, Svoboda K (2009) The subcellular organization of neocortical excitatory connections. Nature 457(7233):1142–1145
Poirazi P, Brannon T, Mel BW (2003) Pyramidal neuron as two-layer neural network. Neuron 37(6):989–999
Rall W (1964) Theoretical significance of dendritic trees for neuronal input–output relations. In: Reiss R (ed) Neural theory and modeling. Stanford Univ. Press, Stanford, CA, pp 73–97
Richardson RJ, Blundon JA, Bayazitov IT, Zakharenko SS (2009) Connectivity patterns revealed by mapping of active inputs on dendrites of thalamorecipient neurons in the auditory cortex. J Neurosci 29(20):6406–6417
Schiller J, Schiller Y, Stuart G, Sakmann B (1997) Calcium action potentials restricted to distal apical dendrites of rat neocortical pyramidal neurons. J Physiol 505(Pt 3):605–616
Schiller J, Major G, Koester HJ, Schiller Y (2000) NMDA spikes in basal dendrites of cortical pyramidal neurons. Nature 404(6775):285–289
Simons DJ (1978) Response properties of vibrissa units in rat SI somatosensory neocortex. J Neurophysiol 41(3):798–820
Watts J, Thomson AM (2005) Excitatory and inhibitory connections show selectivity in the neocortex. J Physiol 562(Pt 1):89–97
Wehr M, Laurent G (1996) Odour encoding by temporal sequences of firing in oscillating neural assemblies. Nature 384(6605):162–166
Williams SR, Stuart GJ (2000) Site independence of EPSP time course is mediated by dendritic I(h) in neocortical pyramidal neurons. J Neurophysiol 83(5):3177–3182
Zador AM, Agmon-Snir H, Segev I (1995) The morphoelectrotonic transform: a graphical approach to dendritic function. J Neurosci 15(3 Pt 1):1669–1682
Zhang LI, Tan AY, Schreiner CE, Merzenich MM (2003) Topography and synaptic shaping of direction selectivity in primary auditory cortex. Nature 424(6945):201–205

Chapter 16
Modelling the Cellular Mechanisms of Fly Optic Flow Processing

Hermann Cuntz, Juergen Haag, and Alexander Borst

Abstract Understanding the computational power of individual neurons is one of the major tasks in neuroscience. Complex calculations in dendrites and axons were identified in recent years in a great number of different systems. The small set of around 60 lobula plate tangential cells (LPTCs) in the fly visual system is a prime example where such computations and their underlying mechanisms are well understood. In this chapter we review recent findings resulting from detailed modelling studies based on experimental data from LPTCs. These studies have shown that the network connectivity is sufficient to explain the morphology of LPTCs to a large extent. Furthermore, an extensive network ubiquitously but highly selectively connects LPTC dendrites and axons with each other and is responsible for sophisticated optic flow calculations. The electrotonic features of LPTCs are affected by this network. We describe how a selective dendritic network between Horizontal System (HS) and Centrifugal Horizontal (CH) cells implements a specialised spatial filter operating on the moving images encoded in the dendrites. We also show how an axonal network renders the axon terminals of Vertical System (VS) cells in the fly selective to rotational optic flow. In summary, fly LPTCs are a prime example to show that the embedding and wiring within the network is crucial to understand morphology and computation.

H. Cuntz (✉)
Ernst Strüngmann Institute (ESI) for Neuroscience in Cooperation with Max Planck Society, Frankfurt/Main, Germany

Institute of Clinical Neuroanatomy, Goethe University, Theodor-Stern-Kai 7, Frankfurt/Main, D-60590, Germany
e-mail: hermann.neuro@gmail.com

J. Haag • A. Borst
Department of Systems and Computational Neurobiology, Max Planck Institute of Neurobiology, Martinsried, Germany

H. Cuntz et al. (eds.), *The Computing Dendrite: From Structure to Function*, Springer Series in Computational Neuroscience 11, DOI 10.1007/978-1-4614-8094-5_16,

16.1 Introduction

Ever since the 1960s when Wilfrid Rall showed that the spatially distributed electric load on the dendrite plays a prominent role in synaptic integration (Rall 1959; Rall and Rinzel 1973), a vast number of complex calculations has been associated with the spatial distribution of inputs in the dendrites of single cells (for review, see London and Häusser 2005). At the same time, a number of computations were explored at the level of a network of neurons which can be understood without specifically modelling the dendritic tree (e.g., Hubel and Wiesel 1962; Seung et al. 2000). In recent years, however, it has increasingly become clear that computation in the nervous system originates from precise arrangements in both the network wiring and the single cell morphological and electrotonic composition (e.g., Bock et al. 2011; Briggman et al. 2011; Watt et al. 2009). Effort has been made to integrate sophisticated, anatomically and electrically realistic compartmental models into larger scale neural networks (Lang et al. 2011; Markram 2006). In most cases, however, the precise wiring in such models is not known, and morphological reconstructions in the network models cannot be connected according to their real biological counterpart. Here, we review one circuit in the fly lobula plate for which such knowledge exists.

A number of invertebrate neural circuits stand out for being particularly well studied both at the level of electrophysiology and anatomy of the single cells involved. For example, the looming sensitivity of the lobula giant movement detector (LGMD) cell in the locust has been successfully associated with a multiplicative computation in its dendrite between the angular velocity and the object size (Gabbiani et al. 2002). Also, the precise circuitry of the locust sky compass (Homberg et al. 2011), cricket cerci (Jacobs et al. 2008) and the central pattern generator of the lobster (Marder and Bucher 2001) have been dissected in terms of electrophysiology and morphology of the individual cells involved. Among neural circuits in general, only few are so amenable to detailed analyses as the network of tangential cells in the lobula plate of the fly (Borst 2012; Borst et al. 2010; Borst and Haag 2002). The lobula plate lies flat in full view when the cuticle is removed at the back of the fly's head. Inserting a tungsten electrode into the lobula plate enables the recording of extracellular spikes from the so-called H1 cell for many hours in the living fly (De Ruyter et al. 1988). This has been a model system to study neural coding and information theory (e.g., Bialek et al. 1991; Borst and Haag 2001) at the level of the single neuron using real-life visual stimulation. The H1 cell is one of only around 60 lobula plate tangential cells (LPTCs) many of which are meanwhile well characterised both electrophysiologically and morphologically.

Details of the underlying computations are known for many LPTCs. The lobula plate, a neural control centre for course corrections during flight (Borst and Haag 2002), encodes visual motion information retinotopically, that is, neighbouring points in the visual field correspond to neighbouring points on the lobula plate. Dendrites of LPTCs that integrate local motion information to produce large-field responses therefore cover large areas of the lobula plate. In agreement with their location within the lobula plate and thus the anatomical pendant of their receptive field, calcium signals measured in dendrites of LPTCs conserve the retinotopy of the visual input (Borst and Egelhaaf 1992). This literally means that motion images are directly mapped onto the dendrites of LPTCs. Via descending neurons, LPTCs then connect to motor neurons in the thoracic ganglia to control flight manoeuvres (Geiger and Nässel 1981; Haag et al. 2007; Wertz et al. 2008). Some LPTCs even connect directly to neck motor neurons responsible for head movements of the fly.

Horizontal System (HS) and Vertical System (VS) cells (Fig. 16.1a) are among the most accessible LPTCs for intracellular electrophysiology in the large blowfly, *Calliphora vicina*, where sharp recording electrodes are positioned in the very large axons (~10–25 μm diameter) of these cells (Hausen 1982a, b). Three HS cells exist that respond to horizontal motion in three horizontal parts of the visual field: the HSN cell responds to large-field motion in the northern part, and the HSE and HSS in the equatorial and southern parts, respectively. The dendrites of the ten VS cells are positioned sequentially along the rostro-caudal axis of the lobula plate jointly covering it with some degree of overlap. VS cells respond preferably to vertical motion in ten vertical stripes arranged from frontal to lateral corresponding to the location of their dendrites.

HS and VS cells respond to visual stimulation with graded membrane potential deflections, depolarising when a stimulus is moved in the preferred direction and hyperpolarising when the stimulus is moved in the anti-preferred direction. A number of computations were identified by studying the single cell electrophysiology of HS and VS cells. Firstly, it was shown that in HS and VS cells voltage-gated ion channels do not produce full-blown action potentials but rather elicit smaller spikelets that enhance the transmission of high temporal frequency signals that would otherwise be filtered out by the time constant of the membrane (Haag and Borst 1996). On the other hand, local irregularities in the motion signal get filtered out by spatial integration in the dendrite, a prime example for a functional interpretation of dendritic morphology (Single and Borst 1998). To better understand the sophisticated computations in single neurons, it has proven useful to design realistic computer models of the cells involved. LPTCs were modelled successfully with respect to their morphology, passive (Borst and Haag 1996) and active membrane properties (Borst and Single 2000; Haag et al. 1997) as well as with respect to their visual responses (Haag et al. 1999). It could be shown that the models reproducing the current flow within the realistic morphology of the individual cells were consistent with the dendritic integration principles mentioned above (Haag and Borst 1996; Single and Borst 1998).

Here, we summarise the results of modelling studies on morphology, electrophysiology and connectivity in HS and VS cells that shed light on their respective computations in the network. It becomes clear that in the case of these particular cells (1) the morphology mainly reflects the wiring in the network and (2) the specialised network connectivity between these neurons plays a decisive role for computation. The morphology and function of the single dendrite can therefore not be fully understood outside of the context of their wiring in the circuitry.

16.2 Modelling the Morphology of HS and VS Cells

Are the morphology and electrophysiology of HS and VS cells indeed specialised for the intricate computations described above? This could be the case similar to locust LGMDs that have a dedicated dendritic structure favouring the response of the cells to looming optic flow stimuli (Gabbiani et al. 2002; Peron et al. 2009; see also Chap. 17). Indeed, early work on LPTCs has shown that morphological features influence dendritic integration properties (Egelhaaf et al. 1994).

The first step in modelling the LPTC circuit therefore concerns the morphology of the cells. In insects, not only individual neuron types but actually individual neurons can be identified accurately because of their stereotypic anatomy and electrophysiology. As such, LPTCs are uniquely identifiable and, thus, represent an ideal subject for investigating the structure–function relationship in dendrites. While their anatomical structures are unique and identifiable by visual inspection, it is also obvious that for one particular cell, e.g., the left HSE cell, no one single dendritic branch from one animal can be associated with one particular branch of the same cell in another animal (Hausen 1982a). It was therefore interesting to study the inter-individual constancy and variability among LPTCs in a quantitative way, i.e., what factors make LPTCs different from each other and what factors are common to all cells. This section summarises this work (Cuntz et al. 2008), where we chose two members of the horizontal system (HSE and HSN, Fig. 16.1a black and red), and two members of the vertical system (VS2 and VS4, Fig. 16.1a green and blue), each of them represented by at least ten individuals from different flies to answer these questions. After the cells were filled with fluorescent dye, the anatomy of each neuron—including soma, axon and dendrite—was manually traced from two-photon image stacks and described by a set of connected cylinders.

We first parameterised the area covered by the dendritic tree, the so-called "dendritic field" (Hausen 1982a). We defined the spanning field of a dendrite by drawing a contour around it at a distance of 25 μm after orienting each reconstructed neuron along its axonal axis (Fig. 16.1b). Parameters relating to the spanning field plainly reflected cell-type-specific differences: All four cells could be easily discriminated by eye by their relative position and the shape of their dendrite spanning fields. Consequently, parameters pertaining to the spanning field readily grouped individual cells into their respective cell types. This is illustrated in Fig. 16.1c where two arbitrarily chosen parameters are plotted against each other: (1) a quantitative measure of convexity of the dendrite spanning field versus and (2) a contextual measure, the relative location of the dendrite off the axonal axis. This shows that the shape of the spanning field of LPTC dendrites determines their identity. On the other hand, using branch-specific statistics (Fig. 16.1d) a qualitative distinction was possible only by detailed examination of distributions of parameters such as dendrite complexity, path lengths to the root and branch order. Clustering the cells according to these branching statistics separated HS from VS cells but not the individual members of the two families (Cuntz et al. 2008).

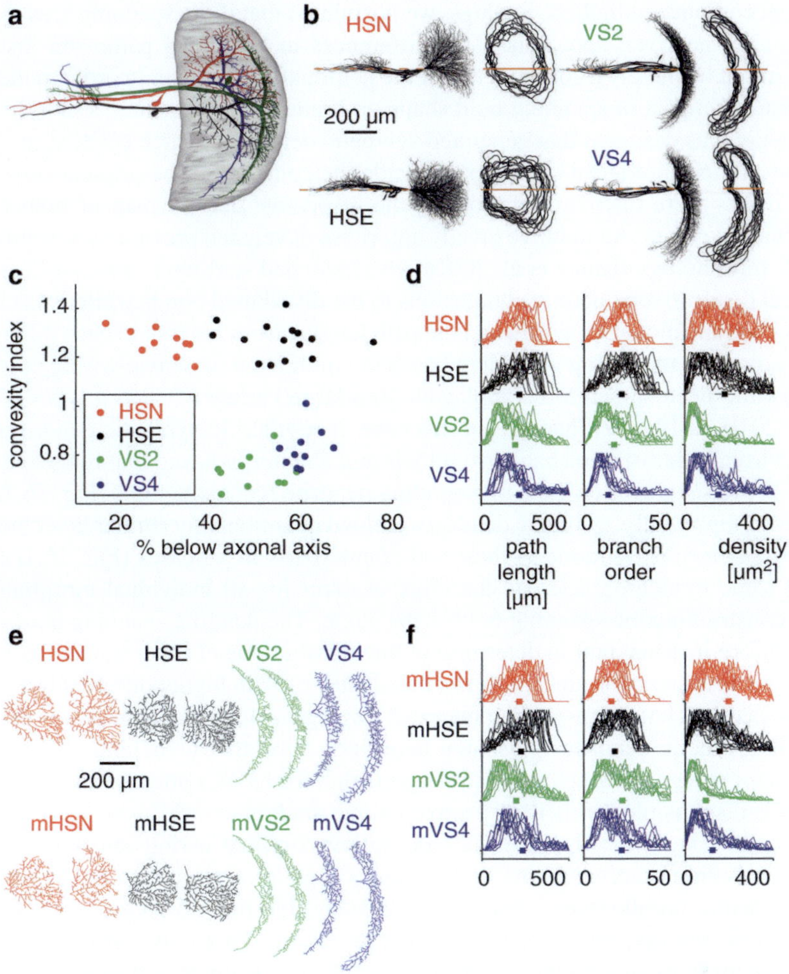

Fig. 16.1 Morphology of HS and VS cells. (**a**) Sketch showing HSE (*black*), HSN (*red*), VS2 (*green*) and VS4 (*blue*) in the context of the lobula plate (*grey*). (**b**) Superimposed full anatomies of all individual cells sorted according to their respective cell type. Cells were aligned along their axonal axis (*orange lines*). To the right, the corresponding dendrite spanning fields are outlined. (**c**) Two dendrite spanning field parameters plotted against each other: the convexity and the relative location to the axonal axis. (**d**) Statistics specifically related to dendrite branching. Statistics are represented as superimposed distribution histograms, *filled squares* show mean values and error bars correspond to standard deviation between individual dendrites. Statistics shown are (1) path length to root values for all topological points, (2) topological branch order values, counting for any point on the dendrite the number of branch points along the tree towards the dendrite root and (3) surface area values assigned to each topological point after Voronoi segmentation indicating topological point density and distribution homogeneity. (**e**) Synthetic dendrites: two examples of each cell type. *Upper row*: real dendrites. *Lower row*, marked by preceding "m": synthetic dendrites corresponding to each of the spanning field of the real dendrites. (**f**) Synthetic dendrite parameter distributions as in (**d**) showing the similarity to the real dendrites. Modified from (Cuntz et al. 2008)

In accordance with these findings, we postulated that if the spanning area readily determines neuronal appearance, the differences in branching parameter distributions might be merely a consequence of the neuronal target zone. In order to identify the critical impact of spanning field shape on branching parameters, a morphological model was designed that generated synthetic dendrites, which covered the same region as their biological counterpart. Inside the contours of the original cells, random points were distributed following the respective density map of branch and termination points. An iterative greedy algorithm developed previously to reproduce LPTC morphology (Cuntz et al. 2007a) was launched starting at the coordinates of the real dendrite root, adding connections to the distributed points while housekeeping both total amount of wiring and the path lengths from the root to each point. The morphological modelling procedure based on optimising these two wiring costs has meanwhile been shown to be applicable to a large variety of neuron types (Cuntz et al. 2010, 2012; see Chap. 6). In this case, it resulted in synthetic dendrites that were virtually indistinguishable from their real counterparts and that remained confined to the same area as the corresponding dendrite reconstructions (Fig. 16.1e).

Most importantly, synthetic dendrites yielded quantitatively similar inner branching parameter distributions as their real counterparts in all cases (Fig. 16.1f). The exact same branching rule can therefore account for all individual morphologies after constraining the spanning field shape alone. The dendrite spanning field shape is therefore instrumental in determining the distributions of inner branching statistics. This is remarkable since it shows that dendrites with highly different branching statistics might have grown using the exact same growth programme as was the case for the synthetic dendrites presented here, indicating that branching statistics are flawed for categorising cells. It also suggests that for LPTCs only the spanning field shape is responsible for their respective unique anatomies, while the inner branching structure is similar for all LPTCs and follows optimal wiring constraints. Since no optimisation criteria relating to dendritic computation were required to generate the dendritic morphologies, there is no evidence that the morphologies of LPTCs are specifically optimised for optic flow processing. This is in agreement with a more recent study that shows that optimising total cable length and path length costs using genetic algorithms results in morphologies with a similar fitness as real LPTCs while adding computational requirements attributed to LPTCs results in synthetic dendrites that perform these tasks better than their real counterparts (Torben-Nielsen and Stiefel 2010). However, dendrites satisfying wiring constraints in such a way might in turn be restricted in the computations that they can implement (Cuntz et al. 2012; see also Chap. 6).

16.3 Dendritic Network Model with HS Cells

We have therefore demonstrated that network wiring plays an important role for dendritic morphology in two different ways: (1) Wiring constraints determine the inner branching pattern and (2) the location of presynaptic input axons, i.e., the anatomical receptive field, determines the dendritic field. We now proceed to discuss the role of network wiring for computation on visual inputs. It was proposed that associating responses from various LPTCs could be useful to provide specific optic flow selectivity (Hausen 1982a, b). The possibility of using dual or more simultaneous intracellular recordings in conjunction with calcium imaging has recently opened up the possibilities of studying the connections between LPTCs and therefore the computational power of the highly specialised networks. In the following two sections we briefly summarise the results from realistic network model studies based on these types of experiments concerning the network of HS cells (Cuntz et al. 2003) and the network of VS cells (Cuntz et al. 2007b).

Apart from HS and VS cells, LPTCs comprise among others two centrifugal horizontal (CH) cells, a dorsal (dCH) and a ventral (vCH) one, whose receptive fields also correspond to the location of their dendrites. CH cells possess inhibitory (Meyer et al. 1986) dendritic output synapses (Gauck et al. 1997). Their output therefore has a spatial component not present in axo-dendritic communication. In contrast to HS cells, CH cells do not receive visual input directly from local motion elements distributed in the lobula plate but indirectly via electrical synapses from the overlapping dendritic trees of HS cells (Haag and Borst 2002). This indicates that the information from HS cells is processed spatially and passed on within a dendritic network of LPTCs. By measuring the dendritic calcium distribution while stimulating the fly with a small moving object, it was previously shown that signals are blurred in CH cells compared to HS cells (Dürr and Egelhaaf 1999; Haag and Borst 2002); this could represent an important processing step made possible only by the dendritic network. It is also known that, after selective laser ablation of the CH cell, the small-field selectivity of so-called figure-detection (FD) cells (Egelhaaf 1985a, b, c) is eliminated (Warzecha et al. 1993). Thus, CH cells and the underlying dendritic network must be part of the neural pathway for the detection of small moving objects.

To understand how the network performs this computation, we arranged compartmental models of a vCH, an HSE and an HSS cell (Borst and Haag 1996) appropriately (Fig. 16.2a) and connected them to each other with 760 spatially distributed linear electrical synapses of 210 pS at the overlapping sections of their dendritic trees. As a consequence, the model required slight adjustments in both the membrane resistance and the axial resistance compared to the disconnected case. Active properties remained unchanged. The model then accurately reproduced the experimental results at many levels. (1) The model revealed a significant outward rectification in the transfer function between the HS and the CH model as had been observed experimentally (Haag and Borst 2002). Since the electrical synapse was represented as a linear conductance, outward rectification resulted from the active membrane currents in both cells (Haag et al. 1997). (2) Dual recordings had shown that spikelets did not propagate from HS to CH cells (Haag and Borst 2002). In our network model, using electrical synapses without any dynamic properties, we saw the same effect. (3) In agreement with experimental results (Haag and Borst 2002), simulated axonal current injections into either HS model cell resulted in a membrane response only in the CH model dendritic area that overlapped with the stimulated HS model. (4) A realistic visual response was found in the CH model when only the HS models received a simulated visual input. The input resistance change was only 5 % in the CH model and, as such, was close to the experimental results. In contrast to our network model, previous simulations where CH cells were modelled to receive visual input directly were not able to reproduce most of the above characteristics faithfully (Haag et al. 1999).

To examine the spatial low-pass characteristics (blurring) found in CH cells experimentally, currents were injected into different dendritic branches of the HS models and the voltage distribution was compared between the two model cell types. The potential spread was consistently broader in the vCH model. This was not an effect of the vCH cell anatomy alone since it was not observed when currents were injected into the dendrites of the isolated models (Fig. 16.2b). These simulation results demonstrate that it is the electrical connection between the two neurons, and not their geometry, that is responsible for the blurring of signals passing from HS to CH neurons. In a simplified model connecting two identical cylinders (2.5 mm long and 3 μm diameter, $R_m = 2{,}500\ \Omega\ \text{cm}^2$ and $R_a = 100\ \Omega\ \text{cm}$) by five conductances of 2.5 nS at 100 μm distance from each other surrounding the current injection site, we found that after point-like current injection in the first cylinder, a broadening of the signal was observed in the second compartment (Fig. 16.2c). The current feedback being negligible, the potential in the second cylinder approximates a simple summation of individual currents through the synapses. This can be examined analytically by convolving the signal with a kernel determined by the filtering properties of the cells. With $f_1(x)$ and $f_2(x)$ describing the potential distribution in the first and in the second cylinder, respectively, we find

$$f_1(x) = e^{-|x|/\lambda} \tag{16.1}$$

$$f_2(x) = \left(1 + |x|/\lambda\right) e^{-|x|/\lambda}, \tag{16.2}$$

where x is the location and λ is the length constant. By solving (16.1) and (16.2) at a distance λ from the point of current injection, the attenuation in the second cylinder is much smaller than in the first one: $f_1(x=\lambda) = 1/e$ and $f_2(x=\lambda) = 2/e$ (see Cuntz et al. 2003 for more details). Hence, the amount of blurring is determined by the length constant λ of both cells which in turn depends on the dendritic anatomy as well as on the axial and transmembrane resistance (Rall 1959).

We have shown that a dendro-dendritic electric coupling leads to a spatial blurring of the signals being passed from one cell to the other. This means that in addition to their integrative properties (Haag and Borst 1996; Single and Borst 1998),

Fig. 16.2 (continued) unconnected cells; Current was injected in HSS (*bottom left*) and vCH (*bottom right*). (**c**) Two cylinders (HS and CH) are connected by five linear conductances surrounding the location of current injection. In HS the signal spreads following an exponential decay, while the CH spread is broader (*bottom traces*). Since the CH signal is smaller than the HS signal, both were normalised to the maximum value. The CH spread can be approximated by the sum of passive spread through each individual conductance (*top traces*). Modified from (Cuntz et al. 2003)

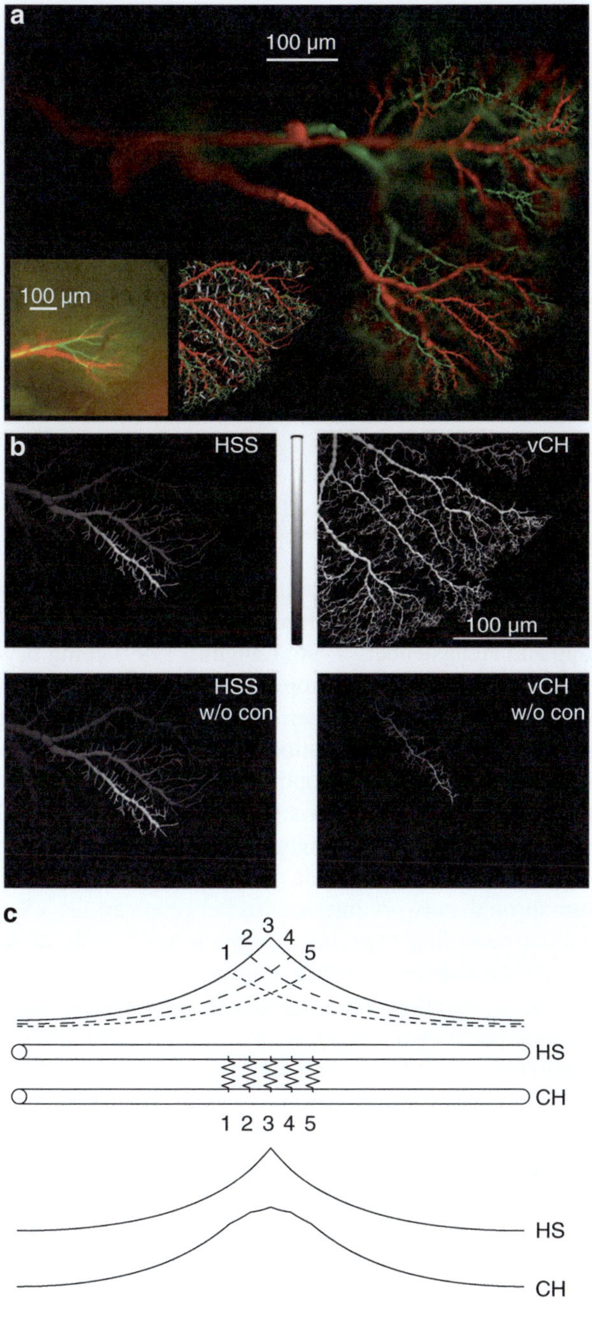

Fig. 16.2 Computation in dendritic network of LPTCs. (**a**) Realistic compartmental models of a vCH cell (*green*), connected to an HSS and an HSE cell (*red*). (*Left Inset*) Original fluorescent staining for comparison. (*Right Inset*) Location of electrical synapses (*white*) used for these analyses. (**b**) Membrane potential spread in HSS model (*top left*) and vCH model (*top right*) after local current injection into HSS. In the *bottom* panels, potential spread is shown similar to *top* panels, but with

LPTCs form a dendritic network allowing for complex image processing with only a few cells. The blurred motion image, as represented in the CH cell dendrites, is passed onto FD cells via inhibitory dendritic synapses. This puts the unique properties of the CH cell into a functional context. Subtracting the CH cell's blurred motion image from the FD cell's own local motion detector input leads to a sharpening of the original image, equivalent to an enhanced motion contrast. Similar interactions of large-field sensitive tangential cells in motion contrast calculation had been suggested before (Egelhaaf 1985c; Reichardt et al. 1983). At that time, however, the spatial dimension of the dendritic output had not been considered (dendrites were considered compact) and an extensive reconnection of the pooled signal to columnar elements was consequently proposed to underlie this computation. We show here an alternative solution where motion contrast is enhanced by a small dendritic network, performing stepwise image processing at a minimal cost with respect to neuronal expenditure.

16.4 Axonal Network Models with VS Cells

As opposed to the comparably simple responses from HS cells to large-field horizontal motion, the ten VS cells (Fig. 16.3a) exhibit more complex receptive fields (Krapp et al. 1998; Krapp and Hengstenberg 1996; Laughlin 1999) going beyond the receptive field size expected from the retinotopic location of their dendrites within the lobula plate. Krapp and colleagues furthermore showed that the orientation preference depends strongly on the spatial location in the visual field. The complex receptive fields of VS cells roughly match the optic flow that occurs during rotational flight manoeuvres of the fly around its own axis (see inset in Fig. 16.3a). Such large and specialised complex receptive fields might either be the result of connections to unknown small-field elements in the lobula plate, or an alteration of the simple receptive field structure through network interactions between various VS cells.

Intracellular dual recording experiments between VS cells clarified this issue (Haag and Borst 2004): Current injection into one VS cell led to a change in the membrane potential in all other VS cells and vice versa. The coupling strength decreased as distance between the dendritic trees of the partner VS cells increased. Moreover, the coupling strength did not tend towards zero for larger distances: for neurons where the dendrites and the receptive fields were clearly separated (like VS1 and VS9), a sign reversal occurred: Depolarisation of VS1 led to a hyperpolarisation of VS9 and hyperpolarisation of VS1 to a depolarisation of VS9. These data strongly suggest that the VS cells are coupled via electrical synapses, which would explain the broad receptive fields of VS cells. Furthermore, to account for the sign reversal between distant VS cells a reciprocal inhibitory connection between VS1 and VS10 was postulated. The gradient in the coupling strength can be explained by two different wiring schemes of the VS cells: (1) each VS cell is directly connected to all other VS cells but the connection strength varies according to the location of the dendritic tree or (2) each VS cell is connected only to its immediate neighbour in a chain-like manner (Fig. 16.3b). The idea of a chain-like

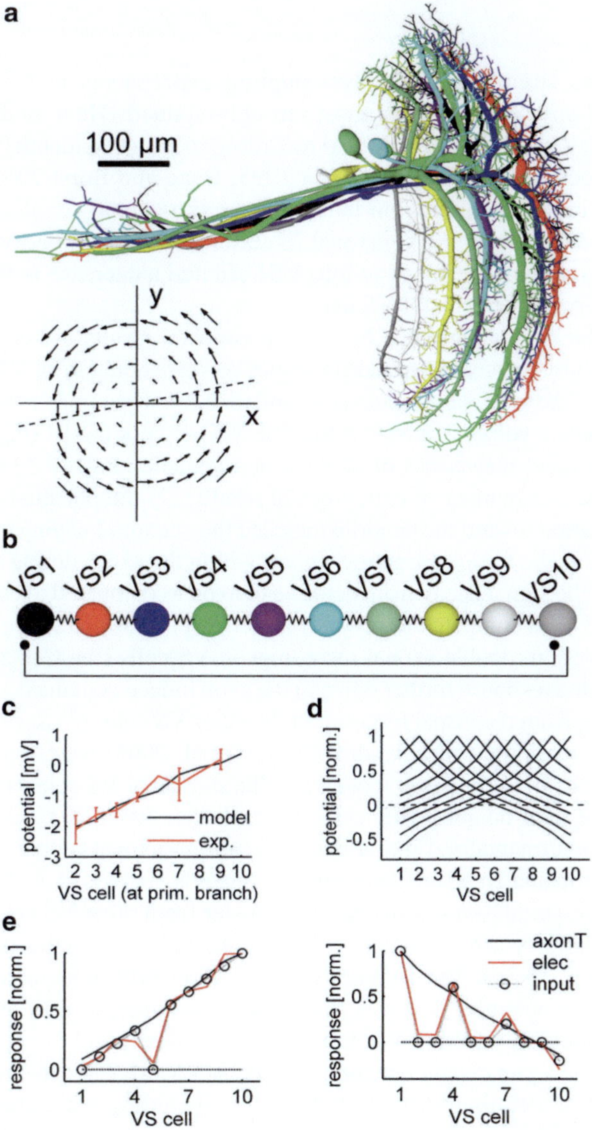

Fig. 16.3 Computation in axonal network between VS cells. (**a**) Ten VS cells as obtained from 2-photon image stacks for which detailed compartmental models were reconstructed. Cells were placed manually according to their position in the lobula plate with neighbouring dendritic arborisations slightly overlapping. *Inset* shows the idealised optic flow field for self-rotation. The vertical components scale linearly with the *x*-axis (*dashed line*). (**b**) Suggested connectivity scheme from (Haag and Borst 2004). Adjacent cells are coupled electrically, and distal cells inhibit each other (schematic *filled spots*). (**c**) Potential responses at the primary dendrite in VS2–10 after current injection of −10 nA in the VS1 primary dendrite; *black line*—model, *red line*—experimental counterpart. (**d**) Voltage transfer from all different VS cells at the location of the synapses in the axon terminal. Normalised amplitudes, *dotted horizontal line* represents 0 mV. (**e**) Examples of the response in the VS cell network to different input signals (*dotted line* with *circles*), measured in the dendrite (*red*) and in the axon terminal (*black*). Traces are normalised to maximum value, *dashed line* represents zero. Modified from (Cuntz et al. 2007b)

connection was strengthened by dye-coupling experiments in which the direct neighbours of any VS cell were most strongly stained (Haag and Borst 2005). Measuring calcium signals with optical recording methods to identify the site of the synaptic connection (Egelhaaf and Borst 1995; Haag and Borst 2000) revealed an accumulation in the axonal region indicating an axo-axonal coupling between the cells. In accordance with an axo-axonal inhibitory connection between distant VS cells, depolarising current injection into VS1 elicited a decrease in fluorescence in the terminal region of the distal VS cell.

To match the circuit suggested by the experimental results, we connected model cells based on fully reconstructed morphologies corresponding to all VS cells via their axon terminals. The synaptic conductances and membrane parameters were then tuned to render a realistic voltage transfer within the VS network model (Fig. 16.3c) while conserving the input resistances of all cells in a realistic range (3.7 ± 0.7 MΩ). This model reproduced a number of experimental results: (1) the potential gradient backward from the axon toward the dendrite matched the calcium imaging results. (2) As a model prediction, the backpropagating signals within the axons during visual stimulation decreased the signal decay from dendrite to axon as compared to current injection (a decay to ~75 % as compared to ~50 %, respectively). This prediction was then verified quantitatively in double axonal recordings in VS cells (see Cuntz et al. 2007b). Consequently, the response further down in the axon indeed contained a larger component originating from the visual response in the other VS cells. (3) Receptive field size alterations due to single cell ablations (Farrow et al. 2005) were reproduced under similar conditions in the network model. (4) The signals of VS cells did not penetrate the dendrites of their neighbouring cells. VS cells can therefore be considered to be efficiently compartmentalised into a dendritic and an axonal region. This was later validated experimentally, again using calcium imaging (Elyada et al. 2009).

Interestingly, simulations of the current transfer from other VS cell models in the network revealed a symmetrical almost linear potential decay at the axonal synapses in all cases (Fig. 16.3d). The network thereby effectively implements a triangular filter. The linear potential decay is the result of a superposition of the potential decaying exponentially from VS1 to more distal VS cells and the inverse potential via the inhibition again decaying exponentially backwards from the distal VS cell to the VS1 cell (see details in Cuntz et al. 2007b; including a simplified model). A triangular filter possesses the interesting characteristic of linearly interpolating the inputs. Through the highly selective coupling, the VS cell network therefore is able to perform linear interpolation on its inputs (Fig. 16.3e): The axonal membrane potential of VS cells whose signals are close to zero, e.g., because no texture is present in their visual receptive field, is set to the linearly interpolated values of their neighbouring cells in the chain. In this way, the VS network interestingly incorporates prior knowledge about the linear relationship of rotational optic flow vectors with horizontal disparity: Fig. 16.3a inset clearly shows this linear relationship in the rotational optic flow field (dashed line). A VS cell without motion response will exhibit in its axonal membrane potential deflection the linearly interpolated value between its neighbours because of the specialised circuit. In this way, the axis of rotation can faithfully be read out by the VS cell whose membrane potential is closest to zero, i.e., the resting potential. This is the case since the line connecting the amplitudes of the vertical components of the optic flow crosses zero in the axis of rotation (compare again dashed line in Fig. 16.3a inset). Using movies of rotational optic flow in natural images, we showed that the jitter due to low-contrast or textureless areas in natural image scenes prevented a correct read-out of the axis of rotation in the VS dendrites. On the other hand the axonal signal was perfectly useful after the linear interpolation through the VS cell network (see details in Cuntz et al. 2007b). This prediction of the network model was later validated experimentally (Elyada et al. 2009). Also, descending neurons that collect the signals from VS cells have been identified that robustly encode the axis of rotation (Wertz et al. 2009).

In summary, we showed that the simple connection scheme between VS cells proposed in Fig. 16.3b could well bring simulations and experiments to close agreement. The benefit of coupling the signals from neighbouring VS cells after dendritic integration is to reliably and robustly represent the centre of rotation in the activity distribution of the VS cell network.

16.5 Conclusions and Outlook

The work summarised above demonstrates that the network structure of the tangential cells within the lobula plate needs to be considered if one is to fully understand computation in LPTCs. We first showed that the dendrite spanning field shape of LPTCs strongly reflects their respective specific connectivity in the network. The dendrite spanning field is what makes LPTCs individually identifiable. Their inner branching properties are similar even between different cell types (HS vs. VS cells), indicating that the wiring constraints are similar. A computational model that describes the branching structure as a function of the dendrite spanning field opens up a number of possibilities. Assuming generality of this principle, the morphology and function of cells, which have not yet been reconstructed, can be inferred and studied based on the contours of their dendrites alone. Moreover, the fly is an animal in which the molecular components that determine neural growth are currently being unravelled, mainly through genetic tools (Grueber et al. 2005). Our framework therefore allows a quantitative study of the impact of genetic modifications far beyond basic statistics. In particular, molecular principles guiding neuronal self-avoidance during development (Hughes et al. 2007), and others can now be put in relation with the branching constraints presented here. Eventually, studying molecular factors shaping dendritic spanning fields separately from a specific branching rule should elucidate a fundamental organisational element in the brain, i.e., the neuron's branching structure. In any case, morphology does not seem to reflect computational requirements for optic flow processing; instead it was useful to study functional aspects in the network context.

Incorporating experimental evidence from multiple simultaneous intracellular electrophysiological recordings into network models including the realistic morphologies of HS or VS cells turned out to be a successful approach to understand many features of optic flow computation. We found that a dendritic network comprises several LPTCs as part of the circuit for the detection of small moving objects. Since the blurring effect could also be reproduced in simplified models, we can conclude that spatial blurring is a feature of any dendro-dendritic connection. This might be relevant for other neural circuits for which dendro-dendritic connections exist, such as in the retina (Cook and Becker 1995). We also found that a specialised combination of electrical and inhibitory synapses can lead to a spatial, triangular filter performing a linear interpolation between neighbouring cells. A model reproducing a more complete connectivity of the lobula plate but using simplified morphologies has meanwhile confirmed that the complex receptive fields measured in real VS cells are a direct result from the specific network connections (Borst and Weber 2011). These findings show how important it is to know precisely the location of the synapses in a given circuit, if one wants to understand the underlying computation.

It has become widely acknowledged that selective connectivity (e.g., Ohki et al. 2005) and the precise location of a synaptic input on a given neuron (e.g., Branco et al. 2010; Gidon and Segev 2012; see Chaps. 15 and 18) are important for computation. The studies presented here serve as an example for including both types of information into network models with realistic morphologies. These models were useful to explain discrepancies between the electrophysiology and the single cell compartmental models, e.g., concerning the visual responses in LPTCs (Haag et al. 1999). The electrical synapses between LPTCs led to a necessary revision of the dendritic integration properties in those cells. The models furthermore provided testable predictions, many of which have been validated for both network models discussed above. This study should be useful for further large-scale network model simulations in other systems and points out the importance of electrical synapses for computation in neural circuits. For the insect brain it shows that the synapses connecting the different neurons are incredibly selective regarding connection partners, location of synapses and type of synapses involved. The resulting models were highly sensitive to those particular features of the connection.

Meanwhile a large number of new findings have continued to expand the known circuitry in the lobula plate (Borst et al. 2010). In particular, the way with which optic flow signals are read out via descending neurons and conveyed to the flight control circuitry has been studied in great detail. These descending neurons have been described in depth, and their connection with LPTCs, partly mediated again by electrical synapses, is now well characterised (Haag et al. 2007, 2010; Wertz et al. 2008, 2012), although a model that incorporates these data and explains the read-out mechanism of the axis of rotation does not yet exist. Also, a large number of heterolateral connections linking both brain halves in the fly involving spiking neurons have been characterised in more detail (Farrow et al. 2006; Haag and Borst 2001, 2008). The results of the computational studies presented here therefore have to be revised in the light of these new findings. Also, more data is needed on the connectivity between LPTCs. This also concerns details of the circuits described in this chapter since the inhibitory connections between VS1 and lateral VS cells as well as between CH cells and FD cells have not yet been identified. This is likely to be solved when the first connectome data (Kleinfeld et al. 2011) will become available for the lobula plate. Morphological models for all existing LPTCs and the descending neurons in combination with that data will provide a more complete and definitive understanding of the lobula plate's role in optic flow computation.

Acknowledgement The work presented here was a collaborative effort together with Friedrich Förstner and Idan Segev.

References

Bialek W, Rieke F, Ruyter D, van Steveninck RR, Warland DK (1991) Reading a neural code. Science 252:1854–1857

Bock DD, Lee W-CA, Kerlin AM et al (2011) Network anatomy and in vivo physiology of visual cortical neurons. Nature 471:177–182

Borst A (2012) Fly motion vision: from optic flow to visual course control. e-Neuroforum 3:59–66

Borst A, Egelhaaf M (1992) In vivo imaging of calcium accumulation in fly interneurons as elicited by visual motion stimulation. Proc Natl Acad Sci USA 89:4139–4143

Borst A, Haag J (2002) Neural networks in the cockpit of the fly. J Comp Physiol A 188:419–437

Borst A, Haag J (2001) Effects of mean firing on neural information rate. J Comput Neurosci 10:213–221

Borst A, Haag J (1996) The intrinsic electrophysiological characteristics of fly lobula plate tangential cells: I. Passive membrane properties. J Comput Neurosci 3:313–336

Borst A, Haag J, Reiff DF (2010) Fly motion vision. Annu Rev Neurosci 33:49–70

Borst A, Single S (2000) Local current spread in electrically compact neurons of the fly. Neurosci Lett 285:123–126

Borst A, Weber F (2011) Neural action fields for optic flow based navigation: a simulation study of the fly lobula plate network. PLoS ONE 6:e16303

Branco T, Clark BA, Häusser M (2010) Dendritic discrimination of temporal input sequences in cortical neurons. Science 329:1671–1675

Briggman KL, Helmstaedter M, Denk W (2011) Wiring specificity in the direction-selectivity circuit of the retina. Nature 471:183–188

Cook JE, Becker DL (1995) Gap junctions in the vertebrate retina. Microsc Res Tech 31:408–419

Cuntz H, Borst A, Segev I (2007a) Optimization principles of dendritic structure. Theor Biol Med Model 4:21

Cuntz H, Forstner F, Borst A, Häusser M (2010) One rule to grow them all: a general theory of neuronal branching and its practical application. PLoS Comput Biol 6:e1000877

Cuntz H, Forstner F, Haag J, Borst A (2008) The morphological identity of insect dendrites. PLoS Comput Biol 4:e1000251

Cuntz H, Haag J, Borst A (2003) Neural image processing by dendritic networks. Proc Natl Acad Sci USA 100:11082–11085

Cuntz H, Haag J, Forstner F et al (2007b) Robust coding of flow-field parameters by axo-axonal gap junctions between fly visual interneurons. Proc Natl Acad Sci USA 104:10229–10233

Cuntz H, Mathy A, Häusser M (2012) A scaling law derived from optimal dendritic wiring. Proc Natl Acad Sci USA 109:11014–11018

Dürr V, Egelhaaf M (1999) In vivo calcium accumulation in presynaptic and postsynaptic dendrites of visual interneurons. J Neurophys 82:3327–3338

Egelhaaf M (1985a) On the neuronal basis of figure-ground discrimination by relative motion in the visual system of the fly. I. Behavioural constraints imposed on the neuronal network. Biol Cybern 52:123–140

Egelhaaf M (1985b) On the neuronal basis of figure-ground discrimination by relative motion in the visual system of the fly. II. Figure-detection cells, a new class of visual interneurones. Biol Cybern 52:195–209

Egelhaaf M (1985c) On the neuronal basis of figure-ground discrimination by relative motion in the visual system of the fly. III. Possible input circuitries and behavioural significance of the FD-cells. Biol Cybern 52:267–280

Egelhaaf M, Borst A (1995) Calcium accumulation in visual interneurons of the fly: stimulus dependence and relationship to membrane potential. J Neurophys 73:2540–2552
Egelhaaf M, Haag J, Borst A (1994) Processing of synaptic information depends on the structure of the dendritic tree. Neuroreport 6:205–208
Elyada YM, Haag J, Borst A (2009) Different receptive fields in axons and dendrites underlie robust coding in motion-sensitive neurons. Nat Neurosci 12:327–332
Farrow K, Borst A, Haag J (2005) Sharing receptive fields with your neighbors: tuning the vertical system cells to wide field motion. J Neurosci 25:3985–3993
Farrow K, Haag J, Borst A (2006) Nonlinear, binocular interactions underlying flow field selectivity of a motion-sensitive neuron. Nat Neurosci 9:1312–1320
Gabbiani F, Krapp HG, Koch C, Laurent G (2002) Multiplicative computation in a visual neuron sensitive to looming. Nature 420:320–324
Gauck V, Egelhaaf M, Borst A (1997) Synapse distribution on VCH, an inhibitory, motion-sensitive interneuron in the fly visual system. J Comp Neurol 381:489–499
Geiger G, Nässel DR (1981) Visual orientation behaviour of flies after selective laser beam ablation of interneurones. Nature 293:398–399
Gidon A, Segev I (2012) Principles governing the operation of synaptic inhibition in dendrites. Neuron 75:330–341
Grueber WB, Yang C-H, Ye B, Jan Y-N (2005) The development of neuronal morphology in insects. Curr Biol 15:R730–R738
Haag J, Borst A (1996) Amplification of high-frequency synaptic inputs by active dendritic membrane processes. Nature 379:639–641
Haag J, Borst A (2002) Dendro-dendritic interactions between motion-sensitive large-field neurons in the fly. J Neurosci 22:3227–3233
Haag J, Borst A (2004) Neural mechanism underlying complex receptive field properties of motion-sensitive interneurons. Nat Neurosci 7:628–634
Haag J, Borst A (2005) Dye-coupling visualizes networks of large-field motion-sensitive neurons in the fly. J Comp Physiol A 191:445–454
Haag J, Borst A (2000) Spatial distribution and characteristics of voltage-gated calcium signals within visual interneurons. J Neurophys 83:1039–1051
Haag J, Borst A (2008) Electrical coupling of lobula plate tangential cells to a heterolateral motion-sensitive neuron in the fly. J Neurosci 28:14435–14442
Haag J, Borst A (2001) Recurrent network interactions underlying flow-field selectivity of visual interneurons. J Neurosci 21:5685–5692
Haag J, Theunissen F, Borst A (1997) The intrinsic electrophysiological characteristics of fly lobula plate tangential cells: II. Active membrane properties. J Comput Neurosci 4:349–369
Haag J, Vermeulen A, Borst A (1999) The intrinsic electrophysiological characteristics of fly lobula plate tangential cells: III. Visual response properties. J Comput Neurosci 7:213–234
Haag J, Wertz A, Borst A (2007) Integration of lobula plate output signals by DNOVS1, an identified premotor descending neuron. J Neurosci 27:1992–2000
Haag J, Wertz A, Borst A (2010) Central gating of fly optomotor response. Proc Natl Acad Sci USA 107:20104–20109
Hausen K (1982a) Motion sensitive interneurons in the optomotor system of the fly – I. The horizontal cells: structure and signals. Biol Cybern 45:143–156
Hausen K (1982b) Motion sensitive interneurons in the optomotor system of the fly – II. The horizontal cells: receptive field organization and response characteristics. Biol Cybern 46:67–79
Homberg U, Heinze S, Pfeiffer K et al (2011) Central neural coding of sky polarization in insects. Philos Trans R Soc Lond B Biol Sci 366:680–687
Hubel DH, Wiesel TN (1962) Receptive fields, binocular interaction and functional architecture in the cat's visual cortex. J Physiol 160:106–154
Hughes ME, Bortnick R, Tsubouchi A et al (2007) Homophilic Dscam interactions control complex dendrite morphogenesis. Neuron 54:417–427
Jacobs GA, Miller JP, Aldworth Z (2008) Computational mechanisms of mechanosensory processing in the cricket. J Exp Biol 211:1819–1828

Kleinfeld D, Bharioke A, Blinder P et al (2011) Large-scale automated histology in the pursuit of connectomes. J Neurosci 31:16125–16138

Krapp HG, Hengstenberg B, Hengstenberg R (1998) Dendritic structure and receptive-field organization of optic flow processing interneurons in the fly. J Neurophys 79:1902–1917

Krapp HG, Hengstenberg R (1996) Estimation of self-motion by optic flow processing in single visual interneurons. Nature 384:463–466

Lang S, Dercksen VJ, Sakmann B, Oberlaender M (2011) Simulation of signal flow in 3D reconstructions of an anatomically realistic neural network in rat vibrissal cortex. Neural Netw 24:998–1011

Laughlin SB (1999) Visual motion: dendritic integration makes sense of the world. Curr Biol 9:R15–R17

London M, Häusser M (2005) Dendritic computation. Annu Rev Neurosci 28:503–532

Marder E, Bucher D (2001) Central pattern generators and the control of rhythmic movements. Curr Biol 11:R986–R996

Markram H (2006) The blue brain project. Nat Rev Neurosci 7:153–160

Meyer EP, Matute C, Streit P, Nässel DR (1986) Insect optic lobe neurons identifiable with monoclonal antibodies to GABA. Histochemistry 84:207–216

Ohki K, Chung S, Ch'ng YH et al (2005) Functional imaging with cellular resolution reveals precise micro-architecture in visual cortex. Nature 433:597–603

Peron SP, Jones PW, Gabbiani F (2009) Precise subcellular input retinotopy and its computational consequences in an identified visual interneuron. Neuron 63:830–842

Rall W (1959) Branching dendritic trees and motoneuron membrane resistivity. Exp Neurol 527:491–527

Rall W, Rinzel J (1973) Branch input resistance and steady attenuation for input to one branch of a dendritic neuron model. Biophys J 13:648–687

Reichardt W, Poggio T, Hausen K (1983) Figure-ground discrimination by relative movement in the visual system of the fly. Biol Cybern 46:1–30

Ruyter D, van Steveninck RR, Bialek W (1988) Real-time performance of a movement-sensitive neuron in the blowfly visual system: coding and information transfer in short spike sequences. Proc R Soc Lond B 234:379–414

Seung HS, Lee DD, Reis BY, Tank DW (2000) Stability of the memory of eye position in a recurrent network of conductance-based model neurons. Neuron 26:259–271

Single S, Borst A (1998) Dendritic integration and its role in computing image velocity. Science 281:1848–1850

Torben-Nielsen B, Stiefel KM (2010) Wide-field motion integration in fly vs cells: insights from an inverse approach. PLoS Comput Biol 6:e1000932

Warzecha A-K, Egelhaaf M, Borst A (1993) Neural circuit tuning fly visual interneurons to motion of small objects. I. Dissection of the circuit by pharmacological and photoinactivation techniques. J Neurophys 69:329–339

Watt AJ, Cuntz H, Mori M et al (2009) Traveling waves in developing cerebellar cortex mediated by asymmetrical Purkinje cell connectivity. Nat Neurosci 12:463–473

Wertz A, Borst A, Haag J (2008) Nonlinear integration of binocular optic flow by DNOVS2, a descending neuron of the fly. J Neurosci 28:3131–3140

Wertz A, Gaub B, Plett J et al (2009) Robust coding of ego-motion in descending neurons of the fly. J Neurosci 29:14993–15000

Wertz A, Haag J, Borst A (2012) Integration of binocular optic flow in cervical neck motor neurons of the fly. J Comp Physiol A 198:655–668

Chapter 17
Biophysical Mechanisms of Computation in a Looming Sensitive Neuron

Simon P. Peron

Abstract The lobula giant movement detector (LGMD) is a large-field visual interneuron believed to be involved in collision avoidance and escape behaviors in orthopteran insects, such as locusts. Responses to approaching—or looming—stimuli are highly stereotypical, producing a peak that signals an angular size threshold. Over the past several decades, investigators have elucidated many of the mechanisms underpinning this response, demonstrating that the LGMD implements a multiplication in log-transformed coordinates. Furthermore, the LGMD possesses several mechanisms that preclude it responding to non-looming stimuli. This chapter explores these biophysical mechanisms, as well as highlighting insights the LGMD provides into general principles of dendritic integration.

17.1 Introduction

Insect neurons have a rich history as model systems for dendritic integration (Borst et al. 2010; Gabbiani et al. 2004; London and Häusser 2005; Segev and London 2000). Experimentally, their advantages include the relative robustness of invertebrate preparations, the existence of individual identified neurons with conserved properties across animals, accessibility in in vivo preparations, and the large size of many of the neurons. The lobula giant movement detector (LGMD) possesses all these features. Because its presumed behavioral role—collision avoidance and escape—is highly specific, it is possible to link computation to function. Biophysical mechanisms can therefore be behaviorally contextualized to a degree not possible in many other systems.

S.P. Peron (✉)
Janelia Farm Research Campus, Ashburn, VA, USA
e-mail: perons@janelia.hhmi.org

H. Cuntz et al. (eds.), *The Computing Dendrite: From Structure to Function*, Springer Series in Computational Neuroscience 11, DOI 10.1007/978-1-4614-8094-5_17,

The LGMD is a large-field visual interneuron found in the third neuropil of the optic lobe of Orthopteran insects that responds preferentially to stimuli on a collision course with the animal, or looming stimuli. O'Shea and Williams (1974) discovered the LGMD when searching for neurons that synapsed onto the descending contralateral movement detector (DCMD), a giant neuron present in the locust nerve cord that was, as its name implies, known to be sensitive to visual motion (Rowell 1971). Subsequent work demonstrated that the LGMD and DCMD are linked via a powerful, mixed electrochemical synapse that transmits spikes from the LGMD to the DCMD in one-to-one fashion (Killmann et al. 1999; Killmann and Schürmann 1985; Rind 1984; O'Shea and Rowell 1975b). Coupled with the observation that the DCMD possesses the largest axon among nerve cord neurons (O'Shea et al. 1974), it was likely that this neuronal tandem subserved an ethologically important function. While O'Shea and colleagues extensively characterized many of the LGMD–DCMD's response features (O'Shea and Rowell 1975a, 1976; Rowell and O'Shea 1976a; Rowell et al. 1977), sensitivity to looming stimuli was first documented by Schlotterer (1977). An example looming stimulus response is shown in Fig. 17.1. Schlotterer's (1977) result, and the known DCMD output to motor centers involved in jumping (Burrows and Rowell 1973; O'Shea et al. 1974) and flight (Simmons 1980), suggested the possibility that the LGMD–DCMD subserves escape behavior in stationary animals and collision avoidance in flying animals.

Over the decades following Schlotterer's (1977) observation that the LGMD–DCMD prefers looming stimuli, evidence for the LGMD–DCMD's behavioral role has expanded (for review, see Fotowat and Gabbiani 2011). The preference for looming stimuli has been probed extensively in varying contexts (Rind and Simmons 1992; Simmons and Rind 1992; Judge and Rind 1997; Guest and Gray 2006). The response has been shown to be invariant to many features of the approaching object (Gabbiani et al. 2001, 2004). Perhaps most convincingly, extracellular recordings from intact, behaving locusts (*Schistocerca americana*) demonstrate that the timing of the DCMD's firing rate peak reliably predicts jump times. Furthermore, during the phase of the jump where the locust builds up muscle tension via co-contraction of both flexors and extensors, the extensor motorneuron spike count correlates strongly with DCMD spike count (Fotowat et al. 2011). Ablation of the DCMD in one nerve cord while severing the other nerve cord does not, however, abolish escape. Instead, it increases the latency variability and reliability of the response (Fotowat et al. 2011; see also Santer et al. 2008), implying that the DCMD, though important in shaping the escape response, is not the only unit capable of driving escape. The DCMD is essentially a relay for the output of its presynaptic partner, the LGMD. As such, investigators interested in computations mediating escape and collision avoidance have focused on the LGMD.

The LGMD's role in collision avoidance and escape makes it an ideal model system in the study of dendritic integration and single-neuron computation. The strong selective pressure presumably experienced by neurons mediating behavior essential for survival likely explains the elaborate menagerie of computational mechanisms observed by investigators. While some processing of visual input is performed upstream to the LGMD, a series of integrative mechanisms produce the

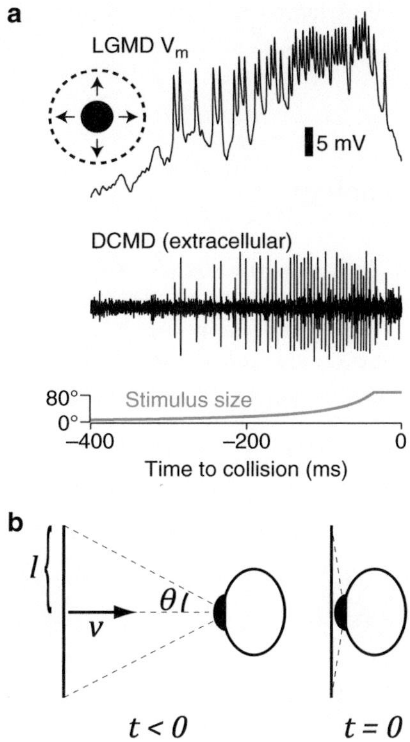

Fig. 17.1 Example response to looming stimulus. (**a**) A disc with $l/|v| = 30$ ms was presented to the animal. The LGMD membrane potential and DCMD spiking activity were recorded concurrently (*top* and *middle traces*, respectively). The *bottom trace* shows the evolution of the stimulus size on the retina with time. Times prior to collision are negative; collision occurs at time 0 (Modified from Peron and Gabbiani 2009a). (**b**) In the context of monocular vision, looming stimuli can be described simply by the angle, θ, that they subtend on the retina of the observer. Trigonometry implies that for an object approaching at a fixed velocity v ($v<0$ for approach) and having a half-height l, the retinal angle at time t ($t<0$ prior to collision) can be described by $\theta(t) = 2 \cdot \tan^{-1} \frac{l}{vt}$. Angular size depends only on time and the ratio of half-height to velocity, $l/|v|$. That is, doubling of an object's size is equivalent to reducing its velocity by half. This effectively limits the parameter space one has to explore when conducting experiments: one has but to sample a reasonable range of $l/|v|$ values. The illustration shows the geometric variables at times $t<0$ and $t=0$

input–output function of this neuron. Some computations are performed proximal to the spike initiation zone (SIZ): summation of excitatory and inhibitory branch input, sodium conductance mediated exponentiation of membrane potential prior to spiking (Gabbiani et al. 2002), and spike-frequency adaptation via a calcium-sensitive potassium conductance (K_{Ca}; Gabbiani and Krapp 2006; Peron and Gabbiani 2009a). Local dendritic computations include orientation preference (Peron et al. 2009) and sublinear summation (Jones and Gabbiani 2012). Additionally, the output from photoreceptors is first processed by interneurons in the lamina and medulla before reaching the LGMD, producing habituation (O'Shea and Rowell 1976) as well as lateral inhibition (O'Shea and Rowell 1975a). This chapter reviews these mechanisms, as well as placing them in the broader context of dendritic computation.

17.2 LGMD Morphology

The LGMD consists of three dendritic fields, one excitatory and two inhibitory (Fig. 17.2a). Like many invertebrate neurons, the soma is attached via a thin process to the rest of the cell and appears to play no role in dendritic integration. Instead, a narrowing of the main process constitutes the spike initiation zone (SIZ; O'Shea 1975). The excitatory branch is connected to the main process at its dendritic end,

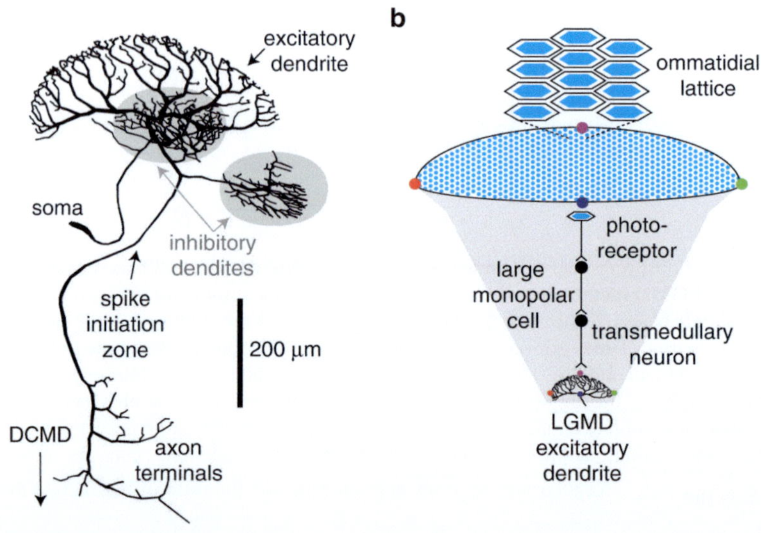

Fig. 17.2 The anatomy of the LGMD. (**a**) The LGMD's excitatory dendrite is shaped like a large ellipsoidal claw, whose shape closely matches that of the locust eye (Peron et al. 2007). The two inhibitory dendrites are highlighted in *grey*. The inhibitory field receiving input from the dorsal uncrossed bundle is the one not overlaid by the excitatory dendritic fan. The spike initiation zone is the narrowest point of the main process. Note the extensive axonal arborization which connects to the DCMD (Modified from Peron and Gabbiani 2009a). (**b**). Visual input in the locust optic lobe originates in ~7,500 individual ommatidia. The ommatidia are arranged in a hexagonal lattice. Individual photoreceptors output onto large monopolar cells (LMCs), which output onto transmedullary neurons. Finally, these output onto the LGMD excitatory dendrites. To illustrate the retinotopy observed in Peron et al. (2009), *red*, *magenta*, *blue*, and *green dots* were placed on the dorsal, anterior, posterior, and ventral edges of the eye, and the recipient portion of the excitatory dendrite. Note that this diagram only includes the pathway believed to be responsible for LGMD excitatory dendritic input; the optic lobe itself is an incredibly complex neural organ, and the connections outlined here are a small subset of those present

while the two inhibitory branches connect to the main process between the excitatory dendrite and the SIZ, a theoretically ideal arrangement for shunting inhibition (Koch et al. 1983).

The LGMD's excitatory dendrite receives a projection of ~15,000 inputs (Strausfeld and Nässel 1981). These inputs form nicotinic cholinergic synapses, as evidenced by immunohistochemistry (Rind and Leitinger 2000) and their block with nicotinic antagonists (Peron et al. 2009). Consistent with observations in other insect neurons (e.g., blowfly lobula plate tangential cells; Borst and Egelhaaf 1992), calcium imaging has revealed that the excitatory input projection onto the LGMD is retinotopic (Peron et al. 2009). In contrast, the inhibitory branches are believed to receive non-retinotopic input (Strausfeld and Nässel 1981). This feed-forward inhibitory input consists of separate ON and OFF pathways responding to increased and decreased luminance, respectively (Rowell et al. 1977). The ON and OFF branches are clearly distinct, as mechanical cutting of the dorsal uncrossed bundle abolishes only the OFF inhibitory responses, leaving ON responses intact (Rowell et al. 1977). Input to both inhibitory branches is likely mediated by $GABA_A$ receptors, as evidenced by their block when the $GABA_A$ antagonist picrotoxin is puffed onto the lobula (Gabbiani et al. 2005). While the excitatory input exhibits a complex response, the inhibitory input appears to simply reflect the overall change in luminance, a feature common in visual feed-forward inhibitory circuits (Rowell et al. 1977; but see Jones and Gabbiani 2012). Indeed, dendritic integration in the excitatory branch appears to be far more mechanistically complex.

17.3 Processing Prior to the LGMD

The processing stream leading to the LGMD is illustrated in Fig. 17.2b. The LGMD's excitatory branch receives its ~15,000 inputs from transmedullary neurons (Strausfeld and Nässel 1981), roughly two per ommatidium (Krapp and Gabbiani 2005). These neurons in turn receive input from large monopolar cells (LMCs) in the lamina, which themselves receive direct photoreceptor input. Though the photoreceptor output is no doubt heavily processed prior to the LGMD, only a few studies have investigated the nature of this processing (e.g., James and Osorio 1996), and the precise projection from photoreceptor to LGMD remains uncharacterized. The first processing stage of photoreceptor input is carried out by the LMCs, and while the second stage cells, the transmedullary neurons, are difficult to record from, some aspects of their activity can be inferred using the voltage clamp technique in the LGMD and looking at synaptic currents. Jones and Gabbiani (2010) employed this approach to study speed sensitivity along the pathway leading to the LGMD, recording from photoreceptors, LMCs, and the LGMD. They showed that the LGMD's motion sensitivity is not due to Reichardt detectors (Reichardt 1987) but is instead due to the rate of luminance change of individual photoreceptors. The LMCs convert the slope of the graded photoreceptor output into a firing rate signal.

Because the slope of a photoreceptor's response will be steeper for a faster luminance change, and because a faster translating edge produces a faster luminance change, the LMC's spiking encodes edge velocity. This luminance rate change sensitivity is present in the LGMD's membrane potential as well as the excitatory postsynaptic current observed under voltage clamp, implying that the impinging transmedullary neurons also share this feature (Jones and Gabbiani 2010). This result is consistent with the LGMD's previously reported sensitivity to the velocity of locally rotating disks (Krapp and Gabbiani 2005).

In addition to speed sensitivity, habituation to local flashes (Palka 1967; O'Shea and Rowell 1976) and looming stimuli (Rogers et al. 2007) is implemented prior to the LGMD's excitatory dendritic field. This habituation is likely implemented at either the synapse itself or the transmedullary neurons, as it can be produced with repeated electrical stimulation of the medulla (O'Shea and Rowell 1976). While separate ON and OFF pathways are common in insect visual systems (Strausfeld and Nässel 1981; Joesch et al. 2010), in the LGMD, habituation to ON and OFF responses transfers to the other response type, suggesting that input on to the excitatory dendrite is of a combined ON–OFF stream (O'Shea and Rowell 1976). Habituation appears to be context dependent: if a locust's leg is stroked, which is believed to induce octopamine release (Bacon et al. 1995), habituation to repeated stimulation can be abolished (Rind et al. 2008). In addition to habituation, experimental evidence has demonstrated that the LGMD's visual response is susceptible to lateral inhibition. Specifically, there is a decline in response to local motion when it is presented along with flaking drifting gratings or other proximal moving stimuli (O'Shea and Rowell 1975a; Rowell and O'Shea 1976b; Rind and Simmons 1998). Electron microscopy suggests that lateral inhibition may result from muscarinic cholinergic synapses between the axon terminals of the adjacent medullary outputs onto the LGMD (Rind and Simmons 1998). If demonstrated conclusively, this would be consistent with habituation taking place at the synapse, as lateral inhibition abolishes local motion habituation (O'Shea and Rowell 1975a).

17.4 Mechanisms Shaping the LGMD's Response to Looming Stimuli

Hatsopoulos et al. (1995) initially made the observation that the LGMD's peak firing rate during looming stimulus presentation occurs with a fixed delay following crossing of an angular threshold. They further found that the firing rate during looming was well fit by the product of the angular velocity of the object and an exponentiation of the object's angular size. This model was refined in Gabbiani et al. (1999) by adding a static nonlinearity to yield:

$$f(t) = g\left(\Psi(t-\delta)\cdot e^{-\alpha\theta(t-\delta)}\right). \tag{17.1}$$

Here, $f(t)$ is the LGMD firing rate at time t (with $t<0$ for times prior to collision), $g(x)$ is a static nonlinearity, $\theta(t)$ and $\psi(t)$ are the angular size and edge velocity of a looming object at time t, respectively, and δ and α are constants. δ is the offset of the timing of the peak relative the crossing of a threshold angle, while α is a scaling factor. Gabbiani et al. (1999) empirically determined the value of the threshold angle, reporting a range of 18.0°–37.0°. Subsequent work showed that the relationship in (17.1) held across several locust species, different looming object approach angles, shapes, textures, and edge counts (Gabbiani et al. 2001), as well as animal temperature (Gabbiani et al. 1999; see Gabbiani et al. 2004 for review).

Behavioral data shows that various phases of the escape response occur at fixed angular thresholds when looming stimuli are presented (Fotowat and Gabbiani 2007), that jump preparation coincides with high frequency LGMD–DCMD spiking (Santer et al. 2008), and that the LGMD–DCMD firing rate peak time predicts the timing of jumps (Fotowat et al. 2011). Because (17.1) implies that the firing rate peak will take place at a fixed delay following the crossing of an angular threshold, the biophysical mechanisms responsible for its implementation can be directly linked to the animal's behavior. Work in the past decade suggests that the multiplication of the exponential term with edge velocity takes place in log-transformed coordinates (Jones and Gabbiani 2012; Gabbiani et al. 2004). That is, ignoring the static nonlinearity $g(x)$ and delay parameter δ, the LGMD appears to implement (17.1) in the following way:

$$\begin{aligned}\Psi(t)\cdot e^{-\alpha\theta(t)} &= e^{\ln\left[\Psi(t)\cdot e^{-\alpha\theta(t)}\right]}\\ &= e^{\ln\left[\Psi(t)\right]+\ln\left[e^{-\alpha\theta(t)}\right]}.\\ &= e^{\ln\left[\Psi(t)\right]-\alpha\theta(t)}\end{aligned} \tag{17.2}$$

Briefly, the logarithm of angular speed is implemented at the excitatory branch (Jones and Gabbiani 2012), the angular size input arrives directly at the inhibitory branches (Rowell et al. 1977; Gabbiani et al. 2005), the subtraction takes place at the level of dendritic integration, and the result is exponentiated at the SIZ (Gabbiani et al. 2002).

The LMCs convert the photoreceptor luminance change signal into a velocity signal, which is relayed to the LGMD via transmedullary neurons (Jones and Gabbiani 2010). Thus, the LGMD excitatory dendrite has ready access to angular velocity information, or $\psi(t)$. The logarithmic transform takes place because the excitatory synapses produce large local depolarizations, bringing membrane potential closer to the synaptic reversal potential. Subsequent input produces less driving force, and therefore a smaller depolarization than predicted from linearity (Johnston and Wu 1995). This sublinear summation is accentuated by the fact that increasing velocity produces larger presynaptic input in an ever-shrinking temporal window (Jones and Gabbiani 2010). It is likely that the precision of the LGMD's retinotopic projection also plays a role (Peron et al. 2009), as sublinear summation should be more pronounced with a more retinotopic mapping when spatially coherent looming stimuli are presented. Thus, dendritic integration results in the LGMD's excitatory branches outputting the logarithm of angular velocity, $\ln[\psi(t)]$.

Two major sources of inhibitory input onto the LGMD's dendrite have been demonstrated: lateral inhibition (O'Shea and Rowell 1975a; Rowell and O'Shea 1976b) and feed-forward inhibition (Rowell et al. 1977; Gabbiani et al. 2005). Lateral inhibition—the suppression of a response if a spatially adjacent stimulus is presented—is presynaptic. Feed-forward inhibition—inhibition in direct response to a given stimulus—is mediated by inputs to the two large inhibitory dendritic fields (see Fig. 17.2a). Gabbiani et al. (2002) showed that it is feed-forward inhibition that dominates during the looming response. First, they demonstrated that a slow drifting background grating, which triggers lateral inhibition, is only effective at suppressing responses during the first ~23° of a looming approach. Second, the sudden onset of a fast moving background grating during the early phase of a looming approach, which drives the large luminance change needed to trigger feed-forward inhibition, abolishes the response prematurely. Finally, this abolition can be prevented with the puffing of picrotoxin onto the lamina. Indeed, under the influence of picrotoxin, the sudden appearance of the rapidly translating background grating elicits a sharp increase in firing. Consistent with this, subsequent experiments have revealed that picrotoxin only influences the later two-third of the looming response (Gabbiani et al. 2005). Simulations incorporating the known features of the LGMD suggest that while inhibitory conductance approximately follows the square of object size, the effect on membrane potential is approximately sigmoidal (Jones and Gabbiani 2012). This is consistent with representing the inhibitory branches as providing an input proportional to angular size or $\alpha\theta(t)$.

Taken together then, the excitatory and inhibitory branches' influence on membrane potential will sum to $\ln[\psi(t)] - \alpha\theta(t)$, with the negative sign indicating inhibitory input. The final step implied by (17.2) is exponentiation of the membrane potential. Gabbiani et al. (2002) employed the sodium channel blocker tetrodotoxin (TTX) to determine the impact of sodium conductances on the membrane potential to firing rate transform. They puffed TTX onto the lobula so as to abolish spiking and unmask the underlying membrane potential response. The transform from membrane potential to spike rate could then be determined, and they showed it to be best approximated by a third-order power law, which is nearly equivalent to an exponential. Thus, it appears that the initial proposal of Hatsopoulos et al. (1995) has been confirmed and that the terms of (17.2) can be explained biophysically. This is summarized in Fig. 17.3.

17.5 Precise Retinotopy in the LGMD

The notion that dendritic input location plays a substantial role in shaping neuronal response dates to Wilfrid Rall and his early modeling studies (e.g., Rall et al. 1967). Evidence suggests that the preservation of retinotopy by the inputs onto the LGMD's

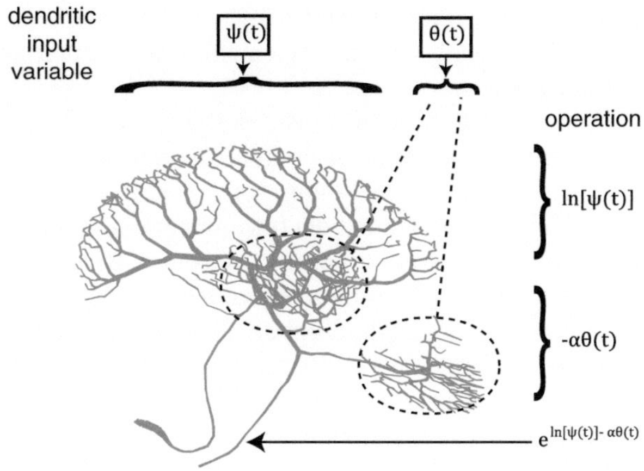

Fig. 17.3 Summary of the core computation performed by the LGMD during looming stimulus presentation. The excitatory dendrite receives as input an angular velocity signal, $\psi(t)$, which it transforms logarithmically, yielding $\ln[\psi(t)]$. The inhibitory fields receive an angular size input, $\theta(t)$, which they scale to yield $-\alpha\theta(t)$. The sum of the two branches is exponentiated at the spike initiation zone, giving $e^{\ln[\psi(t)]-\alpha\theta(t)}$ (17.2)

excitatory branch is highly precise—probably down to the level of individual afferents (Peron et al. 2009). Combined with the ease of visual stimulation and recording of the LGMD's spike output, this makes the neuron an ideal system for studying the influence of synaptic position on integration.

While the projection is retinotopic, the amplitude of the LGMD's response to local motion stimuli across visual space deviates from one predicted by the distribution of individual ommatidia. The locust eye consists of approximately 7,500 ommatidia, and they sample visual space nonuniformly (Krapp and Gabbiani 2005). Specifically, the frontal portion of visual space and the equatorial region are oversampled. In contrast, the receptive field of the LGMD to local rotating disks peaks for posterior stimuli and is fairly uniform azimuthally (Krapp and Gabbiani 2005). This suggests that electrotonic structure—rather than visual input density—dominates the receptive field. Fitting of the LGMD's passive properties reveals the excitatory dendrites to be electrotonically extensive (Peron et al. 2007). The tips are 3–4 log-attenuation units (Zador et al. 1995) away from the SIZ, meaning that only 2–5 % of the local dendritic membrane potential peak is attained at the SIZ. Since posterior inputs project proximally to the SIZ on the excitatory branch, while anterior inputs project distally (Peron et al. 2009; see also Fig. 17.2b), this can partly account for the discrepancy between the sampling of visual space and the response of the neuron.

The "realistic" mapping of ommatidial onto the excitatory dendrites space (realistic in that the dendritic tips receive anterior visual inputs; see Fig. 17.2b) does not, however, explain the lack of response drop off away from the equator (i.e., dorsally or ventrally) observed physiologically. The best-fitting synaptic connectivity model combined the coarse retinotopy of the projection (anterior visual space projecting to dendritic tips) with a uniform sampling of visual space (Peron et al. 2007). This implies some sort of input normalization either prior to or at the LGMD's excitatory dendrite, because dorsal and ventral stimuli should produce responses greater than predicted from ommatidial density, while equatorial stimuli should produce weaker ones. The biophysical implementation of this normalization awaits experimental investigation.

One possible mechanism is presynaptic lateral excitation, whereby activation of an ommatidium spreads to its immediate neighbors. For posterior inputs that activate only a single ommatidium, activation of all adjacent fibers would produce a net activation of seven ommatidia (the ommatidia are organized in a hexagonal array) or a sevenfold increase. In contrast, activation of an anterior input would presumably already activate several ommatidia; if seven were activated initially, this simplistic lateral excitation would produce a net activation of eleven ommatidia, or a ~1.6-fold increase. As the number of directly activated ommatidia increases, the number of additional ommatidia recruited via lateral excitation declines, producing the required normalization. This would also help explain why single ommatidium stimulation produces such strong responses (Peron et al. 2009; Jones and Gabbiani 2010).

An alternative, though not mutually exclusive, explanation could simply be sublinear summation of synaptic inputs (Jones and Gabbiani 2010): such summation would have a greater effect on larger stimuli. The precise retinotopy of the inputs means that such summation will take place more readily with multiple synapses, as this will bring local membrane potential closer to the synaptic reversal potential. However, the use of relatively large stimuli in the modeling study—on the order of those employed by Krapp and Gabbiani (2005) in their electrophysiological study—suggests that sublinear summation is not the only mechanism at play (Peron et al. 2007). Teasing apart the contribution of these various mechanisms would likely require direct stimulation of individual transmedullary afferents to provide accurate estimates of single afferent EPSPs. Additionally, a single ommatidium stimulation apparatus would have to be used to stimulate an ever-larger number of adjacent ommatidia. Combining the results of these experiments in a realistic model should make possible the elucidation of the underlying amplification mechanism.

It is also likely that the precise mapping contributes to the relative decrement of larger stimuli; for instance, if dorsal and ventral inputs are distributed among adjacent branches to a greater degree than equatorial inputs, they would produce less sublinear summation and hence greater responses per ommatidium stimulated. Two-photon calcium imaging in conjunction with local visual stimulation could be employed to experimentally test this hypothesis by producing a more detailed mapping than the one obtained previously. Clearly, active conductances, which are known to be present in the excitatory dendrite (Gabbiani and Krapp 2006), could play a role here (Johnston et al. 1996; Magee 2000; Spruston 2008).

Precise synaptic projections have been observed in several preparations (Borst and Egelhaaf 1992; Ogawa et al. 2006, 2008; Baden and Hedwig 2007; Bollmann and Engert 2009), and novel techniques should make it far easier in diverse preparations (Petreanu et al. 2009). It is generally difficult, however, to investigate complex spatial interactions among inputs because the projection pattern is not amenable to such investigations or the sensory modality makes it difficult. Because the LGMD has a precise retinotopic mapping and synapses can be visually activated, it is an ideal system to study such interactions. In the simplest interaction in passive dendrites, stimulation of proximal synapses should produce sublinear summation and hence reduced responses relative distributed stimulation (Johnston and Wu 1995). This effect has been observed in the LGMD: when presenting vertical bars, which activate synapses in distinct dendritic branches, responses are larger than when presenting horizontal bars, which activate synapses in one or a few branches (Peron et al. 2009).

A second theoretical prediction is that, in passive cells, a sequence of activation that moves toward the soma will, given appropriate timing, elicit a stronger response than an activation sequence moving away from the soma (Rall 1967). This effect has recently been produced in cortical pyramidal neurons using two-photon glutamate uncaging (Branco et al. 2010). In the LGMD, small squares (10°-by-10°) traveling in an anterior–posterior direction elicit stronger responses throughout their trajectory than those traveling in the opposite direction (Peron and Gabbiani 2009a Supp. Fig. 5). Anterior–posterior stimuli are activating synapses that are ever-closer to the SIZ; posterior–anterior stimuli elicit the opposite sequence. This is true throughout the response, across a range of velocities, and occurs independently of spike-frequency adaptation. The only exception is the initial response, which is larger for posterior–anterior stimuli, presumably because posterior inputs activate synapses closer to the SIZ than anterior inputs.

Though the precise retinotopic projection makes the LGMD a useful system for the study of dendritic integration, the retinotopy likely plays only a minor role in the context of looming. First, it is likely to contribute to the logarithmic transform of synaptic input via sublinear summation in the excitatory branch (Jones and Gabbiani 2012). It may also act, in conjunction with presynaptic directional selectivity, to imbue the LGMD with invariance to stimulus position (Gabbiani et al. 2001). Consistent with similar neurons in other insects (Douglass and Strausfeld 2003; Single and Borst 2002; Single et al. 1997), the LGMD's excitatory dendrite receives directionally selective input. The selectivity is such that anterior–posterior drifting grating stimuli produce stronger responses than posterior–anterior gratings (Peron et al. 2009). Because the LGMD has no voltage-gated calcium channels in the excitatory dendrite (Peron and Gabbiani 2009a), the calcium response's directional preference implies the inputs must be directional. Looming stimuli originating in anterior visual space will activate electrotonically distal dendrites, while posterior-originating stimuli activate proximal dendrites. Alone, this would suggest that the response to anterior looming should be smaller. However, because the anterior stimulus is dominated by anterior–posterior motion, while the posterior stimulus is dominated by posterior–anterior motion, the first will be enhanced and the latter mitigated.

17.6 Mechanisms Preventing Responses to Non-looming Stimuli

Circuits mediating escape behaviors must strike a delicate balance: effective escape from real threats is vital, but premature responses can be counterproductive, as they can reveal the animal's presence to potential predators. Lateral inhibition, feed-forward inhibition, and local motion habituation prevent large responses to many classes of stimuli (O'Shea and Rowell 1976; Rowell and O'Shea 1976b; Rowell et al. 1977; Rind and Simmons 1998; Gabbiani et al. 2005). One class of stimuli that would be minimally affected by these mechanisms and should elicit strong responses are constant velocity stimuli. Nevertheless, constant velocity edges and squares moving across visual space elicit weak responses (Rind and Simmons 1992; Simmons and Rind 1992; Peron and Gabbiani 2009a), as do looming approaches with constant edge velocity (i.e., simulating a decelerating approach; Hatsopoulos et al. 1995). A shared feature of many of these responses is a large initial response, followed by rapid decrement to a new steady state (see, e.g., Hatsopoulos et al. 1995, Fig. 17.2b; Peron and Gabbiani 2009a).

One mechanism that could produce this kind of response—large initial response, followed by lower steady state—is spike-frequency adaptation. Spike-frequency adaptation can be elicited in the LGMD by direct current injection (Gabbiani and Krapp 2006; Peron and Gabbiani 2009a). Pharmacological investigation has revealed this adaptation to be mediated by a K_{Ca} conductance. Specifically, Peron and Gabbiani (2009a) showed that the after-hyperpolarization following current injection cessation, a telltale indication of adaptation, as well as the degree of adaptation could both be substantially reduced by cadmium bath application and intracellular BAPTA iontophoresis. Cadmium is a calcium channel blocker (Hille 2001), while BAPTA chelates calcium intracellularly (Tsien 1980), implying that adaptation is calcium mediated. Elevation of extracellular potassium brought the neuron's resting potential to the reversal potential of potassium and abolished the after-hyperpolarization, showing adaptation to be potassium mediated (Peron and Gabbiani 2009a). Combined with additional pharmacology, these results were most consistent with an SK-like K_{Ca} conductance. Subsequent experiments employed intracellular BAPTA iontophoresis, ensuring that only the LGMD would be pharmacologically manipulated. These experiments demonstrated that reducing K_{Ca} produces enhanced responses to constant velocity stimuli (translating edges and squares), while leaving the looming response unaffected. Furthermore, calcium imaging revealed that a large calcium influx took place proximal to the SIZ (Peron et al. 2009), an ideal place to implement shunting inhibition (Koch et al. 1983), and roughly the same location where the inhibitory dendrites join the main process.

Another class of stimulus response that spike-frequency adaptation also mitigates is inverted looming stimuli—that is, receding objects (Peron and Gabbiani 2009b). Receding stimuli produce large onset responses followed by a quick decline in firing (Rind and Simmons 1992; Simmons and Rind 1992; Peron and Gabbiani 2009b). Of course, feed-forward inhibition is likely to be involved in responding to such stimuli.

Nevertheless, simulations show that an LGMD lacking feed-forward inhibition but having adaptation would produce smaller responses to receding than looming stimuli (Peron and Gabbiani 2009b). When feed-forward inhibition is abolished with picrotoxin, sudden rapid background motion during a looming stimulus produces much higher firing rates than normal (Gabbiani et al. 2002, Fig. 17.3b). After the background motion stops, there is very little response to the final phase of the looming stimulus—normally the period of greatest activity. It is likely that adaptation is responsible for this phenomenon.

Fundamentally, spike-frequency adaptation appears to act as a second-derivative filter (Peron and Gabbiani 2009b). That is, it can only be overcome by stimuli having a positive second derivative, such as looming stimuli. Indeed, translating objects having an accelerating velocity profile produce both an onset transient and sustained, increasing activity (Peron and Gabbiani 2009b).

17.7 Outlook

The LGMD is arguably one of the most well-characterized cells in neurobiology. The mathematical description of the looming response (Hatsopoulos et al. 1995; Gabbiani et al. 1999; see (17.2) above) and the linking of underlying biophysical mechanisms to this description (Gabbiani et al. 2002, 2004; Jones and Gabbiani 2012) stand out as a particularly impressive series of experiments elucidating this neuron's function in a behavioral context. It constitutes a novel example of how neurons can perform a multiplicative operation by using log-transformed coordinates. Nevertheless, much remains to be discovered. The presynaptic circuitry remains poorly understood, including the mechanisms responsible for habituation (O'Shea and Rowell 1976) and lateral inhibition (O'Shea and Rowell 1975a; Rowell and O'Shea 1976b; Rind and Simmons 1998). Furthermore, the input onto the inhibitory branches remains under-characterized, especially the ON branches (Rowell and O'Shea 1976a; Gabbiani et al. 2005).

While the presence of sodium conductances is implied by Gabbiani et al.'s (2004) TTX experiments, and Peron and Gabbiani (2009a) demonstrated the presence of K_{Ca}, the precise repertoire of active conductances in the LGMD awaits investigation. Active conductances have a profound role in shaping a neuron's dendritic integration (Johnston et al. 1996; Magee 2000; Stuart et al. 2008; Spruston 2008; Silver 2010). Previous work suggests that an inward rectifying conductance, possibly the h-current, I_h, is present in the excitatory dendrites (Gabbiani and Krapp 2006). Recent work has confirmed that both I_h and an M-current are present (Dewell and Gabbiani 2012). The function of these conductances remains to be elucidated, though they are known to play important roles in pyramidal neurons, including input normalization (Magee 1999).

Additionally, the behavioral role of LGMD–DCMD is still not completely understood. While correlates between behavior and DCMD firing rate have been demonstrated (Fotowat and Gabbiani 2007; Santer et al. 2008; Fotowat et al. 2011), DCMD ablation experiments raise questions about the specific role the neuron plays (Santer et al. 2008; Fotowat et al. 2011), because while ablation increases response latency and variability, it does not abolish the escape response. One hopes that a combination of targeted gene expression (e.g., Judkewitz et al. 2009) and optogenetics could be used to shed light on the subject. Specifically, if excitatory and inhibitory optical transducers could be delivered exclusively to the DCMD, and a method developed to deliver light to the nerve cord in a behaving animal, one could both mimic realistic firing patterns in the absence of visual stimuli as well as prevent DCMD responses in their presence. Clearly, such an experiment would require a novel photostimulation approach and a noninvasive way of delivering the optogenetic construct, but the unprecedented temporal control afforded by such an approach, as well as the ability to both stimulate and silence the DCMD should allow investigators to precisely unmask its behavioral role.

Most work to date has focused on the role of the LGMD in the context of stationary animals. Yet locusts are excellent fliers, and the DCMD projects to flight motorneurons (Simmons 1980). Behavioral work supports the notion that the LGMD–DCMD plays an important role in in-flight collision avoidance and escape via a gliding response (Santer et al. 2005, 2006). Flight should not only produce changes in arousal (Rind et al. 2008) but also present a constant optic flow to the LGMD. The impact of optic flow and flight has been extensively investigated in other insect neurons and is known to be profound (Liang et al. 2011; Jung et al. 2011). At least transiently, it is expected to produce a large increase in synaptic input, which is known to alter electrotonic structure (Bernander et al. 1991). This surge in activity would presumably produce a complex interplay between lateral inhibition, feed-forward inhibition, and habituation (but see Rind et al. 2008), and it is likely that the new steady-state of the LGMD would be distinct from the one observed in stationary animals. While neither flight nor wide-field background motion suppress the LGMD–DCMD response (Santer et al. 2006; Gabbiani et al. 2002), the mechanisms by which the response is maintained remain unstudied. Furthermore, experiments combining flight and optic flow have yet to be performed, and are essential if the goal is to understand the LGMD's behavior under realistic flight conditions.

The LGMD–DCMD system constitutes a rare instance where the link between biophysics and behavior can be made with relative ease. Substantial progress has been made in understanding the mechanisms underlying computation in the LGMD, especially with respect to the response to looming stimuli. Many questions remain, however, and will require novel experimental approaches to be addressed.

References

Bacon JP, Thompson KS, Stern M (1995) Identified octopaminergic neurons provide an arousal mechanism in the locust brain. J Neurophysiol 74:2739–2743

Baden T, Hedwig B (2007) Neurite-specific Ca2+ dynamics underlying sound processing in an auditory interneurone. Dev Neurobiol 67:68–80

Bernander O, Douglas RJ, Martin KA, Koch C (1991) Synaptic background activity influences spatiotemporal integration in single pyramidal cells. Proc Natl Acad Sci USA 88:11569–11573

Bollmann JH, Engert F (2009) Subcellular topography of visually driven dendritic activity in the vertebrate visual system. Neuron 61:895–905

Borst A, Egelhaaf M (1992) In vivo imaging of calcium accumulation in fly interneurons as elicited by visual motion stimulation. Proc Natl Acad Sci USA 89:4139–4143

Borst A, Haag J, Reiff DF (2010) Fly motion vision. Annu Rev Neurosci 33:49–70

Branco T, Clark BA, Häusser M (2010) Dendritic discrimination of temporal input sequences in cortical neurons. Science 329:1671–1675

Burrows M, Rowell CHF (1973) Connections between descending visual interneurons and metathoracic motoneurons in the locust. J Comp Physiol 85:221–234

Dewell RB, Gabbiani F (2012) Role of active conductances for dendritic processing in a looming sensitive neuron. Poster presented at the Society for Neuroscience, New Orleans, 13–17 October 2012

Douglass JK, Strausfeld NJ (2003) Retinotopic pathways providing motion-selective information to the lobula from peripheral elementary motion-detecting circuits. J Comp Neurol 457:326–344

Fotowat H, Gabbiani F (2007) Relationship between the phases of sensory and motor activity during a looming-evoked multistage escape behavior. J Neurosci 27:10047–10059

Fotowat H, Gabbiani F (2011) Collision detection as a model for sensory-motor integration. Annu Rev Neurosci 34:1–19

Fotowat H, Harrison RR, Gabbiani F (2011) Multiplexing of motor information in the discharge of a collision detecting neuron during escape behaviors. Neuron 69:147–158

Gabbiani F, Cohen I, Laurent G (2005) Time-dependent activation of feed-forward inhibition in a looming-sensitive neuron. J Neurophysiol 94:2150–2161

Gabbiani F, Krapp HG (2006) Spike-frequency adaptation and intrinsic properties of an identified, looming-sensitive neuron. J Neurophysiol 96:2951–2962

Gabbiani F, Krapp HG, Hatsopoulos N, Mo CH, Koch C, Laurent G (2004) Multiplication and stimulus invariance in a looming-sensitive neuron. J Physiol Paris 98:19–34

Gabbiani F, Krapp HG, Koch C, Laurent G (2002) Multiplicative computation in a visual neuron sensitive to looming. Nature 420:320–324

Gabbiani F, Krapp HG, Laurent G (1999) Computation of object approach by a wide-field, motion-sensitive neuron. J Neurosci 19:1122–1141

Gabbiani F, Mo C, Laurent G (2001) Invariance of angular threshold computation in a wide-field looming-sensitive neuron. J Neurosci 21:314–329

Guest BB, Gray JR (2006) Responses of a looming-sensitive neuron to compound and paired object approaches. J Neurophysiol 95:1428–1441

Hatsopoulos N, Gabbiani F, Laurent G (1995) Elementary computation of object approach by wide-field visual neuron. Science 270:1000–1003

Hille B (2001) Ionic channels of excitable membranes. Sinauer Associates, Sunderland, MA

James AC, Osorio D (1996) Characterisation of columnar neurons and visual signal processing in the medulla of the locust optic lobe by system identification techniques. J Comp Physiol A 178:183–199

Joesch M, Schnell B, Raghu SV, Reiff DF, Borst A (2010) ON and OFF pathways in Drosophila motion vision. Nature 468:300–304

Johnston D, Magee JC, Colbert CM, Cristie BR (1996) Active properties of neuronal dendrites. Annu Rev Neurosci 19:165–186

Johnston D, Wu SM (1995) Foundations of cellular neurophysiology. Bradford Books, New York, NY

Jones PW, Gabbiani F (2010) Synchronized neural input shapes stimulus selectivity in a collision-detecting neuron. Curr Biol 20:2052–2057

Jones PW, Gabbiani F (2012) Logarithmic compression of sensory signals within the dendritic tree of a collision-sensitive neuron. J Neurosci 32:4923–4934

Judge S, Rind F (1997) The locust DCMD, a movement-detecting neuron tightly tuned to collision trajectories. J Exp Biol 200:2209–2216

Judkewitz B, Rizzi M, Kitamura K, Häusser M (2009) Targeted single-cell electroporation of mammalian neurons in vivo. Nat Protoc 4:862–869

Jung SN, Borst A, Haag J (2011) Flight activity alters velocity tuning of fly motion-sensitive neurons. J Neurosci 31:9231–9237

Killmann F, Gras H, Schürmann F (1999) Types, numbers and distribution of synapses on the dendritic tree of an identified visual interneuron in the brain of the locust. Cell Tissue Res 296:645–665

Killmann F, Schürmann FW (1985) Both electrical and chemical transmission between the lobula giant movement detector and the descending contralateral movement detector neurons of locusts are supported by electron microscopy. J Neurocytol 14:637–652

Koch C, Poggio T, Torre V (1983) Nonlinear interactions in a dendritic tree: localization, timing, and role in information processing. Proc Natl Acad Sci USA 80:2799–2802
Krapp HG, Gabbiani F (2005) Spatial distribution of inputs and local receptive field properties of a wide-field, looming sensitive neuron. J Neurophysiol 93:2240–2253
Liang P, Kern R, Kurtz R, Egelhaaf M (2011) Impact of visual motion adaptation on neural responses to objects and its dependence on the temporal characteristics of optic flow. J Neurophysiol 105:1825–1834
London M, Häusser M (2005) Dendritic computation. Annu Rev Neurosci 28:503–532
Magee JC (1999) Dendritic lh normalizes temporal summation in hippocampal CA1 neurons. Nat Neurosci 2:508–514
Magee JC (2000) Dendritic integration of excitatory synaptic input. Nat Rev Neurosci 1:181–190
O'Shea M (1975) Two sites of axonal spike initiation in a bimodal interneuron. Brain Res 96:93–98
O'Shea M, Rowell CHF (1975a) A spike-transmitting electrical synapse between visual interneurons in the locust movement detector system. J Comp Physiol 97:143–158
O'Shea M, Rowell CHF (1975b) Protection from habituation by lateral inhibition. Nature 254:53–55
O'Shea M, Rowell CHF (1976) The neuronal basis of a sensory analyser, the acridid movement detector system. II. Response decrement, convergence, and the nature of the excitatory afferents to the fan-like dendrites of the LGMD. J Exp Biol 65:289–308
O'Shea M, Rowell CHF, Williams JLD (1974) The anatomy of a locust visual interneurone: the descending contralateral movement detector. J Exp Biol 60:1–12
O'Shea M, Williams JLD (1974) The anatomy and output connection of a locust visual interneurone: the lobular giant movement detector (LGMD) neuron. J Comp Physiol 91:257–266
Ogawa H, Cummins GI, Jacobs GA, Miller JP (2006) Visualization of ensemble activity patterns of mechanosensory afferents in the cricket cercal sensory system with calcium imaging. J Neurobiol 66:293–307
Ogawa H, Cummins GI, Jacobs GA, Oka K (2008) Dendritic design implements algorithm for synaptic extraction of sensory information. J Neurosci 28:4592–4603
Palka J (1967) An inhibitory process influencing visual responses in a fibre of the ventral nerve cord of locusts. J Insect Physiol 13:235–248
Peron SP, Gabbiani F (2009a) Spike frequency adaptation mediates looming stimulus selectivity in a collision-detecting neuron. Nat Neurosci 12:318–326
Peron SP, Gabbiani F (2009b) Role of spike-frequency adaptation in shaping neuronal response to dynamic stimuli. Biol Cybern 100:505–520
Peron SP, Jones PW, Gabbiani F (2009) Precise subcellular input retinotopy and its computational consequences in an identified visual interneuron. Neuron 63:830–842
Peron SP, Krapp HG, Gabbiani F (2007) Influence of electrotonic structure and synaptic mapping on the receptive field properties of a collision-detecting neuron. J Neurophysiol 97:159–177
Petreanu L, Mao T, Sternson SM, Svoboda K (2009) The subcellular organization of neocortical excitatory connections. Nature 457:1142–1145
Rall W (1967) Distinguishing theoretical synaptic potentials computed for different soma-dendritic distributions of synaptic input. J Neurophysiol 30:1138–1168
Rall W, Burke RE, Smith TG, Nelson PG, Frank K (1967) Dendritic location of synapses and possible mechanisms for the monosynaptic EPSP in motoneurons. J Neurophysiol 30:1169–1193
Reichardt W (1987) Computation of optical motion by movement detectors. Biophys Chem 26:263–278
Rind FC (1984) A chemical synapse between two motion detecting neurones in the locust brain. J Exp Biol 110:143–167
Rind FC, Leitinger G (2000) Immunocytochemical evidence that collision sensing neurons in the locust visual system contain acetylcholine. J Comp Neurol 423:389–401
Rind FC, Santer RD, Wright GA (2008) Arousal facilitates collision avoidance mediated by a looming sensitive visual neuron in a flying locust. J Neurophysiol 100:670–680

Rind FC, Simmons PJ (1992) Orthopteran DCMD neuron: a reevaluation of responses to moving objects. I. Selective responses to approaching objects. J Neurophysiol 68:1654–1666
Rind FC, Simmons PJ (1998) Local circuit for the computation of object approach by an identified visual neuron in the locust. J Comp Neurol 395:405–415
Rogers SM, Krapp HG, Burrows M, Matheson T (2007) Compensatory plasticity at an identified synapse tunes a visuomotor pathway. J Neurosci 27:4621–4633
Rowell CHF (1971) Variable responsiveness of a visual interneurone in the free-moving locust, and its relation to behaviour and arousal. J Exp Biol 55:727–747
Rowell CHF, O'Shea M (1976a) The neuronal basis of a sensory analyser, the acridid movement detector system. I. Effects of simple incremental and decremental stimuli in light and dark adapted animals. J Exp Biol 65:273–288
Rowell CHF, O'Shea M (1976b) Neuronal basis of a sensory analyser, the acridid movement detector system. III. Control of response amplitude by tonic lateral inhibition. J Exp Biol 65:617–625
Rowell CHF, O'Shea M, Williams JLD (1977) The neuronal basis of a sensory analyser, the acridid movement detector system. IV. The preference for small field stimuli. J Exp Biol 68:157–185
Santer RD, Rind FC, Stafford R, Simmons PJ (2006) Role of an identified looming-sensitive neuron in triggering a flying locusts escape. J Neurophysiol 95:3391–3400
Santer RD, Simmons PJ, Rind FC (2005) Gliding behaviour elicited by lateral looming stimuli in flying locusts. J Comp Physiol A Neuroethol Sens Neural Behav Physiol 191:61–73
Santer RD, Yamawaki Y, Rind FC, Simmons PJ (2008) Preparing for escape: an examination of the role of the DCMD neuron in locust escape jumps. J Comp Physiol A Neuroethol Sens Neural Behav Physiol 194:69–77
Schlotterer GR (1977) Response of the locust descending movement detector neuron to rapidly approaching and withdrawing visual stimuli. Can J Zool 55:1372–1376
Segev I, London M (2000) Untangling dendrites with quantitative models. Science 290:744–750
Silver RA (2010) Neuronal arithmetic. Nat Rev Neurosci 11:474–489
Simmons PJ (1980) Connexions between a movement-detecting visual interneurone and flight motoneurons of a locust. J Exp Biol 86:87–97
Simmons PJ, Rind FC (1992) Orthopteran DCMD neuron: a reevaluation of responses to moving objects. II. Critical cues for detecting approaching objects. J Neurophysiol 68:1667–1682
Single S, Borst A (2002) Different mechanisms of calcium entry within different dendritic compartments. J Neurophysiol 87:1616–1624
Single S, Haag J, Borst A (1997) Dendritic computation of direction selectivity and gain control in visual interneurons. J Neurosci 17:6023–6030
Spruston N (2008) Pyramidal neurons: dendritic structure and synaptic integration. Nat Rev Neurosci 11:1059–1067
Strausfeld NJ, Nässel DR (1981) Neuroarchitectures serving compound eyes of crustacea and insects. In: Autrum H (ed) Handbook of sensory physiology, vol VII/68. Springer, Berlin
Stuart G, Spruston N, Häusser M (2008) Dendrites. Oxford University Press, New York, NY
Tsien RY (1980) New calcium indicators and buffers with high selectivity against magnesium and protons: design, synthesis, and properties of prototype structures. Biochemistry 19:2396–2404
Zador AM, Agmon-Snir H, Segev I (1995) The morphoelectrotonic transform: a graphical approach to dendritic function. J Neurosci 15:1669–1682

Chapter 18
Biophysics of Synaptic Inhibition in Dendrites

Albert Gidon

Abstract This chapter aims at investigating the functional implications of the biologically realistic and widespread case in which a single inhibitory axon forms multiple (10–20) synaptic contacts on the dendrites of its target neuron. We analyzed the impact of multi-site dendritic inhibition on the neurons' output and, thus, gained several counterintuitive insights into the biophysical and functional implications of such connectivity pattern. In the course of the chapter, we propose a functional role for very distal dendritic inhibition; demonstrate the regional effect of multiple, rather than single, inhibitory synapses in terms of the spread of their collective shunting effect in the dendritic tree; and suggest an explanation as to why, in both cortex and hippocampus, the total number of inhibitory dendritic synapses per pyramidal cell is smaller (about 20 %) than that of excitatory synapses. This chapter, thus, provides a fresh perspective on the biophysical design principles that govern the operation of inhibition in dendrites.

18.1 Introduction

Synaptic inhibition was introduced by Sir Charles Sherrington (in his Nobel lecture in 1932) and discovered later by John Eccles (Brock et al. 1951). Since then, synaptic inhibition was shown to be one of the most fundamental mechanisms that shape the operation of the nervous system. A crucial step towards understanding the inhibitory control over the dynamics of neuronal systems lies in deciphering the functional significance of the very particular connectivity design of inhibitory synapses over their target neurons. Like excitatory synapses, inhibitory synapses are distributed over the whole dendritic tree of their target neuron and, surprisingly, are found

A. Gidon (✉)
Department of Neurobiology, Alexander Silberman Institute of Life Sciences,
The Hebrew University of Jerusalem, Jerusalem 91904, Israel
e-mail: agidon20@gmail.com

H. Cuntz et al. (eds.), *The Computing Dendrite: From Structure to Function*,
Springer Series in Computational Neuroscience 11, DOI 10.1007/978-1-4614-8094-5_18,

even at very distal dendritic branches, far from the output axonal region. Importantly, recent anatomical studies have demonstrated that afferents of inhibitory neurons belonging to a specific interneuron class (e.g., Chandelier, Basket, Martinotti) tend to occupy a particular dendritic subdomain of their target neuron; in these subdomains a single inhibitory axon establishes many (10–20) contacts (Klausberger and Somogyi 2008; Markram et al. 2004). The implications of this highly specific multi-site inhibitory connectivity remain unclear.

In aiming to provide a theoretical understanding for dendritic inhibition, a few (by now classical) analytical studies have utilized Rall's cable theory (Jack et al. 1975; Koch et al. 1982; Rall 1964). Two fundamental insights were provided by these studies. First, the "visibility" of the inhibitory conductance perturbation is highly local. Namely, the effect of local conductance perturbation in dendrites decays very steeply with distance from the inhibitory synaptic contact (Koch et al. 1990; Williams 2004). Second, "on-path" inhibition, located between the excitatory synapse and the soma, is more effective than the respective "off-path" inhibition in dampening the excitatory current that reaches the soma.

These theoretical studies were limited in several ways. First, they adopted a "somato-centric" viewpoint, whereby the impact of dendritic inhibition is considered solely at the soma and/or axon loci. Second, the dendrites were assumed to be passive, and third, inhibition was modeled (in most of the cases) as forming a single synaptic contact onto the dendrites rather than multiple synapses contacting multiple dendritic branches as is the biological case.

Indeed, the existence of multiple contacts between connected pairs of neurons is the rule rather than the exception in many types of cortical cells. Interestingly, single inhibitory interneurons typically make more contacts on PC dendrites than do single excitatory cells (more than ten contacts per connection; see review by Markram et al. 2004). A large number of contacts per connection and the domain-specific innervation of specific axonal afferents as found in the cerebral cortex are also found in the hippocampus for many cell types (Buhl et al. 1994; Miles et al. 1996). Paradoxically, multiple synapses from the same axon reduce the repertoire of inputs potentially obtained by the postsynaptic cell (Wen et al. 2009). Therefore, this common feature in hippocampus, neocortex, and other brain regions [e.g., the striatum (Tepper et al. 2004)] may suggest an underlying common functional principle.

Another important feature of dendrites is that they are decorated by an abundance of voltage-gated excitable ion channels such as NMDA, Ca^{2+}, and Na^{+} inward currents which, in some cases, trigger a dendritic spike. Dendrites of several neuron types were shown to operate in a highly nonlinear fashion (for review see Magee 2007) both in vitro (Larkum et al. 2009) and in vivo (Lavzin et al. 2012; Palmer et al. 2012). In these neurons, inhibition establishes multiple synaptic contacts in a restricted dendritic domain and thus specifically influences regional dendritic nonlinear phenomena. Therefore, an extension of the classical theoretical framework is required to account for conditions that are more realistic. Such a framework which incorporates a "dendro-centric" viewpoint is described in this chapter.

Here, we present an experimentally based theoretical study that solves the cable equation for the case in which multiple synaptic contacts impinge on the dendritic tree. This solution provided several new (sometimes counterintuitive) principles that govern the operation of dendritic inhibition. By introducing a new measure (the "shunt level") for the impact of inhibitory conductance perturbation, we provided an explanation for the role of distal dendritic inhibition and for the domain-specific multi-site inhibitory connectivity pattern and analyzed the strategic placement for synaptic inhibition for optimally controlling dendritic excitability.

Our analytical results enabled us to explain how a relatively small number of strategically placed inhibitory synapses effectively counterbalance the effect of larger number of excitatory synapses found in many central neurons such as cortical (DeFelipe et al. 2002) and hippocampal (Megías et al. 2001) pyramidal cells.

In the modern "age of connectomics" (Bock et al. 2011; Briggman et al. 2011; Seung 2009), new data unraveling the details of specific synaptic connections in neuronal networks is rapidly accumulating. This chapter, which summarizes results from our previous work, serves as a step towards a theoretical understanding that will provide a systematic functional interpretation of the inhibitory connectivity in the neuronal microcircuit.

18.2 Results

Symbols	
X (X_i)	Electrotonic distance (in units of the space constant, λ) from origin to location i (dimensionless)
L	Electrotonic length (in units of λ) of a dendritic branch (dimensionless)
V	Steady membrane potentials, as a deviation from the resting potential (V)
R_i	Input resistance at location i (Ω)
ΔR_i	Change in R_i due to synaptic conductance perturbation (Ω)
g_i	Steady synaptic conductance perturbation at location i (S)
SL	Shunt level ($0 \leq SL \leq 1$; dimensionless)
SL_i	Shunt level $\Delta R_i/R_i$ due to activation of single- or multiple-conductance perturbations ($0 \leq SL \leq 1$; dimensionless)
$R_{i,j}$	Transfer resistance between location i and location j (Ω)
$SL_{i,j}$	Attenuation of SL (SL_j/SL_i) for a single-conductance perturbation at location i ($0 \leq SL_{i,j} \leq 1$; dimensionless)
$A_{i,j}$	Voltage attenuation, V_j/V_i, for current perturbation at location i ($0 \leq A_{i,j} \leq 1$; dimensionless)

18.2.1 The Shunt Level: A Tractable Measure for the Impact of Dendritic Inhibition

When an inhibitory synapse is activated at a dendritic location, i, a local conductance perturbation g_i (a shunt) is induced in the dendritic membrane which is followed by an inhibitory postsynaptic potential (IPSP). "Silent" or "shunting" inhibition occurs when the inhibitory reversal potential is equal to the resting membrane potential (Koch and Poggio 1985); thus, an IPSP is not observed in this case.

Although the membrane shunt due to the activation of the inhibitory synapses at *i* is highly local, its effect spreads to (i.e., is visible at) other dendritic locations (Rall 1967; Koch et al. 1990; Williams 2004). Indeed, this spatial spread is reflected by a change in input resistance, ΔR_d, at location *d*. We defined the shunt level at location *d*, SL_d, as

$$SL_d = \Delta R_d / R_d, \tag{18.1}$$

where R_d is the input resistance in location *d* prior to the activation of g_i. SL_d is thus the relative drop in R_d at location *d* due to the activation of single (or multiple) steady conductance perturbations at arbitrary dendritic locations (see Gidon and Segev 2012 for generalization to the transient case). The value of SL_d ranges from 0 (no shunt) to 1 (infinite shunt) and depends on the particular dendritic distribution of g_is. For example, $SL_d = 0.2$ implies that the inhibitory synapse reduced the input resistance at location *d* by 20 % which is also the relative drop in the steady voltage at *d* due to the inhibition, following the injection of steady current at location *d*.

The solution for ΔR_d (Koch et al. 1990) following the activation of a single steady conductance perturbation, g_i, at location *i* is

$$\Delta R_d = R_d - R_d^* = \frac{g_i R_{i,d}^2}{1 + g_i R_i}, \tag{18.2}$$

where R_d and R_d^* are, respectively, the input resistance prior to and following the activation of g_i (see definitions in the *Symbols* list). The transfer resistance from *i* to *d*, $R_{i,d}$, is (Koch et al. 1983)

$$R_{i,d} = R_{d,i} = R_i A_{i,d} = R_d A_{d,i}. \tag{18.3}$$

Combining (18.2) and (18.3), *SL* at location *d* due to the activation of the conductance perturbation at location *i* is

$$SL_d = SL_i \times SL_{i,d} = \left[\frac{g_i R_i}{1 + g_i R_i} \right] A_{i,d} \times A_{d,i}. \tag{18.4}$$

The bracket denotes, SL_i, the amplitude of *SL* at the input location ($d = i$) and $SL_{i,d} = A_{i,d} \times A_{d,i}$ is the attenuation of *SL*.

Note that the shunt level measure is applicable also for assessing the change in input resistance due to excitatory synapses, which, like inhibition, exert a local membrane conductance change.

In the next sections, we analytically and numerically compute *SL* in idealized dendrites as well as 3D reconstructed dendrites. This provided several new and counterintuitive results regarding the overall impact of multiple inhibitory dendritic synapses in dendrites and explained several experimental and modeling results that were not fully understood prior to the present study.

18.2.2 Distal Dendritic Inhibition Effectively Controls Excitability at Proximal Regions

We started with a geometrically simple case, whereby a single inhibitory synapse impinged on a dendritic cylinder sealed at one end and coupled to an isopotential excitable soma at the other end (Fig. 18.1a). The dendritic cylinder comprised a hotspot (Antic et al. 2010; Larkum et al. 1999; Magee et al. 1995; Schiller et al. 1997, 2000) modeled by a cluster of 20 NMDA synapses, where each synapse was randomly activated at 20 Hz (red circle and red synapse in Fig. 18.1a). We then searched for the strategic placement of the inhibitory synapse that would effectively dampen this local dendritic hotspot.

Using numerical simulations for the nonlinear cable model described above (Fig. 18.1b), we found that when the inhibitory conductance perturbation, g_i, was placed distally ("off-path") to the hotspot, the rate of the soma action potentials was reduced more effectively than when the same inhibitory synapse was placed proximally ("on-path") at the same distance from the hotspot (Fig. 18.1b). The asymmetry of the impact of proximal versus distal inhibition for dampening a local dendritic hotspot holds regardless of the distance from the hotspot (Fig. 18.1c). When we increased the distance (in a given cable length) of both the distal and the proximal synapses from the hotspot the relative advantage of the distal inhibition also increased (Fig. 18.1c).

Indeed, such asymmetry in the impact of proximal versus distal inhibition for dampening a local dendritic hotspot was previously observed in vitro (Jadi et al. 2012; Liu 2004; Lovett-Barron et al. 2012; Miles et al. 1996) and in simulations (Archie and Mel 2000; Rhodes 2006), but the basis for this counterintuitive result has remained unclear.

In order to provide an explanation for this result, we analytically computed the value for SL at the hotspot (h) and thus assessed the impact of inhibition at this location (Fig. 18.1d). In the corresponding passive case, SL_h at the hotspot due to the inhibitory conductance perturbation g_i at location i can be expressed as the product of SL amplitude at location i (SL_i) and the attenuation of SL from i to h ($SL_{i,h}$; see Eq. (18.4)). The asymmetry of the impact of distal versus proximal inhibition on location h (the hotspot) results from the difference in the model's boundary conditions, namely, sealed end boundary at the distal end and an isopotential soma at the proximal end. This difference implies that the input resistance and SL_i (in cases of a fixed g_i) also increase monotonically with distance from the soma. Thus, SL_i at the distal site (depicted by the black synapse) is larger than that at the corresponding proximal site (depicted by the orange synapse). Additionally, the overall $SL_{i,h}$ attenuation from the inhibitory synapses to the hotspot is shallower for the distal synapses than for the proximal synapses, because the latter is more affected by the somatic current sink. The product of these two effects—the initially larger SL_i at the distal synapse and the shallower attenuation of SL_i from the distal synapse to the hotspot—implies that SL at the hotspot (SL_h) is larger for this synapse. The latter conclusion also holds for transient inhibitory synaptic conductance.

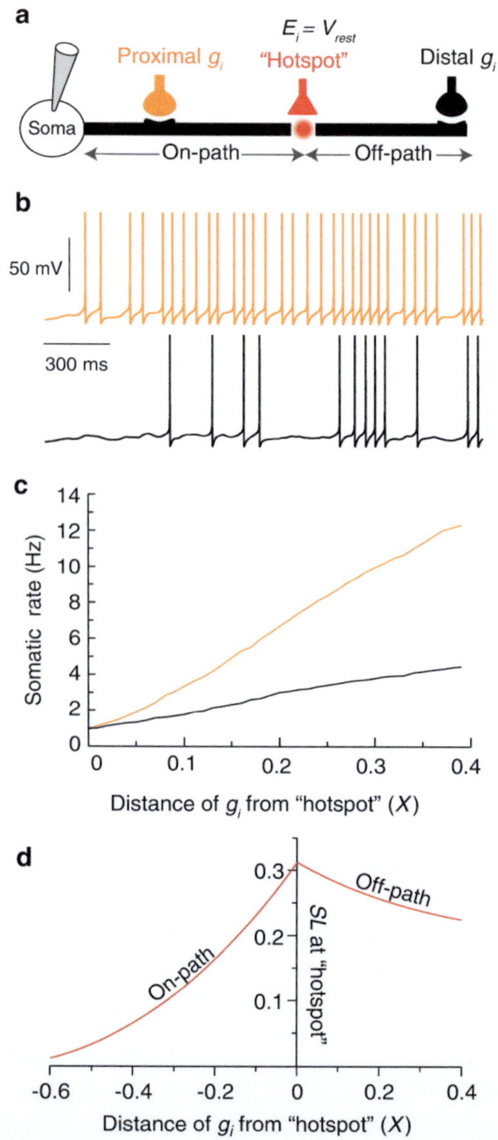

Fig. 18.1 Distal ("off-path") inhibition is more effective than the corresponding proximal ("on-path") inhibition in dampening a local dendritic hotspot. (**a**) A model of a cylindrical cable (sealed end at $L = 1$) coupled to an isopotential excitable soma. Twenty NMDA synapses are clustered at the hotspot located at $X = 0.6$ (*red synapse*); each synapse is randomly activated at 20 Hz. A single inhibitory synapse ($g_i = 1$ nS) is placed either distally or proximally at the same electrotonic distance ($|X| = 0.4$) from the hotspot. (**b**) Inhibition of the somatic Na^+ spikes is more effective when inhibition is placed distally to the hotspot (*black synapse* and corresponding *black somatic spike*s, compared to *orange synapse* and corresponding *orange somatic spikes*). (**c**) Somatic rate as a function of the distance between the synapse and the hotspot. *Black* and *orange lines* correspond to the on-path and off-path synaptic locations, respectively. (**d**) *SL* at the hotspot in the modeled cell depicted in (**a**) as a function of the distance of the inhibitory synapse from the hotspot. Off-path inhibition attenuates less steeply compared to the respective on-path inhibition. Reproduced from Gidon and Segev (2012) with permission from Cell Press

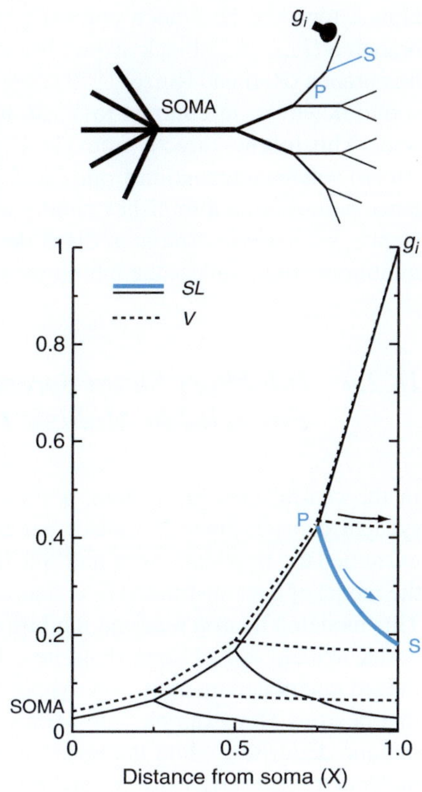

Fig. 18.2 Steep attenuation of *SL* in distal dendrites. *Top*. g_i is located at a single terminal end of an idealized symmetrically branched dendritic tree consisting of six identical stem dendrites of which the structure of only one is fully shown (Rall and Rinzel 1973). *Bottom*. Attenuation of *SL* (*continuous line*) and of steady voltage, *V* (*dotted line*), from the distal input dendritic terminal. Note the steep attenuation of *SL* towards the distal dendritic terminals (*blue arrow*) compared to the attenuation of *V* (*black arrow*). Reproduced from Gidon and Segev (2012) with permission from Cell Press

The results in Fig. 18.1 (in particular Fig. 18.1d) hold *at the dendritic site* for any dendritic region producing inward current, either for voltage-independent conductances (e.g., via an AMPA synapse) or for voltage-dependent conductances (e.g., NMDA currents or active Ca^{+2} or Na^{+} inward currents). But the advantage of distal versus proximal inhibition at that region is amplified in the voltage-dependent (nonlinear) case because inhibition at the hotspot increases the threshold for the activation of regenerative inward currents (Jadi et al. 2012).

18.2.3 SL in Branched Dendrites

The attenuation of *SL* for a single-conductance perturbation, g_i, at location *i* is $SL_{i,d}=A_{i,d}\times A_{d,i}$ (18.4). Therefore, it is straightforward that $SL_{i,d}=SL_{d,i}$, namely, *SL attenuation is symmetrical* for any given two dendritic sites, regardless of the dendritic structure ($A_{i,d}\neq A_{d,i}$ in most cases).

Using (18.4), we compute *SL* attenuation for the case of an idealized dendritic tree receiving a single-conductance perturbation at the distal dendritic terminal (Fig. 18.2). For comparison, the steady voltage attenuation (Rall and Rinzel 1973), *V*, is also shown (Fig. 18.2, dotted line). *V* attenuation is steep from the distal (input) branch towards the branch point (P) but is shallow in the direction of the sibling branch S (Fig. 18.2, black arrow) because of the sealed-end boundary condition in this branch (Rall and Rinzel 1973). Similarly to V, SL attenuates steeply towards the soma; however, in contrast to V, SL attenuates steeply towards terminal S (blue line). This follows directly from (18.4), as SL attenuation from P to S depends on the (steep) voltage attenuation from S to P ($A_{S,P}$). Consequently, the impact of conductance perturbation diminishes rapidly with distance in such thin dendritic branches. Hence, excitatory currents in distal dendrites are electrically "protected" from the inhibitory shunt, unless the inhibitory synapses directly target these branches.

18.2.4 Inhibitory Shunt Spreads Centripetally in Dendrites Encircled by Multiple Inhibitory Synapses

In the central nervous system, a single inhibitory axon typically forms multiple synaptic contacts (up to 20 contacts per axon) on the target dendritic tree. Therefore, we examined the implications of multiple inhibitory synapses for SL in dendrites, using the model of a reconstructed CA1 neuron (Golding et al. 2005) depicted in Fig. 18.3. This modeled neuron received inhibition at three distinct dendritic subdomains: the basal, apical, and oblique dendrites. In CA1, these morphological domains are indeed innervated by inhibitory synapses arising from different classes of inhibitory interneurons [for example, the axon of bistratified cells targets the basal and the oblique dendrites, while the apical dendrite is targeted by the oriens lacunosum-moleculare cells (Klausberger and Somogyi 2008)]. We assumed that each domain receives a cluster of five inhibitory contacts (white dots).

The color-coded SL value induced by the activation of these 15 inhibitory synapses is shown in Fig. 18.3a, b, superimposed on the modeled cell. SL spreads poorly (it attenuates steeply) in the direction of the dendritic terminals (Fig. 18.3a, blue dendrites), but, surprisingly, it spreads effectively (Fig. 18.3a, red region) hundreds of micrometers centripetally to the region delineated (dashed line) by the contact sites themselves. Even more surprising was that SL became larger in regions lacking inhibitory synapses compared to SL at the synaptic sites themselves (Fig. 18.3b). This is in contrast to the prevailing view that the maximal effect of inhibition is always at the synaptic site itself (Jack et al. 1975). This was further demonstrated by simulation, whereby an excitatory synapse in the proximal apical tree, far away from any inhibitory synapse, was more inhibited than an excitatory synapse contacting the oblique branches (top EPSP; see Fig. 18.3b, continuous yellow line—before inhibition; dashed line—following inhibition).

The elevated centripetal increase in SL (red central dendritic regions in Fig. 18.3a) existed under a wide range of conditions. Interestingly, we can show analytically that such elevation in centripetal inhibition required at least three inhibitory synapses encircling a dendritic region consisting of at least three branches (see Sect. 18.2.5 and further analysis in Gidon and Segev 2012).

For comparison, we also computed the impact of dendritic inhibition as observed at the soma. In Fig. 18.3c, d, the same CA1 cell as in Fig. 18.3a, b was modeled, but

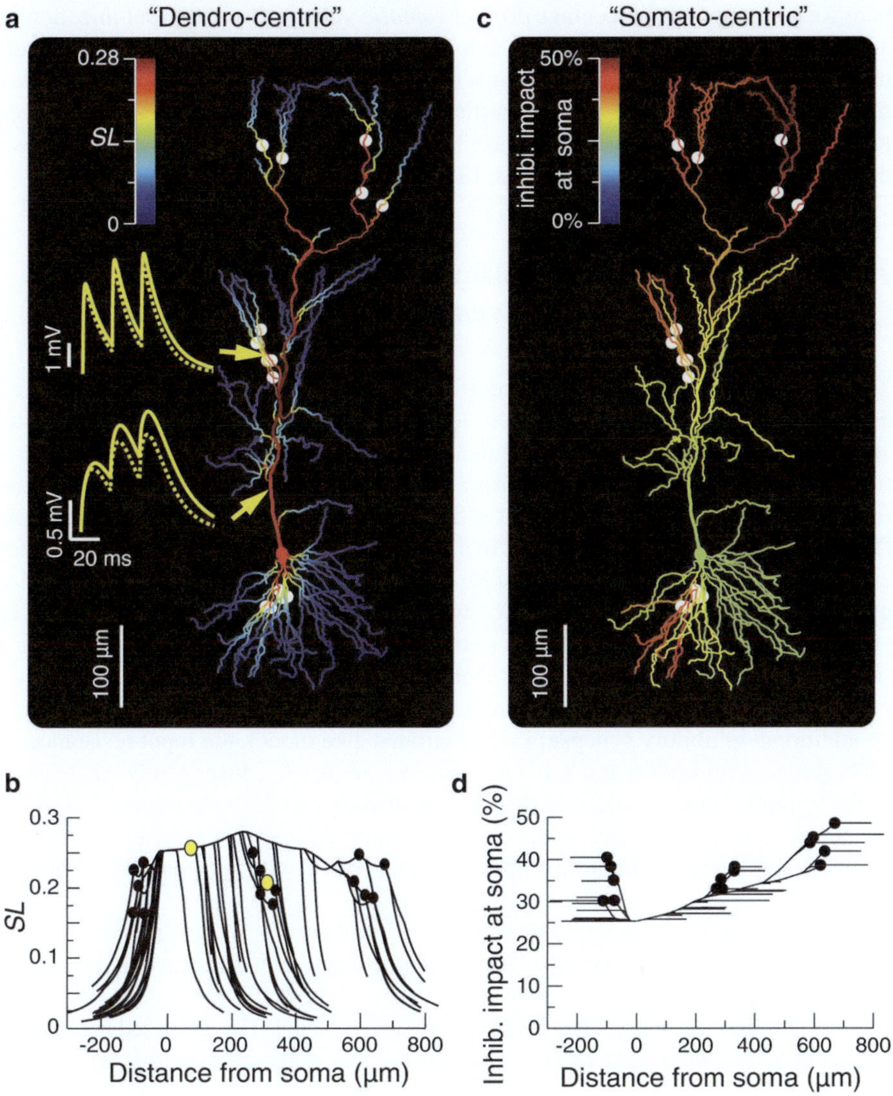

Fig. 18.3 Global spread of inhibitory shunt in trees with multiple inhibitory synapses. (**a**) *SL* (color coded) in a model of a reconstructed CA1 pyramidal neuron receiving a total of 15 inhibitory synapses (*white dots*; g_i=0.5 nS each, at steady state) targeting distinct dendritic subdomains (basal, oblique, and apical dendrites; delineated by *dashed lines*). *Yellow traces*: EPSPs generated and recorded at the sites denoted by *yellow arrows* before (*continuous line*) and following (*dashed line*) the activation of the inhibitory synapses. The excitatory synapse on the oblique branch is co-localized with one of the inhibitory synapses; the EPSP generated by this synapse is less inhibited (*SL*=0.2) than the excitatory synapse located on the apical trunk (*SL*=0.25), far from any one of the inhibitory synapses. The EPSP was simulated using AMPA-like conductance (see Appendix). (**b**) *SL* as a function of distance from the soma for the model shown in (**a**). Inhibitory synapses are marked by *black dots* and excitatory synapses by *yellow dots*. (**c**, **d**) As in **a** and **b**, but the impact of inhibition is measured as the percentage drop of the somatic voltage for steady current injected at any given dendritic site. Reproduced from Gidon and Segev (2012) with permission from Cell Press

here we computed the percentage drop of *somatic* voltage due to current injection at any dendritic location following the activation of the 15 inhibitory synapses. When measured at the soma, the largest impact of inhibition was obtained for depolarization originating at distal dendrites, particularly for distal branches receiving inhibitory synapses (red branches in Fig. 18.3c). In contrast, *SL* was very small in these distal branches (blue branches in Fig. 18.3a).

18.2.5 Maximal Inhibition May Occur in Dendritic Domains Lacking Inhibitory Synapses

To analytically explain the counterintuitive results whereby inhibition accumulates centrally to the synapses which encircle the somatic regions (Fig. 18.3) we constructed a symmetrical starburst-like dendritic model (Fig. 18.4) consisting of multiple identical branches stemming from a common junction ($X=0$). Each of these branches received an identical g_i at a fixed distance ($X=0.4$) from the junction (Fig. 18.4a). In order to solve for *SL* in a symmetrical starburst-like dendritic model consisting of *N* identical branches with *N* synapses we used an equivalent model consisting of two connected branches with two synapses (Fig. 18.4c, inset). It is straightforward to show that in such a structure, *SL* at the junction remains constant (Rall 1959), independent of the number of stem branches (Fig. 18.4b, all curves converge at $X=0$). However, when we increased the number of branches (each with an additional inhibitory synapse) in the starburst-like model, the input resistance at each of the synaptic sites and at the junction was reduced. Consequently, SL_i at each of the synaptic sites was reduced (Fig. 18.4b, arrow), and the attenuation of *SL* from the junction to each of the synaptic sites became steeper. Namely, synapses had progressively smaller shunting impact on each other with increasing number of branches.

Together, these results imply that when the number of branches is large enough, *SL* at the junction (lacking synapses) may become larger than *SL* at each of the synaptic sites. Figure 18.4c summarizes *SL* as a function of the distance of g_i from the soma for different number of branches in the starburst-like model. The enhanced centripetal impact of *SL* becomes more prominent as the number of branches increases.

18.3 Discussion

The shunt level, *SL*, introduced in this study is a simple, intuitive, and analytically tractable measure for assessing the impact of inhibitory conductance perturbation in the dendritic tree. The "dendro-centric" view of inhibition using *SL* in arbitrary passive dendritic trees receiving multiple inhibitory contacts has provided several surprising results. In particular, distal dendritic sites are effective in dampening more proximal nonlinear events. Additionally, we found that with multiple synapses, *SL* spreads very effectively towards dendritic regions encircled by these synapses and that it may become larger in these regions than at the synaptic loci themselves.

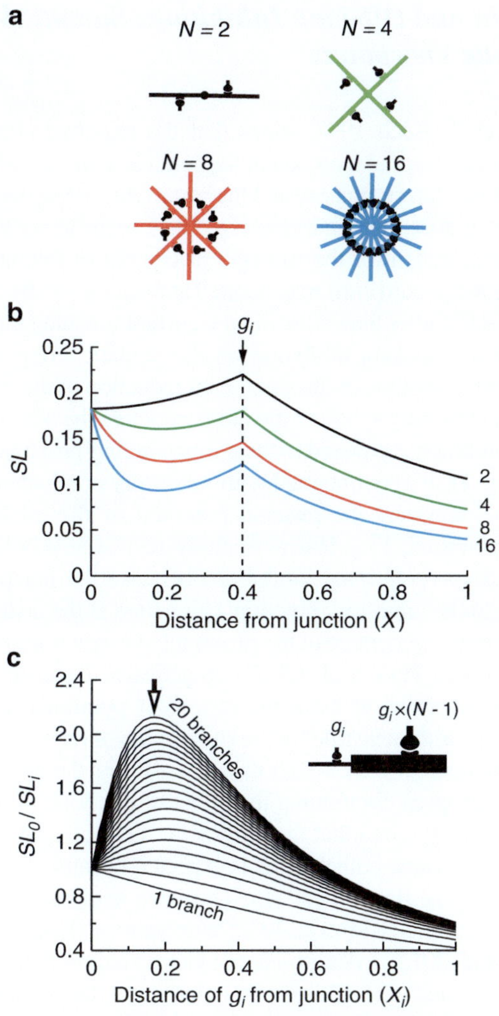

Fig. 18.4 Enhanced centripetal accumulation of inhibitory shunt in trees with multiple inhibitory synapses. (**a**) Symmetrical starburst-like dendritic models consisting of multiple (N=2, 4, 8, 16) identical branches (L=1) stemming from a common junction (X=0). (**b**) SL for the corresponding models depicted in **a**. Each branch receives a single g_i (1 nS) at X=0.4 (*dashed line*). For N=8 (*blue*) and N=16 (*red*), SL at the junction (lacking synapses) is larger than SL at the site of the inhibitory synapse. (**c**) Ratio between SL at the junction (SL_0) and at the synaptic sites (SL_i) as a function of the distance of the synapses from the junction. Models as in **a** but with varying number of branches (1–20), each with one synapse. SL_0/SL_i increases with the number of branches (with ~2.2 for 20 synapses and branches; *arrow*). Inset: Equivalent dendritic model used to solve for SL where N is the total number of branches and synapses in the starburst-like model (see Appendix). Reproduced from Gidon and Segev (2012) with permission from Cell Press

These findings yielded several new insights regarding the functional implications of the unique connectivity pattern of dendritic inhibition. Importantly, although these insights were based on the solution for the steady state case and for passive dendrites, they nevertheless explained the simulated results obtained for corresponding nonlinear and transient cases.

18.3.1 On-Path and Off-Path Inhibition: Somatic Versus Dendritic Viewpoints

The "on-path theorem" (Koch 1998) states that the maximal effect of inhibition in reducing the excitatory potential recorded at the soma is achieved when inhibition is on the path between the excitatory synapse and the soma (Rall 1964; Jack et al. 1975; Koch et al. 1983). At first glance, our findings (Fig. 18.1) seem to contradict this classical result. However, we searched for the strategic placement of inhibition so that it most effectively dampens the inward current generated at the locus of the excitatory synapses (or the "hotspot") itself, rather than reducing the current reaching the soma. Indeed, the powerful impact of the off-path inhibition on the somatic firing as demonstrated in Fig. 18.1 is a secondary outcome of the significant reduction of the inward current in the hotspot by the distal inhibitory synapse: the more excitable the hotspot, the more advantageous the distal inhibition compared to the corresponding proximal inhibition.

The results in Fig. 18.1 were obtained with inhibitory conductance, i.e., the case of a "silent inhibition," whereby the reversal potential of the inhibitory synapse, E_i, equals the resting potential, V_{rest}. However, these results still hold when E_i is more negative than V_{rest} (hyperpolarizing inhibition) because the hotspot would be more hyperpolarized due to the inhibitory potential originated at the distal synapse than due to the inhibitory potential generated at the proximal synapse (Gidon and Segev 2012).

In recent experiments, Hao et al. (2009) co-activated dendritic inhibition, g_i, and excitation, g_e, while recording at the soma of a CA1 pyramidal cell (somato-centric view). They derived an arithmetic rule for the summation of the somatic EPSP and IPSP, confirming the predictions of the on-path theorem also for the case of multiple inhibitory and excitatory synapses. Examining the effect of dendritic inhibition on dendritic spikes invoked by g_e, they found that the arithmetic rule does not hold when g_i and g_e were co-activated on the same branch. This is expected because, in this case, g_i directly inhibits the dendritic spike (large local *SL*). This case demonstrates that for dendrites with active nonlinear currents (Kim et al. 2012; Murayama and Larkum 2009; Murayama et al. 2009; Palmer et al. 2012), a dendro-centric view is required in order to characterize the impact of dendritic inhibition. This is particularly true due to the global and centripetal spread of inhibition in dendrites with multiple inhibitory synapses.

18.3.2 Anatomical Versus Functional Inhibitory Dendritic Subdomains

Typically, axons belonging to particular inhibitory classes contact specific dendritic subdomains of their target dendrites (Klausberger and Somogyi 2008; Markram et al. 2004). For example, Chandelier axo-axonic inhibitory interneurons target the axon initial segment of excitatory neurons (Somogyi et al. 1983; Szentágothai and Arbib 1974), while Basket cells target the soma and proximal dendrites (Wang et al. 2002). Interestingly, individual inhibitory axons often form multiple (10–20) synaptic contacts on the target dendritic tree. Therefore, in most cases, the principles discussed in this chapter are relevant even at the level of a single inhibitory axon and its postsynaptic dendritic target.

The effective spread of *SL* into the dendritic region surrounded by multiple inhibitory synapses (Fig. 18.3) leads to a spatially extended shunted dendritic domain beyond the anatomical domain demarcated by these synapses. Therefore, the spatial extent of *SL* could serve as a way to assess spatial specificity of the *impact* of different classes of interneurons on their postsynaptic targets ("functional subdomain"), rather than relying on the anatomical connectivity alone.

The global spread of the inhibitory shunt implies that the functional subdomain would be less specific than the anatomical subdomain. Namely, in order to dampen excitatory/excitable dendritic currents, it is not necessary to match each excitatory synapse with a corresponding adjacent inhibitory synapse. Rather, by surrounding a dendritic region with a few inhibitory contacts, it is possible to effectively dampen the excitatory/excitable current that would be generated in this region (Figs. 18.3 and 18.4) and thereby effectively control the neuron's output. This may explain why in the neocortex and the hippocampus, only a small number (~20 %) of the synapses are inhibitory (DeFelipe and Fariñas 1992; Megías et al. 2001; Merchán-Pérez et al. 2009).

18.3.3 Why Does Inhibition Target Distal Dendrites?

One surprising analytic result of this study is that distal off-path inhibition is more effective than the corresponding on-path inhibition for dampening a midway dendritic nonlinear hotspot (Fig. 18.1, and see also experimental validation in Miles et al. 1996; Lovett-Barron et al. 2012 and simulated results in Archie and Mel 2000; Rhodes 2006). This result, together with the result showing that *SL* spreads poorly to thin distal branches (Figs. 18.2 and 18.3), implies that in order to control nonlinear processes in distal dendritic branches, inhibitory synapses should directly target the distal end of these branches.

Interestingly, branch-specific off-path (rather than on-path) distal inhibition is expected to powerfully affect the plasticity of excitatory synapses in these branches, as this process depends on the influx of (active) Ca^{2+} current via either NMDA-dependent receptors or voltage-dependent Ca^{2+} channels (MacDonald et al. 2006; Malenka 1991; Malenka and Nicoll 1993). Indeed, a recent study (Bar-Ilan et al. 2013) has shown that inhibitory synapses can fine-tune the transition between synaptic potentiation and synaptic depression with high spatial resolution, even in electrically short branches (despite the global spread of synaptic inhibition).

The result in Fig. 18.1 relies, in part, on the increase of the input resistance R_d in distal branches (Rall and Rinzel 1973; Rinzel and Rall 1974). However, in some cell types the specific membrane resistivity, R_m, along the main stem dendrite decreases with distance from the soma (Ledergerber and Larkum 2010; Magee 1998; Stuart and Spruston 1998), and this could lead to a decrease, rather than an increase, in R_d with distance from the soma (Magee 1998; but see Ledergerber and Larkum 2010). However, it is possible to show in simulations that due to the thin diameter of distal dendritic branches and the effect of the adjacent sealed-end boundary conditions, even with the observed decrease in R_m with distance from the soma, R_d in thin distal branches still increases towards the distal tips and thus the advantage of the off-path versus on-path conditions still holds.

In summary, this work advocates a "dendro-centric" viewpoint for understanding how the neuron's output is first and foremost shaped in the dendrites, whereby excitatory and inhibitory dendritic synapses interact with nonlinear membrane currents before an output is generated at the axon.

Appendix

In all the models we used unless stated otherwise, the axial resistance was $R_a = 100\,\Omega\,\text{cm}$ and the specific membrane capacitance was $C_m = 1\ \mu\text{F/cm}^2$. In Fig. 18.1, the model consisted of a sealed-end passive cylindrical cable ($L = 1$; $R_m = 20{,}000\,\Omega \times \text{cm}^2$) with a diameter of 1 μm, coupled at $X = 0$ to an isopotential soma such that the dendritic-to-somatic conductance ratio was 0.1. In addition to the passive membrane resistance, the somatic conductances in Fig. 18.1 included Na^+ and K^+ channels [model and parameters as previously described (Traub et al. 1991) with activation and inactivation functions shifted by +15 mV]. NMDA synapses were modeled (with $g_{\text{max}} = 0.5$ nS) as previously described (Sarid et al. 2007). In Fig. 18.3 we used the reconstructed morphology of a CA1 pyramidal neuron (Ascoli et al. 2007; Golding et al. 2005) with $R_m = 15{,}000\,\Omega \times \text{cm}^2$. In Fig. 18.3a, the AMPA-like synapse was modeled by voltage-independent conductance with peak value of 0.5 nS and rise and decay time constants of 0.2 and 10 ms, respectively.

The equivalent model as in Fig. 18.4c (inset) consists of two cylindrical branches: the thin branch is identical to any one of the original branches of the starburst-like models in Fig. 18.4a, whereas the thick branch is the equivalent cylinder model for the other $N-1$ branches with diameter $D_{\text{eq}} = D_0 \times N^{2/3}$ (Rall 1959), where D_0 is the diameter of the thin branch. The equivalent cylinder receives conductance of $(N-1) \times g_i$ at a distance X_i (as in the thin branch) from the junction.

References

Antic SD, Zhou WL, Moore AR, Short SM, Ikonomu KD (2010) The decade of the dendritic NMDA spike. J Neurosci Res 88:2991–3001

Archie KA, Mel BW (2000) A model for intradendritic computation of binocular disparity. Nat Neurosci 3:54–63

Ascoli GA, Donohue DE, Halavi M (2007) NeuroMorpho.Org: a central resource for neuronal morphologies. J Neurosci 27:9247–9251

Bar-Ilan L, Gidon A, Segev I (2013) The role of dendritic inhibition in shaping the plasticity of excitatory synapses. Front Neural Circuits 6:118

Bock DD, Lee W-CA, Kerlin AM, Andermann ML, Hood G, Wetzel AW, Yurgenson S, Soucy ER, Kim HS, Reid RC (2011) Network anatomy and in vivo physiology of visual cortical neurons. Nature 471:177–182

Briggman KL, Helmstaedter M, Denk W (2011) Wiring specificity in the direction-selectivity circuit of the retina. Nature 471:183–188

Brock LG, Coombs JS, Eccles JC (1951) Action potentials of motoneurones with intracellular electrode. In: Proc. Univ. Otago Med. Sch, pp 14–15

Buhl EH, Halasy K, Somogyi P (1994) Diverse sources of hippocampal unitary inhibitory postsynaptic potentials and the number of synaptic release sites. Nature 368:823–828

DeFelipe J, Fariñas I (1992) The pyramidal neuron of the cerebral cortex: morphological and chemical characteristics of the synaptic inputs. Prog Neurobiol 39:563–607
DeFelipe J, Alonso-Nanclares L, Arellano JI (2002) Microstructure of the neocortex: comparative aspects. J Neurocytol 31:299–316
Gidon A, Segev I (2012) Principles governing the operation of synaptic inhibition in dendrites. Neuron 75:330–341
Golding NL, Mickus TJ, Katz Y, Kath WL, Spruston N (2005) Factors mediating powerful voltage attenuation along CA1 pyramidal neuron dendrites. J Physiol 568:69–82
Hao J, Wang XD, Dan Y, Poo MM, Zhang XH (2009) An arithmetic rule for spatial summation of excitatory and inhibitory inputs in pyramidal neurons. Proc Natl Acad Sci USA 106:21906–21911
Jack JJB, Noble D, Tsien RW (1975) Electric current flow in excitable cells. Clarendon, Oxford
Jadi M, Polsky A, Schiller J, Mel BW (2012) Location-dependent effects of inhibition on local spiking in pyramidal neuron dendrites. Plos Comput Biol 8:e1002550
Kim S, Guzman SJ, Hu H, Jonas P (2012) Active dendrites support efficient initiation of dendritic spikes in hippocampal CA3 pyramidal neurons. Nat Neurosci 15(4):600–606
Klausberger T, Somogyi P (2008) Neuronal diversity and temporal dynamics: the unity of hippocampal circuit operations. Science 321:53–57
Koch C (1998) Biophysics of computation: information processing in single neurons, 1st edn. Oxford University Press, New York, NY
Koch C, Poggio T (1985) The synaptic veto mechanism: does it underlying direction and orientation selectivity in the visual cortex. In: Dobson VG, Rose D (eds) Models of the visual cortex. Wiley-Blackwell, New York, NY, pp 408–419
Koch C, Poggio T, Torres V (1982) Retinal ganglion cells: a functional interpretation of Dendritic morphology. Philos Trans R Soc Lond B Biol Sci 298:227–263
Koch C, Poggio T, Torre V (1983) Nonlinear interactions in a dendritic tree: localization, timing, and role in information processing. Proc Natl Acad Sci USA 80:2799–2802
Koch C, Douglas R, Wehmeier U (1990) Visibility of synaptically induced conductance changes: theory and simulations of anatomically characterized cortical pyramidal cells. J Neurosci 10:1728–1744
Larkum ME, Kaiser KMM, Sakmann B (1999) Calcium electrogenesis in distal apical dendrites of layer 5 pyramidal cells at a critical frequency of back-propagating action potentials. Proc Natl Acad Sci 96:14600–14604
Larkum ME, Nevian T, Sandler M, Polsky A, Schiller J (2009) Synaptic integration in tuft dendrites of layer 5 pyramidal neurons: a new unifying principle. Science 325:756–760
Lavzin M, Rapoport S, Polsky A, Garion L, Schiller J (2012) Nonlinear dendritic processing determines angular tuning of barrel cortex neurons in vivo. Nature 490:397–401
Ledergerber D, Larkum ME (2010) Properties of layer 6 pyramidal neuron apical dendrites. J Neurosci 30:13031–13044
Liu G (2004) Local structural balance and functional interaction of excitatory and inhibitory synapses in hippocampal dendrites. Nat Neurosci 7:373–379
Lovett-Barron M, Turi GF, Kaifosh P, Lee PH, Bolze F, Sun X-H, Nicoud J-F, Zemelman BV, Sternson SM, Losonczy A (2012) Regulation of neuronal input transformations by tunable dendritic inhibition. Nat Neurosci 15(3):423–430
MacDonald JF, Jackson MF, Beazely MA (2006) Hippocampal long-term synaptic plasticity and signal amplification of NMDA receptors. Crit Rev Neurobiol 18:71–84
Magee JC (1998) Dendritic hyperpolarization-activated currents modify the integrative properties of hippocampal CA1 pyramidal neurons. J Neurosci 18:7613–7624
Magee JC (2007) Voltage-gated ion channels in dendrites. In: Stuart G, Spruston N, Häusser M (eds) Dendrites. Oxford University Press, New York, pp 139–160
Magee JC, Christofi G, Miyakawa H, Christie B, Lasser-Ross N, Johnston D (1995) Subthreshold synaptic activation of voltage-gated Ca^{2+} channels mediates a localized Ca^{2+} influx into the dendrites of hippocampal pyramidal neurons. J Neurophysiol 74:1335–1342
Malenka RC (1991) The role of postsynaptic calcium in the induction of long-term potentiation. Mol Neurobiol 5:289–295
Malenka RC, Nicoll RA (1993) NMDA-receptor-dependent synaptic plasticity: multiple forms and mechanisms. Trends Neurosci 16:521–527

Markram H, Toledo-Rodriguez M, Wang Y, Gupta A, Silberberg G, Wu C (2004) Interneurons of the neocortical inhibitory system. Nat Rev Neurosci 5:793–807
Megías M, Emri Z, Freund TF, Gulyás AI (2001) Total number and distribution of inhibitory and excitatory synapses on hippocampal CA1 pyramidal cells. Neuroscience 102:527–540
Merchán-Pérez A, Rodriguez J-R, Alonso-Nanclares L, Schertel A, Defelipe J (2009) Counting synapses using FIB/SEM microscopy: a true revolution for ultrastructural volume reconstruction. Front Neuroanat 3:18
Miles R, Toth K, Gulyas AI, Hajos N, Freund TF (1996) Differences between somatic and dendritic inhibition in the hippocampus. Neuron 16:815–823
Murayama M, Larkum ME (2009) Enhanced dendritic activity in awake rats. Proc Natl Acad Sci USA 106:20482–20486
Murayama M, Perez-Garci E, Nevian T, Bock T, Senn W, Larkum ME (2009) Dendritic encoding of sensory stimuli controlled by deep cortical interneurons. Nature 457:1137–1141
Palmer LM, Schulz JM, Murphy SC, Ledergerber D, Murayama M, Larkum ME (2012) The cellular basis of GABAB-mediated interhemispheric inhibition. Science 335:989–993
Rall W (1959) Branching dendritic trees and motoneuron membrane resistivity. Exp Neurol 1:491–527
Rall W (1964) Theoretical significance of dendritic trees for neuronal input–output relations. In: Reiss RF (ed) Neural theory and modeling. Stanford University Press, Palo Alto, CA, pp 73–97
Rall W (1967) Distinguishing theoretical synaptic potentials computed for different soma-dendritic distributions of synaptic input. J Neurophysiol 30:1138–1168
Rall W, Rinzel J (1973) Branch input resistance and steady attenuation for input to one branch of a dendritic neuron model. Biophys J 13:648–687
Rhodes P (2006) The properties and implications of NMDA spikes in neocortical pyramidal cells. J Neurosci 26:6704–6715
Rinzel J, Rall W (1974) Transient response in a dendritic neuron model for current injected at one branch. Biophys J 14:759–790
Sarid L, Bruno R, Sakmann B, Segev I, Feldmeyer D (2007) Modeling a layer 4-to-layer 2/3 module of a single column in rat neocortex: interweaving in vitro and in vivo experimental observations. Proc Natl Acad Sci USA 104:16353–16358
Schiller J, Schiller Y, Stuart G, Sakmann B (1997) Calcium action potentials restricted to distal apical dendrites of rat neocortical pyramidal neurons. J Physiol 505(Pt 3):605–616
Schiller J, Major G, Koester HJ, Schiller Y (2000) NMDA spikes in basal dendrites of cortical pyramidal neurons. Nature 404:285–289
Seung HS (2009) Reading the book of memory: sparse sampling versus dense mapping of connectomes. Neuron 62:17–29
Somogyi P, Nunzi MG, Gorio A, Smith AD (1983) A new type of specific interneuron in the monkey hippocampus forming synapses exclusively with the axon initial segments of pyramidal cells. Brain Res 259:137–142
Stuart G, Spruston N (1998) Determinants of voltage attenuation in neocortical pyramidal neuron dendrites. J Neurosci 18:3501–3510
Szentágothai J, Arbib MA (1974) Conceptual models of neural organization. Neurosci Res Program Bull 12:305–510
Tepper JM, Koos T, Wilson CJ (2004) GABAergic microcircuits in the neostriatum. Trends Neurosci 27:662–669
Traub RD, Wong RK, Miles R, Michelson H (1991) A model of a CA3 hippocampal pyramidal neuron incorporating voltage-clamp data on intrinsic conductances. J Neurophysiol 66:635–650
Wang Y, Gupta A, Toledo-Rodriguez M, Wu CZ, Markram H (2002) Anatomical, physiological, molecular and circuit properties of nest basket cells in the developing somatosensory cortex. Cereb Cortex 12:395–410
Wen Q, Stepanyants A, Elston GN, Grosberg AY, Chklovskii DB (2009) Maximization of the connectivity repertoire as a statistical principle governing the shapes of dendritic arbors. Proc Natl Acad Sci 106:12536–12541
Williams SR (2004) Spatial compartmentalization and functional impact of conductance in pyramidal neurons. Nat Neurosci 7:961–967

Chapter 19
Role of Non-uniform Dendrite Properties on Input Processing by GABAergic Interneurons

Anja Matthiä, Marlene Bartos, and Imre Vida

Abstract GABAergic interneurons in cortical circuits control the activation of principal cells and orchestrate network activity patterns, including oscillations at different frequency ranges. Recruitment of interneurons depends on integration of convergent synaptic inputs along the dendro-somatic axis; however, dendritic processing in these cells is still poorly understood.

In this chapter, we summarise our results on the cable properties, electrotonic structure and dendritic processing in "basket cells" (BCs; Nörenberg et al. 2010), one of the most prevalent types of cortical interneurons mediating perisomatic inhibition. In order to investigate integrative properties, we have performed two-electrode whole-cell patch clamp recordings, visualised and reconstructed the recorded interneurons and created passive single-cell models with biophysical properties derived from the experiments. Our results indicate that membrane properties, in particular membrane resistivity, are inhomogeneous along the somato-dendritic axis of the cell. Derived values and the gradient of membrane resistivity are different from those obtained for excitatory principal cells. The divergent passive membrane properties of BCs facilitate rapid signalling from proximal basal dendritic inputs but at the same time increase synapse-to-soma transfer for slow signals from the distal apical dendrites.

Our results demonstrate that BCs possess distinct integrative properties. Future computational models investigating the diverse functions of neuronal circuits need to consider this diversity and incorporate realistic dendritic properties not only of excitatory principal cells but also various types of inhibitory interneurons.

A. Matthiä, Ph.D. (✉) • I. Vida, M.D, Ph.D.
Cluster of Excellence NeuroCure, Institute of Integrative Neuroanatomy,
Charité Berlin, Charitéplatz 1, D-10117 Berlin, Germany
e-mail: anja.matthiae@charite.de

M. Bartos, Ph.D.
Institute for Physiology I, Albert-Ludwigs-Universität Freiburg, Hermann-Herder-Str. 7,
D-79104 Freiburg, Germany

H. Cuntz et al. (eds.), *The Computing Dendrite: From Structure to Function*,
Springer Series in Computational Neuroscience 11, DOI 10.1007/978-1-4614-8094-5_19,

19.1 Introduction

GABAergic inhibition is a crucial factor for proper network function in cortical circuits. The inhibitory system controls the level of neuronal excitation and the timing and frequency of discharge and plays a major factor in the generation of population activity patterns (Buzsáki 2006). A critical change in the balance of inhibition to excitation is thought to underlie many major brain disorders, such as epilepsy and schizophrenia (Lewis et al. 2005).

Inhibitory interneurons, in contrast to excitatory principal cells, are much lower in number and show a high level of heterogeneity with respect to their morphological, physiological and neurochemical properties (Ascoli et al. 2008; Freund and Buzsáki 1996). Interneurons can be classified into two main classes, preferentially innervating either the dendrites or the perisomatic region of their targets (Bartos et al. 2011). BCs are the main type of perisomatic inhibitory interneurons that innervate the soma and the proximal dendrites of principal cells and various types of interneurons. BCs are interconnected by mutually inhibitory synapses and gap junctions and are thought to play a central role in the generation of oscillations at theta (4–12 Hz), gamma (30–90 Hz) and ripple frequencies (14–200 Hz) (Bartos et al. 2007; Buzsáki and Draguhn 2004).

BCs are recruited by different mechanisms, including the tonic drive through metabotropic receptors (Whittington et al. 1995; Fisahn et al. 2002) or the phasic excitation mediated by ionotropic receptors in glutamatergic synapses (Geiger et al. 1997). With respect to circuit structure, excitatory synapses are formed by either (1) afferent pathways originating outside the area or (2) "recurrent" collaterals of local principal cells resulting in feedforward vs. feedback activation of the interneurons, respectively.

19.1.1 Circuitry of the Dentate Gyrus

In order to understand synaptic activation and integration in BCs, we need to consider the specific organisation of the surrounding network.

The dentate gyrus (DG) is considered to be the gateway to the hippocampus, the first station of the so-called "trisynaptic pathway" (Andersen et al. 1971). As a cortical area, the DG shows a strict but simple layered structure (Fig. 19.1a): the cell body layer densely packed with somata of mainly principal cells, the granule cells (GCs), also called as the granule cell layer. GC dendrites emerge from the apical pole of the soma and extend into the neuropil, the "molecular layer", whereas the axon arises from the basal pole and extends into the "hilus". Most interneurons, including BCs, differ from GCs in that they have a bipolar dendritic arbour with apical dendrites in the molecular layer and basal dendrites in the hilus.

The main input to the DG, the perforant path, constitutes fibres from the entorhinal cortex and terminates in the outer two-thirds of the molecular layer forming synapses onto distal dendrites of GCs as well as inhibitory interneurons

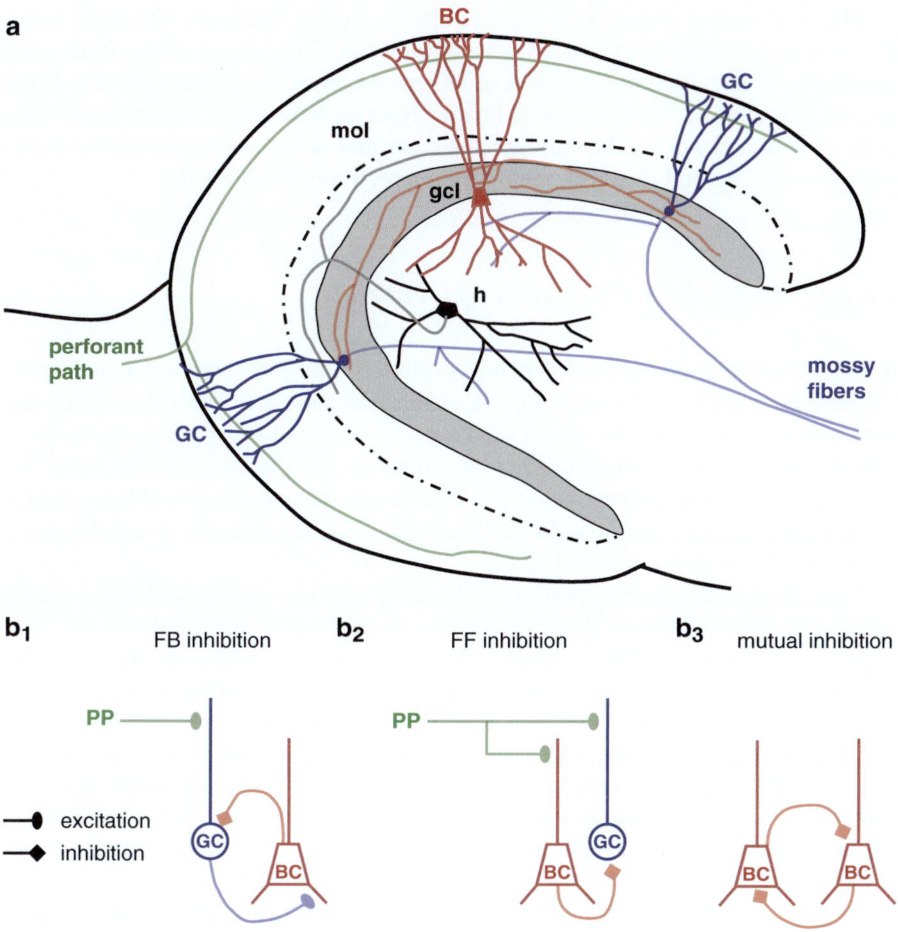

Fig. 19.1 The dentate gyrus circuitry. (**a**) The granule cell layer with the densely packed somata of granule cells (GC, *blue*) is shown in *grey*. These cells are activated mainly by distal excitatory perforant path synapses (in *green*) but also from excitatory mossy cells (*black*). Granule cell axons form the mossy fibres with local collaterals in the hilus. Basket cells (BC, *red*) have a larger cell body located at the border of the granule cell layer (gcl) to the hilus (h). The apical dendrites of basket cells, similar to those of granule cells, extend into the molecular layer (mol) that is subdivided into an outer and inner part (indicated by the *dashed line*). Basket cell dendrites receive distal synaptic input from the perforant path and proximal input from mossy cells within the inner molecular layer. In contrast to granule cells basket cells have basal dendrites which extend into the hilus and receive recurrent excitatory input from granule cells. (**b**) Schematic representations of connectivity between perforant path (PP), basket cells and granule cells. BCs are involved in feedback (FB, b_1) or feedforward inhibition (FF, b_2). At the same time they are interconnected by mutual inhibitory synapses (b_3). Whereas FF inhibition is generated by the afferent activation of BCs, FB inhibition involves the activation of GCs

(Amaral 1993). Dendrites in the inner third of the molecular layer are targeted by commissural/associational fibres originating from excitatory mossy cells in the ipsi- and contralateral hilus. Thereby, dendritic trees of GCs and interneurons receive strictly laminated, segregated inputs (see Förster et al. 2006).

The main output of the DG is conveyed by the mossy fibres, i.e., GC axons being directed to the hilus and the CA3 area. Within the hilus, mossy fibres form axon collaterals contacting basal dendrites of DG interneurons and thereby provide direct excitatory feedback. Thus, many DG interneurons, including BCs, receive input from both afferent pathways and recurrent collaterals, and therefore they are embedded in both feedforward and feedback inhibitory microcircuits (Fig. 19.1b).

19.1.2 Integrative Properties of Dendrites

The dendritic tree of a neuron receives excitatory and inhibitory synaptic inputs. These inputs are transmitted along the dendro-somatic axis until they reach the perisomatic domain and the site of action potential initiation, typically at the axon initial segment. How different types of inputs are integrated to produce specific neuronal output is dependent on the spatio-temporal pattern of inputs (Magee 2000), the morphological structure of the cell (Rall 1959; Vetter et al. 2001) and the membrane properties (Gulledge et al. 2005).

Integrative membrane properties can be determined by passive and active mechanisms. Passive membrane properties together with the morphology define the electrotonic structure of neurons (Rall 1959, 1962) and therefore set the basis for integration of synaptic signals within the dendritic tree. Integration or summation of different inputs over time is enhanced when the cell membrane has low leak conductance and the intracellular ("axial") conductance (Spruston et al. 2008) is high. As an effect, the integration time window becomes larger, and many inputs from different sources, e.g. different brain regions, can be integrated. Thus, the cell will act as an "integrator". In contrast, temporal summation in a neuron is smaller when the leak conductance is large. Synaptic signals would decay rapidly and show more attenuation along the dendrites, and inputs would summate only within a brief time window. These passive properties would, thus, result in a "coincidence' or a "synchrony detector" neuron. Therefore, passive membrane properties can set the basis for different modes of integration, such as strong integration (e.g. DG GC Krueppel et al. 2011) or synchrony detection (e.g. CA1 pyramidal cell Losonczy and Magee 2006).

On the basis of the passive properties, active, i.e. voltage dependent, conductances increase the cell's repertoire of integrating synaptic inputs, for example by counteracting the passive location-dependent attenuation (Magee 2000) or compartmentalising the dendrite into different functional units (e.g. Larkum et al. 2009; Scott et al. 2005).

19.2 Dendritic Properties of Dentate Gyrus Basket Cells

The passive properties of each dendritic cable segment can be characterised by three main parameters: the specific membrane resistance R_m, the specific membrane capacitance C_m and the intracellular (axial) resistivity R_i. R_m describes the leakiness of the membrane and is determined by the presence of leak channels in the membrane. C_m is a measure of the capacitive nature of the membrane. The lipid and protein content is the main determinant of the dielectric attributes of the membrane and thus C_m. Finally, R_i reflects the resistive properties of the intracellular medium, and it may change depending on the intracellular content, including the number and size of intracellular organelles and the fluidity of intracellular medium. Ultimately, the knowledge of R_m, C_m and R_i as well as the detailed morphological structure of a neuron makes it possible to create a realistic computational model of the cell and analyse its integrative properties either in isolation or as part of a network.

The development of a realistic model of BCs includes three major steps: (1) whole-cell patch-clamp recordings of passive voltage responses to current application; (2) visualisation and 3-dimensional reconstruction of cell morphology and (3) derivation of R_m, C_m and R_i by numerical simulations using a model based on the recorded voltage traces and the respective neuronal morphology.

Upon completion of these separate steps the developed model was used to study the integrative properties of BC dendrites. We showed that BC dendrites are optimised for speed and efficacy of signalling properties. BC dendrites exhibit faster signalling properties in basal than apical dendrites, but at the same time the efficacy of slow distal apical inputs is increased. In the final part of the chapter, we discuss the implications of BC signalling properties for the proper function of the DG circuitry.

19.2.1 Electrophysiological Experiments

In the experiments, BCs were selected in 300 μm hippocampal slices of 17–19-day-old Wistar rats on the basis of their large cell body located at the border of the granule cell layer to the hilar region. To confirm their physiological identity, a series of hyperpolarising subthreshold and depolarising suprathreshold current pulses were injected to evaluate input resistance and firing patterns under current clamp conditions. Typically, BCs show low input resistance, produce very brief action potentials followed by a large-amplitude fast afterhyperpolarisation and can discharge at very high frequencies in response to sustained current pulses (119–216 Hz at 0.75–1 nA current amplitude; at 30–34 °C; Fig. 19.2a).

To record passive responses from the neurons, ligand- and voltage-dependent currents need to be eliminated. Therefore, blockers of glutamatergic (10 μM 6-cyano-7-nitroquinoxaline-2,3-dione) and GABAergic (20 μM bicuculline or 1–2 μM gabazine) synaptic transmission and active conductances (sodium channels, 1 μM tetrodotoxin), including the hyperpolarisation-activated current (I_h, 30 μM ZD7288), were bath-applied, and voltage responses were kept as small as possible (see below, Fig. 19.2).

Electrophysiological recordings were performed with two electrodes in two different experimental configurations: (1) dual-somatic recordings and (2) dual-somato-dendritic recordings (Fig. 19.2a, d). In the dual-somatic recording configuration, current injection is separated from voltage recording. Therefore, artefacts that

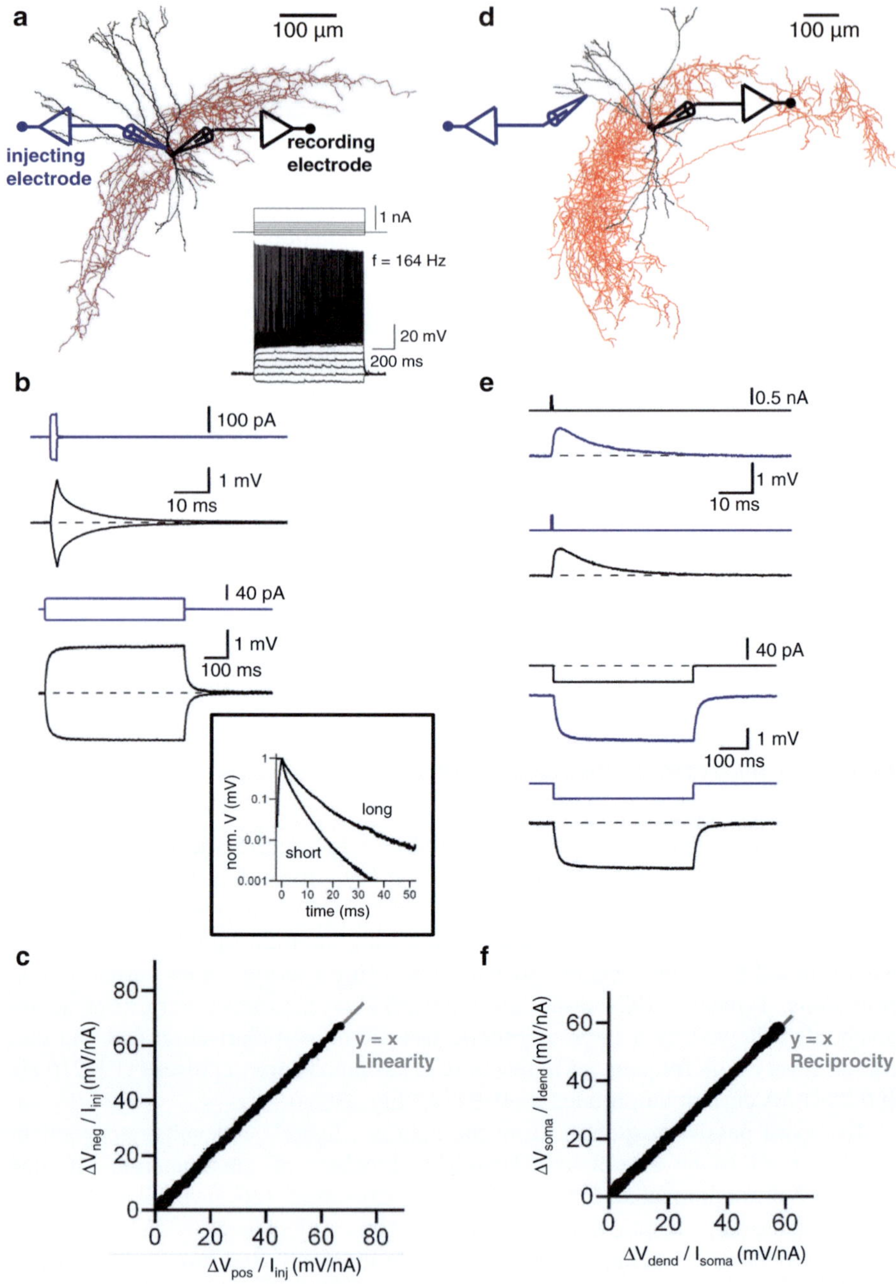

Fig. 19.2 Two-electrode recordings of passive responses to short and long current pulses. (**a**) Detailed reconstruction of a basket cell with a schematic representation of the dual-somatic recording configuration. Injection and recording electrodes were separated, and only voltage responses from the recording site were used for further analysis. *Inset*: Sub- and suprathreshold responses to a family of hyperpolarising and depolarising current pulses. This cell fired action potentials at 164 Hz in

would emerge during current application due to series resistance and capacitance of the electrodes are eliminated from the voltage-recording electrode. This is especially important for the early, fast decay of the voltage transient in response to brief current injections. This early transient reflects a rapid redistribution of charge within the cell after the end of the current pulses. Conceptually, the intracellular medium is a low-resistance path (reflected by R_i) which enables the rapid charge redistribution before the membrane capacitance can fully discharge through the high-resistance path across the membrane (determined by R_m). In case of single-electrode recordings this transient is superimposed by electrode artefacts (Schmidt-Hieber et al. 2007; Thurbon et al. 1998; Major et al. 1994), but dual-somatic electrode configurations as in our experiments enable the reliable recording of this essential determinant of R_i.

In the second somato-dendritic recording configuration, one electrode was positioned at the soma and the other on the apical dendrite between 100 and 300 μm distance from soma (Fig. 19.2d, Nörenberg et al. 2010; Hu et al. 2010). Currents were applied to the soma, and voltage recordings were performed at both sites, somatic and dendritic (Fig. 19.2e, black and blue traces, respectively). In these recordings, the electrode artefact could not be avoided in the somatic responses, and thus the initial fast decay was excluded from analysis (see overlay of simulated and experimental traces in Fig. 19.3a). Nevertheless, voltage responses obtained from an additional point of the dendritic structure served as a spatial constraint for the derivation of the parameters R_m, C_m or R_i. This was particularly important to enable us to assess homogeneity along the somato-dendritic axis.

In both configurations, membrane properties were probed by two sets of current pulses, either long pulses with small amplitude (400–600 ms, 20–50 pA, Fig. 19.2b, e lower traces) or short pulses with large amplitude (0.5 or 2 ms, 100–500 pA, Fig. 19.2b, e, upper traces). Voltage responses to both sets of pulses were always kept smaller than 4 mV to prevent activation of voltage-dependent conductances (Fig. 19.2b, e). The advantage of the dual pulses was that the brief pulses probe membrane properties mainly locally, i.e. at the soma, whereas the long pulses can charge more distal membrane surface, and thereby the kinetics of charging and discharging phases may uncover potential inhomogeneities in R_m or C_m.

Indeed, analysing the decay of voltage responses to short and long current injections revealed a difference in the apparent membrane time constant (Fig. 19.2b, inset).

Fig. 19.2 (continued) response to a 1 nA current pulse. (**b**) Voltage responses (*black*) to short (*top*) and long (*bottom*) current pulses (*blue*) in the dual-somatic recording configuration. Inset: Semilogarithmic plot of the decay phase of short and long voltage responses shows their different time constants. (**c**) Point-by-point comparison between voltage responses to positive and negative current injection indicates linearity of the voltage responses. (**d**) Detailed reconstruction of a basket cell with a schematic representation of the dual-somato-dendritic recording configuration. (**e**) Voltage responses to short (*top*) and long (*bottom*) current pulses recorded from the soma (in *black*, with current injected to the dendrite in *blue*) or the dendrite (in *blue*, with current injected to the soma in *black*). (**f**) Point-by-point comparison of somatic responses to dendritic current injection, and dendritic responses to somatic current injection indicate reciprocity of the response. Reproduced with permission from Nörenberg et al. (2010)

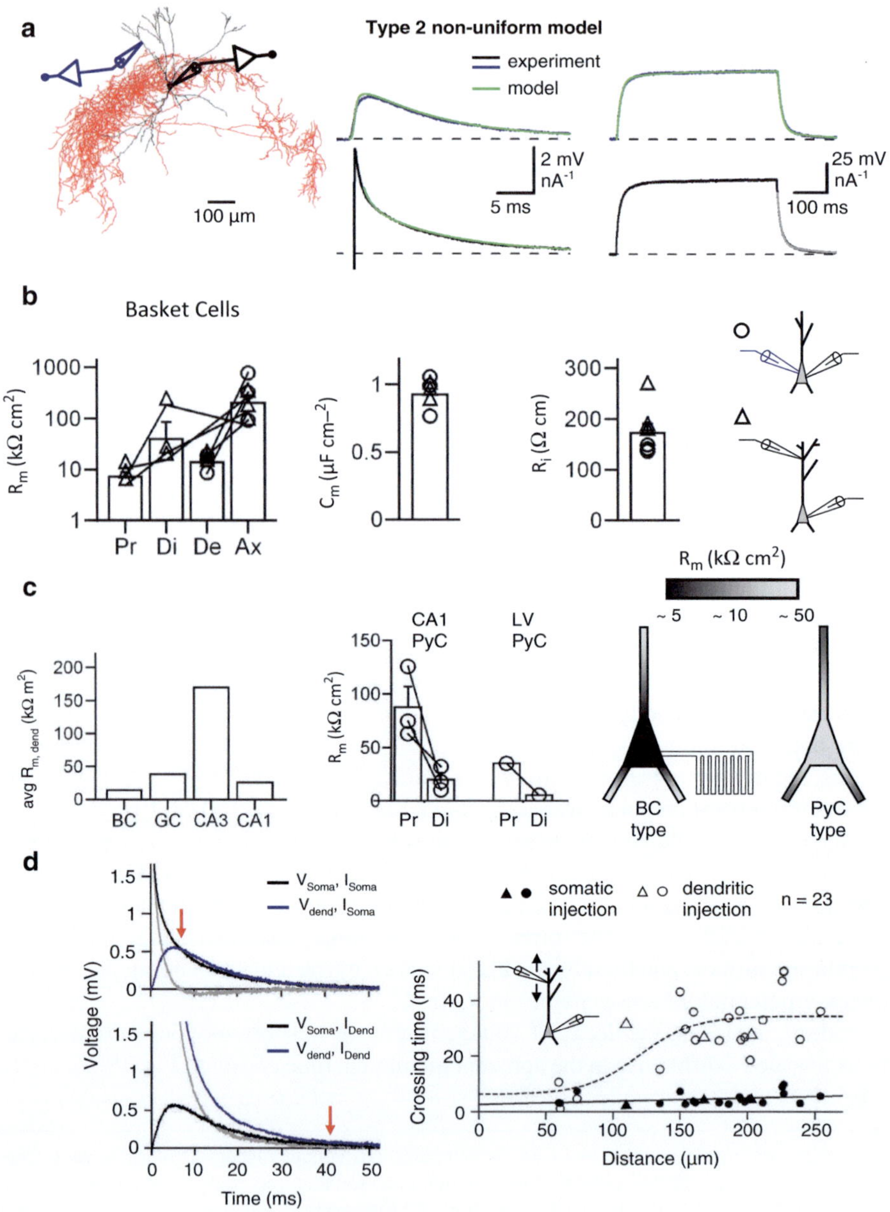

Fig. 19.3 Basket cell cable parameters in comparison to other cell's properties. (**a**) The best-fit parameter combination of R_m, C_m, R_i and the residual pipette capacitance resulted in a good reproduction of the experimental traces (somatic, *black*; dendritic, *blue*) in non-uniform R_m model (five parameters: $R_{m,\ prox}$, $R_{m,\ dist}$, $R_{m,\ axon}$, C_m and R_i). Simulated traces are shown in *green*. *Left*: Reconstruction and recording configuration for the basket cell. (**b**) Summary bar charts of best-fit cable parameters R_m (*left*), C_m (*middle*) and R_i (*right*) for six basket cells, three dual-somatic recordings (*circles*) and three somato-dendritic recordings (*triangles*). Abbr.: Pr, $R_{m,\ prox}$; Di, $R_{m,\ dist}$;

Short voltage responses decayed much faster than long voltage responses. This difference suggested that proximal and distal surface membrane areas may display differences in their passive membrane properties, more specifically a non-uniform membrane resistance.

Altogether, the combination of dual-somatic and somato-dendritic recordings and the analysis of brief and long responses can constrain R_m, C_m and R_i well, offering good estimates of these parameters.

To confirm that our recordings reflected a passive electrotonic system, dual-somatic responses were checked for linearity and somato-dendritic responses were checked for reciprocity, prerequisites for passive systems (Major et al. 1994). To test for linearity, voltage responses to positive and negative current pulses were compared. The point-by-point comparison did not show any significant deviation from the identity line (Fig. 19.2c) irrespective of whether short or long current pulses were used. To test for reciprocity, dendritic current injections were performed and the resulting somatic voltage responses were compared to the dendritic ones evoked by somatic current injections. The points did not deviate from the identity line in this test either, indicating that reciprocity held true also for our somato-dendritic recordings (Fig. 19.2f).

19.2.2 Morphological Analysis and Reconstruction

During the experiment, recorded cells were filled with biocytin and subsequently visualised for reconstruction and morphological analysis. Reconstructions were only performed of those neurons whose extensive dendritic and axonal arborisation was considered complete.

On the basis of stable electrophysiological recording conditions and the morphological criteria, six cells (three dual-soma recordings and three somato-dendritic recordings) were selected for further analysis. Full 3-dimensional reconstructions were made using a Neurolucida reconstruction system at x63 magnification. First, the soma was traced as a closed contour by following its perimeter. Second, dendrites and axon were reconstructed as a series of virtual cylinders giving rise to a set of interconnected local compartments. Dendritic branches had larger (0.68–2.52 μm) and slowly changing diameter, whereas axon collaterals had thin diameter

Fig. 19.3 (continued) De, average dendritic R_m; Ax, $R_{m,\,axon}$. (**c**) Comparison between R_m of basket cells and principal cells such as granule cells (GCs), CA3 pyramidal cells (CA3), CA1 pyramidal cells (CA1) and layer V neocortical pyramidal cells (LV PyC). (**d**) Somatic (*black*) and dendritic (*blue*) responses to short somatic (*top left*) or dendritic (*bottom left*) current injections. The difference between both traces is shown in *grey*. Note that in case of somatic injection somatic and dendritic responses cross each other at an early time point of the decay phase (6.6 ms), but both responses to dendritic injections do not cross but converge at a later time point (50.6 ms). The summary graph for all 23 somato-dendritic recordings (*right*) shows that this is true for all recordings with the dendritic recording site <120 μm from soma. Reproduced with permission from Nörenberg et al. (2010)

(0.32–1.33 μm at distances of 0–500 μm from soma) and regularly spaced varicosities ("boutons"). Dendrites extended either into the molecular layer or the hilus corresponding to apical or basal, respectively. In contrast, the axon arborised extensively in and near the granule cell layer. Finally, dendrites and axons could also be distinguished on the basis of their branching pattern: The angle between two children branches was narrow for dendrites (<90°), whereas for the axon the angle between children branches could be up to 180°. Because the diameters of dendrites appeared unchanged by fixation, the reconstructed morphology was not corrected for any shrinkage.

19.2.3 Detailed Biophysical Model of BCs

To determine the three cable parameters, R_m, C_m and R_i, the reconstructed morphology was imported to the simulation environment NEURON (Hines and Carnevale 1997; Carnevale and Hines 2006) and used as the spatial framework for subsequent single-cell simulations. In the NEURON model, current pulses identical to those in the experiment were used to simulate voltage responses of the imported cell assuming certain values for R_m, C_m and R_i. These values were then systematically changed using a random-walk procedure and NEURON's built-in Brent's principal axis algorithm (Brent 2002) in order to best reproduce voltage responses measured in the experiments. The best fit between experimental and simulated traces was quantified by the minimal sum of their pointwise squared errors (SSE) for both short and long pulses (Fig. 19.3a).

Initially, R_m, C_m and R_i were assumed to be uniform throughout the entire cell (data not shown), but no parameter combination was found that reproduced well the difference in the decay time course of short and long pulses. Therefore, alternative models with non-uniform parameters were tested next. While no substantial improvement in SSE was found for non-uniform C_m or R_i, implementing non-uniform R_m improved the fit and reduced the SSE markedly. In the non-uniform R_m model, three domains were defined: soma/proximal dendrites (≤120 μm), distal dendrites (>120 μm) and axon. On average, the SSE was reduced by 84.7 % in the best-fit non-uniform model in comparison to the uniform one.

The best-fit value for C_m was 0.93 μF cm^{-2}; for R_i it was 172.1 Ω cm. In the final non-uniform model, R_m was low in proximal dendrites with $R_{m, prox} = 7.6$ kΩ cm^2 ($x \leq 120$ μm) and high in distal dendrites with $R_{m, dist} = 74.3$ kΩ cm^2. The highest R_m, however, was found to be in the axon ($R_{m, axon} = 281.6$ kΩ cm^2; Fig. 19.3b; Nörenberg et al. 2010).

This result was in strong contrast with the non-uniformity observed in principal cells (Stuart and Spruston 1998; Golding et al. 2005). To obtain further evidence for the gradient of R_m along the somato-dendritic axis in BCs, the time course of voltage transients evoked by brief current pulses to either the dendritic or the somatic recording sites was analysed (Fig. 19.3d). Theoretically, if brief current pulses are injected into compartments with differing R_m, the voltage transients will decay more rapidly in the low-R_m region (in our case at the BC soma) than in the high-R_m region (BC distal dendrites). This leads to a crossing of superimposed voltage responses in the two recording sites (London et al. 1999). Conversely, injecting current into the high-R_m region, local signals will decay more slowly than remote ones; thus both traces would not cross but converge to a common level at a later time point. To compare this theory with the experiment, somatic and dendritic voltage transients to short current pulses were compared in somato-dendritic recordings from BCs (Fig. 19.3d). For somatic stimulation, somatic transients showed faster decay than dendritic ones and the two transients crossed within 10 ms after the pulse (23 somato-dendritic recordings) indicating that a lower R_m region is located close to the soma. For dendritic stimulation 100 μm or more from soma, the two transients did not cross but converged to a common level.

In summary, this analysis confirmed a decreasing gradient of R_m towards the soma in BCs. The value of R_m in BCs differed substantially from values in other cell types in the hippocampus and the neocortex (Fig. 19.3c). The average dendritic R_m ($R_{m,\,dend}$ = 14.2 kΩ cm^2, in the presence of I_h channel blocker ZD 7288) was lower in BCs than in hippocampal principal cells, including GCs (Schmidt-Hieber et al. 2007), CA3 (Major et al. 1994) and CA1 pyramidal cells (Golding et al. 2005: weighted mean $R_{m,\,dend}$ = 26 kΩ cm^2, assuming a distal:proximal membrane area ratio of 2:1). Furthermore, BCs showed a gradient of R_m which decreases from distal dendrites towards the perisomatic domain, opposite to that observed in both CA1 and neocortical layer V pyramidal cells (Golding et al. 2005; Stuart and Spruston 1998).

In contrast to R_m, values for both C_m and R_i in BCs were comparable to those in other cell types in the hippocampus and the neocortex. The best-fit non-uniform model revealed a C_m of 0.93 μF cm^{-2}, consistent with previous capacitive measurements of nucleated patch recordings from hippocampal neurons (0.90 μF cm^{-2}; Gentet et al. 2000). Furthermore, similar values of C_m were obtained using whole-cell recordings of GCs (1.01 μF cm^{-2}; Schmidt-Hieber et al. 2007), CA3 pyramidal neurons (0.75 μF cm^{-2}; Major et al. 1994) and CA1 pyramidal neurons (1.50 μF cm^{-2}; Golding et al. 2005). The similarity of C_m between different cell types quite plausibly suggests comparable lipid and protein composition of cellular membranes in these distinct cell populations.

The mean value of R_i in BCs was 172 ± 18.5 Ω cm. First studies of interneuron cable properties either failed to determine R_i (Chitwood et al. 1999) or obtained highly variable values (52–484 Ω cm, Thurbon et al. 1998). Early estimates of R_i in hippocampal principal cells were also more variable and higher (160–340 Ω cm, Major et al. 1994). These values were based on single-electrode recordings and were very sensitive to recording conditions, such as electrode parameters affecting the fast initial decay of voltage responses. More recent studies using a dual-electrode recording configuration, in contrast, report values of R_i in GCs (194 Ω cm; Schmidt-Hieber et al. 2007) and CA1 pyramidal neurons (178 Ω cm; Golding et al. 2005) which are closer to our estimate for BCs but markedly lower for layer 5 neocortical pyramidal neurons (70–100 Ω cm; Stuart and Spruston 1998). As R_i is highly dependent on the density of intracellular organelles as well as cytoplasmic ion concentrations, it may vary between different cell types. For example, in BCs the density of mitochondria is higher due to a higher metabolic rate (Gulyás et al. 2006) and thus R_i could be increased. On the other hand, due to a more positive Cl^- equilibrium potential in BCs compared to principal neurons (Vida et al. 2006), the intracellular Cl^- concentration will be higher and result in a decreased R_i because of the high mobility of Cl^- (Hille 2001).

In the following section, we show what effects BC cable parameters have on synaptic integration in these cells.

19.2.4 *Integration of Synaptic Inputs in Basket Cells*

How are the different inputs onto apical versus basal dendrites integrated in BCs? How do these passive cable properties contribute to fast signalling?

To answer these questions, simulations were run in one single-cell model out of our sample[1] using its best-fit parameters of the non-uniform R_m model (Fig. 19.4). To better reproduce the low input resistance and the physiological conditions in BCs, a uniformly distributed I_h conductance with 6.8 pS μm^{-2} was inserted into the membrane of the model cell. This value was chosen from the experimental measurement of input resistance of the same BC before ZD 7288 application.

To study the generation and propagation of synaptic inputs, an excitatory synapse was placed on apical or basal dendrites at different locations corresponding to the perforant path input distally, the commissural-associational path input onto proximal apical dendrites and the recurrent GC input onto basal dendrites (Fig. 19.4a, left). These synapses were modelled as conductance-based excitatory synapses with a rise time of 0.2 ms, a decay time constant of 1 ms and a reversal potential of −5 mV (Geiger et al. 1997).

Since the action potential initiation site is close to the soma (Hu et al. 2010), it is important to know how excitatory postsynaptic potentials (EPSPs) are shaped at the synapse and the soma. Figure 19.4a shows the EPSP amplitude plotted against the distance of the synapse that was activated on basal dendrites (negative *x*-values) or apical dendrites (positive *x*-values). Distal synapses evoke somatic EPSPs with small amplitude because of strong attenuation along the dendrites, whereas proximal synapses showed less attenuation and were thus larger (Fig. 19.4a, centre). Attenuation of EPSPs was similar for basal and apical dendrites, even though the local EPSP amplitude was larger for basal than for apical sites (Fig. 19.4a, right).

BCs are well known for their fast and precise timing of their intrinsic and synaptic signalling. To study temporal aspects of dendritic integration with realistic specific cable parameters, latency, decay time constant and half-duration of EPSPs were analysed (Fig. 19.4b). The latency was calculated between the onset of synapse activation and the peak of somatic EPSP waveform, and thus it is a measure of propagation velocity. Somatic EPSPs induced by inputs onto basal dendrites showed

[1]All model BCs can be downloaded from ModelDB (#140789; http://senselab.med.yale.edu/modeldb/).

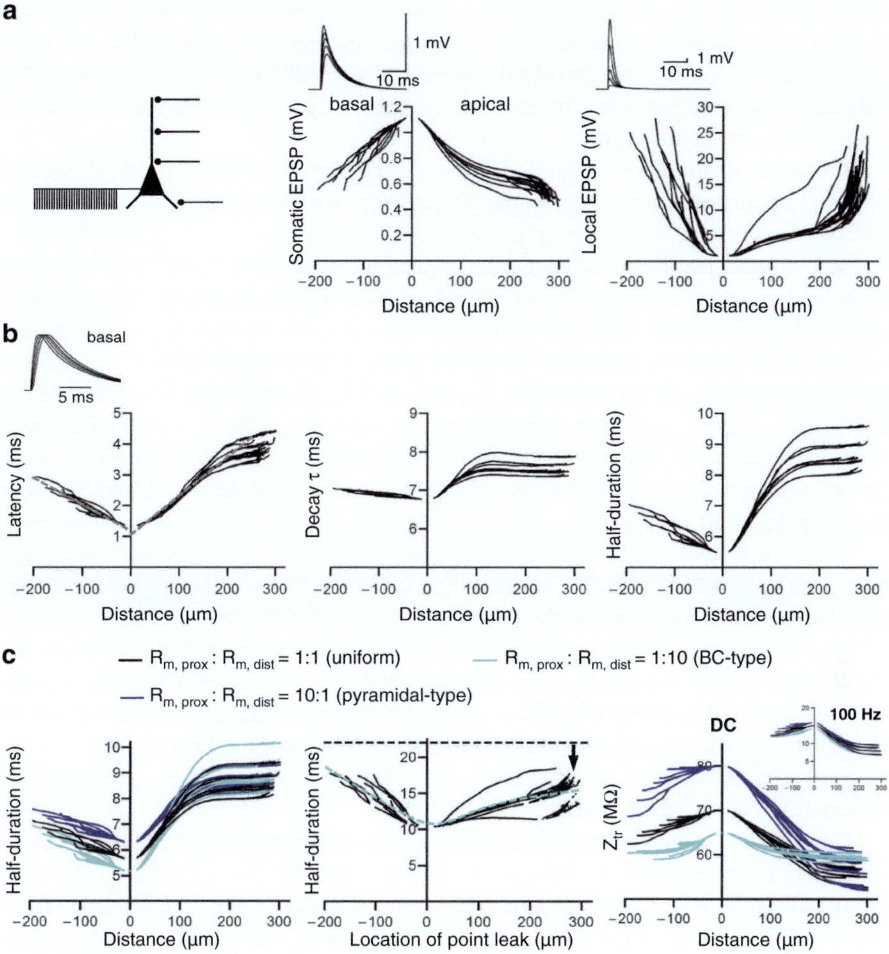

Fig. 19.4 Timing and efficacy in basket cells. (**a**) For each graph one conductance-based synapse was placed at different locations (scheme, *left*) along the somato-dendritic axis, with negative *x*-values for basal and positive ones for apical dendrites. For each synapse, the somatic (*middle*) and local (*right*) excitatory postsynaptic potential (EPSP) was analysed and plotted against the distance of the synaptic site from soma. (**b**) Latency (*left*), decay time constant τ (*middle*) and half-duration (*right*) of somatic EPSPs were plotted against the distance of the activated synapse from soma. Inset, *left*: Representative somatic EPSP waveforms for the activation of basal synapses at different distances from soma. (**c**) The ratio between $R_{m,\,prox}$ and $R_{m,\,dist}$ was subsequently changed from uniform (*black*) to either pyramidal type (*dark blue*) or basket cell type (*light blue*). The half-duration (*left*) and transfer resistance Z_{tr} (*right*) were then computed for the three different scenarios. Inset: Transfer impedance for 100 Hz signals. Reproduced with permission from Nörenberg et al. (2010)

shorter latencies than those in response to apical dendritic inputs, indicating faster signal propagation on basal dendrites (116 μm ms^{-1} vs. 79 μm ms^{-1} for distances ≤200 μm; Fig. 19.4b, left). Similarly, the decay of EPSPs as well as somatic EPSP half-durations were faster and shorter, respectively, for basal than for apical dendrites (Fig. 19.4b, centre and right).

In summary, the fastest somatic EPSPs were generated if the synapse is located on basal dendrites close to the soma. This difference gives a kinetic advantage to feedback over feedforward synapses (see Sect. 19.3.2).

How do the specific cable properties contribute to the fast signalling phenotype of BCs? To understand the effect of the R_m gradient in BCs on timing, we compared the effect of three different scenarios with the same cell morphology keeping the total leak conductance constant. Either no gradient was implemented, i.e. R_m was equal for all dendritic segments ($R_{m,\,prox}$:$R_{m,\,dist}$ = 1:1), the gradient was either of the BC type ($R_{m,\,prox}$:$R_{m,\,dist}$ = 1:10) or pyramidal cell type ($R_{m,\,prox}$:$R_{m,\,dist}$ = 10:1). For basal and proximal apical dendrites, the BC-type gradient produced faster EPSP half-durations than the other two, the model with no gradient and the pyramidal cell-type model (Fig. 19.4c left).

However, somatic EPSP half-durations for distal apical synapses were heterogeneous, independent of which scenario was used. To systematically analyse the relation between the location of the low-R_m region and the kinetics of the somatic EPSPs, the R_m in the model with BC-type gradient was uniformly increased to the distal, high R_m and the low R_m was implemented in a single segment with varying location along the somato-dendritic axis. Activating a distal synapse (arrow in Fig. 19.4c, centre) the corresponding somatic EPSPs were the fastest if the low-R_m region was implemented close to the soma. Thus, a low somatic R_m produced the maximal EPSP speeding for distal inputs, too.

These results indicate that BC cable properties are optimised for EPSP speeding. However, a low R_m leads to stronger attenuation and thus lower efficacy of distal inputs. To study the efficacy of synaptic inputs, transfer impedance for fast inputs and transfer resistance (Z_{tr}) were computed (Fig. 19.4c, right). The transfer impedance denotes the degree of coupling between a synaptic location on the dendritic tree and the soma, i.e. the "efficacy" of signal transmission (Koch 1999). The smaller the transfer impedance the larger the attenuation, as it is the case for distal inputs, revealed in our analysis (Fig. 19.4c, right).

The BC-type non-uniform model led to an increase in Z_{tr} for distal inputs, whereas it remained low in the model with no R_m gradient (Fig. 19.4c, right). Furthermore, compared to the pyramidal cell-type non-uniformity, the BC-type model equalises Z_{tr} for all inputs along the apical dendrite. Therefore, interpreting Z_{tr} as a measure for efficacy of intracellular signalling, the high R_m on distal dendrites enhances signal transfer for slow distal inputs in BCs.

In summary, the distinct non-uniformity of R_m in BCs, in particular the low perisomatic R_m, ensures fast kinetics of somatic EPSPs generated by fast inputs albeit at the cost of substantial attenuation for distal inputs. At the same time, the high R_m on distal dendrites increases the efficacy for distal inputs at low frequencies.

19.3 Discussion

19.3.1 *Inhomogeneity of* $\mathbf{R}_m$ *in BCs Enhances Transfer of Distal Slow Input to the Soma*

Our results show that BCs exhibit distinct passive membrane properties; in particular R_m is non-uniform along the somato-dendritic axis, with a high value on the distal dendrites and low in the perisomatic region. This finding deviates from the uniform R_m in DG GCs (Schmidt-Hieber et al. 2007) and CA3 pyramidal cells (Major et al. 1994) and is in stark contrast to the gradient observed in CA1 or neocortical pyramidal cells (Golding et al. 2005; Stuart and Spruston 1998).

How do these divergent distributions of R_m influence synaptic integration? We showed that in BCs the transfer resistance and thus the somatic response to synaptic inputs onto the dendrites are equalised over a wide range of distances (Fig. 19.4c, right, light blue). In contrast, if a uniform R_m is implemented the difference of transfer resistance between proximal and distal compartments is stronger (Fig. 19.4c, right, black). This effect is further enhanced if an opposite gradient of R_m is used, leading to an electrotonic separation of the distal dendritic tree from the soma as for example in the thin and distal apical tufts of pyramidal cell dendrites (London et al. 1999). Voltage-dependent conductances are required for local nonlinear integration and transmission of distal synaptic inputs to the soma, as observed in neocortical pyramidal neurons (Larkum et al. 2009). Since BCs show only minimal voltage-dependent conductances along their dendritic tree (Hu et al. 2010), the BC-type non-uniformity with high R_m in the distal dendrites appears to be optimised for the transmission of distal synaptic inputs down to the soma.

Consistent with the slow integrative properties of distal apical dendrites, the entorhinal input is primarily tuned to the low theta frequency range. Recent experimental evidence further suggests that theta activity from the entorhinal cortex at 10 Hz activates GCs most efficiently in the functional DG network (Ewell and Jones 2010). The proposed mechanism is that lower as well as higher frequencies lead to increased inhibition, whereas a frequency around 10 Hz leads to EPSP facilitation, efficient temporal summation and reduced inhibition in GCs. The non-uniform R_m in BCs could support such a mechanism. For low frequencies, inhibition in GCs by BCs is enhanced because of the more efficient signal transfer (see Fig. 19.4c, right) at distal apical dendrites. In contrast, at higher frequencies, proximal and basal inputs may preferentially recruit BCs and increase inhibition onto GCs.

19.3.2 *Fast Signalling in the Basal Dendrites of BCs*

BCs are well known for their fast synaptic and intrinsic properties. They have fast excitatory (Geiger et al. 1997) and inhibitory synaptic inputs (Bartos et al. 2001, 2002), fast membrane time constant (Jonas et al. 2004), fast discharge pattern (Pike et al.

2000) and fast synaptic output (Bucurenciu et al. 2008). These fast properties are important for the precise timing of principal discharge and the generation of gamma oscillations within the DG (Bartos et al. 2007). Our results further demonstrate that membrane properties, in particular the R_m, and the electrotonic structure appear to be also optimised to serve fast signalling in the basal dendrites and the perisomatic domain of these interneurons.

In fact, there are marked differences between apical and basal dendrites as well as proximal and distal synapses. These differences are consistent with the divergent properties of the synapses themselves: Inputs of the perforant path onto the distal apical BC dendrites express slower calcium-impermeable glutamate receptors of the AMPA type, whereas mossy fibre inputs onto the proximal basal BC dendrites express fast calcium-permeable AMPA receptors (Sambandan et al. 2010). Our results further show that EPSPs on apical dendrites propagate slower and will have slower kinetics when reaching the soma than those on basal dendrites (Fig. 19.4b). Thus, the recurrent inputs onto the basal dendrites have a kinetic advantage over those onto the apical ones.

What might be the network impact of this kinetic advantage of the feedback circuit? The DG consists of ~10^3–10^4 times more GCs than BCs, and the divergent GC–BC connectivity is smaller than the convergent BC–GC connection (Patton and McNaughton 1995). As soon as GCs are activated by entorhinal inputs, they can recruit BCs in a fast and reliable manner, thereby producing strong feedback inhibition to the GC population. The strong inhibitory feedback and the lack of recurrent excitatory synapses (i.e. GC–GC synapses) make DG circuits a competitive network (Rolls and Treves 1998).

In such a network, sparse activity and a high signal-to-noise ratio can be achieved. Specifically in the DG, information flow could be as follows. The main input, the perforant path, is weak and slow and will not activate BCs but rather put them into a ready-to-fire subthreshold state. However, when a small set of highly excited GCs discharge, they can recruit BCs. Feedback inhibition will suppress any further activity in the GC population, maintaining the sparse activity pattern and a good signal-to-noise ratio. Synapses onto the "winner" GC could be potentiated under these conditions, whereas synapses onto suppressed GCs will not undergo potentiation or may even show depression. Thus, potentiated GCs can emerge as a cell assembly which transmits the given information further to CA3, consistent with a winner-takes-all mechanism.

The synaptic and dendritic properties of the basal dendrites involved in feedback inhibitory loop are central to these mechanisms, because they enable the fast, temporally precise and efficient activation of BCs by low number of coincident inputs (Geiger et al. 1997; Nörenberg et al. 2010). Furthermore, the GC-to-BC synapse might itself undergo potentiation in an associative manner under these conditions (Sambandan et al. 2010). The precise role of associative plasticity is not known, but one can speculate that on the one hand it will help to preserve the balance of excitation and inhibition in the circuit. On the other hand, it may facilitate the emergence and stabilise the "winner" cell assembly.

Although main aspects of computational function of the DG are becoming apparent, several details of the underlying microcircuit mechanisms are still unknown. In particular the role of the various types of interneurons, e.g. the distinct operation of dendrite-inhibiting versus perisomatic-inhibiting interneurons, remains elusive. Our results provide some insights into how the divergent properties of synaptic inputs and surface membranes of different subcellular compartments in BCs contribute to microcircuit operation in the DG. However, further systematic analysis of synaptic and integrative properties as well as the pattern and timing of activity of the various interneuron types will be required to better understand how inhibitory neurons contribute to the control of flow and storage of information in the DG and other cortical networks.

References

Amaral DG (1993) Emerging principles of intrinsic hippocampal organization. Curr Opin Neurobiol 3(2):225–229

Andersen P, Bliss TV, Skrede KK (1971) Lamellar organization of hippocampal pathways. Exp Brain Res 13(2):222–238

Ascoli GA et al (2008) Petilla terminology: nomenclature of features of GABAergic interneurons of the cerebral cortex. Nat Rev Neurosci 9(7):557–568

Bartos M, Vida I, Frotscher M et al (2001) Rapid signaling at inhibitory synapses in a dentate gyrus interneuron network. J Neurosci 21(8):2687–2698

Bartos M, Vida I, Frotscher M et al (2002) Fast synaptic inhibition promotes synchronized gamma oscillations in hippocampal interneuron networks. Proc Natl Acad Sci USA 99(20): 13222–13227

Bartos M, Vida I, Jonas P (2007) Synaptic mechanisms of synchronized gamma oscillations in inhibitory interneuron networks. Nat Rev Neurosci 8(1):45–56

Bartos M, Alle H, Vida I (2011) Role of microcircuit structure and input integration in hippocampal interneuron recruitment and plasticity. Neuropharmacology 60(5):730–739

Brent RP (2002) Algorithms for minimization without derivatives. Dover Publications, Mineola, NY

Bucurenciu I, Kulik A, Schwaller B et al (2008) Nanodomain coupling between Ca2+ channels and Ca2+ sensors promotes fast and efficient transmitter release at a cortical GABAergic synapse. Neuron 57(4):536–545

Buzsáki G (2006) Rhythms of the brain. Oxford University Press, New York, NY

Buzsáki G, Draguhn A (2004) Neuronal oscillations in cortical networks. Science 304(5679):1926–1929

Carnevale NT, Hines ML (2006) The NEURON book. Cambridge University Press, Cambridge

Chitwood RA, Hubbard A, Jaffe DB (1999) Passive electrotonic properties of rat hippocampal CA3 interneurones. J Physiol 515(Pt 3):743–756

Ewell LA, Jones MV (2010) Frequency-tuned distribution of inhibition in the dentate gyrus. J Neurosci 30(38):12597–12607

Fisahn A, Yamada M, Duttaroy A et al (2002) Muscarinic induction of hippocampal gamma oscillations requires coupling of the M1 receptor to two mixed cation currents. Neuron 33(4):615–624

Förster E, Zhao S, Frotscher M (2006) Laminating the hippocampus. Nat Rev Neurosci 7(4): 259–267

Freund TF, Buzsáki G (1996) Interneurons of the hippocampus. Hippocampus 6(4):347–470

Geiger JR, Lübke J, Roth A et al (1997) Submillisecond AMPA receptor-mediated signaling at a principal neuron-interneuron synapse. Neuron 18(6):1009–1023
Gentet LJ, Stuart GJ, Clements JD (2000) Direct measurement of specific membrane capacitance in neurons. Biophys J 79(1):314–320
Golding NL, Mickus TJ, Katz Y et al (2005) Factors mediating powerful voltage attenuation along CA1 pyramidal neuron dendrites. J Physiol 568(Pt 1):69–82
Gulledge AT, Kampa BM, Stuart GJ (2005) Synaptic integration in dendritic trees. J Neurobiol 64(1):75–90
Gulyás AI, Buzsáki G, Freund TF et al (2006) Populations of hippocampal inhibitory neurons express different levels of cytochrome c. Eur J Neurosci 23(10):2581–2594
Hille B (2001) Ion channels of excitable membranes. Sinauer Associates Inc, Sunderland
Hines ML, Carnevale NT (1997) The NEURON simulation environment. Neural Comput 9(6):1179–1209
Hu H, Martina M, Jonas P (2010) Dendritic mechanisms underlying rapid synaptic activation of fast-spiking hippocampal interneurons. Science 327(5961):52–58
Jonas P, Bischofberger J, Fricker D et al (2004) Interneuron diversity series: fast in, fast out–temporal and spatial signal processing in hippocampal interneurons. Trends Neurosci 27(1):30–40
Koch C (1999) Biophysics of computation. Oxford Univ Press, Oxford
Krueppel R, Remy S, Beck H (2011) Dendritic integration in hippocampal dentate granule cells. Neuron 71(3):512–528
Larkum ME, Nevian T, Sandler M et al (2009) Synaptic integration in tuft dendrites of layer 5 pyramidal neurons: a new unifying principle. Science 325(5941):756–760
Lewis DA, Hashimoto T, Volk DW (2005) Cortical inhibitory neurons and schizophrenia. Nat Rev Neurosci 6(4):312–324
London M, Meunier C, Segev I (1999) Signal transfer in passive dendrites with nonuniform membrane conductance. J Neurosci 19(19):8219–8233
Losonczy A, Magee JC (2006) Integrative properties of radial oblique dendrites in hippocampal CA1 pyramidal neurons. Neuron 50(2):291–307
Magee JC (2000) Dendritic integration of excitatory synaptic input. Nat Rev Neurosci 1(3): 181–190
Major G, Larkman AU, Jonas P et al (1994) Detailed passive cable models of whole-cell recorded CA3 pyramidal neurons in rat hippocampal slices. J Neurosci 14(8):4613–4638
Nörenberg A, Hu H, Vida I et al (2010) Distinct nonuniform cable properties optimize rapid and efficient activation of fast-spiking GABAergic interneurons. Proc Natl Acad Sci USA 107(2):894–899
Patton PE, McNaughton B (1995) Connection matrix of the hippocampal formation: I. The dentate gyrus. Hippocampus 5(4):245–286
Pike FG, Goddard RS, Suckling JM et al (2000) Distinct frequency preferences of different types of rat hippocampal neurones in response to oscillatory input currents. J Physiol 529(Pt 1): 205–213
Rall W (1959) Branching dendritic trees and motoneuron membrane resistivity. Exp Neurol 1:491–527
Rall W (1962) Electrophysiology of a dendritic neuron model. Biophys J 2(2 Pt 2):145–167
Rolls ET, Treves A (1998) The hippocampus and memory. In: Rolls ET, Treves A (eds) Neural networks and brain function. Oxford University Press, New York, NY, pp 118–120
Sambandan S, Sauer JF, Vida I et al (2010) Associative plasticity at excitatory synapses facilitates recruitment of fast-spiking interneurons in the dentate gyrus. J Neurosci 30(35):11826–11837
Schmidt-Hieber C, Jonas P, Bischofberger J (2007) Subthreshold dendritic signal processing and coincidence detection in dentate gyrus granule cells. J Neurosci 27(31):8430–8441
Scott LL, Mathews PJ, Golding NL (2005) Posthearing developmental refinement of temporal processing in principal neurons of the medial superior olive. J Neurosci 25(35):7887–7895
Spruston N, Stuart G, Häusser M (2008) Dendritic integration. In: Stuart G, Spruston N, Häusser M (eds) Dendrites, 2nd edn. Oxford University Press, Oxford, pp 231–270

Stuart G, Spruston N (1998) Determinants of voltage attenuation in neocortical pyramidal neuron dendrites. J Neurosci 18(10):3501–3510
Thurbon D, Lüscher HR, Hofstetter T et al (1998) Passive electrical properties of ventral horn neurons in rat spinal cord slices. J Neurophysiol 79(5):2485–2502
Vetter P, Roth A, Häusser M (2001) Propagation of action potentials in dendrites depends on dendritic morphology. J Neurophysiol 85(2):926–937
Vida I, Bartos M, Jonas P (2006) Shunting inhibition improves robustness of gamma oscillations in hippocampal interneuron networks by homogenizing firing rates. Neuron 49(1):107–117
Whittington MA, Traub RD, Jefferys JG (1995) Synchronized oscillations in interneuron networks driven by metabotropic glutamate receptor activation. Nature 373(6515):612–615

Chapter 20
Subthreshold Resonance and Membrane Potential Oscillations in a Neuron with Nonuniform Active Dendritic Properties

Ekaterina Zhuchkova, Michiel W.H. Remme, and Susanne Schreiber

Abstract Synaptic input to neurons is subject to cell-intrinsic filtering. In the subthreshold membrane potential range this filtering can have either low-pass or resonant characteristics and thereby have a key role in the frequency-dependent information flow in neuronal networks. Experimental classification of neurons as resonant versus nonresonant is usually based on somatic measurements, which, as we demonstrate here, may not accurately reflect neuronal filter properties because of nonuniform distributions of active membrane processes. Using cable theory, we identify conditions under which dendritic currents, in particular I_{h}, can generate somatic resonances. We find that even a strong dendritic resonance may not be detectable somatically in pyramidal cells with a high density of HCN channels in the distal parts of the dendrites. In addition, we show that noise-driven membrane potential oscillations caused by dendritic resonance can propagate to the soma where they can be recorded in the absence of somatic resonance.

20.1 Introduction

Experimental studies have demonstrated in the last decades that dendrites can express a large variety of voltage-gated channels (see for example Migliore and Shepherd, 2002). Often, these channels are not distributed uniformly across the dendrite, but can either be found at specific dendritic locations, or arranged in a gradient (Nusser, 2009). This fine-tuned expression of channels suggests a role for the spatial distribution of dendritic ion channels in neural computation.

E. Zhuchkova • M.W.H. Remme • S. Schreiber (✉)
Institute for Theoretical Biology, Humboldt-Universität zu Berlin,
Invalidenstrasse 43, 10115 Berlin, Germany
e-mail: ekaterina.zhuchkova@hu-berlin.de; michiel.remme@hu-berlin.de;
s.schreiber@hu-berlin.de

H. Cuntz et al. (eds.), *The Computing Dendrite: From Structure to Function*,
Springer Series in Computational Neuroscience 11, DOI 10.1007/978-1-4614-8094-5_20,

In this chapter we focus on a subset of conductances that activate at subthreshold membrane potentials and are involved in the generation of so-called membrane potential resonances (Hutcheon and Yarom, 2000). Subthreshold membrane potential resonance characterizes the property of certain neurons to respond best, i.e., with maximal voltage amplitude, to subthreshold inputs at a specific (nonzero) frequency. Such resonances hence lead to band-pass-like filtering of neuronal inputs and are therefore relevant for the frequency-dependent information flow in the brain (Gloveli et al., 1997). Here, we specifically aim to understand how the spatial localization of such conductances affects somatic versus dendritic input filtering.

Hyperpolarization-activated, cyclic nucleotide-gated HCN channels (h-channels) have been identified as key players in shaping subthreshold resonance in many neurons: the current flowing through activated h-channels (designated as I_h) underlies membrane potential resonance in various cortical and hippocampal neurons (Hutcheon et al., 1996a; Ulrich, 2002; Wang et al., 2006; Nolan et al., 2007; Wahl-Schott and Biel, 2009; Zemankovics et al., 2010; Gastrein et al., 2011). It produces a membrane potential resonance as it actively opposes changes in membrane voltage and activates slowly relative to the membrane time constant (Hutcheon and Yarom, 2000). The biophysical properties of h-channels can give rise to a resonance within the theta range (4–12 Hz; see for example Klink and Alonso, 1993; Pike et al., 2000; Hu et al., 2002; Wahl-Schott and Biel, 2009), and could thus play a central role in the generation of local-field theta oscillations (Nolan et al., 2004; Rotstein et al., 2005).

The h-channels are distributed in a nonuniform fashion across the soma and dendrites in various types of neurons (Nusser, 2009). A subcellular distribution was first described in CA1 pyramidal cells (Magee, 1998). Subsequent work demonstrated that the density of h-channels increases up to 60-fold with distance from the soma along the apical dendrites of hippocampal and neocortical layer 5 pyramidal cells (Williams and Stuart, 2000; Berger et al., 2001; Nusser, 2009; Lörincz et al., 2002). An important consequence of this location-dependence is that properties of the membrane potential resonance typically also vary across the neuron (Narayanan and Johnston, 2007; Hu et al., 2009), and may hence be expected to affect the processing of synaptic inputs in a location-dependent manner, which we explore in this chapter.

In addition we consider the occurrence of subthreshold membrane potential oscillations (MPOs), which have been associated with membrane potential resonance and are thought to contribute to the generation of various brain rhythms (Buzsaki and Draguhn, 2004). Modeling work suggests that self-sustained, highly regular MPOs arise from an instability of the membrane's resting state leading to subthreshold limit-cycle oscillations of the membrane potential (Lampl and Yarom, 1997; Hutcheon and Yarom, 2000; see also Chap. 21 in this book). Alternatively, MPOs can also result from the interplay between resonance-generating active conductances and noise that arises from, e.g., ion-channel stochasticity (Dorval and White, 2005; Erchova et al., 2004). The latter MPOs are noise-driven and hence less regular, but show a prominent peak in the voltage power spectrum. They can, for example, arise from the h-channel-mediated, resonant filtering of cell-intrinsic noise. Interestingly, we find that MPOs and resonance may have distinct "visibility" in somatic recordings.

In this chapter we show that neurons can express a strong dendritic resonance that filters the somatic response to dendritic inputs without affecting the response to somatic inputs. It is important to note that, as a consequence, experimental classification of resonant versus nonresonant cells may be misleading when it is exclusively based on somatic recordings. We use a simple, analytically tractable model to identify morphological and physiological parameters that foster a masking of dendritic resonance when measuring somatically. In addition, we demonstrate that dendritically generated MPOs may still propagate to the soma where they can be picked up by somatic measurements under circumstances where dendritic resonance itself is "invisible" to somatic recordings. Note that this chapter is a modified and abbreviated version of Zhuchkova et al. (2013).

20.2 Mathematical Analysis of the Impedance Amplitude Profiles

In this section we briefly lay the ground for the analytical derivation of dendritic and somatic resonance properties in a simple model of a spatially extended neuron with h-conductances that are distributed nonuniformly along the dendrite. We consider a passive dendritic cable of length l attached to a single compartment representing the soma, similar to the Rall model of the motoneuron (Rall, 1964). Unless stated otherwise, the model is extended with a lumped distal dendritic compartment where all h-conductances are located (Fig. 20.1a). This simplification gives us an analytically treatable model, capturing the steep asymmetry of h-channel density along the apical dendrite of pyramidal neurons.
The passive cable equation satisfies

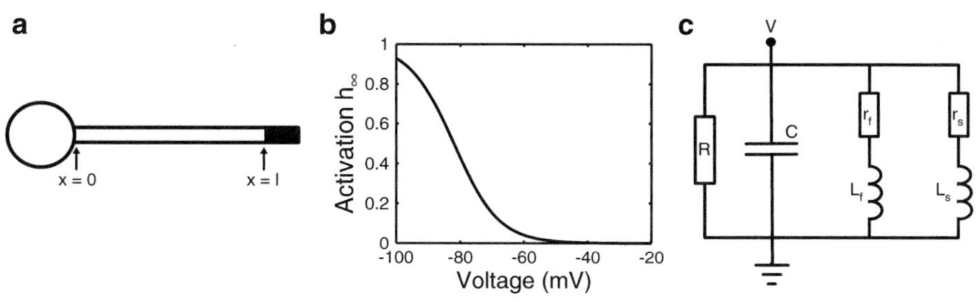

Fig. 20.1 Modeling a dendritic neuron with nonuniform membrane properties. (**a**) The analytically treatable model of a pyramidal cell consists of a finite passive cable, with a soma connected at $x = 0$ and a distal dendritic segment with h-channels at $x = l$. (**b**) Steady-state activation curve of the h-current (see Appendix). (**c**) Resonant dynamics is represented by an LRC-circuit: for perturbations around holding potential V_R, the h-current responds as if the total membrane resistance is in parallel with two inductive branches

$$\lambda^2 \frac{\partial^2 V(x,t)}{\partial x^2} - \tau_m \frac{\partial V(x,t)}{\partial t} - (V(x,t) - E_L) = 0, \tag{20.1}$$

where $V(x,t)$ is the membrane voltage along the cable, λ is the space constant, τ_m is the membrane time constant, and E_L is the leak reversal potential. The soma, represented by a single isopotential compartment, is attached at $x = 0$:

$$C_s \frac{\partial V(0,t)}{\partial t} = -\frac{V(0,t) - E_L}{R_s} + \frac{1}{r_a} \frac{\partial V(0,t)}{\partial x}, \tag{20.2}$$

where C_s is the capacitance and R_s is the membrane resistance of the soma membrane, and r_a is the axial resistance. We typically consider that the h-conductances are concentrated in a lumped compartment at the distal end of the dendritic cable:

$$C_d \frac{\partial V(l,t)}{\partial t} = -\frac{V(l,t) - E_L}{R_d} - \frac{1}{r_a} \frac{\partial V(l,t)}{\partial x} - I_h + I_{inj}, \tag{20.3}$$

where C_d is the capacitance and R_d the membrane resistance of the dendritic compartment, and I_{inj} is an arbitrary current input.

The description of the h-current, I_h, is based on recordings from cortical neurons (Spain et al., 1987) and consists of a fast and a slow component: $I_h = \bar{g}_h(0.8h_f + 0.2h_s)(V - E_h)$, where $\bar{g}_h$ is the peak conductance and E_h the reversal potential of the h-current. The gating variables h_f and h_s have steady-state activation function $h_\infty(V)$ (Fig. 20.1b) and evolve according to standard first order differential equations with time constants $\tau_f = 40$ ms and $\tau_s = 300$ ms (for details see Appendix).

In order to analyze the frequency-dependent input filtering of the neuron model, we first linearize the h-current about a holding membrane potential V_R (here, – 60 mV, see Mauro et al., 1970). The linear system can be described as an LRC electric circuit consisting of two phenomenological inductances, three resistances, and a capacitance (Fig. 20.1c; see Appendix); for perturbations around V_R the h-current responds as if the total membrane resistance is in parallel with two inductive branches. A passive cell membrane is represented by an RC circuit and always shows low-pass filtering, i.e., the membrane impedance decreases with increasing frequency. In contrast, the LRC circuit may behave as a band-pass filter and its impedance can have a prominent maximum. In this case, the RC part of the circuit forms a low-pass filter, while the inductive branches act as high-pass filters. To produce a band-pass filter, a voltage-gated current should actively oppose changes in membrane voltage and activate slowly relative to the membrane time constant (Hutcheon and Yarom, 2000). The h-current, considered in this study, satisfies both conditions and, hence, can cause resonant behavior.

To quantify the effect of the filtering of inputs at various locations, we compute the transfer impedance. It relates the current injected in one location to the voltage response that this current elicits in a different location. As the majority of excitatory inputs arrives at the dendritic tree, but action potentials are usually generated in the axon initial segment close to the soma, the transfer impedance between the distal dendritic and the somatic compartment is most interesting for our work (note that a transfer impedance is always identical in both directions). To calculate this transfer impedance, we rewrite (20.1) with the boundary conditions given by (20.2) and (20.3) in the Fourier domain, and solve for a δ-current impulse at the distal dendritic end. Details are given in Appendix. In addition to the transfer impedance, we compute the input impedances for somatic and dendritic compartments (current injection and response measurement in the same location).

To describe the quality of a band-pass filter we compute the Q-value (Koch, 1984; Hutcheon et al., 1996b): the ratio of the impedance amplitude at its resonant frequency to the input resistance (i.e., the impedance at zero frequency). If $Q > 1$, the membrane shows resonant dynamics. Note that in experimental studies one should generally rely on larger Q-values (usually more than 1.2) to identify membrane resonances in order to surpass the intrinsic noise level (Erchova et al., 2004). For completeness, we also compare our analytical results with numerical simulations of the nonlinear conductance-based model using the NEURON simulation environment (Hines and Carnevale, 1997; see Appendix). In these simulations we consider ZAP current stimuli or injection of white noise.

20.3 Dendritic Resonance Can be Invisible in a Somatic Assessment of Input Impedance

To set the reference frame, we first illustrate the frequency filtering in a neuron without dendritic h-conductances. We hence compute the local membrane impedance in a purely passive neuron model as well as in a neuron model with somatic (but no dendritic) expression of I_h. The passive neuron behaves as a low-pass filter (Fig. 20.2a, blue curve), while the neuron with somatic I_h displays a resonance (magenta curve). Both behaviors are not surprising, as long as h-channels are located in the soma and one measures the somatic input impedance.

However, a different situation occurs if we analyze the input impedance when the h-channels are distributed in pyramidal-like manner, where h-conductances are concentrated in the distal dendritic end (Fig. 20.2b). When measuring the input impedance at the distal dendritic end, a strong resonance is seen ($Q = 1.36$, gray curve). However, the input impedance at the soma demonstrates a low pass filter, as if there were no resonant current present in the neuron (red curve). Note that we used the same number of h-channels in the distal end as in the soma in Fig. 20.2a. The thick curves were calculated analytically and coincide with the thin curves obtained from numerical simulations of the full nonlinear model stimulated with a ZAP current (see Appendix). This shows that the nonlinear models are very well approximated by the analytically treatable linear ones. These results also demonstrate that a somatic measurement of the input impedance could misleadingly suggest that a neuron cannot show band-pass filtering despite dendritic resonance. In principle, a distal dendritic location of a resonant current can be ineffective for somatic synaptic inputs, which in this scenario remain low-pass filtered, whereas synaptic inputs impinging directly on the distal dendrites are locally band-pass filtered.

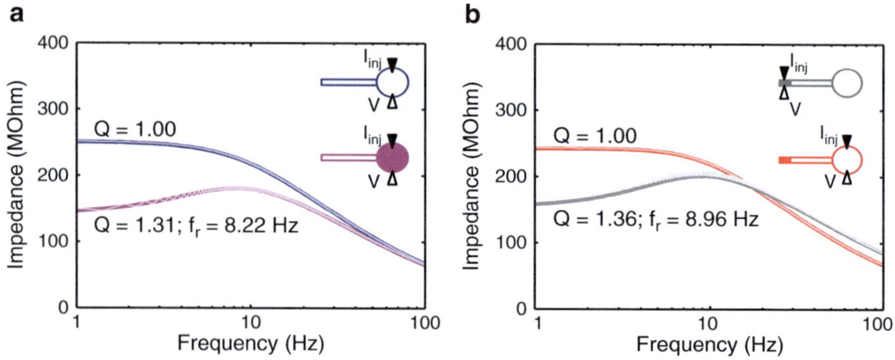

Fig. 20.2 Dendritic resonance may not be detectable in somatic measurements. (**a**) The somatic input impedance of a passive neuron (*blue*) demonstrates low-pass behavior. When I_h is concentrated in the soma, the interplay between the resonant current and the passive properties results in a band-pass filter (*magenta*). (**b**) If h-channels are located in the distal dendritic end, resonance is found in the dendrite (*gray*), but is not visible in a somatic recording (*red*). *Thin curves* in panels **a** and **b** give numerical results from the nonlinear model

20.4 Dendritic Resonance-Induced MPOs Can Propagate to the Soma

Next, instead of determining the local response to external inputs arriving at the soma or distal dendritic end (i.e., the input impedance), we describe the filtering along the dendritic cable (i.e., the transfer impedance). We consider the transfer impedance between the distal dendritic end and the soma. We want to relate the transfer impedance to the characterization of somatic membrane potential fluctuations. In experiments, one can measure such fluctuations under in vitro or in vivo conditions and determine their spectral properties (Dudman and Nolan, 2009). Both the spectral properties and the amplitude of membrane potential fluctuations are shaped by the presence of h-channels. If there is a prominent peak at a nonzero frequency in the voltage power spectrum, one may talk about the presence of MPOs.

As in Sect. 20.3, we again consider the three neuron models: a passive neuron, a model with h-channels in the soma, or a model with h-channels in the distal dendritic end (Fig. 20.3, parts a, b, and c, respectively). Now, we inject white noise (representing synaptic or channel noise) in the distal dendritic segment and measure the membrane voltage response at the soma. It is not clear from the voltage traces themselves whether the subthreshold somatic voltage has any oscillatory component (Fig. 20.3, top traces; see also Dudman and Nolan, 2009). However, the voltage power spectra demonstrate maxima in the theta-range (f_{MPO} = 6.58 Hz and f_{MPO} = 6.84 Hz, when h-channels are in the soma or in the distal end, respectively) for both neurons with h-channels, while the passive neuron does not show a preferred frequency. The spectra of the nonlinear models are well approximated by the squared

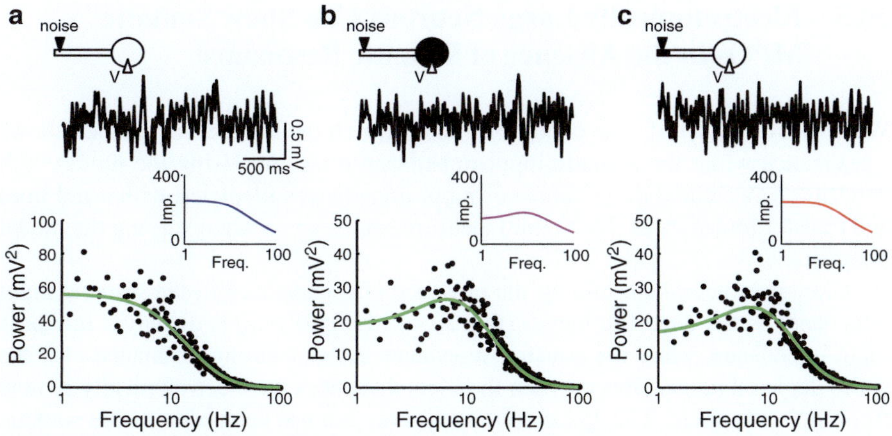

Fig. 20.3 MPOs resulting from interplay of dendritic I_h and synaptic/channel noise may propagate to the soma. White noise is injected at the distal dendritic end. Voltage traces recorded at the soma (*solid black curves*) and their power spectra (*black dots*) approximated by squared dendro-somatic transfer impedances (*green curves*) are shown for three models. (**a**) The neuron is entirely passive. (**b**) h-Channels are only present in the soma. (**c**) h-Channels are localized in the distal end of the dendrite. *Insets* show local somatic impedances from Fig. 20.2. Even if the somatic resonance is invisible, dendritic resonance-induced MPOs can "show up" in somatic measurements

transfer impedances (multiplied by the noise power spectrum) calculated for the linear models (Fig. 20.3, black dots versus green curves in bottom panels).

The Q-values of the transfer impedances for current injected at the distal end and measured at the soma are 1.00 (passive), 1.25 (somatic I_h), and 1.28 (dendritic I_h) for the three models. Although low-pass properties are expected for the passive neuron, it is perhaps surprising that the two models with I_h show similar Q-values. In fact, the transfer impedances of these two models are very similar because the total number of h-channels is the same. In accordance with results from Angelo et al. (2007), the transfer impedance is hardly affected by the precise distribution of I_h between input and output locations. While both distributions have similar somatic impedance profiles when the current is injected at the distal end, this is not the case for somatic input (insets in Fig. 20.3). Presence or absence of local resonances strongly depends on the h-channel distribution along the neuron.

Comparing Q-values of the somatic input impedance (inset in Fig. 20.3c) and transfer impedance in the case of a dendritic neuron with a pyramidal cell-like distribution of h-conductances, we come to the conclusion that even if the somatic resonance is absent, there still can be somatic MPOs resulting from dendritic resonances. Note that both noise sources, synaptic background activity and stochastic opening of ion channels, could introduce filtering by themselves (e.g., they could have band-pass power spectra, see White et al., 2000), which could modify the amplitude and frequency of MPOs.

20.5 Electrotonically Large Neurons Can Show Somatic MPOs in the Absence of Somatic Resonance

We next aim to identify the conditions under which the transfer impedance shows a resonance while the somatic input impedance does not. Using the model with dendritic I_h, we compare *Q*-values of input impedances (depicted with a red line) and transfer impedances (black line) measured at the soma when varying the model parameters.

A key parameter determining the presence of a resonance in the somatic input impedance is the dendritic length (Fig. 20.4a). Long dendrites display a low-pass input impedance, while the transfer impedance shows a strong resonance over the entire depicted range. Hence, when the h-conductances are electrotonically remote from the soma (here, 1–2 space constants), they are not detectable in the somatic input impedance.

In our default parameter set (indicated by vertical dashed lines) the somatic input impedance does not show a resonance. This does not change significantly when increasing the dendritic diameter (Fig. 20.4b) or soma surface area (Fig. 20.4c). Both parameters do decrease the *Q*-value of the transfer impedance, because they decrease the relative contribution of the distal dendritic segment to the total membrane surface area.

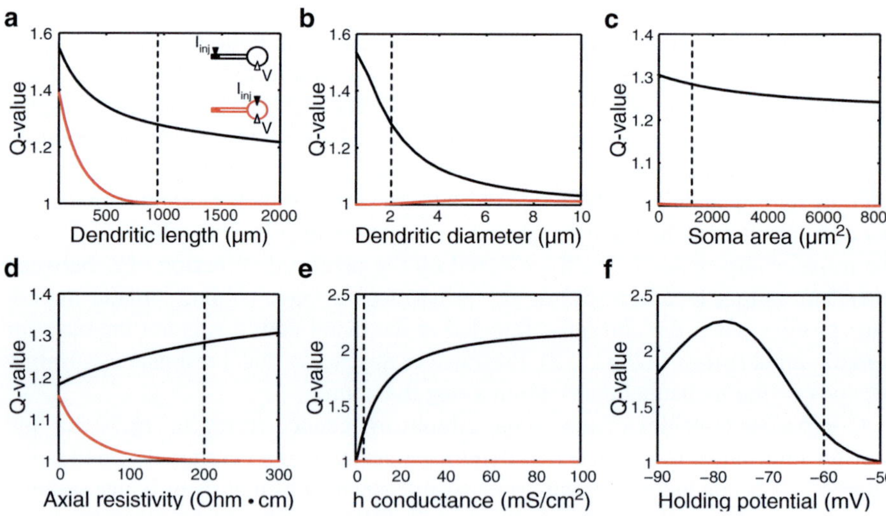

Fig. 20.4 Despite the low-pass somatic filtering, somatic MPOs can be recorded in electrotonically large cells. *Q*-values of somatic input impedance (*red*) and transfer impedance (*black*) under variation of various model parameters are shown. *Vertical dashed lines* indicate the default parameter values used in this study. Electrotonic separation of the dendritic segment with I_h from the soma determines the strength of somatic resonance

Another important parameter controlling the input impedance resonance is the axial resistivity, R_a. Decreasing this parameter reveals the dendritic resonance in the soma (Fig. 20.4d), since it decreases the electrotonic separation between dendrite and soma. It is important to realize that there is a large uncertainty about the actual values of axial resistivity for different neurons. For example, Golding et al. (2005) found that the axial resistivity of CA1 pyramidal neurons lies within the range of 139–218 Ωcm, which is approximately twice as high as estimates for cortical layer V pyramidal neurons (70–100 Ωcm, see Stuart and Spruston, 1998).

Finally, parameters that affect the resonance directly, such as the density of h-channels in the distal dendritic end (Fig. 20.4e) and the cell's holding potential (Fig. 20.4f), also control the Q-values of the transfer impedance. However, it does not change the low-pass nature of the somatic input impedance, since these parameters do not affect the electrotonic separation between the soma and the active, distal dendritic segment.

20.6 Discussion

We have used mathematical descriptions of spatially extended neurons to analyze the effect of dendritic resonance on dendritic and somatic input. We have found that a dendritic resonance may strongly affect the somatic response to dendritic inputs, without affecting somatic input, or indeed, without being detectable with somatic current clamp recordings.

The extent to which dendritic resonance affects somatic input depends on the electrotonic separation of the resonant dendritic segments from the soma. If the resonant membrane is close to the soma, the somatic input will activate the resonant current, which in turn will affect the somatic response. However, if the resonant membrane is more distant, the attenuated somatic input will perhaps activate the resonant currents somewhat, but these currents will be further attenuated on the way back to the soma, thereby becoming negligible. In contrast, distal dendritic input will be locally filtered by the resonance, and the response, though attenuated, will still be detectable at the soma. Indeed, we show that MPOs that are generated by distal, resonant dendritic segments may propagate to the soma, leading to a situation where such cells, when measured somatically, do exhibit subthreshold MPOs in the apparent absence of somatic resonance.

Usually, it is assumed that if subthreshold MPOs can be detected, also a subthreshold membrane potential resonance should be present. Our results show, however, that depending on the spatial location of the resonating mechanism, MPOs can be picked up by somatic recordings while recordings in this compartment do not show a pronounced resonant peak in the frequency-dependent membrane impedance. This is the case if the resonating currents are located in electrotonically distant compartments, like the distal ends of apical dendrites, such that the local resonance is not strongly reflected in the voltage responses to currents injected somatically. Noise-driven MPOs of distal dendritic origin (caused either by channel or synaptic noise) may nevertheless still reach the somatic compartment and reflect the local dendritic filtering in a peaked voltage power spectrum.

It should be noted that at least two alternative mechanisms exist that lead to a scenario where somatic MPOs can occur in the apparent absence of somatic membrane potential resonance. The first mechanism relies not on a spatial, but a depolarization-dependent separation of resonance. For example, the M-type potassium current leads to resonance at membrane potentials that are more depolarized than the membrane potentials at which the h-current mediated resonance is prevalent. If tested at resting potential, a neuron with the M-type current may be classified as nonresonant, as the current is not sufficiently activated. Nevertheless, such a cell may show pronounced MPOs in the subthreshold range close to threshold. In analogy to the spatial separation presented in this chapter, the underlying cause of MPOs is also a resonance in this case. The resonance, however, may be missed if not measured at the right level of depolarization. Experimentally, these measurements may not always be straightforward because the amplitude of (the time-varying component of) the ZAP current has to be small in order to avoid spiking. Such a constellation of depolarization-dependent resonance has been reported in hippocampal CA1 pyramidal cells (Hu et al., 2009) and has been related to differences in the spiking dynamics of these cells—integrator versus resonator—under in vitro and in vivo conditions (Prescott et al., 2008).

A second mechanism for MPOs without subthreshold resonance relates to the properties of resonance itself. A theoretical study by Richardson et al. (2003) showed that even in point neurons there is no equivalence between subthreshold resonance and damped oscillations in response to temporary stimulus changes. These authors identified a restricted parameter range, for which the subthreshold dynamics of a simple neuronal model exhibited damped oscillations despite the absence of membrane potential resonance. In this regime, a cell is likely to exhibit noise-driven MPOs without showing a resonant peak in its impedance.

It should also be mentioned that while MPOs can occur somatically without somatic resonance, as we argue here, the opposite is also possible as outlined by Erchova et al. (2004). They show that a resonance may not be sufficient to cause MPOs of substantial size. Only if the cell-intrinsic noise is of sufficient size, MPOs can become large enough to be detected.

The spatial distribution of resonant currents as analyzed in this chapter has important implications for the filtering properties of neurons. A high density of h-channels in the distal parts of the dendrite will lead to band-pass-like filtering of synaptic inputs impinging on this part of the dendritic tree. Our work implies that somatic inputs to the same cell, however, may—depending on the electrotonic separation—be low-pass filtered, as the dendritic resonance does not substantially affect inputs to the somatic compartment. Thus the spatial segregation of resonance allows for differential filtering of inputs arriving at different neuronal sites. In CA1 pyramidal cells, inputs at distal dendrites stem from entorhinal projections, while the proximal inputs come from hippocampal CA3 cells. The large I_h gradient of h-channels along the dendrites of CA1 pyramidal cells suggests that inputs arriving from these two pathways are subject to differential frequency filtering (see also Narayanan and Johnston, 2007). Moreover, spike initiation may be subject to a third frequency-dependent filter process despite its near-somatic location due to local resonant currents that activate at more depolarized levels (Engel et al., 2008; Hu et al., 2009; Schreiber et al., 2009). Altogether, spatial separation of resonance is likely to contribute to frequency-dependent information routing in the brain.

Previous mathematical work on active dendritic currents has typically considered uniform membrane properties, using the so-called quasi-active description of dendritic cables (Sabah and Leibovic, 1969; Koch, 1984; Coombes et al., 2007; Remme and Rinzel, 2011). An important result comes from the study by Koch (1984), who showed that if a small patch of dendritic membrane acts as a band-pass filter, both the local input impedance and the transfer impedance of the entire uniform cable will have the same response characteristics. Here, to mathematically analyze the effects of a strongly polarized distribution of active conductances on the filtering of dendritic and somatic input, we extended the Rall model of a passive dendritic neuron (Rall, 1964) by including an active distal dendritic segment. With this compact description we showed that one can find a band-pass somatic response to distal dendritic stimuli and a low-pass filter response when stimulating proximally. Hence, nonuniform ion channel distributions can underlie differential filtering of segregated synaptic inputs. In such cases, it may not be sufficient to assess fundamental properties of neuronal dynamics based on somatic recordings alone.

Acknowledgements This work was funded by the German Federal Ministry of Education and Research (Grants No. 01GQ0901, No. 01GQ1001A, and No. 01GQ0972), DFG (Grants No. SFB 618 and No. GRK1589), and the Einstein Foundation Berlin.

Appendix

A.1 Model Parameters

Leak-conductance $g_L = 0.09\,\text{mS/cm}^2$, capacitance $C = 1\,\mu\text{F/cm}^2$, passive membrane time constant $\tau_m = C / g_L = 11$ ms. Length of passive dendritic cable $l = 900\,\mu\text{m}$, length of the active distal end $l_d = 100\,\mu\text{m}$, dendritic diameter $d = 2\,\mu\text{m}$, surface area of the distal end $S_d = \text{p}dl_d = 628\ \mu\text{m}^2$, surface area of the soma $S_s = 1{,}257\,\mu\text{m}^2$. Axial resistivity $R_a = 200\,\Omega\,\text{cm}$, space constant $1 = \sqrt{d/(4R_a g_L)} = 527\,\mu\text{m}$.

A.2 h-Channel Kinetics

The equations describing the h-current are based on Spain et al. (1987), see also Bernander et al. (1994) and Hutcheon et al. (1996b):

$$I_{\mathrm{h}} = \bar{g}_{\mathrm{h}}(0.8h_{\mathrm{f}} + 0.2h_{\mathrm{s}})(V - E_{\mathrm{h}})$$
$$\tau_{\mathrm{f}}\frac{dh_{\mathrm{f}}}{dt} = h_{\infty}(V) - h_{\mathrm{f}}$$
$$\tau_{\mathrm{s}}\frac{dh_{\mathrm{s}}}{dt} = h_{\infty}(V) - h_{\mathrm{s}}$$
$$h_{\infty}(V) = \frac{1}{1+\exp((V+82)/7)}.$$

Here, peak conductance of the active distal dendritic segment $\bar{g}_{\mathrm{h}} = 3.8$ mS/cm², reversal potential $E_{\mathrm{h}} = -43$ mV, fast time constant $\tau_{\mathrm{f}} = 40$ ms, and slow time constant $\tau_{\mathrm{s}} = 300$ ms. The steady-state activation curve $h_{\infty}(V)$ is shown in Fig. 20.1b. Note that in the case of an active soma $\bar{g}_{\mathrm{h}} = 1.9$ mS/cm² to ensure the same number of h-channels as in the distal dendritic end.

A.3 Equations Describing the LRC Circuit

When an isopotential compartment containing a leak current and the h-current described above is linearized around a holding voltage V_{R}, we obtain the equations describing an LRC electric circuit (Fig. 20.1c):

$$C\frac{dV}{dt} = -\frac{V}{R} - I_{\mathrm{f}} - I_{\mathrm{s}} + I_{\mathrm{inj}}$$
$$L_{\mathrm{f}}\frac{dI_{\mathrm{f}}}{dt} = -r_{\mathrm{f}}I_{\mathrm{f}} + V$$
$$L_{\mathrm{s}}\frac{dI_{\mathrm{s}}}{dt} = -r_{\mathrm{s}}I_{\mathrm{s}} + V.$$

The resistances and inductances are computed as

$$R = \frac{1}{g_{\mathrm{L}} + \bar{g}_{\mathrm{h}}h_{\infty}(V_{\mathrm{R}})}$$
$$r_{\mathrm{f}} = \frac{1}{0.8\bar{g}_{\mathrm{h}}(V_{\mathrm{R}} - E_{\mathrm{h}})\frac{\partial}{\partial V}h_{\infty}(V_{\mathrm{R}})}$$
$$r_{\mathrm{s}} = \frac{1}{0.2\bar{g}_{\mathrm{h}}(V_{\mathrm{R}} - E_{\mathrm{h}})\frac{\partial}{\partial V}h_{\infty}(V_{\mathrm{R}})}$$
$$L_{\mathrm{f}} = r_{\mathrm{f}}\tau_{\mathrm{f}}$$
$$L_{\mathrm{s}} = r_{\mathrm{s}}\tau_{\mathrm{s}}.$$

For our default parameters we have $R = 4{,}019\,\Omega\text{cm}^2$, $r_f = 3{,}381\,\Omega\text{cm}^2$, $r_s = 13{,}522\,\Omega\text{cm}^2$, $L_f = 135\,\text{H}\,\text{cm}^2$, and $L_s = 4{,}057\,\text{H}\,\text{cm}^2$.

A.4 Calculation of the Local and Transfer Impedances

The impedance amplitude profile of the spatially extended neuron equals the absolute value of the (complex-valued) transfer function (i.e., the impulse response function in the frequency domain). Here, we provide the transfer functions for the passive cable with a passive soma attached at $x = 0$ and the active distal dendritic segment at $x = l$. The current is injected at the distal dendritic end. The passive cable equation (20.1) can be expressed in the frequency domain as

$$\frac{d^2\tilde{V}(x,\omega)}{dx^2} - \gamma^2(\omega)\tilde{V}(x,\omega) = 0 \tag{20.4}$$

with $\omega = 2\pi f$ (where frequency f is in Hz) and with propagation constant

$$\gamma^2(\omega) = \frac{1 + i\omega\tau_{\text{m}}}{\lambda^2}.$$

The boundary conditions defined by (20.2) and (20.3) with an impulse current $I_{\text{inj}} = \delta(t)$ in the distal dendritic segment can be written in the frequency domain as

$$\frac{1}{r_{\text{a}}}\frac{\partial\tilde{V}(0,\omega)}{\partial x} = \gamma_{\text{s}}(\omega)\tilde{V}(0,\omega) \tag{20.5}$$

$$\frac{1}{r_{\text{a}}}\frac{\partial\tilde{V}(l,\omega)}{\partial x} = 1 - \gamma_{\text{d}}(\omega)\tilde{V}(l,\omega), \tag{20.6}$$

where for the passive soma $\gamma_{\text{s}}(\omega) = S_{\text{s}}\left(i\omega C + g_{\text{L}}\right)$ and for the active dendritic segment $\gamma_{\text{d}}(\omega) = S_{\text{d}}\left(i\omega C + g_{\text{L}} + \bar{g}_{\text{h}}h_{\infty}(V_{\text{R}}) + (1/(r_{\text{f}} + i\omega L_{\text{f}})) + (1/(r_{\text{s}} + i\omega L_{\text{s}}))\right)$. By solving (20.4) with boundary conditions given by (20.5) and (20.6) we obtain the transfer function of the neuron model

$$\tilde{G}(x,\omega) = \tilde{V}(x,\omega) = A\cosh(\gamma x) + B\sinh(\gamma x), \tag{20.7}$$

where coefficient $A = A(\omega) = \gamma K$ and $B = B(\omega) = r_{\text{a}}\gamma_{\text{s}}K$, with

$$K = K(\omega) = \frac{r_{\text{a}}}{r_{\text{a}}\gamma(\gamma_{\text{s}} + \gamma_{\text{d}})\cosh(\gamma l) + (\gamma_{\text{s}}\gamma_{\text{d}}r_{\text{a}}^2 + \gamma^2)\sinh(\gamma l)},$$

and $\gamma = \gamma(\omega)$, $\gamma_{\text{s}} = \gamma_{\text{s}}(\omega)$, $\gamma_{\text{d}} = \gamma_{\text{d}}(\omega)$. To compute the dendritic input impedance or the transfer impedance between the distal dendritic end and the soma one must let $x = l$ or $x = 0$ in (20.7), respectively. When the current input is injected somatically we have $A = (\gamma\cosh(\gamma l) + r_{\text{a}}\gamma_{\text{d}}\sinh(\gamma l))K$ and $B = -(r_{\text{a}}\gamma_{\text{d}}\cosh(\gamma l) + \gamma\sinh(\gamma l))K$ and one can compute the somatic input impedance by letting $x = 0$. Input and transfer impedances for the neuron with active soma and/or passive distal dendritic segment can be obtained by setting γ_{s} and γ_{d} appropriately.

A.5 Computation of the Impedances from the Simulations

To compute somatic and dendritic input impedances, we inject a so-called ZAP current $I_{ZAP}(t) = I_0\sin(2\pi f(t)t)$, with frequency $f(t) = f_m t / 2T$, input amplitude I_0 = 0.01 nA, maximum frequency f_m = 100 Hz, and stimulus length T = 150 s. At the same location we measure the membrane potential $V(t)$ and compute the impedance as $\tilde{G}(f) = \text{FFT}(V(t)) / \text{FFT}(I_{ZAP}(t))$, where FFT is the fast Fourier transform. To compute transfer impedances and power spectra, we inject a white noise current (with a duration of 100 s and standard deviation of 0.1 nA) at the distal dendritic end and measure the somatic voltage $V(t)$. Impedance amplitude profile is determined as $|\tilde{G}(f)| = \text{FFT}(V(t)) / \text{FFT}(I_{noise}(t))$.

References

Angelo K, London M, Christensen SR, Häusser M (2007) Local and global effects of I(h) distribution in dendrites of mammalian neurons. J Neurosci 27(32):8643–8653

Berger T, Larkum ME, Lüscher HR (2001) High I(h) channel density in the distal apical dendrite of layer V pyramidal cells increases bidirectional attenuation of EPSPs. J Neurophysiol 85(2):855–868

Bernander O, Koch C, Douglas RJ (1994) Amplification and linearization of distal synaptic input to cortical pyramidal cells. J Neurophysiol 72(6):2743–2753

Buzsaki G, Draguhn A (2004) Neuronal oscillations in cortical networks. Science 304(5679): 1926–1929

Coombes S, Timofeeva Y, Svensson CM, Lord GJ, Josić K, Cox SJ, Colbert CM (2007) Branching dendrites with resonant membrane: a "sum-over-trips" approach. Biol Cybern 97(2):137–149

Dorval AD, White JA (2005) Channel noise is essential for perithreshold oscillations in entorhinal stellate neurons. J Neurosci 25(43):10025–10028

Dudman JT, Nolan MF (2009) Stochastically gating ion channels enable patterned spike firing through activity-dependent modulation of spike probability. PLoS Comput Biol 5(2):e1000290

Engel TA, Schimansky-Geier L, Herz AVM, Schreiber S, Erchova I (2008) Subthreshold membrane-potential resonances shape spike-train patterns in the entorhinal cortex. J Neurophysiol 100(3):1576–1589

Erchova I, Kreck G, Heinemann U, Herz AVM (2004) Dynamics of rat entorhinal cortex layer II and III cells: characteristics of membrane potential resonance at rest predict oscillation properties near threshold. J Physiol 560(Pt 1):89–110

Gastrein P, Campanac E, Gasselin C, Cudmore RH, Bialowas A, Carlier E, Fronzaroli-Molinieres L, Ankri N, Debanne D (2011) The role of hyperpolarization-activated cationic current in spike-time precision and intrinsic resonance in cortical neurons in vitro. J Physiol 589 (Pt 15):3753–3773

Gloveli T, Schmitz D, Empson RM, Heinemann U (1997) Frequency-dependent information flow from the entorhinal cortex to the hippocampus. J Neurophysiol 78(6):3444–3449

Golding NL, Mickus TJ, Katz Y, Kath WL, Spruston N (2005) Factors mediating powerful voltage attenuation along CA1 pyramidal neuron dendrites. J Physiol 568(Pt 1):69–82

Hines ML, Carnevale NT (1997) The NEURON simulation environment. Neural Comput 9(6): 1179–1209

Hu H, Vervaeke K, Storm JF (2002) Two forms of electrical resonance at theta frequencies, generated by M-current, h-current and persistent Na^+ current in rat hippocampal pyramidal cells. J Physiol 545(Pt 3):783–805

Hu H, Vervaeke K, Graham LJ, Storm JF (2009) Complementary theta resonance filtering by two spatially segregated mechanisms in CA1 hippocampal pyramidal neurons. J Neurosci 29(46):14472–14483

Hutcheon B, Yarom Y (2000) Resonance, oscillation and the intrinsic frequency preferences of neurons. Trends Neurosci 23(5):216–222

Hutcheon B, Miura RM, Puil E (1996a) Models of subthreshold membrane resonance in neocortical neurons. J Neurophysiol 76(2):698–714

Hutcheon B, Miura RM, Puil E (1996b) Subthreshold membrane resonance in neocortical neurons. J Neurophysiol 76(2):683–697

Klink R, Alonso A (1993) Ionic mechanisms for the subthreshold oscillations and differential electroresponsiveness of medial entorhinal cortex layer II neurons. J Neurophysiol 70(1):144–157

Koch C (1984) Cable theory in neurons with active, linearized membranes. Biol Cybern 50(1): 15–33

Lampl I, Yarom Y (1997) Subthreshold oscillations and resonant behavior: two manifestations of the same mechanism. Neuroscience 78(2):325–341

Lörincz A, Notomi T, Tamás G, Shigemoto R, Nusser Z (2002) Polarized and compartment-dependent distribution of HCN1 in pyramidal cell dendrites. Nat Neurosci 5(11):1185–1193

Magee JC (1998) Dendritic hyperpolarization-activated currents modify the integrative properties of hippocampal CA1 pyramidal neurons. J Neurosci 18(19):7613–7624

Mauro A, Conti F, Dodge F, Schor R (1970) Subthreshold behavior and phenomenological impedance of the squid giant axon. J Gen Physiol 55:497–523

Migliore M, Shepherd GM (2002) Emerging rules for the distributions of active dendritic conductances. Nat Rev Neurosci 3(5):362–370

Narayanan R, Johnston D (2007) Long-term potentiation in rat hippocampal neurons is accompanied by spatially widespread changes in intrinsic oscillatory dynamics and excitability. Neuron 56(6):1061–1075

Nolan MF, Malleret G, Dudman JT, Buhl DL, Santoro B, Gibbs E, Vronskaya S, Buzsaki G, Siegelbaum SA, Kandel ER, Morozov A (2004) A behavioral role for dendritic integration: HCN1 channels constrain spatial memory and plasticity at inputs to distal dendrites of CA1 pyramidal neurons. Cell 119(5):719–732

Nolan MF, Dudman JT, Dodson PD, Santoro B (2007) HCN1 channels control resting and active integrative properties of stellate cells from layer II of the entorhinal cortex. J Neurosci 27(46):12440–12451

Nusser Z (2009) Variability in the subcellular distribution of ion channels increases neuronal diversity. Trends Neurosci 32(5):267–274

Pike FG, Goddard RS, Suckling JM, Ganter P, Kasthuri N, Paulsen O (2000) Distinct frequency preferences of different types of rat hippocampal neurones in response to oscillatory input currents. J Physiol 529(Pt 1):205–213

Prescott SA, Ratté S, De Koninck Y, Sejnowski TJ (2008) Pyramidal neurons switch from integrators in vitro to resonators under in vivo-like conditions. J Neurophysiol 100(6):3030–3042

Rall W (1964) Theoretical significance of dendritic trees for neuronal input–output relations. In: Reis R (ed) Neural theory and modeling. Stanford University Press, Stanford, pp 73–97

Remme MWH, Rinzel J (2011) Role of active dendritic conductances in subthreshold input integration. J Comput Neurosci 31(1):13–30

Richardson MJE, Brunel N, Hakim V (2003) From subthreshold to firing-rate resonance. J Neurophysiol 89(5):2538–2554

Rotstein HG, Pervouchine DD, Acker CD, Gillies MJ, White JA, Buhl EH, Whittington MA, Kopell N (2005) Slow and fast inhibition and an H-current interact to create a theta rhythm in a model of CA1 interneuron network. J Neurophysiol 94(2):1509–1518

Sabah NH, Leibovic KN (1969) Subthreshold oscillatory responses of the Hodgkin–Huxley cable model for the squid giant axon. Biophys J 9(10):1206–1222

Schreiber S, Samengo I, Herz AVM (2009) Two distinct mechanisms shape the reliability of neural responses. J Neurophysiol 101(5):2239–2251

Spain WJ, Schwindt PC, Crill WE (1987) Anomalous rectification in neurons from cat sensorimotor cortex in vitro. J Neurophysiol 57(5):1555–1576

Stuart G, Spruston N (1998) Determinants of voltage attenuation in neocortical pyramidal neuron dendrites. J Neurosci 18(10):3501–3510

Ulrich D (2002) Dendritic resonance in rat neocortical pyramidal cells. J Neurophysiol 87(6):2753–2759

Wahl-Schott C, Biel M (2009) HCN channels: structure, cellular regulation and physiological function. Cell Mol Life Sci 66(3):470–494

Wang WT, Wan YH, Zhu JL, Lei GS, Wang YY, Zhang P, Hu SJ (2006) Theta-frequency membrane resonance and its ionic mechanisms in rat subicular pyramidal neurons. Neuroscience 140(1):45–55

White JA, Rubinstein JT, Kay AR (2000) Channel noise in neurons. Trends Neurosci 23(3): 131–137

Williams SR, Stuart GJ (2000) Site independence of EPSP time course is mediated by dendritic I(h) in neocortical pyramidal neurons. J Neurophysiol 83(5):3177–3182

Zemankovics R, Káli S, Paulsen O, Freund TF, Hájos N (2010) Differences in subthreshold resonance of hippocampal pyramidal cells and interneurons: the role of h-current and passive membrane characteristics. J Physiol 588(Pt 12):2109–2132

Zhuchkova E, Remme MWH, Schreiber S (2013) Somatic versus dendritic resonance: differential filtering of inputs through non-uniform distributions of active conductances. PLoS ONE (in press)

Chapter 21
A Trade-Off Between Dendritic Democracy and Independence in Neurons with Intrinsic Subthreshold Membrane Potential Oscillations

Michiel W.H. Remme, Máté Lengyel, and Boris S. Gutkin

Abstract Membrane potential oscillations are ubiquitous in neurons and have been proposed to underly important neuronal computations. As a paradigmatic example, the periodic spatial tuning of stellate cells from medial entorhinal cortex neurons is thought to be generated by the interference patterns arising from multiple, independent dendritic oscillators, each controlled by direction-selective input. We analyzed how multiple dendritic oscillators embedded in the same neuron integrate inputs separately and determine somatic membrane voltage jointly. We found that the interaction of dendritic oscillations leads to phase locking, which sets an upper limit on the time scale for independent input integration. Factors that increase this time scale also decrease the influence that the oscillations exert on somatic voltage. In stellate cells, inter-dendritic coupling dominates and causes these cells to act as single oscillators. Our results suggest a fundamental trade-off between local and global processing in dendritic trees integrating ongoing signals.

M.W.H. Remme (✉)
Institute for Theoretical Biology, Humboldt-Universität zu Berlin, Invalidenstrasse 43, 10115 Berlin, Germany
e-mail: michiel.remme@hu-berlin.de

M. Lengyel
Department of Engineering, Computational and Biological Learning Lab, University of Cambridge, Cambridge, UK
e-mail: m.lengyel@eng.cam.ac.uk

B.S. Gutkin
Département d'Études Cognitives, Group for Neural Theory, École Normale Supérieure, Paris, France
e-mail: boris.gutkin@ens.fr

H. Cuntz et al. (eds.), *The Computing Dendrite: From Structure to Function*, Springer Series in Computational Neuroscience 11, DOI 10.1007/978-1-4614-8094-5_21,

21.1 Introduction

Dendritic trees possess a wide variety of voltage-dependent processes (Stuart et al., 2007) that render them into sophisticated computing devices (Poirazi et al., 2003). Many studies characterized how synaptic inputs are mapped into dendritic membrane potentials, for example into dendritic spikes (Golding and Spruston, 1998), and how the local membrane potential signals from several such dendritic units then jointly determine the somatic membrane potential and ultimately the action potential output of cells (Rudolph and Destexhe, 2003). Two key features of neuronal computation emerged from these studies. First, the dendritic tree can consist of several functional compartments, each processing its inputs locally and largely independently from the other compartments ("dendritic independence", Polsky et al., 2004). Second, the soma integrates the outputs of these compartments in a way that even distant compartments exert an influence on it ("dendritic democracy," Magee and Cook, 2000; Häusser, 2001).

The issue of dendritic independence and democracy lies at the heart of a prominent theory for the spatial tuning of neurons in the entorhinal cortex. "Grid cells" of the rat medial entorhinal cortex respond by forming characteristic grid patterns when the animal is navigating through its environment (Hafting et al., 2005; Sargolini et al., 2006). These hexagonal grid patterns remain stable over long periods of time and even persist in the dark for as long as 30 min (Hafting et al., 2005). This persistence is a signature of a path integration mechanism which computes the spatial position of the animal by the continuous integration of self-motion-generated cues (McNaughton et al., 2006). Intrinsic membrane potential oscillations have been proposed to be particularly well suited for integrating synaptic inputs on long time scales (Huhn et al., 2005), and thus to play a key role in path integration (Lengyel et al., 2003). This is because the phase of an oscillator naturally integrates inputs modulating its frequency. Consequently, a dominant theory to provide a mechanistic explanation for the firing pattern of grid cells posits the existence of several independent oscillatory units in the dendritic tree—each integrating the animal's velocity along a different direction—and a "democratic" summation of the signals contributed by these dendritic oscillations at the soma (O'Keefe and Burgess, 2005; Burgess et al., 2007; Hasselmo, 2007).

The "multiple oscillator" theory of grid cells is supported by several lines of evidence. First, entorhinal spiny stellate cells show subthreshold membrane potential oscillations (Alonso and Llinás, 1989; Alonso and Klink, 1993), which appear to result from the interaction between a persistent sodium current and the hyperpolarization-activated inward current (Alonso and Llinás, 1989; Dickson et al., 2000; Fransén et al., 2004; Rotstein et al., 2006). Second, on top of the spatial tuning, the theory can also explain the dynamics of the phase relationship between the grid cell firings and the local field potential (theta) oscillations as the animal passes through each peak of the grid field (Hafting et al., 2008). Finally, the theory can account for the correlation between the intrinsic oscillator frequencies and the spacing and size of grid fields (Giocomo et al., 2007). However, while holding considerable explanatory power, the multiple oscillator theory does not consider realistic membrane potential dynamics. Critically, it relies on the assumption that dendritic independence and democracy can coexist on long time scales in an oscillatory regime. Therefore, understanding the requirements for these two features to emerge in oscillating dendrites would offer important insights into the mechanisms underlying grid field formation, as well as into the nature of ongoing dendritic computations in general.

In this chapter, we summarize and extend our study (Remme et al., 2010); see the original study for further details that have been omitted from this chapter.

21.2 Results

21.2.1 The Multiple Dendritic Oscillators Model of Grid Cells

We implemented the multiple oscillator model of grid cells (O'Keefe and Burgess, 2005; Burgess et al., 2007; Hasselmo, 2007) using three dendritic oscillators (Fig. 21.1a). The frequency of each of these oscillators is modulated by external input that reflects the movement speed $s(t)$ of the animal in a particular direction $\phi(t)$. The input to oscillator i has a preferred direction ϕ_i (differing by multiples of 120° for the three oscillators) and is linearly dependent on speed, such that the input to oscillator i is proportional to $s(t)\cos(\phi(t) - \phi_i)$. In turn, the oscillator frequency is linearly dependent on the input current. Hence, the phase ϑ_i of oscillator i evolves as

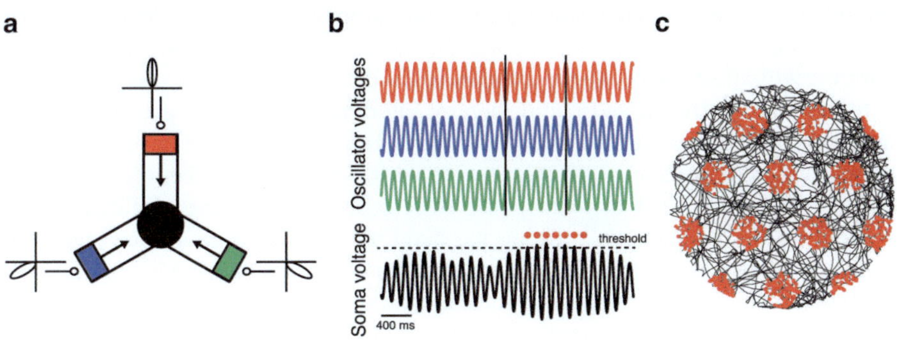

Fig. 21.1 Independent dendritic processing of velocity signals yields stable grid-fields. (**a**) Schematic of a neuron implementing the interference model showing the soma (*black*) and three dendritic oscillators (*red*, *blue*, *green*). The oscillator frequencies are modulated by velocity-tuned external input. The preferred direction of the external input to the three oscillators differs by multiples of 120° (as indicated by the panels next to each oscillator). (**b**) Membrane potentials of the three dendritic oscillators sum at the soma, thereby producing somatic voltage interference patterns. Spikes (*red dots*) are determined by threshold crossings. (**c**) Example of a 10-min simulated trajectory of a rat where the velocity-tuned inputs control the oscillator frequencies of the model in panel **a**: spike output (*red dots*) is organized in a hexagonal grid pattern. Modified from Remme et al. (2010)

$$\frac{d\vartheta_i}{dt} = \omega + \beta s(t)\cos(\phi(t) - \phi_i)$$

with oscillator frequency $\omega = 2\pi / T$ and period $T = 0.125$ s, and constant $\beta > 0$. Somatic voltage is simply determined by the sum of the dendritic voltages: $V_{\text{soma}}(t) = \frac{1}{3}(\cos\vartheta_1 + \cos\vartheta_2 + \cos\vartheta_3)$, i.e., each oscillator contributes equally to the somatic response. Variation over time of the dendritic oscillator frequencies lead to interference patterns in the somatic membrane potential; spikes are generated when the somatic membrane potential crosses a threshold (Fig. 21.1b). Using this model, we simulated the activity of a grid cell as an animal explored an environment, and reproduced the hexagonal grid-like firing rate fields of entorhinal cells (Fig. 21.1c). Thus, independent dendritic processing of continuous input signals in the idealized multiple oscillator model could produce stable grid field patterns.

21.2.2 A Trade-Off Between Independence and Democracy for Dendritic Oscillators

One of the key assumptions in the multiple oscillator model is that the interaction between the dendritic oscillators and somatic voltage is unidirectional (arrows in Fig. 21.1a): somatic voltage does not affect the oscillators, thus ensuring their perfect independence. However, within a real neuron, electrotonic coupling prevents dendritic compartments from being completely independent. The coupling results from the voltage gradient between the soma and the oscillators, and it is bidirectional: intracellular currents are not only propagated from the dendrites to the soma, but also propagate from the soma back to the dendrites. These interdendritic currents will affect the oscillator frequencies in the same way as the velocity-tuned external input currents, and could thereby interfere with correct path integration and, hence, with the formation of stable grid fields. In fact, we have previously demonstrated (Remme et al., 2009) that the periodic interactions between dendritic oscillators tend to lead to fixed phase differences between the oscillators. The strength of the interactions between the dendritic oscillators determines on what time scale this phase locking takes place and, therefore, whether this interaction still allows for stable grid formation on the time scale of minutes. We therefore formulated a mathematical theory that established a direct relationship between the time scale of phase locking of dendritic oscillations and the biophysical properties of the dendritic membrane to assess whether a biophysically realistic time scale of phase locking is sufficient for maintaining stable grid fields.

In order to estimate the time scale on which the dendritic oscillators phase lock, we mathematically analyzed the interactions between two sinusoidal oscillators connected by a segment of membrane that itself did not generate intrinsic oscillations (Fig. 21.2a, top panel; see Appendix). The phase difference between two oscillators that are not perturbed by external input exponentially decays to zero with time

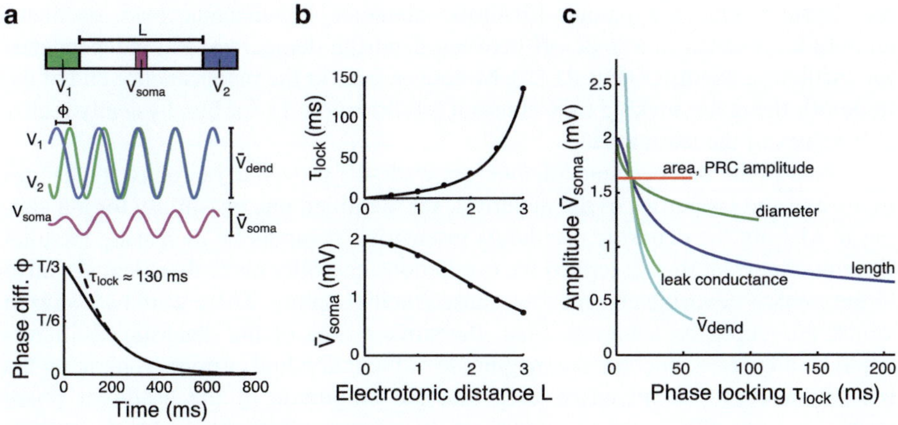

Fig. 21.2 Analysis of dendritic oscillator interactions shows a trade-off between the speed of dendritic phase locking and somatic signal amplitude. (**a**) Two dendritic oscillators with voltages V_1 (*green*) and V_2 (*blue*), oscillation amplitude $\bar{V}_{dend}$ and phase difference ϕ are coupled via a cable with electrotonic length L. The "somatic" voltage (*magenta*) at halfway point of the cable, V_{soma}, has maximal oscillation amplitude $\bar{V}_{soma}$. *Bottom panel* illustrates the phase locking of coupled dendritic oscillators. In this example, where $L = 3$ and the oscillator period $T = 125$ ms, the phase difference ϕ approaches 0 with time constant $\tau_{lock} \approx 130$ ms as indicated by the exponential fit (*dashed line*). (**b**) Phase locking time constant τ_{lock} (*top*) increases, while the somatic oscillation amplitude $\bar{V}_{soma}$ (*bottom*) decreases as a function of L. (**c**) Trade-off between phase locking time constant τ_{lock} and somatic oscillation amplitude $\bar{V}_{soma}$ persists across biophysical parameter ranges. *Curves* show the relationship between τ_{lock} and $\bar{V}_{soma}$ when each of the parameters indicated is varied over a realistic range while keeping the other parameters at their standard values. Reproduced from Remme et al. (2010) with permission from Elsevier

constant τ_{lock} (Fig. 21.2a, bottom panel). Our analysis focussed on deriving τ_{lock} to determine how it can be slowed down to the time scale of minutes. The electrotonic separation, L, of the dendritic oscillators is the most obvious parameter to affect their independence: the larger it is, the less the oscillators should influence each other. This intuition was confirmed by our analysis (Fig. 21.2b, top panel), although τ_{lock} still remained in the sub-second range for realistic values of L (up to 3 length constants). Importantly, increasing electrotonic separation also resulted in a decreasing effect of the dendritic oscillators on somatic voltage, $\bar{V}_{soma}$ (Fig. 21.2b, bottom panel). This trade-off between the independence (large phase locking time constant) and democracy (strong signal propagation to the soma) of dendritic oscillations is easily seen when plotting $\bar{V}_{soma}$ against τ_{lock} for different lengths of the connecting cable (Fig. 21.2c, blue curve).

In order for the dendritic tree to be able to realize more than one functional oscillatory integrator, as required by the multiple oscillator model of grid cells, democracy and independence must coexist. Thus, we analyzed the effects of various biophysical properties of dendritic membrane potential dynamics on τ_{lock} and $\bar{V}_{soma}$ to see what conditions may loosen the democracy-independence trade-off. We found that changing most of the relevant parameters characterizing dendritic membrane potential dynamics (dendritic diameter, conductance load, oscillator amplitude) resulted in a trade-off between dendritic democracy and independence for oscillating dendrites (Fig. 21.2c). Moreover, even at the independence end of the trade-off, the phase locking time constant fell far below 1 s for biophysically realizable values of the parameters.

Although the independence-democracy trade-off proved to be robust to changes in most of the properties of the dendrites, we identified one possibility for alleviating it. Making the dendritic oscillators insensitive to inputs by increasing the total amount of current that generated the oscillations naturally made them less sensitive to the somato-dendritic currents that caused their coupling. There were two ways in which this could be achieved. First, the surface area of the dendritic oscillators could be increased. Second, the magnitude of the individual currents generating the oscillation could be increased such that the amplitude of the oscillator phase response curve (PRC), describing the amount of phase shift obtained by unit external perturbations (see Appendix), was decreased. The effects of these changes were formally equivalent: both increased τ_{lock} without affecting $\overline{V}_{\text{soma}}$ (Fig. 21.2c, red line). Nevertheless, τ_{lock} still remained well below 1 s in a realistic range of these parameters. Further analyses showed that increasing τ_{lock} to an extent that allows for stable grid fields on a time scale of many minutes would require membrane properties that lie far outside the physiological range (results not shown). Moreover, an increased insensitivity of the oscillators to the coupling currents also makes them increasingly insensitive to synaptic inputs since the response to such inputs is also determined by the PRC amplitude and the oscillator surface area.

21.2.3 Simulations of a Reconstructed Stellate Cell Show Dendritic Phase Locking and Failure of Grid-Field Formation

Our mathematical theory pointed to a general trade-off between independence and democracy for oscillating dendrites and made specific predictions about how biophysical properties of the dendritic tree influence the phase locking behavior of dendritic oscillators and their effects on somatic voltage. We next constructed a detailed biophysical model of stellate cells—the cell type believed to produce the grid field responses—to test whether successful path integration in such cells is limited by the lack of independence of the dendritic oscillators. In this model we set all parameters to values that allowed for the most independence within a realistic range.

The subthreshold oscillations were generated by an interaction between a persistent sodium current, I_{NaP}, and a hyperpolarization-activated cation current, I_{h} (Dickson et al., 2000). These oscillator currents were moved to the distal dendritic tips (>150 μm from the soma) to promote independence of the oscillators (Fig. 21.3a). We first determined the electrotonic structure of the stellate cell model, computing the distribution of electrotonic distances between all dendritic oscillators. The average electrotonic

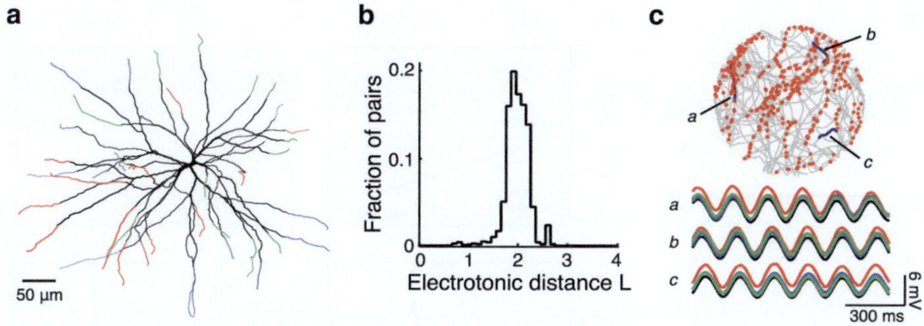

Fig. 21.3 Simulations of stellate cell activity with a detailed compartmental model with realistic electrotonic scales show strong dendritic coupling. (**a**) Stellate cell model with oscillations-generating conductances in distal dendritic segments (>150 μm from soma) are grouped into three clusters (*red*, *blue*, *green*) each of which receives external input with a different preferred direction. (**b**) Electrotonic distances between all pairs of dendritic oscillators. (**c**) Activity of the stellate cell during a simulated 5-min exploration. *Top*, trajectory with threshold crossings (*red dots*). *Bottom traces* show membrane potentials for three indicated 1-s trajectories (*a*,*b*,*c*). The *colors* of the traces each correspond to one dendrite from the three clusters of oscillators in panel **a**. *Black trace* shows somatic voltage. Reproduced from Remme et al. (2010) with permission from Elsevier

distance between all pairs was found to be ∼2 (Fig. 21.3b). Our theoretical analysis predicted fast coupling in this range of electrotonic distances (see Fig. 21.2b).

We next attempted to generate grid field activity with the compartmental model in the same way as in the idealized oscillator interference model (Fig. 21.1). We subdivided the dendritic tree into three groups of dendrites (see Fig. 21.3a) that received external input with preferred head directions differing by multiples of 120°. Despite setting up the simulation in favor of dendritic independence, we found it to be impossible to obtain grid field activity (Fig. 21.3c, top). The oscillating dendritic segments locked so strongly that the cell always showed global synchronized membrane potential oscillations. Tracking the membrane potential during three representative 1-s segments of the exploration episode revealed complete synchronization of the soma and the oscillating segments in the dendrites (Fig. 21.3c, bottom).

These results demonstrate that even with optimistic estimates of neuronal properties for independence, the coupling between dendritic segments is too strong to maintain several independent oscillators within one stellate cell. As a consequence, these cells act as single oscillators.

21.2.4 Alleviating the Trade-Off Through Large Dendritic Conduction Delays and High Oscillation Frequencies

Since the above analysis focussed on the multiple oscillator model of grid field formation by stellate cells, the results were restricted to the relevant parameter range regarding electrotonic size of the cell and oscillation frequency (i.e., within the theta range). We next explore the phase locking dynamics of dendritic

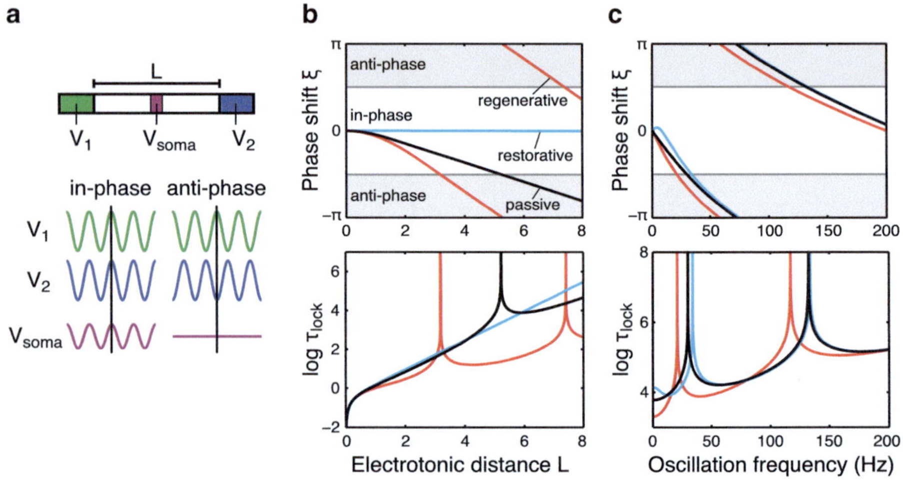

Fig. 21.4 The stable phase-locked solution of dendritic oscillators is controlled by their electrotonic separation and oscillation frequency. (**a**) The oscillators with voltage trajectories $V_1(t)$ and $V_2(t)$ determine the membrane potential at the ends of a cable with electrotonic length L. When the oscillators are synchronized (in-phase, *left traces*), a membrane potential oscillation is also visible halfway along the cable (V_{soma}). When the oscillators lock in anti-phase (*right traces*), no oscillation is visible at the soma. (**b**) *Top panel*, the value of phase shift ξ determines where the in-phase (*white area*) or anti-phase solution (*gray area*) is stable. It is plotted as a function of the electrotonic distance L between the oscillators. The cable is passive (*black*) or includes a regenerative (*red*) or restorative (*cyan*) current. *Bottom panel*, the logarithm of the phase locking time constant τ_{lock} (arbitrary units) increases with L and shows peaks for values of L where the stable phase-locked solution switches from in-phase to anti-phase or vice versa. The oscillator frequency is 8 Hz. (**c**) As in panels **b**, but showing ξ and the logarithm of τ_{lock} as a function of the oscillation frequency. The electrotonic distance between the oscillators is $L = 2$

oscillators when considering a larger range of both electrotonic length L and the oscillation frequency (Fig. 21.4).

In the parameter regime we explored above, dendritic oscillators always phase lock with zero phase difference. However, other stable phase-locked solutions are possible, depending on the properties of the oscillators and the connecting dendritic cables. Using the mathematical analysis of two cable-coupled dendritic oscillators (see Fig. 21.2) we can also analyze the stable phase-locked solution (see Appendix). The sinusoidal oscillations can either synchronize or lock in anti-phase (Fig. 21.4a). Which of the two states is stable has great consequences for the membrane potential fluctuations measured in between the dendritic oscillators. When considering that the soma is located halfway between the oscillators, oscillators that are locked in-phase will lead to (attenuated) somatic oscillations, while dendritic oscillations that lock in anti-phase will be invisible at the soma.

Coupling of the oscillators via a dendritic cable introduces conduction delays. The stable phase-locked solution is uniquely determined by this cable-induced phase shift, ξ. When the absolute dendritic phase shift is smaller than $\pi/2$, the oscillators will synchronize. However, when it is larger than $\pi/2$, the in-phase state is unstable and the oscillators will lock in anti-phase instead (Fig. 21.4b, top panel). The phase shift naturally increases with the electrotonic separation L between the oscillators. For large values of L, the phase shift ξ increases linearly with L (see Remme et al., 2009). When oscillators with an intrinsic frequency of 8 Hz are coupled through a passive dendrite, the transition to anti-phase locking occurs at $L \sim 5$ (black curve). However, the dendritic delays can be strongly affected by voltage-dependent membrane currents that are active in the subthreshold range (Remme and Rinzel, 2011). Such currents can be divided into two classes: restorative currents (e.g., the hyperpolarization-activated h-type current) that actively counteract changes of the membrane potential, and regenerative currents (e.g., the persistent sodium current) that amplify membrane potential changes. Restorative currents decrease dendritic delays, whereas regenerative currents increase delays. As a consequence, the transition from in-phase to anti-phase locking occurs at a much smaller value of L ($\sim$3.2) in dendrites with regenerative currents (red curve) than in passive dendrites. In contrast, a restorative current (cyan curve) can stabilize the in-phase locking.

Interestingly, the transition from in-phase to anti-phase locking also has great consequences for the phase locking time constant (Fig. 21.4b, bottom panel). In the cable with a restorative current (which does not show such transitions) the time constant simply increases exponentially with L. However, the time constant goes to infinity at parameter values where the stability transitions occur (e.g., at $L \sim 3.2$ in the regenerative cable).

The oscillation frequency is also a key parameter determining the transitions between in-phase and anti-phase locking, since it is really the dendritic delay relative to the oscillation period that is the critical factor. Hence, when increasing the frequency of the dendritic oscillators, a critical value will be reached where a transition in the stability of the phase-locked solution occurs (Fig. 21.4c, top panel). Consequently, the phase locking time constant can become very large at realistic values of L when the oscillation frequency is much higher than the theta frequency that we considered for the grid cell model (Fig. 21.4c, bottom panel). Note that the oscillation frequency will also affect the dynamics of the voltage-dependent dendritic currents, which will modulate the oscillator interactions less and less as the frequency increases, such that active dendritic cables behave as passive ones for high frequencies.

In summary, dendritic conduction delays that are sufficiently large relative to the oscillation period can lead to anti-phase locking of dendritic oscillators. The dendritic oscillators can be independent at the transition point, thereby avoiding the trade-off between dendritic democracy and independence.

21.3 Discussion

We report mathematical analyses and numerical simulations of interacting dendritic oscillations. Intrinsic subthreshold membrane potential oscillations have been demonstrated in various types of neurons: in stellate cells from entorhinal cortex layer 2 (Alonso and Llinás, 1989; Alonso and Klink, 1993), neurons from the frontal cortex (Gutfreund et al., 1995), neurons from the amygdala complex (Pape et al., 1998; Sanhueza and Bacigalupo, 2005), and pyramidal cells and interneurons from the hippocampal CA1 area (Leung and Yim, 1991; Chapman and Lacaille, 1999). Our results suggest that in such an oscillatory regime, there is a trade-off between dendritic "democracy" (Häusser, 2001), expressing how much each oscillator can influence the somatic membrane potential and hence the spiking output of the cell, and dendritic "independence," that is how much each oscillator can integrate its inputs independently of the other oscillators. This is because the same electrotonic coupling that is necessary for dendritic signals to reach the soma also promotes phase locking of dendritic oscillations.

We find that the time constant of phase locking of dendritic oscillators in biophysically realistic regimes can be on the order of hundreds of milliseconds. This defines two different modes of operation for multiple dendritic oscillations. On time scales shorter than that of phase locking, inputs are integrated in each oscillator locally and independently, and somatic firing is determined by their joint effect. Importantly, this integration can still take place on a time scale that is considerably longer than that suggested simply by membrane time constants. Once phase locking occurs, it causes cells to act as single oscillators. In this mode, synaptic inputs throughout the dendritic tree are integrated in the phase of this single "global" oscillator, which in turn determines somatic firing. Thus, the main difference between local (shorter time scale) and global (longer time scale) dendritic integration of inputs is in the way dendritic nonlinearities and summation act on incoming signals. This difference closely parallels that found between traditional (McCulloch and Pit ts, 1943; Rosenblatt, 1958; Hopfield, 1982; Rumelhart and McClelland, 1986) and more recent accounts (Poirazi et al., 2003) of near instantaneous dendritic processing, treating the dendritic tree as a single global computational unit, or as a "network" of multiple local computational units, respectively (Fig. 21.5).

Our results also predict neuronal morphologies that promote independence of dendritic oscillators by slowing down their phase locking. First, a key parameter is the electrotonic distance between the dendritic oscillators, typically related to the spatial extent of the neuron. The larger the electrotonic separation between the oscillators, the weaker their interactions and the slower the phase locking. Second, when the oscillation generating currents are present over a large stretch of dendritic membrane, more current will be needed to shift the phase of this "large" oscillator and, hence, such an oscillator will phase lock more slowly than a "small" oscillator (i.e., having weak currents or small membrane area). Tufted dendritic terminal branches seem particularly well suited for creating large but electrotonically separated oscillators. This is because each tuft can contain multiple branches, thus creating an oscillator with a large total surface area, but different tufts can be placed at the ends of different dendrites thus ensuring the separation of their oscillators. Certain heavily branching cells, such as Purkinje neurons, could exploit such an arrangement to slow down phase locking of dendritic oscillations.

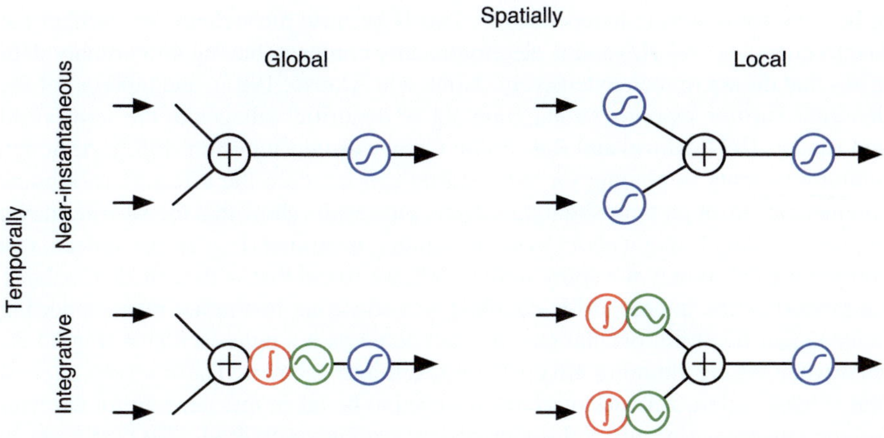

Fig. 21.5 Near-instantaneous versus integrative modes of dendritic operation. During near-instantaneous processing of inputs the net signal to the soma depends only on the current level of the inputs driving different dendritic subunits. Classical neural network models assumed that dendritic signals are first summed globally and then passed through a nonlinearity (*blue*) that determines the firing rate of the cell (*left*; McCulloch and Pitts, 1943). More recent results indicate that dendritic subunits perform local nonlinear operations before their signals are summed at the soma (*right*; Poirazi et al., 2003). When inputs are integrated by dendritic oscillations, somatic voltage depends on the history of the inputs. In particular, the information in the inputs is integrated by the oscillation phase. Beyond the time scale on which the dendritic oscillators phase lock, the dendritic tree acts as a single global oscillator integrating all inputs in its phase (*red*, *left*). The somatic membrane potential is a (sinusoid-like) nonlinear function of the phase of this global oscillator (*green*). Below the time scale of phase locking, each dendritic subunit integrates its inputs locally before the dendritic signals are summed at the soma (*right*). Reproduced from Remme et al. (2010) with permission from Elsevier

Our results showed that besides the morphology, also the oscillation frequency is a central parameter determining the phase locking dynamics of dendritic oscillators. Not only do higher oscillation frequencies undergo stronger attenuation, therefore leading to slower phase locking, but the oscillation frequency is also important for the stable phase difference that the oscillators move towards. For low frequencies, phase locking typically leads to synchronization of the oscillators, but as oscillation frequency increases, a point will be reached where the synchronized state loses stability and the oscillators lock in anti-phase. When oscillators reach this state, the oscillation amplitude will strongly decrease along the dendrites connecting the oscillators, and even be invisible exactly halfway between the oscillators. Importantly, at the oscillation frequency where the stability of the phase-locked state switches, the phase locking time constant becomes very large, yielding a possibility for the coexistence of dendritic independence and democracy.

The oscillation frequency and electrotonic structure of cortical stellate cells—the cell type thought to produce the grid field activity—are quite unlike that predicted to be ideal for dendritic independence. This is because the stellate cell oscillates at low frequencies (~8 Hz) and is electrotonically compact, having 4–6 primary dendrites that do not branch extensively (Klink and Alonso, 1997), and tapering of the dendrites further supports strong coupling of dendritic voltages to the soma (Rall and Rinzel, 1973; Rinzel and Rall, 1974; Carnevale and Johnston, 1982). Although inhibitory inputs impinging on the dendrite can increase the effective membrane conductance through their shunting effects, our results show that the slowing-down of phase locking brought about by such shunting is limited (Fig. 21.2c). Indeed, in a compartmental model of a spiny stellate cell, we found that with realistic biophysical properties the interdendritic coupling was so strong that rather than supporting independent dendritic oscillations, the cell acted as a single oscillator (Fig. 21.3), even when strong shunting effects were taken into account. These results suggest that at least certain network mechanisms need to be taken into account for explaining the emergence of grid fields (for review see Giocomo et al., 2011). Recent in vivo intracellular recordings from entorhinal cortex cells in navigating mice (in virtual-reality environments) have demonstrated slow, large depolarizing ramps during traversals of firing fields, supporting attractor network mechanisms rather than oscillator interference mechanisms to underly the spatial tuning of these cells (Schmidt-Hieber and Häusser, 2013; Domnisoru et al., 2013).

The dendritic democracy-independence trade-off we identified is unique to ongoing dendritic processing, such as that achieved by oscillations: if inputs are integrated only on short time scales, then the dendritic processing is essentially over by the time different dendritic branches would start interacting. Previous studies of dendritic integration were focusing on such near-instantaneous transformations of dendritic inputs into somatic outputs and thus did not address this issue (Poirazi et al., 2003). More generally, the contributions of active ionic conductances to dendritic processing have almost exclusively been studied in the context of near-instantaneous processing (Wei et al., 2001; Gasparini et al., 2004; Losonczy and Magee, 2006; Nevian et al., 2007; Larkum et al., 2009). The analysis presented here represents the first step towards understanding dendritic computation in another and hitherto scarcely studied dynamical regime in which active processes may play a crucial role, that of ongoing dendritic oscillations, and identifies computationally relevant features that are unique to this mode of operation.

Appendix

A.1 Derivation of Phase Locking Time Constant and Stable Phase-Locked Solution

We performed a mathematical analysis to determine the dynamics of two subthreshold dendritic oscillators. We determined how fast phase locking of dendritic oscillators occurs as a function of the oscillator properties and the properties of the membrane segment connecting the oscillators, and what is the phase difference ϕ between the oscillators in the stable phase-locked solution. Consider a system of two identical oscillators with natural frequency f (in Hz) that are coupled via a cable of length l (in cm), with oscillator $i = 1,2$ located at $x = 0$ and $x = l$, respectively (see Figs. 21.2a and 21.4a). The membrane potential $V_i(t)$ (in millivolts) of each dendritic oscillator is described by a sinusoidal function:

$$V_i(t) = \bar{V}_{\text{dend}} \cos(\omega(t + \theta_i)) + V_{\text{R}}, \tag{21.1}$$

with dendritic oscillator amplitude $\bar{V}_{\text{dend}}$ (in millivolts), angular frequency $\omega = 2\pi f$, phase shift θ_i, and resting membrane potential V_{R}. We consider the oscillators are weakly coupled (i.e., the interactions only affect the oscillators' phases). We can then write the changes in the phase shifts of the oscillators as

$$\dot{\theta}_i = \varepsilon Z(t) p_i(t), \tag{21.2}$$

where the positive parameter $\varepsilon \ll 1$, and $Z(t)$ is the infinitesimal PRC which is assumed to be identical for both oscillators. It describes the change of the oscillator's phase shift in response to an infinitesimally small and short perturbation at a particular phase (Izhikevich, 2007). Here we consider $Z(t) = -Q\sin(\omega t)$, where Q is the amplitude of the PRC (in seconds per millivolt). Note that when $Q = 1/\omega\bar{V}_{\text{dend}}$ we obtain the PRC of Andronov–Hopf oscillators, the minimal dynamical system to produce the sinusoidal limit cycle oscillations in (21.1). The perturbations $p_i(t)$ result from the axial currents that flow between the cable and oscillator i. The passive properties of the cable are determined by membrane time constant τ (in seconds), leak reversal potential E_{L} (in millivolts), and length constant λ (in centimeters), giving the cable an electrotonic length $L = l/\lambda$. The cable also expresses a voltage-dependent conductance with reversal potential E_{m}. The dynamics of this conductance are determined by a single gating variable $m(x,t)$ with activation function $m_\infty(V)$ and time constant $\tau_{\text{m}}(V)$. The equations governing the membrane potential $V(x,t)$ and gating variable $m(x,t)$ along the cable are

$$\begin{aligned}
\tau \frac{\partial}{\partial t} V(x,t) &= \lambda^2 \frac{\partial^2}{\partial x^2} V(x,t) - (V(x,t) - E_{\text{L}}) - \gamma_{\text{m}}\, m(x,t)(V(x,t) - E_{\text{m}}), \\
\tau_{\text{m}}(V(x,t)) \frac{\partial}{\partial t} m(x,t) &= m_\infty(V(x,t)) - m(x,t),
\end{aligned} \tag{21.3}$$

where γ_{m} is the ratio of the maximal conductance of the active current to the passive membrane conductance. In order to determine the perturbations to the oscillators we need to solve (21.3) with the oscillators at the ends of the cable giving the periodically forced end conditions of the cable. For this, we first linearize (21.3) about the membrane voltage V_{R} around which the membrane potential oscillates, leading to the quasi-active approximation for the cable (Sabah and Leibovic, 1969; Koch, 1984; Remme and Rinzel, 2011). We define $U(x,t)$ as the difference between the oscillating solution and the resting membrane potential V_{R}, i.e., $U(x,t) \equiv V(x,t) - V_{\text{R}}$ and we define $w(x,t)$ analogously as $w(x,t) \equiv m(x,t) - m_\infty(V_{\text{R}})$. The equations describing the quasi-active cable read

$$\tau \frac{\partial}{\partial t} U(x,t) = \lambda^2 \frac{\partial^2}{\partial x^2} U(x,t) - \gamma_R U(x,t) - \gamma_m (V_R - E_m) w(x,t),$$
$$\tau_m \frac{\partial}{\partial t} w(x,t) = \frac{\partial}{\partial V} m_\infty (V_R) U(x,t) - w(x,t), \tag{21.4}$$

where $\gamma_R = 1 + \gamma_m m_\infty (V_R)$ and $\tau_m = \tau_m(V_R)$. Using the oscillators from (21.1) as the periodically forced end conditions for (21.4) we can write the solution as

$$U(x,t) = \bar{V}_{dend} \cdot \text{Re}\left[\frac{\sinh(b(L - x/\lambda))}{\sinh(bL)} e^{i\omega(t+\theta_1)} + \frac{\sinh(bx/\lambda)}{\sinh(bL)} e^{i\omega(t+\theta_2)} \right], \tag{21.5}$$

where Re[z] is the real part of the complex number z and where

$$b = \sqrt{\gamma_R + \frac{\mu}{1 + (\omega\tau_m)^2} + i\omega\left(\tau - \frac{\mu\tau_m}{1 + (\omega\tau_m)^2}\right)}, \tag{21.6}$$

with

$$\mu = \gamma_m (V_R - E_m) \frac{\partial}{\partial V} m_\infty (V_R). \tag{21.7}$$

Note that the sign of μ determines whether the active current is restorative ($\mu > 0$) or regenerative ($\mu < 0$; the current is passive when $\mu = 0$). We can now show that the perturbation to oscillator $i = 1$ reads

$$p_1(t) = \frac{\partial}{\partial x} U(0,t) = \frac{\bar{V}_{dend}}{\lambda} \times \text{Re}\left[\frac{b}{\sinh(bL)} e^{i\omega(t+\theta_2)} - \coth(bL) e^{i\omega(t+\theta_1)} \right]. \tag{21.8}$$

We want to describe the evolution of the phase difference $\phi(t) = \theta_2(t) - \theta_1(t)$. For this we first need to determine the phase interaction function $H_i(\phi)$ that describes the average effect of perturbation $p_i(t)$ on the phase of oscillator i over a cycle of period $T = 2\pi / \omega$. For oscillator $i = 1$ this interaction function reads

$$H_1(\phi) = \frac{1}{T} \int_0^T Z(t) p_1(t + \phi) dt = \frac{Q\bar{V}_{dend}}{2\lambda} \rho \sin(\omega\phi + \xi) + \upsilon, \tag{21.9}$$

where

$$\rho = \left| \frac{b}{\sin \text{h}(bL)} \right|, \tag{21.10}$$

$$\xi = \arg\left[\frac{b}{\sinh(bL)} \right], \tag{21.11}$$

and υ is a constant, and where $|z|$ and arg[z] are the absolute value and the angle of the complex number z, respectively. Since we consider two identical oscillators the interaction function $H_2(\phi) = H_1(-\phi)$. We now obtain an equation describing the evolution of the phase difference between the two oscillators:

$$\dot{\phi} = \dot{\theta}_2 - \dot{\theta}_1 = \varepsilon(H_1(-\phi) - H_1(\phi)) = -\varepsilon \frac{Q\bar{V}_{\text{dend}}}{\lambda} \rho \cos(\xi) \sin(\omega\phi) \qquad (21.12)$$

The fixed points of this differential equation (i.e., the points where $\dot{\phi} = 0$) are $\phi = k \cdot T / 2$, where k is an integer. The stable fixed points are those points where $d\dot{\phi} / d\phi < 0$. The synchronous solution $\phi = 0$ is thus stable when $\cos(\xi) > 0$. When this solution is stable the anti-phase solution $\phi = T / 2$ is unstable and vice versa.

Close to the stable fixed point of (21.12), the term $\sin(\omega\ \phi)$ is approximately linear and the phase locking time constant can be written as

$$\tau_{\text{lock}} = \frac{\lambda}{\varepsilon\omega Q\bar{V}_{\text{dend}}} \frac{1}{\rho\,|\cos\xi\,|}. \qquad (21.13)$$

The parameter ε can be described in terms of cable and oscillator parameters:

$$\varepsilon = \frac{\pi d^2}{4R_{\text{i}}C_{\text{m}}a} = \frac{\pi d\lambda^2}{a\tau} \qquad (21.14)$$

where R_{i} is the intracellular resistivity (in Ωcm), a is the surface area of one oscillator (in cm^2), and C_{m} is the specific membrane capacitance (in μF/cm^2). The second equality uses $\lambda = \sqrt{d / 4R_{\text{i}}g_{\text{L}}}$ and $\tau = C_{\text{m}} / g_{\text{L}}$, where g_{L} is the specific membrane conductance (in S/cm^2). Hence we can write the phase locking time constant as

$$\tau_{\text{lock}} = \frac{a\tau}{\pi d\lambda\omega Q\bar{V}_{\text{dend}}} \frac{1}{\rho|\cos\xi|}. \qquad (21.15)$$

A.2 *Amplitude of the Voltage Oscillations Along the Cable*

Using (21.5) we can easily obtain explicit expressions for the amplitude of the voltage oscillations along the cable as a function of the cable parameters. For synchronized dendritic oscillators (i.e., $\phi = 0$) the oscillation amplitude $\bar{V}(x)$ at any point x along the cable of length l is

$$\bar{V}(x) = \bar{V}_{\text{dend}} \left| \frac{\sinh(b(L - x/\lambda)) + \sinh(bx/\lambda)}{\sinh(bL)} \right|. \qquad (21.16)$$

Considering the voltage at the middle of the cable as the "somatic" voltage $V_{\text{soma}}(t) = U(l/2, t)$, we find the somatic oscillation amplitudes for synchronized oscillators:

$$\bar{V}_{\text{soma}} = \bar{V}\left(\frac{l}{2}\right) = \frac{\bar{V}_{\text{dend}}}{|\cosh(bL/2)|}. \qquad (21.17)$$

Acknowledgments This work was supported by a Marie Curie Team of Excellence Grant (BIND MEXT-CT-2005-024831; MR, BSG), the Gatsby Charitable Foundation (ML), Alliance France-UK collaborative grant (MR, ML, BSG), the CNRS (BSG), the Wellcome Trust (ML), German Federal Ministry of Education and Research (No. 01GQ0972; MR), and the Einstein Foundation Berlin (MR).

References

Alonso AA, Klink R (1993) Differential electroresponsiveness of stellate and pyramidal-like cells of medial entorhinal cortex layer II. J Neurophysiol 70(1):128–143

Alonso AA, Llinás RR (1989) Subthreshold Na^+-dependent theta-like rhythmicity in stellate cells of entorhinal cortex layer II. Nature 342(6246):175–177

Burgess N, Barry C, O'Keefe J (2007) An oscillatory interference model of grid cell firing. Hippocampus 17(9):801–812

Carnevale NT, Johnston D (1982) Electrophysiological characterization of remote chemical synapses. J Neurophysiol 47(4):606–621

Chapman CA, Lacaille JC (1999) Intrinsic theta-frequency membrane potential oscillations in hippocampal CA1 interneurons of stratum lacunosum-moleculare. J Neurophysiol 81(3): 1296–1307

Dickson CT, Magistretti J, Shalinsky MH, Fransén E, Hasselmo ME, Alonso AA (2000) Properties and role of I_h in the pacing of subthreshold oscillations in entorhinal cortex layer II neurons. J Neurophysiol 83(5):2562–2579

Domnisoru C, Kinkhabwala AA, Tank DW (2013) Membrane potential dynamics of grid cells. Nature 495(7440):199–204

Fransén E, Alonso AA, Dickson CT, Magistretti J, Hasselmo ME (2004) Ionic mechanisms in the generation of subthreshold oscillations and action potential clustering in entorhinal layer II stellate neurons. Hippocampus 14(3):368–384

Gasparini S, Migliore M, Magee JC (2004) On the initiation and propagation of dendritic spikes in CA1 pyramidal neurons. J Neurosci 24(49):11046–11056

Giocomo L, Zilli E, Fransén E, Hasselmo M (2007) Temporal frequency of subthreshold oscillations scales with entorhinal grid cell field spacing. Science 315(5819):1719–1722

Giocomo LM, Moser MB, Moser EI (2011) Computational models of grid cells. Neuron 71(4):589–603

Golding NL, Spruston N (1998) Dendritic sodium spikes are variable triggers of axonal action potentials in hippocampal CA1 pyramidal neurons. Neuron 21(5):1189–1200

Gutfreund Y, Yarom Y, Segev I (1995) Subthreshold oscillations and resonant frequency in guinea-pig cortical neurons: physiology and modelling. J Physiol 483:621–640

Hafting T, Fyhn M, Molden S, Moser MB, Moser EI (2005) Microstructure of a spatial map in the entorhinal cortex. Nature 436(7052):801–806

Hafting T, Fyhn M, Bonnevie T, Moser MB, Moser EI (2008) Hippocampus-independent phase precession in entorhinal grid cells. Nature 453(7199):1248–1252

Hasselmo ME (2007) Arc length coding by interference of theta frequency oscillations may underlie context-dependent hippocampal unit data and episodic memory function. Learn Mem 14(11):782–794

Häusser M (2001) Synaptic function: dendritic democracy. Curr Biol 11(1):R10–R12

Hopfield JJ (1982) Neural networks and physical systems with emergent collective computational abilities. Proc Natl Acad Sci 79(8):2554–2558

Huhn Z, Orbán G, Érdi P, Lengyel M (2005) Theta oscillation-coupled dendritic spiking integrates inputs on a long time scale. Hippocampus 15(7):950–962

Izhikevich E (2007) Dynamical systems in neuroscience: the geometry of excitability and bursting. MIT, Cambridge

Klink R, Alonso AA (1997) Morphological characteristics of layer II projection neurons in the rat medial entorhinal cortex. Hippocampus 7(5):571–583

Koch C (1984) Cable theory in neurons with active, linearized membranes. Biol Cybern 50(1):15–33

Larkum ME, Nevian T, Sandler M, Polsky A, Schiller J (2009) Synaptic integration in tuft dendrites of layer 5 pyramidal neurons: a new unifying principle. Science 325(5941):756–760

Lengyel M, Szatmáry Z, Érdi P (2003) Dynamically detuned oscillations account for the coupled rate and temporal code of place cell firing. Hippocampus 13(6):700–714

Leung LW, Yim CY (1991) Intrinsic membrane potential oscillations in hippocampal neurons *in vitro*. Brain Res 553(2):261–274

Losonczy A, Magee JC (2006) Integrative properties of radial oblique dendrites in hippocampal CA1 pyramidal neurons. Neuron 50(2):291–307

Magee JC, Cook EP (2000) Somatic EPSP amplitude is independent of synapse location in hippocampal pyramidal neurons. Nat Neurosci 3(9):895–903

McCulloch WS, Pitts W (1943) A logical calculus of the ideas immanent in nervous activity. Bull Math Biophys 5:155–133

McNaughton BL, Battaglia FP, Jensen O, Moser EI, Moser MB (2006) Path integration and the neural basis of the 'cognitive map'. Nat Rev Neurosci 7(8):663–678

Nevian T, Larkum ME, Polsky A, Schiller J (2007) Properties of basal dendrites of layer 5 pyramidal neurons: a direct patch-clamp recording study. Nat Neurosci 10(2):206–214

O'Keefe J, Burgess N (2005) Dual phase and rate coding in hippocampal place cells: theoretical significance and relationship to entorhinal grid cells. Hippocampus 15(7):853–866

Pape HC, Paré D, Driesang RB (1998) Two types of intrinsic oscillations in neurons of the lateral and basolateral nuclei of the amygdala. J Neurophysiol 79(1):205–216

Poirazi P, Brannon T, Mel BW (2003) Pyramidal neuron as two-layer neural network. Neuron 37(6):989–999

Polsky A, Mel BW, Schiller J (2004) Computational subunits in thin dendrites of pyramidal cells. Nat Neurosci 7(6):621–627

Rall W, Rinzel J (1973) Branch input resistance and steady attenuation for input to one branch of a dendritic neuron model. Biophys J 13(7):648–687

Remme MWH, Rinzel J (2011) Role of active dendritic conductances in subthreshold input integration. J Comput Neurosci 31(1):13–30

Remme MWH, Lengyel M, Gutkin BS (2009) The role of ongoing dendritic oscillations in single-neuron dynamics. PLoS Comput Biol 5(9):e1000493

Remme MWH, Lengyel M, Gutkin BS (2010) Democracy-independence trade-off in oscillating dendrites and its implications for grid cells. Neuron 66(3):429–437

Rinzel J, Rall W (1974) Transient response in a dendritic neuron model for current injected at one branch. Biophys J 14(10):759–790

Rosenblatt F (1958) The perceptron: a probabilistic model for information storage and organization in the brain. Psychol Rev 65(6):386–408

Rotstein HG, Oppermann T, White JA, Kopell NJ (2006) The dynamic structure underlying subthreshold oscillatory activity and the onset of spikes in a model of medial entorhinal cortex stellate cells. J Comput Neurosci 21(3):271–292

Rudolph M, Destexhe A (2003) A fast-conducting, stochastic integrative mode for neocortical neurons in vivo. J Neurosci 23(6):2466–2476

Rumelhart DE, McClelland JL (eds) (1986) Parallel distributed processing: explorations in the microstructure of cognition. MIT, Cambridge

Sabah NH, Leibovic KN (1969) Subthreshold oscillatory responses of the Hodgkin–Huxley cable model for the squid giant axon. Biophys J 9(10):1206–1222

Sanhueza M, Bacigalupo J (2005) Intrinsic subthreshold oscillations of the membrane potential in pyramidal neurons of the olfactory amygdala. Eur J Neurosci 22(7):1618–1626

Sargolini F, Fyhn M, Hafting T, McNaughton BL, Witter MP, Moser MB, Moser EI (2006) Conjunctive representation of position, direction, and velocity in entorhinal cortex. Science 312(5774):758–762

Schmidt-Hieber C, Häusser M (2013) Cellular mechanisms of spatial navigation in the medial entorhinal cortex. Nat Neurosci 16(3):325–331

Stuart G, Spruston N, Häusser M (2007) Dendrites, 2nd edn. Oxford University Press, New York

Wei DS, Mei YA, Bagal A, Kao JP, Thompson SM, Tang CM (2001) Compartmentalized and binary behavior of terminal dendrites in hippocampal pyramidal neurons. Science 293(5538):2272–2275

Chapter 22
Dendrites Enhance Both Single Neuron and Network Computation

Romain D. Cazé, Mark D. Humphries, and Boris S. Gutkin

Abstract In a single dendritic branch of a neuron, multiple excitatory inputs can locally interact in a nonlinear fashion. The local sum of multiple excitatory post-synaptic potentials (EPSPs) can be inferior or superior to their arithmetic sum; in these cases summation is respectively sublinear or supralinear. While this experimental observation can be well explained by conductance-based models, the computational impact of these local nonlinearities remains to be elucidated. Are there any examples of computation that are only possible with nonlinear dendrites? What is the impact of nonlinear dendrites at the network scale? We show here that both supralinear summation and sublinear summation enhance single neuron computation. We use Boolean functions, whose input and output consists of zeros and ones, and demonstrate that a few local dendritic nonlinearities allow a single neuron to compute new functions like the well-known exclusive OR (XOR). Furthermore, we show that these new computational capacities help resolve two problems faced by network composed of linearly integrating units. Certain functions require (1) that at least one unit in the network have an arbitrarily large receptive field and (2) that the range of synaptic weights be large. This chapter demonstrates that both of these limitations can be overcome in a network of nonlinearly integrating units.

R.D. Cazé
INSERM U960, École Normale Supérieure, 29 rue d'Ulm, 75005 Paris, France

Paris Diderot 7, 5 rue Thomas Mann, 75013 Paris, France
e-mail: romain.caze@ens.fr

M.D. Humphries
Faculty of Life Sciences, University of Manchester, Manchester M60 1QD, UK

B.S. Gutkin (✉)
INSERM U960, École Normale Supérieure, 29 rue d'Ulm, 75005 Paris, France
e-mail: boris.gutkin@ens.fr

H. Cuntz et al. (eds.), *The Computing Dendrite: From Structure to Function*, Springer Series in Computational Neuroscience 11, DOI 10.1007/978-1-4614-8094-5_22,

> It would be better to think of the brain not as a single network whose elements operate in accord with a uniform set of principles but as a network *whose components are themselves networks having a large variety of different architectures* and control systems.
>
> (Minsky and Papert, 1988)

22.1 Introduction

In a single dendrite, the *measured* excitatory post-synaptic potential (EPSP) resulting from multiple stimulations can be different from the *expected* algebraic sum of the EPSPs resulting if each stimulation arrives independently, as shown in Fig. 22.1a.

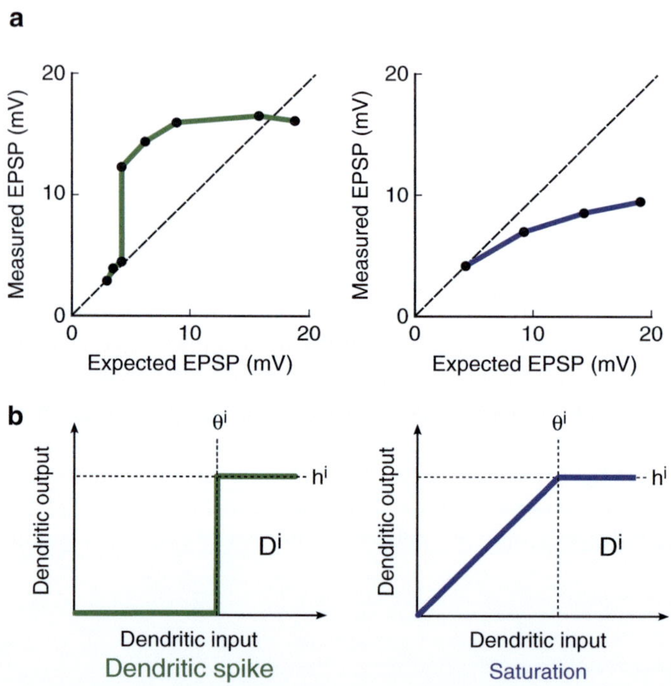

Fig. 22.1 Spiking and saturating dendrites. (**a**) The *x*-axis (Expected EPSP) is the arithmetic sum of two EPSPs induced by two separate stimulations and the *y*-axis (Measured EPSP) is the measured EPSP when the stimulations are simultaneous; the *dashed line* corresponds to linear summation. *Left*, observations from cortical pyramidal neurons (redrawn from Polsky et al., 2004). Summation is both supralinear and sublinear due to the occurrence of a dendritic spike. *Right*, observations from cerebellar interneurons (redrawn from Abrahamsson et al., 2012). In this case summation is purely sublinear due to a saturation caused by a reduction in driving force. (**b**) The activation function of a dendritic subunit D_i (*left*) modeling dendritic spike-type nonlinear summation: both supralinear and sublinear on $[0,\infty]$, or (*right*) the saturation type nonlinear summation: strictly sublinear on $[0,\infty]$. This function has two parameters: a threshold Θ_i and a height h_i. These figures are copied from Cazé et al. (2013) (under CCAL)

Within a dendrite, summation can either be supralinear, sublinear, or both. Supralinear local summation of EPSPs can be due to the opening of NMDA, Ca^{2+}, or Na^{+} voltage-gated channels (Wei et al., 2001; Ariav et al., 2003; Nevian et al., 2007; Polsky et al., 2004; Losonczy and Magee, 2006; Polsky et al., 2009), which cause a sharp deflection in the local membrane voltage called a dendritic spike. Sublinear local summation of EPSPs is most often due to a reduction in driving force: i.e., a reduction in the potential difference between inside and outside the membrane which drives the ionic current (Haag et al., 1992; Heck, 2003; Abraham sson et al., 2012). These two experimental findings can be explained using conductance-based models (see Rall, 1967; Koch, 2005 for details). Nevertheless, the impact of the nonlinear summation of EPSPs on single neuron computation remains to be elucidated. This question requires the use of a different kind of model, i.e., binary neuron models. In this chapter, we use such a model to demonstrate that local nonlinear summation of EPSPs enhances the computational capacity of both single neurons and networks.

The first binary models of neurons by McCulloch and Pitts (1943) and Rosenblatt (1958) assumed linear integration of excitatory inputs, and these models have informed much subsequent thinking about how neurons compute, both individually and collectively. The limitations of linear integration models—or threshold linear units (TLU)—are well known: TLUs cannot compute linearly nonseparable functions (see example just below, and Results section for a formal definition). By contrast, we show that nonlinear summation of EPSPs enables a binary neuron model, schematized in Fig. 22.2a, to compute linearly non-separable functions.

We show in the first part of this chapter that a binary neuron model with a single nonlinear dendrite can compute a well-known example of linearly non-separable function: the famous exclusive OR (XOR). Suppose that a neuron has two sources of presynaptic inputs: this neuron computes the XOR if it fires when only one of the input sources is active, but remains silent if no or both sources are active simultaneously. The truth table of an XOR and its implementation are shown in Fig. 22.2b. The reason why a TLU cannot compute XOR is an intuitive one. A neuron that computes XOR needs to fire when at least one of two excitatory input sources (i.e., sets of excitatory synapses) is active, but it must stay silent when both sets are active at once. This is impossible when the EPSPs add up linearly, but when excitatory inputs interact nonlinearly then it becomes possible (Zador et al., 1993). This last theoretical study demonstrates that a neuron can compute XOR because activity in two excitatory input sources can depolarize the soma less than activity in a single excitatory input source, due to the local activation of potassium voltage-gated channels. In this chapter, we propose a way to implement this function using classic inhibition and dendritic saturation. In this case, this capacity is due to the local nonlinear summation which had been theoretically conjectured by Koch et al. (1982) and has since been experimentally demonstrated by Polsky et al. (2009). Local summation means that the dendritic tree can be segmented into independent subunits, meaning, for instance, that two EPSPs generated in two different dendritic branches sum linearly and that two EPSPs generated on the same branch sum nonlinearly,

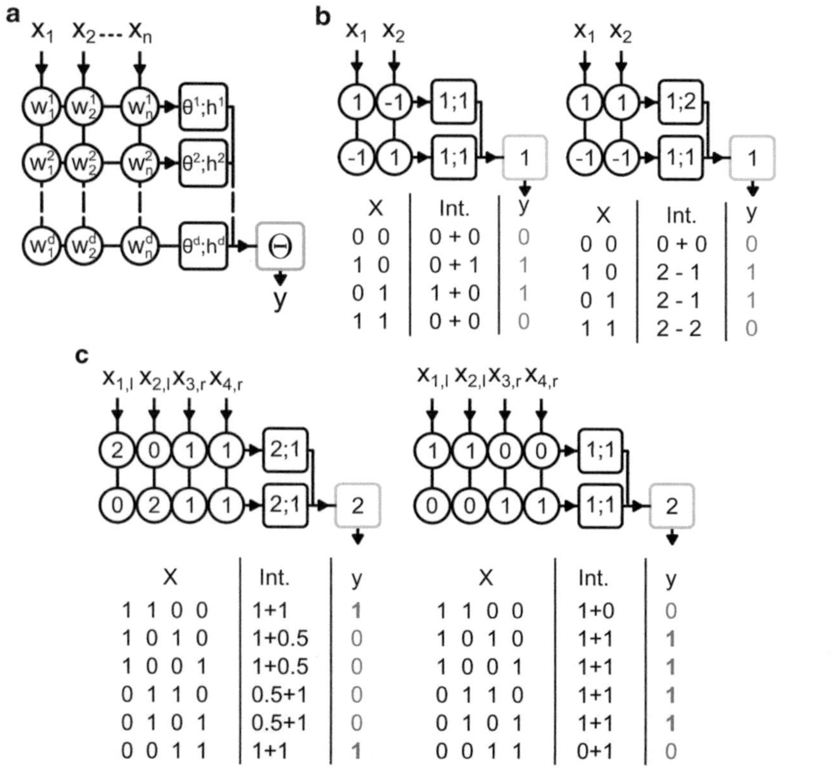

Fig. 22.2 The two-stage neuron model and its computing capacity. (**a**) is a general definition of this model, (**b**) and (**c**) are specific implementations of linearly non-separable functions. (**a–c**) Each x_i is a binary variable, corresponding to the activity of the cell assembly i (0 = inactive; 1 = active). y is a binary variable, representing post-synaptic activity. In *circles* are independent sets of integer-valued synaptic weights, corresponding to the amplitude of the EPSP or the IPSP recorded locally when a presynaptic source is active. In *black squares*, θ^j and h^j are the integer-valued thresholds and heights which are the parameters of the activation function D^j of dendrites j—this function is either spiking (Fig. 22.1b, left) or saturating (Fig. 22.1b, right). In the *gray square*, Θ is a positive integer-valued threshold determining whether the post-synaptic neuron is active or silent. (**b**) Implementation of the XOR function: the lower part is the truth table. The *numbers* in the second column are the result of dendritic integration. The two parts of the sum in the second column correspond to the activation states of the two dendritic subunits. *Left*, an implementation where the dendritic subunit can only be of the spiking type. *Right*, an implementation where the dendritic subunit can either be spiking or saturating. (**c**) *Left*, the implementation of the object-feature binding problem (FBP). *Right*, the implementation of the dual version (dFBP). Note that either spiking or saturating dendritic activation functions can be used for both implementations

e.g., by inducing a dendritic spike (Polsky et al., 2009) or a dendritic saturation (Heck, 2003). This capacity enables a single neuron to compute linearly non-separable functions. To conclude this first part, we refer to our previous publications (Cazé et al., 2012, 2013) which introduce a set of linearly non-separable functions computable with a small number of dendritic subunits.

We go on to discuss the impact of such new computational capacities on *networks* of neurons possessing nonlinear dendrites. Our opening quotation is taken from the edition of Perceptron published after the Parallel Distributed Processing book (McClelland et al., 1986). This manifesto of connectionism argued that much of what at first sight seem to be bafflingly complex learning and memory processes in cognition can be explained by layered networks of simple interconnected units such as TLUs. Nonetheless, networks of TLUs may, in practice, be unable to implement certain functions (Minsky and Papert, 1988). Why? Minsky and Papert's book demonstrated at least two limitations of networks of TLUs. First, a function can require a unit with an arbitrarily large receptive field. For instance, the parity function, a generalization of the XOR operation, requires at least one unit of the network with a receptive field covering the whole input space. Second, the implementation of a function can require synapses with a huge number of distinct synaptic weight values. Our quotation from their book proposes a way to overcome these two limitations: the use of networks made up of more complex units. This proposition fits with the recent observation that single neurons are computationally complex entities capable of linearly non-separable computations. In the second part of this chapter, we demonstrate that a network made up of complex neurons with small receptive fields can implement the parity function. Finally, we show that dendritic nonlinearities make it possible for intrinsic parameters to store information, reducing the range of synaptic weight values needed to implement the parity function.

22.2 Impact of Nonlinear Summation in Dendrites of Single Neurons

In this section, we present a simple neuron model that takes into account nonlinear dendritic summation of EPSPs. These nonlinearities are due to voltage-gated channels (Schiller et al., 2000; Polsky et al., 2004, 2009) or to the passive electrical properties of the dendritic cable, as shown in Fig. 22.1. Depending on the intensity of stimulation, the sum of multiple EPSPs can be higher or lower than their arithmetic sum—respectively supralinear and sublinear—or both (see Poirazi et al., 2003b for a first definition, and Cazé et al., 2012, 2013 for a formal definition). Moreover, a neuron's dendrites can be decomposed into a set of quasi-independent subunits based on some measure of relative electrical compartmentalization, such as between branch points (Koch et al., 1982; Hendrickson et al., 2011). Each subunit may thus independently support nonlinear integration of its inputs. The existence of so-called dendritic spikes and of independent subunits has led to the proposal that a single neuron behaves as a two-layer feed-forward neural network (Gurney, 2001a,b; Poirazi et al., 2003a).

Figure 22.2a introduces this two-layer model. We distinguish between models composed of spiking subunits, which are both supralinear and sublinear (Fig. 22.1b, left), and those composed of saturating subunits, which are strictly sublinear (Fig. 22.1b, right). In the former case, we use spiking activation functions for the dendritic subunits to model the generation of local dendritic spikes, and in the latter we use saturating activation functions to model local dendritic saturations. We call this class of models threshold nonlinear unit (TNLU), to contrast with the classical TLU.

22.2.1 Dendrites Enable the Computation of Linearly Non-separable Functions

To quantify the difference between the TLU and the TNLU, we used Boolean functions. These functions are central in the study of computation (Wegener, 1987; Muroga, 1971), and they can be defined as follows:

Definition 1. A Boolean function is a function of n input sources on $0,1^n$ into $0,1$.

In biological terms, the n input sources correspond to the activity of n assemblies of pre-synaptic neurons. 0 means that the pre-synaptic assembly is inactive and 1 means that it is active. If the post-synaptic neuron responds then the output of the Boolean function is 1, if it remains silent then the output is 0.

Using this framework, it can be demonstrated that the computational capacities of a TNLU are both quantitatively superior to and qualitatively different from those of a TLU. A TLU is limited to linearly separable functions which are defined below (Crama and Hammer, 2011; Muroga, 1971).

Definition 2. f is a *linearly separable* Boolean function if and only if there exists at least one vector W and a number Θ such that for all $f(X) = 0$ we have $W \cdot X < \Theta$ and for all $f(X) = 1$ we have $W \cdot X \geq \Theta$.

A TNLU can compute every Boolean function provided a sufficient number of spiking dendritic subunits (see Anthony, 2010, Theorem 13.9). This proof can be explained intuitively: all Boolean functions can be decomposed into a "disjunction of conjunctions," an OR of ANDs; for instance the XOR function can be expressed as $(x_1 AND (NOT x_2)) OR ((NOT x_1) AND x_2)$. A spiking dendritic subunit can implement an AND term through its supralinear summation: only some combination of inputs can elicit the subunit's output, not a single input alone. Moreover, the soma with a sufficiently low threshold can implement the OR term by firing whenever one of the spiking subunits has an output. Thus, with one spiking subunit per AND term, a TNLU can theoretically implement any Boolean function. Figure 22.2b (left) shows how the TNLU with spiking subunits can implement the XOR function.

In biological terms, these results hold only for neurons with spiking dendrites like cortical and hippocampal pyramidal neurons (Polsky et al., 2009; Katz et al., 2009). Extending the generality of this result, we have found that even when dendrites are passive, as illustrated in Fig. 22.1b, the distinction between TLUs and TNLUs remains (Cazé et al., 2012, 2013): a sufficient number of saturating subunits can implement any

Boolean function. This proof can be understood intuitively in a similar way. All Boolean functions can be expressed as a "conjunction of disjunctions," an AND of ORs; for instance, the XOR function can be expressed as $(x_1 OR x_2) AND((NOT x_1) OR (NOT x_2))$. A saturating dendritic subunit can implement an OR term through its sublinear summation: each input elicits an output, but the output does not grow with further active inputs. A soma with a sufficiently high threshold can implement the AND term by only firing when a sufficient number of saturating subunits have an output. Thus, with one saturating subunit per OR term, a TNLU can theoretically implement any Boolean function. Figure 22.2b (right) shows how the TNLU with saturating subunits can implement the XOR function.

The linearly separable functions represent a tiny proportion of all Boolean functions: as the number of input variables x_i increases the proportion of functions that are linearly separable tends to zero: if n is the number of input variables, there are at most 2^{n^2} (Hastad, 1994) linearly separable functions, whereas the total number of Boolean functions is 2^{2^n} (Wegener, 1987). Consequently, the mere existence of dendrites increases the computational capacity of a neuron beyond that bestowed by linear integration, even if the dendrites are not spiking. The newly accessible functions include the feature-binding problems, reviewed in the next section, which map onto a wide range of computations performed by vertebrate and invertebrate neural systems.

The previous formal propositions are valid for an infinite number of dendritic subunits, but the number of independent nonlinear subunits in a real neuron is finite. To show that these formal propositions are relevant to computation by real neurons we needed to answer two fundamental questions. What functions are computable with a limited number of dendritic subunits? And how a single neuron can implement these functions?

22.2.2 A Finite Number of Dendritic Subunits Expand the Computational Capacity of a Single Neuron

We used large systematic parameter searches for the TNLU model (described in "Method" section of Cazé et al., 2013) to find functions computable given a limited number of dendritic subunits. We found that, with a number of subunits linearly proportional to the number of input sources, a TNLU can compute an interesting set of linearly non-separable functions. Contrary to XOR, inhibition is not necessary for these functions; in formal terms, such functions are linearly non-separable and monotone (Shmulevich et al., 1995; Korshunov, 2002). Some of these functions are already known as object-feature binding problems (FBPs) (Legenstein and Maass, 2011). Two subsets of FBPs are particularly notable as they correspond to biological problems solved by single neurons in vertebrates (Agmon-Snir et al., 1998; Archie and Mel, 2000). One of these functions is already well known as the object-feature binding problem (FBP) (Legenstein and Maass, 2011), we call

Table 22.1 Feature-object binding problems expressed by their truth table

$x_{l,1}$	$x_{l,2}$	$x_{r,1}$	$x_{r,2}$	FBP	dFBP
1	1	0	0	1	0
1	0	1	0	0	1
1	0	0	1	0	1
0	1	1	0	0	1
0	1	0	1	0	1
0	0	1	1	1	0

The partial truth tables for the Feature Binding Problem (FBP) and its dual version (dFBP) for $n = 4$ variables. All input vectors containing fewer 1s than in this truth table yield $y = 0$; all input vectors containing more 1s than in this truth table yield $y = 1$

the other the "dual" FBP (dFBP); Table 22.1 gives the truth tables defining these functions. The FBP is a well-known problem in visual (Treisman and Gelade, 1980) and auditory perception (Hall and Wieberg, 2003; for review see Treisman, 1998; Roskies and Binding, 1999; Taraborelli, 2002; Holcombe, 2009; F eldman, 2012). An example of this problem is to correctly signal distinct objects that are each characterized by disjoint sets of features, for instance shape and color, e.g., to detect red Xs or blue Os but not blue Os or red Xs. Binocular disparity (Archie and Mel, 2000) is another example of a FBP problem: consider input in the form of four binary variables $[x_{l,1}, x_{l,2}, x_{r,1}, x_{r,2}]$, two each from the left (l) and right (r) eye. The neuron responds maximally if all inputs from a single eye are active, i.e., [1, 1, 0, 0] and [0, 0, 1, 1], but not when inputs from both eyes are partially active, e.g., [0, 1, 0, 1]. The problem that we have called the dFBP is also a binding problem, important in binaural coincidence detection (Agmon-Snir et al., 1998). Using the same notation with l and r referring to the left and right ear, this function consists in responding more to inputs from both ears simultaneously than to inputs from only one ear (Segev, 1998).

Figure 22.2c shows how these functions can be implemented in a TNLU with two nonlinear dendritic subunits. In these examples there are two objects made of two features, but one can implement the FBP or dFBP of o objects made of f features with only o dendritic subunits (Cazé et al., 2012).

22.2.3 Different Strategies for Implementing Functions in Nonlinear Dendrites

The decomposition of Boolean functions into a "disjunction of conjunctions" (called the disjunctive normal form) and "conjunction of disjunctions" (called the conjunctive normal form) provide a guaranteed method for implementing functions using a decomposition into spiking or saturating dendritic subunits. They are in turn examples of two broader classes of function implementations: a *local* strategy where a single dendritic subunit is capable of eliciting a somatic spike (like the "disjunction of conjunctions"); and a *global* strategy where no single dendritic subunit can generate a somatic spike thus somatic spiking always requires simultaneous activity in at least two dendritic subunits (like the "conjunction of disjunctions").

We have shown that using a local strategy with saturating subunits cannot implement any linearly non-separable Boolean function. By contrast, using a local strategy with spiking dendritic subunits can implement any linearly non-separable Boolean function (Cazé et al., 2012, 2013). This in turn contrasts with the global strategy, which can be used with both types of subunits to implement any linearly non-separable Boolean function. We illustrate in Fig. 22.4 how the FBP and dFBP can be computed using a global strategy. In these implementations, each dendritic subunit can reach its maximum possible level of activity and yet the neuron stays silent. The existence of a global strategy could explain why Layer 2/3 pyramidal neurons have different input tunings between their dendrites and the output of the whole neuron: inputs which elicit the largest calcium response in a local region of dendrite did not cause the neuron to fire, while inputs that caused the neuron to fire did not elicit the maximum calcium response in the dendrite (Jia et al., 2010).

22.3 Impact of Nonlinear Summation in Dendrites on a Network of Neurons

If an individual neuron (a TNLU) is computationally more powerful than a TLU, yet multi-layered networks of TLUs can implement any Boolean function, what further advantages do we gain from a network of TNLUs? In this section, we demonstrate that networks made of TNLUs are computationally more efficient than networks of TLUs; first we show that a network of TNLUs can better parallelize a function than a network of TLUs; we then show that they allow for a more flexible and larger storage capacity.

22.3.1 Dendrites Enable the Efficient Parallelization of Functions

To show that networks of TNLUs are more efficient in parallelizing certain functions we have first to understand the limitations of a network of TLUs, the topic of the famous book *Perceptrons* by Minsky and Papert (1988). A Perceptron is a two layer feed-forward network: an *input layer* made of units with unlimited computation capacity projecting to a *processing layer* which is a single TLU. Their book starts with a simple question: Why does the Perceptron fail at computing certain functions *in practice*? Indeed, provided a sufficiently large number of units or sufficiently complex units in the input layer, the Perceptron can compute every Boolean

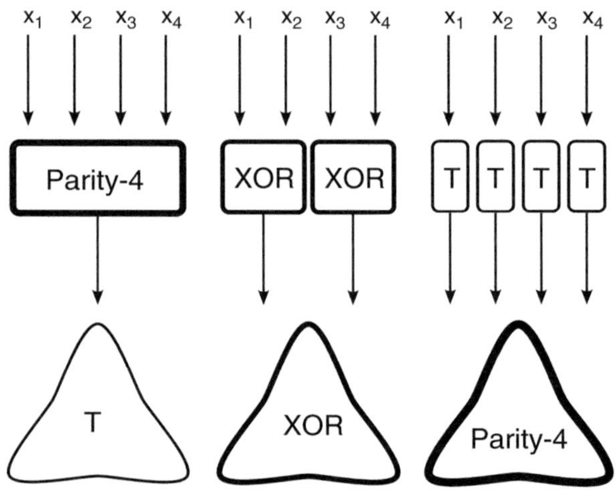

Fig. 22.3 Parallelizing the computation of the parity function. These three feed-forward networks compute the 4-parity function: they determine whether a retina made of 4 pixels contains an odd number of black pixels. In *rectangles*, the computing capacity of units in the input layer; in *rounded triangles*, the computing capacity of the unit in the processing layer. T is the identity function, 4-Parity is the parity function with 4 input variables. In the network in the middle the computational burden is distributed equally, i.e. every unit has the same receptive field and computing capacity

functions *in theory*, but *in practice* a Perceptron can fail to implement certain of these functions.

The parity function is one generalization of the XOR: it requires signalling 1 for input vectors containing an odd number of 1s and remaining silent (signalling 0) for input vectors containing an even number of 1s. Minsky and Papert found that the parity function can be implemented in a Perceptron only if it obeys the biologically implausible constraint that at least one of the units in the input layer takes as input the whole input space (see Rojas, 1996, Chap. 3 for a detailed explanation). This is true even if the input layer of a Perceptron is made of universal units, which can compute every Boolean functions—including the parity function itself (Fig. 22.3, left). To put this into perspective, if the input is a retina, the Perceptron can compute the parity function only if at least one neuron of the network has a receptive field as large as the retina. This necessary convergence of inputs creates a processing bottleneck.

What happens if the processing layer of the Perceptron is a TNLU instead of a TLU? If we assume a situation where a TNLU is capable of computing the parity function then, while the receptive field of the units in the input layer can be arbitrary small, as is shown on Fig. 22.3, the processing later still needs to receive as many inputs from those units as there are inputs. But two layers of TNLUs can remove the necessary convergence by decomposing the parity function. We saw that a TNLU is capable of computing the XOR function, as illustrated in Fig. 22.2b. If the unit of the processing layer is capable of such computation, then the parity function can be computed with further XOR units in the input layer, as shown in Fig. 22.3. Each unit receives a limited number of inputs in this network of TNLUs, whereas this type of implementation is impossible with TLUs.

The implementation of some functions can also be impossible in practise because they require a large range of synaptic weight values between the input and the processing layer. For instance the range of synaptic weight values necessary to compute the parity function can grow exponentially with the number of pixels n of the retina if the network is made of TLUs. But a network of TNLUs can implement the parity function with binary synaptic weights, in this case the implementation is easy.

22.3.2 Dendrites Expand the Information Storage Capacity of Intrinsic Parameters

Information storage is a vaporous notion, difficult to define for a neuron. This is surprising because defining memory for a computer is easy. Memory and its processing are independent in a computer based on von Neumann architecture. This means that n bits of information can be stored in n free bits of memory. Memory and its processing are interwoven in a neuron, thus memory cannot be measured directly. To overcome this difficulty there is an indirect but quantitative way to measure memory storage capacity in a neuron (used in Brunel et al., 2004; Poirazi and Mel, 2000), by estimating the number of functions a neuron can potentially compute in a given range of parameter values. Counting the number of functions a neuron can implement enables us to directly quantify the information storage capacity of a neuron.

Before looking at the general case, let us look again at the parity function example. This function cannot be implemented by a single TLU; however, a network of TLUs can implement it in theory. Nevertheless, if the input layer is made of ORs and the processing layer is a TLU, then the synapses between the input and the processing layer need to have integer values ranging between 1 and 2^{n-1}, where n is the number of binary input variables (Minsky and Papert, 1988). If $n = 4$, such Perceptron would require eight different distinct synaptic weight values to compute the parity function whereas a network made of XORs, which are TNLUs, only requires binary weights. Other functions can be found where such a disparity between networks of TLUs and networks TNLUs exists (Hastad and Goldmann, 1991).

We can push this observation much farther by looking at the capacity advantage gained by having a single dendritic subunit with nonlinear summation. Our exhaustive parameter searches demonstrated that a TNLU with one nonlinear dendritic subunit can implement tenfold more functions than a single TLU (Table 22.2) (Cazé et al., 2013). Notably this tenfold increase in memory capacity necessitates a smaller range of synaptic weight values for a TNLU than for a TLU, as shown in Table 22.2. For $n = 6$, a

Table 22.2 Information storage capacity of TLUs and TNLUs as a function of the number of input variables

n	# of computable functions	# of possible synaptic weights
6	[1113;13505]	[10;5]
5	[119;208]	[6;4]
4	[30;30]	[4;3]

The numbers of implementable functions with or without a nonlinear dendritic subunit [TLU;TNLU] are in the first column. The numbers of synaptic weight values required to reach this capacity are given in the second column. This tablewas built using large parameter searches (see Cazé et al., 2012 for moredetails)

TNLU needs a number of distinct synaptic weight values twofold smaller to reach a memory storage capacity tenfold higher than a TLU. Thus, given practical, biological restrictions on possible synaptic weight values, a TNLU neuron can have a dramatically greater storage capacity.

The root of this advantage is in the availability of intrinsic plasticity for the TNLU. While the only intrinsic storage space of a TLU is its threshold, a single parameter which is negligible compared to the multidimensional space of synaptic weights, TNLUs have a much larger number of intrinsic parameters than TLUs to store information, namely the subcellular placement of synapses and the dendritic local excitability (here modeled as the height and threshold of the dendritic activation function *D*).

Mainstream theories of memory and its use focus on synapses as the main information reservoir in the brain, learnt through synaptic plasticity (Abbott, 1990; Bliss et al., 1973; Sjöström et al., 2008), but plasticity of intrinsic neuron properties has long been described experimentally (Brons and Woody, 1980; Frick and Johnston, 2005; Schulz, 2006; Williams et al., 2007). Intrinsic plasticity works by modifying the properties of the neuron itself, i.e., the property of the voltage-gated ionic channels contained in the membrane (Alkon, 1984; Desai et al., 1999; Frick and Johnston, 2005; Zhang and Linden, 2003) or the subcellular placement of synapses (Chklovskii et al., 2004). As has been suggested in an experimental study in invertebrate (Alkon, 1984), local intrinsic plasticity is possible in dendrites (Losonczy et al., 2008).

Consequently, a brain made of TLNU-like neurons—those with local nonlinear dendritic summation—has a greater storage capacity than has traditionally been recognized (Fig. 22.4): in a network made of TLUs, learning occurs through synaptic plasticity because information is only stored in the connections between nodes (McClelland et al., 1986); but a significant amount of information can also be stored within the nodes of a networks of TNLUs.

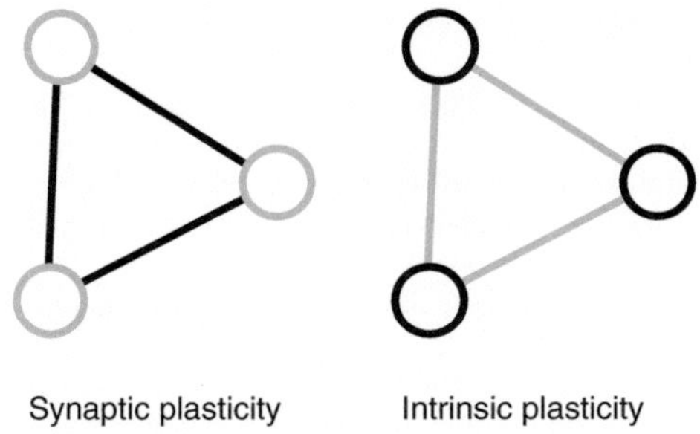

Fig. 22.4 Two methods for storing information in a network of neurons. The nodes stand for the neuron and the edges for their connections. The *black part* of the graph is plastic while the *gray part* is static. *Left*: synaptic plasticity is a change in the intensity of the link between two nodes. *Right*: intrinsic plasticity is a change in the computation performed by a node

22.4 Conclusion

In this chapter, we have discussed three implications of nonlinear summation in dendrites for single neurons and networks.

We showed, first, that a neuron can use its analogue spiking or saturating dendrites to perform new sets of computations like XOR, FBP, or dFBP, suggesting that neurons are complex computing units rather than networks of simple units. We labeled these units TNLUs. This observation challenges the connectionist hypothesis that the brain is a network made of simple computing elements, either neuronal or dendritic.

Second, we demonstrated that networks of TNLUs can do better at parallelizing a difficult function than networks of TLUs. In the former type of networks some computations, such as the parity function, require pre-synaptic units with large receptive fields; this requirement can be weakened or eliminated in networks of TNLUs. This observation can be generalized to recurrent networks if the input layer is considered to receive both feed-forward and recurrent inputs.

Third, we have shown that a TNLU can store a substantial amount of information in its intrinsic parameters. Consequently, intrinsic plasticity plays a more important role in a network of TNLUs, and such network can store more information with a smaller range of synaptic weight values than a network of TLUs containing the same number of units. This last property allows networks of TNLUs to store information more efficiently than networks of TLUs.

References

Abbott LF (1990) Learning in neural network memories. Physics 1:105–122

Abrahamsson T, Cathala L, Matsui K, Shigemoto R, DiGregorio DA (2012) Thin dendrites of cerebellar interneurons confer sublinear synaptic integration and a gradient of short-term plasticity. Neuron 73(6):1159–1172

Agmon-Snir H, Carr C, Rinzel J (1998) The role of dendrites in auditory coincidence detection. Nature 393(6682):268–272

Alkon D (1984) Calcium-mediated reduction of ionic currents: a biophysical memory trace. Science 226(4678):1037

Anthony M (2010) Neural networks and boolean functions. In: Crama Y, Hammer PL (eds) Boolean models and methods in mathematics, computer science, and engineering. Cambridge University Press, Cambridge, pp 554–595

Archie KA, Mel BW (2000) A model for intradendritic computation of binocular disparity. Nat Neurosci 3(1):54–63

Ariav G, Polsky A, Schiller J (2003) Submillisecond precision of the input–output transformation function mediated by fast sodium dendritic spikes in basal dendrites of CA1 pyramidal neurons. J Neurosci 23(21):7750

Bliss TV, Lomo T, Lø mo T (1973) Long-lasting potentiation of synaptic transmission in the dentate area of the anaesthetized rabbit following stimulation of the perforant path. J Physiol 232(2):331

Brons JF, Woody CD (1980) Long-term changes in excitability of cortical neurons after Pavlovian conditioning and extinction. J Neurophysiol 44(3):605–615

Brunel N, Hakim V, Isope P, Nadal JP, Barbour B (2004) Optimal information storage and the distribution of synaptic weights: perceptron versus Purkinje cell. Neuron 43(5):745–757

Cazé RD, Humphries M, Gutkin BS (2012) Spiking and saturating dendrites differentially expand single neuron computation capacity. In: Advances in neural information processing systems, vol 25, pp 1079–1087

Cazé RD, Humphries M, Gutkin B (2013) Passive dendrites enable single neurons to compute linearly non-separable functions. PLoS Comput Biol 9(2):e1002867

Chklovskii DBDB, Mel BW, Svoboda K (2004) Cortical rewiring and information storage. Nature 431(7010):782–788

Crama Y, Hammer PL (2011) Boolean functions: theory, algorithms, and applications (Encyclopedia of mathematics and its applications). Cambridge University Press, Cambridge

Desai NS, Rutherford LC, Turrigiano GG (1999) Plasticity in the intrinsic excitability of cortical pyramidal neurons. Nat Neurosci 2(6):515–520

Feldman J (2012) The neural binding problem (s). Cogn Neurodyn 7:1–11

Frick A, Johnston D (2005) Plasticity of dendritic excitability. J Neurobiol 64(1):100–115

Gurney KN (2001a) Information processing in dendrites: I. Input pattern generalisation. Neural Netw 14(8):991–1004

Gurney KN (2001b) Information processing in dendrites: II. Information theoretic complexity. Neural Netw 14(8):1005–1022

Haag J, Egelhaaf M, Borst A (1992) Dendritic integration of motion information in visual interneurons of the blowfly. Neurosci Lett 140(2):173–176

Hall MD, Wieberg K (2003) Illusory conjunctions of musical pitch and timbre. Acoust Res Lett Online 4(3):65

Hastad J (1994) On the size of weights for threshold gates. SIAM J Discrete Math 7(3):484–492

Hastad J, Goldmann M (1991) On the power of small-depth threshold circuits. Comput Complex 1(2):113–129

Heck D (2003) Passive spatial and temporal integration of excitatory synaptic inputs in cerebellar Purkinje cells of young rats. Neurosci Lett 341(1):79–83

Hendrickson EB, Edgerton JR, Jaeger D (2011) The capabilities and limitations of conductance-based compartmental neuron models with reduced branched or unbranched morphologies and active dendrites. J Comput Neurosci 30(2):301–321

Holcombe AO (2009) The binding problem. In: Goldstein EB (ed) The Sage encyclopedia of perception. Sage, Los Angeles

Jia H, Rochefort NL, Chen X, Konnerth A (2010) Dendritic organization of sensory input to cortical neurons in vivo. Nature 464(7293):1307–1312

Katz Y, Menon V, Nicholson DA, Geinisman Y, Kath WL, Spruston N (2009) Synapse distribution suggests a two-stage model of dendritic integration in CA1 pyramidal neurons. Neuron 63(2): 171–177
Koch C (2005) Biophysics of computation: information processing in single neurons. Oxford University Press, New York
Koch C, Poggio T, Torres V (1982) Retinal ganglion cells: a functional interpretation of dendritic morphology. Philos Trans R Soc Lond B Biol Sci 298(1090):227
Korshunov A (2002) On the distribution of the number of monotone Boolean functions relative to the number of lower units. Discrete Math 257(2–3):463–479
Legenstein R, Maass W (2011) Branch-specific plasticity enables self-organization of nonlinear computation in single neurons. J Neurosci 31(30):10787–10802
Losonczy A, Magee J (2006) Integrative properties of radial oblique dendrites in hippocampal CA1 pyramidal neurons. Neuron 50(2):291–307
Losonczy A, Makara JK, Magee JC (2008) Compartmentalized dendritic plasticity and input feature storage in neurons. Nature 452(7186):436–441
McClelland J, Rumelhart D, Hinton GE (1986) The appeal of parallel distributed processing. In: The parallel distributed processing: explorations in the microstructure of cognition: foundations, vol 1, MIT press
McCulloch WS, Pitts W (1943) A logical calculus of the ideas immanent in nervous activity. Bull Math Biol 52(1–2):99–115 [discussion 73–97]
Minsky M, Papert S (1988) Perceptrons: an introduction to computational geometry, expanded edn. MIT, Cambridge
Muroga S (1971) Threshold logic and its applications. Wiley, New York
Nevian T, Larkum M, Polsky A, Schiller J (2007) Properties of basal dendrites of layer 5 pyramidal neurons: a direct patch-clamp recording study. Nature 200(2):7
Poirazi P, Mel B (2000) Memory capacity of linear vs. nonlinear models of dendritic integration. In: Advances in neural information processing system, vol 12, pp 157–163
Poirazi P, Brannon T, Mel B (2003a) Pyramidal neuron as two-layer neural network. Neuron 37(6):989–999
Poirazi P, Brannon T, Mel BW (2003b) Arithmetic of subthreshold synaptic summation in a model CA1 pyramidal cell. Neuron 37(6):977–987
Polsky A, Mel BW, Schiller J (2004) Computational subunits in thin dendrites of pyramidal cells. Nat Neurosci 7(6):621–627
Polsky A, Mel B, Schiller J (2009) Encoding and decoding bursts by NMDA spikes in basal dendrites of layer 5 pyramidal neurons. J Neurosci 29(38):11891
Rall W (1967) Distinguishing theoretical synaptic potentials computed for different soma-dendritic distributions of synaptic input. J Neurophysiol 30(5):1138–1168
Rojas R (1996) Neural networks: a systematic introduction. Springer, Berlin
Rosenblatt F (1958) The perceptron: a probabilistic model for information storage and organization in the brain. Psychol Rev 65(6):386–408
Roskies AL, Binding WI (1999) The binding problem. Neuron 24:7–9
Schiller J, Major G, Koester H, Schiller Y (2000) NMDA spikes in basal dendrites of cortical pyramidal neurons. Nature 404(6775):285–289
Schulz DJ (2006) Plasticity and stability in neuronal output via changes in intrinsic excitability: it's what's inside that counts. J Exp Biol 209(24):4821–4827
Segev I (1998) Sound grounds for computing dendrites. Nature 393(6682):207–207
Shmulevich I, Sellke T, Gabbouj M, Coyle E (1995) Stack filters and free distributive lattices. In: Proceedings of 1995 IEEE workshop on nonlinear signal processing, Halkidiki, Greece
Sjöström PJ, Rancz EAE, Roth A, Häusser M (2008) Dendritic excitability and synaptic plasticity. Physiol Rev 88(2):769–840
Taraborelli D (2002) Feature binding and object perception. In: European Society for Philosophy and Psychology, Lyon, pp 1–16

Treisman A (1998) Feature binding, attention and object perception. Philos Trans R Soc Lond B Biol Sci 353(1373):1295–1306

Treisman A, Gelade G (1980) A feature-integration theory of attention. Cogn Psychol 136:97–136

Wegener I (1987) The complexity of Boolean functions. Wiley, Stuttgart

Wei DS, Mei YA, Bagal A, Kao JP, Thompson SM, Tang CM (2001) Compartmentalized and binary behavior of terminal dendrites in hippocampal pyramidal neurons. Science 293(5538):2272–2275

Williams SR, Wozny C, Mitchell SJ (2007) The back and forth of dendritic plasticity. Neuron 56(6):947–953

Zador AAM, Claiborne B, Brown TTH, Clairborne BJ (1993) Nonlinear pattern separation in single hippocampal neurons with active dendritic membrane. In: Advances in neural information processing system, p 51

Zhang W, Linden DJ (2003) The other side of the engram: experience-driven changes in neuronal intrinsic excitability. Nat Rev Neurosci 4(11):885–900

Chapter 23
Dendritic Size and Topology Influence Burst Firing in Pyramidal Cells

Arjen van Ooyen and Ronald A.J. van Elburg

Abstract Neurons have highly branched dendrites that form characteristic tree-like structures. The morphology of these dendritic arborizations is not fixed and can undergo significant alterations in many pathological conditions. However, little is known about the impact of morphological changes on neuronal activity. Using computational models of pyramidal cells, we study the influence of dendritic tree size and branching structure on burst firing. Burst firing is the generation of two or more action potentials in close succession, a form of neuronal activity that is critically involved in neuronal signaling and synaptic plasticity. We show that there is only a range of dendritic tree sizes that supports burst firing, and that this range is modulated by the branching structure of the tree. Shortening as well as lengthening the dendritic tree, or even just modifying the pattern in which the branches in the tree are connected, can shift the cell's firing pattern from bursting to tonic firing. The influence of dendritic morphology on burst firing is attributable to the effect that dendritic size and branching pattern have on the average spatial extent of the dendritic tree and the spatiotemporal dynamics of the dendritic membrane potential. Our results suggest that alterations in pyramidal cell morphology, such as those observed in Alzheimer's disease, mental retardation, epilepsy, and chronic stress, can change neuronal burst firing and thus ultimately affect information processing and cognition.

A. van Ooyen (✉)
Computational Neuroscience Group, Department of Integrative Neurophysiology, Center for Neurogenomics and Cognitive Research, VU University, Amsterdam, The Netherlands
e-mail: arjen.van.ooyen@cncr.vu.nl

R.A.J van Elburg
Institute of Artificial Intelligence and Cognitive Engineering (ALICE), University of Groningen, 9700 AK, Groningen, The Netherlands

H. Cuntz et al. (eds.), *The Computing Dendrite: From Structure to Function*, Springer Series in Computational Neuroscience 11, DOI 10.1007/978-1-4614-8094-5_23,

23.1 Introduction

Neurons exhibit a wide range of intrinsic firing patterns (Connors and Gutnick 1990; Contreras 2004). A distinct firing pattern that is critically involved in neuronal signaling and synaptic plasticity is burst firing, the generation of clusters of spikes with short interspike intervals (Krahe and Gabbiani 2004). Bursts improve the signal-to-noise ratio (Eggermont and Smith 1996), are transmitted more reliably than isolated spikes (Swadlow and Gusev 2001), are more effective than single spikes in inducing synaptic long-term potentiation (LTP) (Yun et al. 2002), and can even determine whether LTP or LTD (long-term depression) occurs (Birtoli and Ulrich 2004).

Two main classes of ionic mechanisms underlying intrinsic neuronal burst firing have been identified (Krahe and Gabbiani 2004). In so-called dendrite-independent mechanisms—responsible for bursting in thalamic relay neurons (McCormick and Huguenard 1992), for example—the fast, spike-generating conductances and the slow, burst-controlling conductances are co-localized in the soma. Conversely, in dendrite-dependent mechanisms—involved in pyramidal cell burst firing—these conductances are distributed across the soma and dendrites, with the interaction between somatic and dendritic conductances playing an essential role in burst generation. Dendritic voltage-gated Na^+ and K^+ channels, which promote propagation of action potentials from the soma into the dendrites, cause the dendrites to be depolarized when, at the end of a somatic spike, the soma is hyperpolarized, leading to a return current from dendrites to soma. The return current gives rise to a depolarizing afterpotential at the soma, which, if strong enough, produces another somatic spike (Williams and Stuart 1999; Wang 1999). This whole process was described by Wang (1999) as "ping-pong" interaction between soma and dendrites.

Although ion channels play a central role in burst firing, dendritic morphology also appears to be an important modulating factor. In many cell types, including neocortical and hippocampal pyramidal cells (Mason and Larkman 1990; Chagnac-Amitai et al. 1990; Bilkey and Schwartzkroin 1990), neuronal firing patterns and the occurrence of bursts are correlated with dendritic morphology. Results from modeling studies also suggest a relationship between dendritic morphology and firing pattern (Mainen and Sejnowski 1996; Sheasby and Fohlmeister 1999; Van Ooyen et al. 2002; Krichmar et al. 2002). However, these studies are mainly correlative, focus on morphologically very distinct cell classes, use only the physiologically less appropriate stimulation protocol of somatic current injection, and do not investigate the impact of topological structure of dendritic arborizations.

In this chapter, we summarize our study (Van Elburg and Van Ooyen 2010) in which we used computational models of neocortical pyramidal neurons to investigate the impact of a cell's dendritic morphology (both size and topological structure) on the ping-pong mechanism of burst firing, under either somatic current injection or synaptic stimulation of the apical dendritic tree.

23.2 Methods

23.2.1 Pyramidal Cell

We used a morphologically and biophysically realistic model of a bursting layer 5 pyramidal cell (Mainen and Sejnowski 1996) implemented in NEURON (Hines and Carnevale 1997). The pyramidal cell was activated by either somatic or dendritic stimulation. For somatic stimulation, the cell was continuously stimulated with a fixed current injection of 0.2 nA. For dendritic stimulation, the cell was stimulated by synapses that were regularly distributed across the apical dendrite. Each synapse was randomly activated.

We investigated the effect of both dendritic size and dendritic topology on burst firing. The size of a dendritic tree is the total length of all its dendritic segments. The topology of a dendritic tree is the way in which the dendritic segments are connected to each other. For example, a tree with a given number of terminal segments can be connected in a fully asymmetrical (Fig. 23.4a) or a fully symmetrical way (Fig. 23.4b).

To investigate how the dendritic size of the pyramidal cell influences burst firing, we varied the total length of the cell's apical dendrite according to two methods. In the first method, we successively pruned terminal segments from the apical dendritic tree. Starting with the full pyramidal cell morphology, in each round of pruning we randomly removed a number of terminal segments from the apical dendritic tree. From the reduced dendritic tree, we again randomly cut terminal segments, and so on, until the whole apical dendrite was eliminated. This whole procedure of starting with the complete apical dendritic tree and sequentially stripping it of all its segments was repeated 20 times. The density of synapses was kept constant during pruning, so with dendritic stimulation pruning also changed the total input to the cell. With somatic simulation, the total input to the cell did not change when the apical dendrite was pruned.

In the second method, we kept the dendritic arborization intact and changed the size of the apical dendrite by multiplying the lengths of all its segments by the same factor. Thus, the entire apical dendritic tree was compressed or expanded. For dendritic stimulation, we kept the total synaptic input to the cell constant by adapting the density of the synapses. So, both with somatic and dendritic stimulation, the total input to the cell did not change when the size of the apical dendrite was modified.

To examine the impact of the cell's dendritic branching structure on burst firing, we varied the topology of the apical dendritic tree by swapping branches within the tree. The apical dendritic trees that were generated in this way had exactly the same total dendritic length and other metrical properties such as total dendritic surface area and differed only in their topological structure. The total input to the cell, both with somatic and dendritic stimulation, did not change when the topological structure was altered.

23.2.2 Morphologically Simplified Cells

We also studied a set of morphologically simplified neurons, consisting of all the 23 topologically different trees with eight terminal segments (two examples are shown in Fig. 23.4a, b), but with the same ion channel composition as in the full pyramidal cell model. All dendritic segments had the same length, so the tree topologies did not differ in total dendritic length. All terminal segments were given the same diameter (0.7 μm; Larkman 1991), while the diameters of intermediate segments were calculated using Rall's power law (Rall 1959), with a branch power of 1.5. This implied that asymmetrical topologies had a higher total dendritic surface area than symmetrical topologies. We therefore also considered the case in which all segments in the tree had the same diameter (3 μm) and all tree topologies thus had the same dendritic surface area. The neurons were continuously stimulated with a fixed current injection of 0.03 nA (0.1 nA for the non-Rall neurons) at the soma, or by uniformly distributed synapses across the dendritic tree.

To examine how the size of the dendritic tree influences firing pattern, we changed the total dendritic length of a given tree topology by multiplying the lengths of all its segments by the same factor. For dendritic stimulation, the total number of synapses on the tree was thereby kept constant. Thus, both with somatic and dendritic stimulation, the total input to the cell did not change when the size of the dendritic tree was modified.

23.2.3 Quantifying Burst Firing

Bursting is defined as the occurrence of two or more consecutive spikes with short interspike intervals followed by a relatively long interspike interval. To quantify bursting, we used the burst measure B developed in Van Elburg and Van Ooyen (2004). This measure is based solely on spike times and detects the correlated occurrence of one or more short (intraburst) interspike intervals followed by a long (interburst) interspike interval. It quantifies the extent of bursting in the whole spike train; it does not try to identify individual bursts. The higher the ratio of interburst to intraburst interspike intervals, the stronger the bursting and the higher the value of B.

23.2.4 Input Conductance and Mean Electrotonic Path Length

The input conductance of a pyramidal cell was determined by applying a subthreshold current injection at the soma. The ratio of the magnitude of the injected current to the resulting change in membrane potential at the soma is defined as the input conductance of the cell.

To quantify the electrotonic extent of a dendritic tree, we used the mean electrotonic path length (MEP) (Van Elburg and Van Ooyen 2010). For a given terminal segment, the electrotonic path length is the length (normalized to the electrotonic length constant) of the path from the tip of the segment to the soma. This electrotonic path length is determined for each terminal segment, and the sum of all electrotonic path lengths is divided by the total number of terminal segments to obtain the MEP of the dendritic tree.

23.3 Results

Employing a standard model of a bursting pyramidal cell (Mainen and Sejnowski 1996), we investigated how dendritic morphology influences burst firing by varying either the size or the topology of the apical dendrite. We also examined a set of morphologically simplified cells with systematic differences in dendritic topology.

23.3.1 Pyramidal Cell

Dendritic size. To investigate how pyramidal cell size influences burst firing, we changed the total length of the apical dendrite according to two methods. In the first method, we successively pruned terminal branches off the apical dendrite. Both with somatic and with dendritic stimulation, the degree of bursting decreased as the dendritic tree became shorter (Fig. 23.1a). Reducing the size of the apical dendrite ultimately transformed the bursting pyramidal cell into a tonically firing cell. The removal of only a few small terminal segments could be enough to completely change the firing state of the cell (Fig. 23.1b). Because of the random activation of synapses, the changes in the degree of bursting were more gradual with dendritic than with somatic stimulation.

In the second method, we kept the dendritic arborization intact and varied the size of the apical dendritic tree by multiplying the lengths of all its segments by the same factor. Both with somatic and with dendritic stimulation, and in line with the previous results, burst firing disappeared as the total dendritic length was decreased (Fig. 23.2a). Interestingly, the pyramidal cell also did not exhibit burst firing when the apical dendrite became too large. Only when the length of the apical dendrite remains within a certain range are bursts generated. Figure 23.2b shows the firing patterns of the pyramidal cell at increasing lengths of its apical dendrite.

Dendritic topology. To examine whether dendritic branching structure, or topology, could influence burst firing, we varied the topology of the apical dendritic tree by swapping branches within the tree. Thus, all the dendritic trees generated in this way had exactly the same metrical properties (total length, total surface area) and differed only in the way their branches were connected. Within this set, we found pyramidal

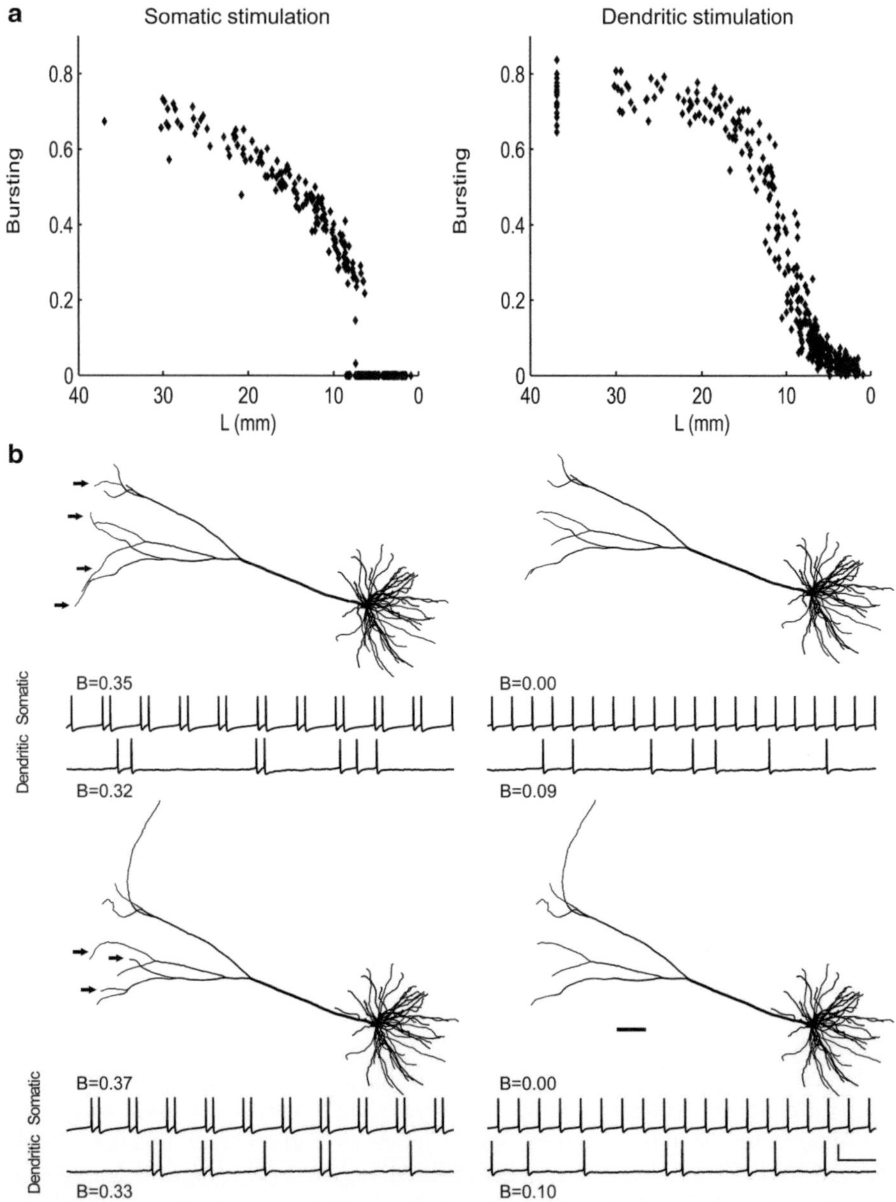

Fig. 23.1 Both with somatic and with dendritic stimulation, pyramidal cell burst firing decreased as the apical dendrite became shorter. (**a**) The degree of bursting, as measured by the burst measure *B*, against the total length of the apical dendritic tree. The total length of the apical dendritic tree was gradually reduced by successively removing terminal segments. Results are shown of 20 times starting with the complete apical dendritic tree and randomly removing terminal segments until the whole tree was eliminated. The complete apical dendritic tree had a total length of 36,865 μm. (**b**) Examples from the experiment in (**a**) showing that the removal of only a few small terminal segments from the apical dendritic tree could change the firing state of the cell. Morphology of pruned pyramidal cells, and voltage traces for both somatic and dendritic stimulation. *Left*, Bursting cells (*Top*, 9,772 μm; *Bottom*, 8,925 μm). *Right*, Non-bursting cells (*Top*, 8,184 μm; *Bottom*, 6,927 μm). Scale bar: 100 ms, 50 mV. Scale bar (anatomy): 100 μm. *Arrows* in the bursting cells indicate the branches that were shorter or absent in the non-bursting cells. (Reproduced from Van Elburg and Van Ooyen 2010)

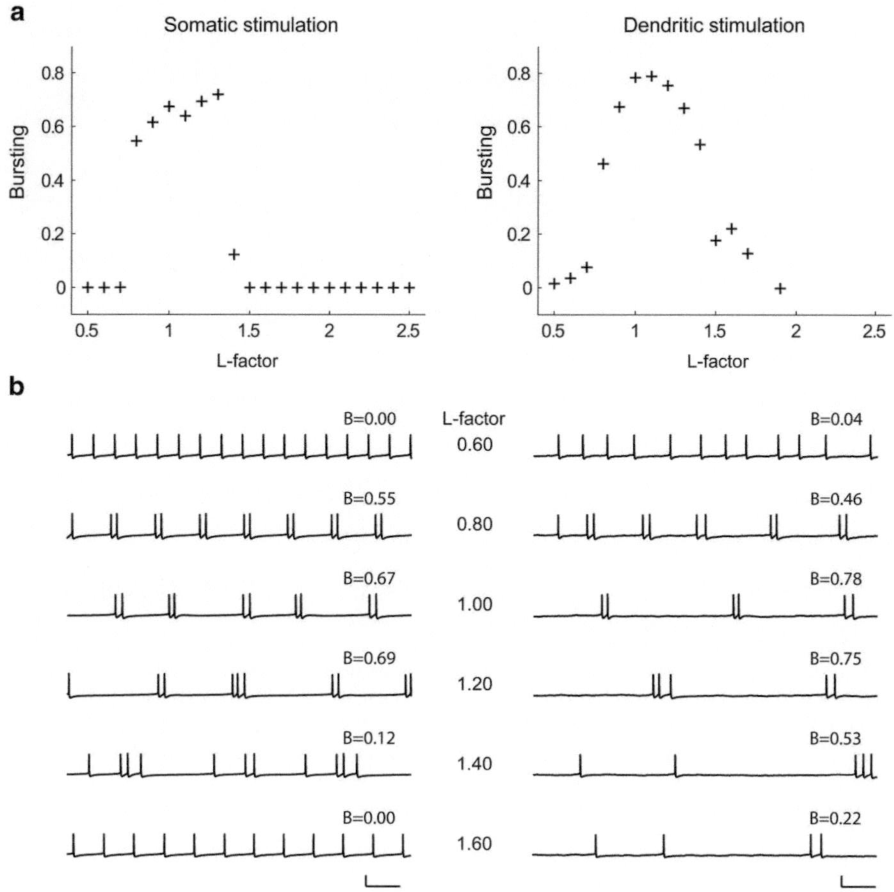

Fig. 23.2 Both with somatic and with dendritic stimulation, pyramidal cell burst firing disappeared when the apical dendrite became either too large or too small. We varied the size of the apical dendrite by scaling the entire apical dendrite, thus keeping the dendritic arborization intact. (**a**) The degree of bursting against the factor by which the length of all the apical dendritic segments was multiplied. (**b**) Voltage traces obtained for different sizes of the apical dendrite. *Left*, somatic stimulation. *Right*, dendritic stimulation. Scale bars: 100 ms, 50 mV. (Reproduced from Van Elburg and Van Ooyen 2010)

cells that produced firing patterns ranging from tonic firing to strongly bursting (Fig. 23.3). Just remodeling the topology of the apical dendrite could completely change the firing state of the cell and turn a bursting cell into tonically firing cell or vice versa. Interestingly, dendritic topology not only affected whether a cell exhibited bursting or not (Fig. 23.3a–c vs. Fig. 23.3d–f, but also influenced the fine structure or degree of bursting. The cell displayed in Fig. 23.3a generated (with somatic stimulation) two-spike bursts alternating with single spikes. The cells in Fig. 23.3b, c both produced a pattern of two-spike bursts, but the relative sizes of the interspike

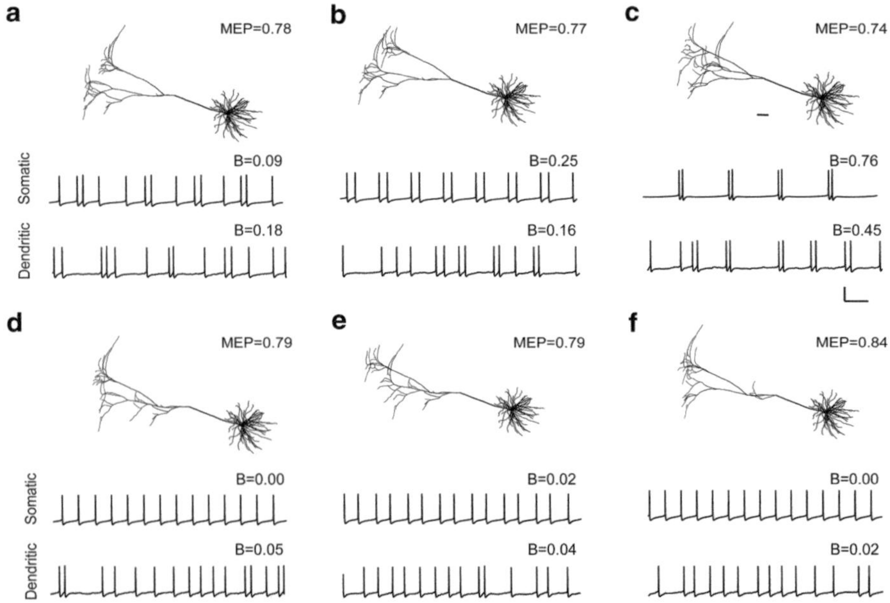

Fig. 23.3 Both with somatic and with dendritic stimulation, dendritic topology affected pyramidal cell burst firing. We varied the topology of the apical dendritic tree by swapping branches within the tree. Thus, all the pyramidal cells shown had exactly the same total dendritic length and dendritic surface area and differed only in the topology of their apical dendrite (basal dendrites were the same). Voltage traces for three bursting cells (**a–c**), and three non-bursting cells (**d–f**). Scale bar: 100 ms, 50 mV. Scale bar (anatomy): 100 μm. MEP values indicate the mean electrotonic path length of the apical dendritic tree. (Reproduced from Van Elburg and Van Ooyen 2010)

intervals between and within bursts were different. Although dendritic stimulation introduced irregularities in firing pattern because of the stochastic nature of the activation of synapses, somatic and dendritic stimulation yielded comparable results.

23.3.2 Morphologically Simplified Cells

Dendritic size and topology. To analyze more precisely the effect of dendritic morphology on burst firing, we also investigated (Van Elburg and Van Ooyen 2010) a set of 23 morphologically simplified neurons consisting of all the topologically different trees with eight terminal segments (two examples are shown in Fig. 23.4a, b). The total length of a given tree topology was varied by changing the lengths of all the segments in the tree by the same factor. Under all conditions (somatic or dendritic simulation, trees with uniform segment diameters or trees whose segments diameters obeyed Rall's power law), bursting occurred only for a certain range of tree sizes, as in the full pyramidal cell model. Interestingly, this range depended on

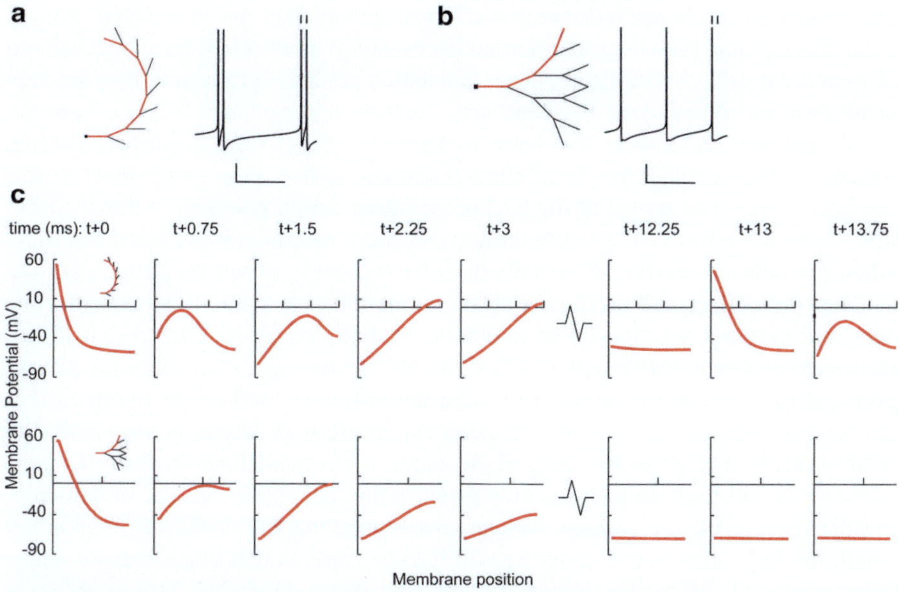

Fig. 23.4 The importance of electrotonic distance for burst firing and the impact of dendritic topology illustrated with a fully asymmetrical and a fully symmetrical tree (**a**, **b**). At this dendritic size, the asymmetrical tree (**a**) generated bursts, whereas the symmetrical tree (**b**) produced single spikes. The segment diameters in the trees obeyed Rall's power law, and both trees had the same total dendritic length (1,600 μm). The cell was stimulated at the soma. Scale bars: 100 ms, 20 mV. The ticks on top of the action potentials in (**a**) and (**b**) indicate the spikes that are shown at t+0 and t+13 in (**c**). (**c**) The membrane potential evolution over time in the asymmetrical tree (*top row*) and the symmetrical tree (*bottom row*) along the dendritic paths indicated in (**a**) and (**b**). Time is relative to the first spike (at 0 ms), and membrane position on the *x*-axis runs from soma to the tip of the terminal segment. Because the distance between soma and terminal segment is smaller in the symmetrical than in the asymmetrical tree, the membrane potential evolution in the symmetrical tree had less spatial differentiation, the membrane potential reached a lower value at the distal end, and the distal membrane potential started decreasing earlier in time so that the return current from dendrites to soma reached the soma at a time when the delayed-rectifier K^+ channels were still open, preventing the generation of a second spike. (Reproduced from Van Elburg and Van Ooyen 2010)

the topology of the dendritic tree: trees with an asymmetric branching structure started bursting at a lower total dendritic length than trees with a symmetric branching structure, and also stopped bursting at a lower total dendritic length.

Also under all conditions, we found that both the onset and the cessation of bursting were strongly correlated not to the input conductance but to the mean electrotonic path length of the dendritic tree (see Sect. 23.2), with burst firing occurring only within a certain range of path lengths. Burst firing occurred when the mean electrotonic path length was higher than a certain critical value. At the same total dendritic length, asymmetrical trees have a higher mean electrotonic path length than symmetrical trees, and consequently reached this critical value earlier than symmetrical trees as dendritic length was increased.

Importance of electrotonic distance and dendritic topology for burst firing. Why is mean electrotonic path length important for bursting? In the burst firing mechanism of pyramidal cells, a somatically generated action potential propagates into the dendritic tree and depolarizes the dendrites, creating a potential difference between distal dendrites and soma. This leads to a return current from dendrites to soma, which, if strong enough, produces another somatic spike ("ping-pong" mechanism; see Sect. 23.1). The arrival of the backpropagating action potential in the dendritic tips marks the onset of the return currents. If these return currents reach the soma when the delayed-rectifier K^+ conductance is still high, it will be difficult for the soma to depolarize and produce a spike. Since the propagation velocity of voltages and currents is proportional to the electrotonic length constant (Koch 1999), the mean electronic path length is a measure for the average time it takes for a backpropagating action potential to travel to the dendritic tips (and for the return current to move to the soma). Thus, if the mean electronic path length is too small, the return current will arrive too early at the soma, when the delayed-rectifier K^+ conductance is still high, so that it cannot produce another spike—that is, no bursting. Furthermore, if the electrotonic distance between soma and distal dendrites is too small, the large conductive coupling will lead to currents that quickly annul membrane potential differences between soma and distal dendrites. This prohibits a strong and long-lasting differentiation in membrane potential between soma and distal dendrites, which is the generator of the return current.

However, if the electrotonic distance between soma and distal dendrites is too large, bursting will also not occur. Note that even in the absence of a return current, the cell will generate a next spike as a result of the external (somatic or dendritic) stimulation. So, what the return current in fact does when it causes bursting is to advance the timing of the next spike. If the electronic distance is too large, the return current will arrive too late—that is, not before the external stimulation has already caused the cell to spike. Furthermore, if the electrotonic distance is too large, the potential gradient between distal dendrites and soma will become too shallow for a strong return current.

Importantly, the mean electrotonic path length depends also on the topology of the dendritic tree, which accounts for the influence of dendritic topology on burst firing. In asymmetrical trees, the terminal segments are on average further away from the soma than in symmetrical trees. Consequently, at the same dendritic size, asymmetrical trees have a higher mean electrotonic path length—as well as "normal" mean path length—than symmetrical trees (both in trees with uniform segment diameters and in trees whose segments diameters obey Rall's power law). As a result, asymmetrical trees reach the critical values of mean electrotonic path length from where bursting starts, and from where it stops, at lower dendritic sizes than symmetrical trees. Figure 23.4 illustrates the importance of electrotonic distance and the impact of topological structure on burst firing.

In general, the results obtained with dendritic stimulation are comparable to those produced with somatic stimulation (Figs. 23.1–23.3). In the ping-pong mechanism of burst firing, the sequence and timing of events start when a somatic action potential propagates into the dendritic tree. How this action potential is generated in the first place, by current injection into the soma or as a result of summation of dendritic synaptic inputs, appears not to be crucial.

23.4 Discussion

Burst firing is critically involved in synaptic plasticity and neuronal signaling, and it is therefore important to know what factors might affect bursting. We showed that either shortening or lengthening the apical dendrite tree beyond a certain range can transform a bursting pyramidal cell into a tonically firing cell. Remarkably, altering only the topology of the dendritic tree, whereby the total length of the tree remains unchanged, can likewise shift the firing pattern from bursting to non-bursting or vice versa. Moreover, both dendritic size and dendritic topology not only influence whether a cell is bursting or not, but also affect the number of spikes per burst and the interspike intervals between and within bursts.

The influence of dendritic morphology on burst firing is attributable to the effect dendritic length and dendritic topology have, not on input conductance, but on the spatial extent of the dendritic tree, as measured by the mean electrotonic path length between soma and distal dendrites. For the spatiotemporal dynamics of dendritic membrane potential to generate burst firing, this electrotonic distance should be neither too small nor too large. Because the degree of symmetry of the dendritic tree also determines mean electrotonic path length, with asymmetrical trees having larger mean path lengths than symmetrical trees, dendritic topology as well as dendritic size affects the occurrence of burst firing.

In Mainen and Sejnowski's (1996) two-compartment model for explaining the role of dendritic morphology in shaping firing pattern, the spatial dimension of morphology was completely reduced away. Although the model is able to reproduce a wide range of firing patterns, it does not capture the essential influence of dendritic morphology on burst firing, for which, as we have shown here, the spatial extent of the dendritic tree and the resulting spatiotemporal dynamics of the dendritic membrane potential are important.

Our results are in accord with empirical observations suggesting that pyramidal cells should have reached a minimal size to be capable of burst firing. In weakly electric fish, the tendency of pyramidal cells to fire bursts is positively correlated with the size of the cell's apical dendritic tree (Bastian and Nguyenkim 2001). In rat prefrontal cortex (Yang et al. 1996) and visual cortex (Mason and Larkman 1990), the classes of pyramidal cells that exhibit burst firing have a greater total dendritic length than the other classes.

In addition, the developmental time course of bursting shows similarities with that of dendritic morphology. In rat sensorimotor cortex, the proportion of bursting pyramidal cells progressively increases from postnatal day 7 onwards, while at the same time the dendritic arborizations become more complex (Franceschetti et al. 1998). In pyramidal cells from rat prefrontal cortex, the total lengths of apical and basal dendrites increase dramatically between postnatal days 3 and 21, with neurons capable of burst firing appearing only from postnatal day 18 onwards (Zhang 2004; Dégenètais et al. 2002).

Dendritic morphology can undergo significant alterations in many pathological conditions, including chronic stress (Sousa et al. 2000; Radley et al. 2004; Cook and

Wellman 2004; Magariños et al. 1996), epilepsy (Teskey et al. 2006), hypoxic ischemia (Ruan et al. 2006), Alzheimer (Yamada et al. 1988, Moolman et al. 2004), and disorders associated with mental retardation (Kaufmann and Moser 2000). Functional consequences of these morphological changes are usually interpreted in terms of loss or formation of synaptic connections as a result of a diminished or expanded post-synaptic surface area. Our modeling results indicate that alterations in dendritic morphology can directly modify neuronal firing, irrespective of changes in total synaptic input.

Chronic stress, as well as daily administration of corticosterone, induces extensive regression of pyramidal apical dendrites in hippocampus (Sousa et al. 2000; Magariños et al. 1996; Woolley et al. 1990) and prefrontal cortex (Radley et al. 2004; Cook and Wellman 2004). As a result of a decrease in the number and length of terminal branches, the total apical dendritic length can reduce by as much as 32 % (Cook and Wellman 2004). Similarly large alterations have been observed in response to mild, short-term stress (Brown et al. 2005). Our results predict that stress and the accompanying reduction in apical dendritic length could turn a bursting neuron into a non-bursting one. Indeed, Okuhara and Beck (1998) found that 2 weeks of high corticosterone treatment caused a decrease in the relative number of intrinsically bursting CA3 pyramidal cells. Since burst firing of CA3 pyramidal cells is critically involved in LTP (Bains et al. 1999), this could have profound functional consequences for hippocampal information processing (Pavlides et al. 2002).

With regard to epilepsy, a significant decrease in total dendritic length and number of branches has been found in pyramidal cells following neocortical kindling (Teskey et al. 2006). In line with our results, Valentine et al. (2004) reported that cells recorded from the primary auditory cortex of kindled cats showed a reduction in the amount of burst firing and a decrease in the number of spikes per burst.

In Alzheimer's disease, various aberrations in dendritic morphology have been observed—including a reduction in total dendritic length and number of dendritic branches (Yamada et al. 1988; Moolman et al. 2004) and alterations in the pattern of dendritic arborization (Arendt et al. 1997)—which may contribute to the abnormal neurophysiological properties of Alzheimer pyramidal cells (Stern et al. 2004). The anomalies in morphology could influence the cells' ability to produce burst, and, because of the role of burst firing in LTP and LTD (Yun et al. 2002; Birtoli and Ulrich 2004), ultimately affect cognition. In disorders related with mental retardation, the observed alterations in dendritic length and pattern of dendritic branching (Kaufmann and Moser 2000), e.g., changes in the degree of symmetry of the apical dendrite (Belichenko et al. 1994), may likewise be hypothesized to contribute to impaired cognition.

In conclusion, our results show that alterations in either the size or the topology of dendritic arborizations, such as those observed in many pathological conditions, could have a marked impact on pyramidal cell burst firing and, because of the critical role of bursting in neuronal signaling and synaptic plasticity, ultimately affect cognition.

23.5 Outlook

One way to test experimentally the importance of mean electrotonic path length for burst firing is to study whether the occurrence of bursting in a large set of electrophysiologically similar pyramidal cells correlates with the mean electrotonic path length (or alternatively, "normal" path length) of their apical dendrites. Direct experimental testing of the influence of dendritic morphology on burst firing could be done by physically manipulating the shape or size of the dendritic tree, e.g., by using techniques developed by Bekkers and Häusser (2007). In line with our results, they showed that dendrotomy of the apical dendrite abolished bursting in layer 5 pyramidal cells.

Since firing patterns characteristic of different classes of neurons may in part be determined by total dendritic length, we expect on the basis of our results that a neuron may try to keep its dendritic size within a restricted range in order to maintain functional performance. Indeed, Samsonovich and Ascoli (2006) have shown that total dendritic size appears to be under intrinsic homeostatic control. Statistically analyzing a large collection of pyramidal cells, they found that, for a given morphological class and anatomical location, fluctuations in dendritic size in one part of a cell tend to be counterbalanced by changes in other parts of the same cell so that the total dendritic size of each cell is conserved. We predict that dendritic topology may similarly be protected from large variations. In fact, there could be a trade-off between dendritic size and dendritic topology. In a set of bursting pyramidal cells, we expect that apical dendritic trees with a lower degree of symmetry are shorter in terms of total dendritic length or have thicker dendrites to reduce electrotonic length than those with a higher degree of symmetry.

We investigated the impact of dendritic morphology on the ping-pong mechanism of burst firing (Williams and Stuart 1999; Wang 1999) in pyramidal cells. Future computational studies could also examine the role of morphology in other mechanisms of burst firing (McCormick and Huguenard 1992).

References

Arendt T, Schindler C, Brückner MK, Eschrich K, Bigl V, Zedlick D, Marcova L (1997) Plastic neuronal remodeling is impaired in patients with Alzheimer's disease carrying apolipoprotein epsilon 4 allele. J Neurosci 17:516–529

Bains JS, Longacher JM, Staley KJ (1999) Reciprocal interaction between CA3 network activity and strength of recurrent collateral synapses. Nat Neurosci 2:720–726

Bastian J, Nguyenkim J (2001) Dendritic modulation of burst-like firing in sensory neurons. J Neurophysiol 85:10–22

Bekkers JM, Häusser M (2007) Targeted dendrotomy reveals active and passive contributions of the dendritic tree to synaptic integration and neuronal output. Proc Natl Acad Sci USA 104:11447–11452

Belichenko PV, Oldfors A, Hagberg B, Dahlström A (1994) Rett syndrome: 3-D confocal microcospy of cortical pyramidal dendrites and afferents. Neuroreport 5:1509–1513

Bilkey D, Schwartzkroin P (1990) Variation in electrophysiology and morphology of hippocampal CA3 pyramidal cells. Brain Res 514:77–83
Birtoli B, Ulrich D (2004) Firing mode-dependent synaptic plasticity in rat neocortical pyramidal neurons. J Neurosci 24:4935–4940
Brown SM, Henning S, Wellman CL (2005) Mild, short-term stress alters dendritic morphology in rat medial prefrontal cortex. Cereb Cortex 15:1714–1722
Chagnac-Amitai Y, Luhmann HJ, Prince DA (1990) Burst generating and regular spiking layer 5 pyramidal neurons of rat neocortex have different morphological features. J Comp Neurol 296:598–613
Connors BW, Gutnick MJ (1990) Intrinsic firing patterns of diverse neocortical neurons. Trends Neurosci 13:99–104
Contreras D (2004) Electrophysiological classes of neocortical neurons. Neural Netw 17:633–646
Cook SC, Wellman CL (2004) Chronic stress alters dendritic morphology in rat medial prefrontal cortex. J Neurobiol 60:236–248
Dégenètais E, Thierry AM, Glowinski J, Gioanni Y (2002) Electrophysiological properties of pyramidal neurons in the rat prefrontal cortex: an in vivo intracellular recording study. Cereb Cortex 12:1–16
Eggermont JJ, Smith GM (1996) Burst-firing sharpens frequency-tuning in primary auditory cortex. Neuroreport 7:753–757
Franceschetti S, Sancini G, Panzica F, Radici C, Avanzini G (1998) Postnatal differentiation of firing properties and morphological characteristics in layer V pyramidal neurons of the sensorimotor cortex. Neuroscience 83:1013–1024
Hines ML, Carnevale NT (1997) The NEURON simulation environment. Neural Comput 9:1179–1209
Kaufmann WE, Moser HW (2000) Dendritic anomalies in disorders associated with mental retardation. Cereb Cortex 10:981–991
Koch C (1999) Biophysics of computation. Oxford University Press, New York; Oxford
Krahe R, Gabbiani F (2004) Burst firing in sensory systems. Nat Rev Neurosci 5:13–23
Krichmar JL, Nasuto SJ, Scorcioni R, Washington SD, Ascoli GA (2002) Effects of dendritic morphology on CA3 pyramidal cell electrophysiology: a simulation study. Brain Res 941:11–28
Larkman AU (1991) Dendritic morphology of pyramidal neurones of the visual cortex of the rat: I. Branching patterns. J Comp Neurol 306:307–319
Magariños AM, McEwen BS, Flügge G, Fuchs E (1996) Chronic psychosocial stress causes apical dendritic atrophy of hippocampal CA3 pyramidal neurons in subordinate tree shews. J Neurosci 16:3534–3540
Mainen Z, Sejnowski T (1996) Influence of dendritic structure on firing patterns in model neocortical neurons. Nature 382:363–366
Mason A, Larkman A (1990) Correlations between morphology and electrophysiology of pyramidal neurons in slices of rat visual cortex. II. Electrophysiology. J Neurosci 10:1415–1428
McCormick DA, Huguenard JR (1992) A model of the electrophysiological properties of thalamocortical relay neurons. J Neurophysiol 68:1384–1400
Moolman DL, Vitolo OV, Vonsattel J-P G, Shelanski ML (2004) Dendrite and dendritic spine alterations in Alzheimer models. J Neurocytol 33:377–387
Okuhara DY, Beck SG (1998) Corticosteroids influence the action potential firing pattern of hippocampal subfield CA3 pyramidal cells. Neuroendocrinology 67:58–66
Pavlides C, Nivon LG, McEwen BS (2002) Effects of chronic stress on hippocampal long-term potentiation. Hippocampus 12:245–257
Radley JJ, Sisti HM, Hao J, Rocher AB, McCall T, Hof PR, McEwen BS, Morrison JH (2004) Chronic behavioral stress induces apical dendritic reorganization in pyramidal neurons of the medial prefrontal cortex. Neuroscience 125:1–6
Rall W (1959) Branching dendritic trees and motoneuron membrane resistivity. Exp Neurol 1:491–527
Ruan Y-W, Zou B, Fan Y, Li Y, Lin N, Zeng Y-S, Gao T-M, Yao Z, Xu ZC (2006) Dendritic plasticity of CA1 pyramidal neurons after transient global ischemia. Neuroscience 140:191–201

Samsonovich AV, Ascoli GA (2006) Morphological homeostasis in cortical dendrites. Proc Natl Acad Sci USA 103:1569–1574

Sheasby BW, Fohlmeister JF (1999) Impulse encoding across the dendritic morphologies of retinal ganglion cells. J Neurophysiol 81:1685–1698

Sousa N, Lukoyanov NV, Madeira MD, Almeida OFX, Paula-Barbosa NM (2000) Reorganization of the morphology of hippocampal neurites and synapses after stress-induced damage correlates with behavioral improvement. Neuroscience 97:253–266

Stern EA, Bacskai BJ, Hickey GA, Attenello FJ, Lombardo JA, Hyman BT (2004) Cortical synaptic integration in vivo is disrupted by amyloid-beta plaques. J Neurosci 24:4535–4540

Swadlow HA, Gusev AG (2001) The impact of 'bursting' thalamic impulses at a neocortical synapse. Nat Neurosci 4:402–408

Teskey GC, Monfils M-H, Silasi G, Kolb B (2006) Neocortical kindling is associated with opposing alterations in dendritic morphology in neocortical layer V and striatum from neocortical layer III. Synapse 59:1–9

Valentine PA, Teskey GC, Eggermont JJ (2004) Kindling changes burst firing, neural synchrony and tonotopic organization of cat primary auditory cortex. Cereb Cortex 14:827–839

Van Elburg RAJ, Van Ooyen A (2004) A new measure for bursting. Neurocomputing 58–60:497–502

Van Elburg RAJ, Van Ooyen A (2010) Impact of dendritic size and dendritic topology on burst firing in pyramidal cells. PloS Comput Biol 6(5):e1000781

Van Ooyen A, Duijnhouwer J, Remme MWH, Van Pelt J (2002) The effect of dendritic topology on firing patterns in model neurons. Network 13:311–325

Wang X-J (1999) Fast burst firing and short-term synaptic plasticity: a model of neocortical chattering neurons. Neuroscience 89:347–362

Williams SR, Stuart GJ (1999) Mechanisms and consequences of action potential burst firing in rat neocortical pyramidal neurons. J Physiol (Lond) 521:467–482

Woolley C, Gould E, McEwen BS (1990) Exposure to excess glucocorticoids alters dendritic morphology of adult hippocampal pyramidal neurons. Brain Res 531:225–231

Yamada M, Wada Y, Tsukagoshi H, Otomo E, Hayakawa M (1988) A quantitative Golgi study of basal dendrites of hippocampal CA1 pyramidal cells in senile dementia of Alzheimer type. J Neurol Neurosurg Psyhiatry 51:1088–1090

Yang CR, Seamans JK, Gorelova N (1996) Electrophysiological and morphological properties of layers V-VI principal pyramidal cells in rat prefrontal cortex in vitro. J Neurosci 16:1904–1921

Yun SH, Mook-Jung I, Jung MW (2002) Variation in effective stimulus patterns for induction of long-term potentiation across different layers of rat entorhinal cortex. J Neurosci 22:RC214

Zhang Z-W (2004) Maturation of layer V pyramidal neurons in the rat prefrontal cortex: intrinsic properties and synaptic function. J Neurophysiol 91:1171–1182

Chapter 24
Stochastic Ion Channel Gating and Probabilistic Computation in Dendritic Neurons

Cian O'Donnell and Matthew F. Nolan

Abstract The electrical signals underlying neural computations are mediated by membrane ion channels. Although these ion channels are well known to operate stochastically, most computational models of dendritic neurons instead make the approximation that ionic conductances are deterministic. We review the basic mathematical considerations underlying this approximation and new efficient simulation tools that allow it to be evaluated systematically. We show how this approximation breaks down for dendritic neurons, with the relative functional influence of stochastic ion channel gating likely to depend strongly on neuron type. An important consequence of stochastic gating of ion channels may be that it causes dendritic neurons to integrate synaptic inputs probabilistically, rather than in the all or nothing fashion predicted by deterministic models.

24.1 Introduction

Neural circuits are computational devices. Increasing experimental evidence indicates that key neural computations involve biochemical and electrical signals that are localized to subcellular compartments such as synapses and dendrites. In very large compartments, such as the cell soma, biochemical and electrical signaling is mediated by many thousands of molecules and is typically reliable, with fluctuations that are small relative to the signals generated. However, many neural computations take place in smaller subcellular compartments, such as dendrites, in which relatively few molecules mediate biochemical and electrical signaling. Because molecular reactions are stochastic, signaling at the subcellular level may therefore

C. O'Donnell • M.F. Nolan (✉)
Centre for Integrative Physiology, University of Edinburgh,
Hugh Robson Building, 15 George Square, Edinburgh EH8 9XD, UK
e-mail: mattnolan@ed.ac.uk

H. Cuntz et al. (eds.), *The Computing Dendrite: From Structure to Function*,
Springer Series in Computational Neuroscience 11, DOI 10.1007/978-1-4614-8094-5_24,

be unreliable. While the consequences of this stochasticity have been overlooked in many studies, recent theoretical evidence suggests that stochastic effects may fundamentally shape computation in neural circuits. Here, we will address the influence on neural computations of stochastic gating of ion channels found in dendrites. Other recent work has addressed consequences of stochastic biochemical signaling including molecular diffusion, probabilistic biochemical reactions, and stochastic gene expression (Raj and van Oudenaarden 2008; Bhalla 2004a, b).

We will address three questions. How can the impact of stochastic ion channel gating on dendritic computation be predicted? What are the effects of stochastic gating of dendritic ion channels on neuronal membrane potential dynamics? Does unreliability of signaling, particularly by stochastic dendritic ion channels, have functional consequences for computations carried out by neurons? We will first review basic principles of ion channel gating and theoretical considerations for the impact of stochastic ion channels in dendrites. We will highlight practical difficulties in simulating stochastic ion channel gating in dendrites and review new simulation tools that solve many of these difficulties. We will then show how these tools reveal influences of dendritic morphology on the functional impact of stochastic ion channel gating and how this stochastic gating leads to probabilistic modes of dendritic communication. Finally, we will suggest key areas for future research.

24.2 Stochastic Gating of Dendritic Ion Channels

Cells generate electrical signals by opening and closing of ion channels that selectively conduct specific ions. Some of these ion channels are sensitive to the membrane voltage (voltage-gated ion channels) whereas others are influenced by neurotransmitters (ligand-gated ion channels). The resting membrane potential of most neurons is determined primarily by leak potassium channels, while the rapid depolarization of an action potential is due to opening of voltage-gated sodium channels. Fast excitatory synaptic transmission typically involves opening of ligand-gated ion channels that selectively conduct sodium and potassium ions, while fast inhibitory transmission is typically through ligand-gated ion channels that selectively conduct chloride ions. In almost all models of electrical signaling in neurons the properties of single ion channels are not accounted for. Instead, electrical signals are modeled as arising from macroscopic conductances that represent the collective behavior of large populations of ion channels (Hodgkin and Huxley 1952; Hines and Carnevale 1997). In these models, conductances change smoothly over time and react deterministically to the model's voltage dynamics, while ion channels are assumed to be continuously distributed in the cell membrane rather than localized to particular points in space.

These assumptions are in contrast to data from patch-clamp recordings which show that macroscopic conductances are mediated through opening and closing of many single ion channels (Neher and Sakmann 1976). If the behavior of a single ion channel is recorded in stationary conditions, its conductance is not constant, but

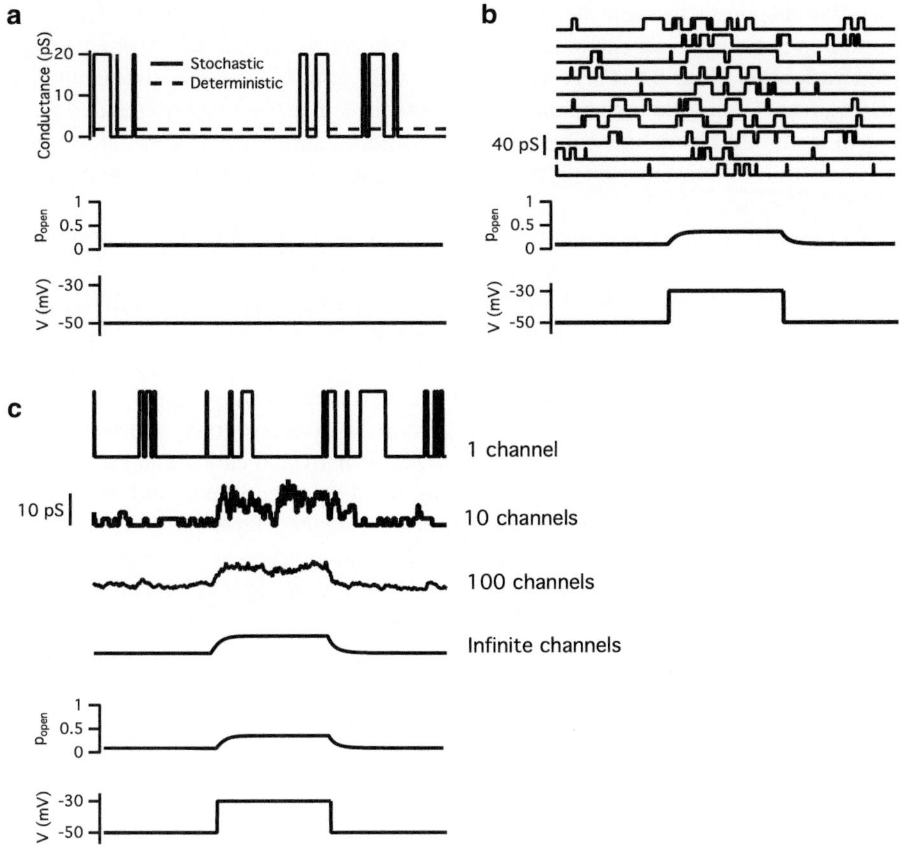

Fig. 24.1 Examples of fluctuating currents from simulated Hodgkin–Huxley potassium channels. (**a**) A single channel shows a fluctuating conductance (*top, solid trace*) by opening and closing randomly even when subject to a stationary voltage (*bottom*) at a fixed open channel probability (*middle*). The equivalent deterministic conductance is also plotted (*top, dashed trace*). (**b**) The channel responds to a depolarizing voltage step (*bottom*) with an elevated open probability (*middle*). Notably, the conductance response (*top*) is different on each trial (*sequential rows*) despite an identical stimulus. (**c**) The conductance response (*top four traces*) to a voltage stimulus (*bottom trace*) is less noisy for larger ion channel populations. For this illustration, single channel conductances were scaled inversely to channel number in order to normalize total conductance amplitudes

instead jumps rapidly between discrete closed and open states (Fig. 24.1a). Typically the open states of single ion channels have conductances from <1 pS to as high as 150 pS (Hille 2001). Stimuli that modulate macroscopic ionic conductances usually do so not by modulating the single channel conductance, but by influencing the probability that single channels will be in their open state. For example, changes in membrane potential influence voltage-gated ion channels by modulating the probability of their opening. As a result, a single ion channel generates unpredictable step-like responses to a repeated stimulus (Fig. 24.1b). In contrast, the responses of large populations of ion channels are predictable and show relatively little trial-to-trial variability (Fig. 24.1c). A key question then is to what extent the variability arising from stochastic ion channel gating influences neuronal function and to what extent it should be included in models of neuronal signaling. Before addressing this question we consider the properties of single ion channels found in dendrites of central neurons.

Cell-attached patch-clamp recordings and antibody labeling data demonstrate localization of voltage-gated ion channels to dendrites of many neuron types (Spruston 2008; Magee 2000; Johnston et al. 2003; Reyes 2001; Nusser 2009). The estimated densities of channels in dendritic membrane range from fewer than 1 μm^{-2} to greater than 500 μm^{-2} (Kole et al. 2006; Engel and Jonas 2005; Chen and Johnston 2004). In cortical pyramidal neurons dendritically located voltage-gated Na^+ and K^+ channels support propagation of action potentials in anterograde and retrograde directions in the dendritic tree. Hyperpolarization-activated cation channels (HCN channels), which mediate the hyperpolarization-activated current I_h, and some other K^+ channel types, for example mediating the A current, have little influence on action potential propagation, but instead influence integration of synaptic responses (Robinson and Siegelbaum 2003; Magee 2000; Wahl-Schott and Biel 2009; Johnston and Narayanan 2008). Most electrophysiological investigations of dendritic ion channels have relied on macroscopic currents that reflect gating of many ion channels within the patch of recorded membrane (Hoffman et al. 1997; Johnston et al. 2003), but single channel recordings from dendrites have been reported in a few studies (Chen and Johnston 2004; Magistretti et al. 1999; Magee and Johnston 1995; Bittner et al. 2012). These studies reveal the kinetics of dendritic ion channel gating and show that macroscopic dendritic conductances can be explained by changes in single channel open probability. Typical single channel conductances of dendritic voltage-gated Na^+ and K^+ channels are 15–30 pS (Chen and Johnston 2004; Magee and Johnston 1995). In contrast, the single channel openings of dendritic HCN channels have not been observed directly, but have been inferred from noise analysis to be less than 1 pS (Kole et al. 2006). The reasons for this diversity of single channel conductances are not yet clear, but a possibility raised by the work we describe below is that it enables fine tuning of the functional consequences of stochastic ion channel gating.

24.3 Theoretical Considerations for Impact of Stochastic Dendritic Ion Channels

Are deterministic models sufficient to explain dendritic computation? Or are there circumstances in which it is necessary also to take account of stochastic gating of single channels? We first address these questions by examining a simple model. Consider a population of N stochastically gating ion channels in a cell membrane, each with the same open probability p and single channel current i. The mean current through this population of channels is:

$$I = iNp$$

In a deterministic model the current flowing through these channels is always equal to the mean current. In contrast, for real populations of channels, and for models that account for their stochastic opening and closing, the actual population current fluctuates around the mean value, hence generating electrical *noise* around the mean *signal*. The standard deviation of these current fluctuations is a natural measure of the amplitude of this noise, and is given from binomial statistics as:

$$\sigma = i\sqrt{Np(1-p)}.$$

This equation leads to a number of predictions. First, the absolute amplitude of noise is maximal when $p=0.5$, and minimal when $p=0$ or $p=1$. Below action potential threshold, most ion channels in the nervous system have open probabilities less than 0.5. In this range, increasing the open probability will increase the current fluctuations. Because for many ion channels p increases upon depolarization (Hille 2001), this suggests that the absolute amplitude of ion channel noise will increase with depolarization. In contrast to the absolute amplitude, the relative amplitude of fluctuations, when defined as the standard deviation of the current divided by its mean, is proportional to $\sqrt{\frac{(1-p)}{p}}$. Because this quantity decreases with increasing p, the *relative* amplitude of fluctuations usually decreases upon depolarization. Second, because the standard deviation is proportional to $\sqrt{N}$, the absolute amplitude of current fluctuations will be greater for larger populations of ion channels. On the other hand, because the mean current is proportional to N, the relative amplitude of current fluctuations will decrease for larger populations of ion channels. Third, larger single channel currents will also lead to greater amplitude current fluctuations (both absolutely and relatively). Because single channel currents can differ by three or more orders of magnitude between channel types, both because of differences in single-channel conductances and differences in reversal potentials (Hille 2001), this could lead to considerable diversity in the impact of stochastic ion channel gating.

How will current fluctuations from the ion channel populations influence membrane potential dynamics? When current flows through an ion channel it can modify the membrane potential by charging or discharging the membrane capacitance or it can flow axially to other parts of the cell (Rall 1959, 1962). The functional impact of stochastic ion channel gating therefore depends not only on the properties of stochastic currents, but also on the electrical properties of the cell in which they take place. The cell membrane's parallel capacitance and resistance act as a low-pass filter—they preferentially attenuate high frequency components of membrane currents. The product of the capacitance and resistance determines the time constant of filtering, while the resistance sets the amplitude of membrane potential changes. The low pass filter characteristics of the membrane suggest that current fluctuations from ion channels with faster gating kinetics will be more attenuated than those from channels with slower gating kinetics (DeFelice 1981). This prediction is confirmed by simulations comparing channels with different physiological gating kinetics (Cannon et al. 2010). In addition to membrane properties, the cytoplasmic resistivity will influence the axial flow of ionic currents. Because axial flow of current is determined by axial resistance, membrane resistance and membrane capacitance throughout a cell, the cell's morphology and the spatial arrangement of the cell's compartments influences both the local and distal impact of stochastic current fluctuations. Finally, the cell's eventual voltage dynamics will also depend on its entire complement of ion channels, and on the cell's level of depolarization. For example, in most cells near resting potential small voltage transients are dampened and quickly decay back to rest. In contrast, at membrane potentials nearer to spike threshold the voltage dynamics become nonlinear. In this region, small voltage fluctuations from channel noise can be quickly amplified by the macroscopic dynamics, leading to large amplitude transients or oscillations (White et al. 1995, 1998). If the fluctuations are large enough, they may even trigger action potentials (Skaugen and Walloe 1979; Johansson and Arhem 1994). Hence, microscopic fluctuations from stochastic channel gating can have large macroscopic consequences (Schneidman et al. 1998; Dudman and Nolan 2009; Faisal et al. 2005). This sensitivity to noise near threshold reflects a general property of all nonlinear excitable systems near a bifurcation (Lindner et al. 2004).

In summary, a number of factors may determine the influence of stochastic ion channel gating on neuronal computation. Together, single channel conductance, the number of channels and their open probability determine the amplitude of current fluctuations that arise from a population of channels. The impact of these current fluctuations on membrane potential dynamics is determined by local membrane capacitance, axial current flow, and presence of other ion channels. This suggests that in principle stochastic ion channel gating could profoundly influence membrane potential dynamics, particularly in small electrically isolated structures. However, because many of the properties that will determine the influence of stochastic ion channel gating vary both within and between cells, predicting the functional consequences of stochastic ion channel gating requires models that account for these details in a particular neuron.

24.4 Simulation of Stochastic Ion Channels in Complex Neuronal Morphologies

The computational demands of accurately simulating model neurons in which all ion channels gate stochastically have until recently made it impractical to carry out detailed investigations of the consequences of stochastic ion channel gating. The scale of this challenge is illustrated by the fact that a single neuron has on the order of one million ion channels, each at a different location on its membrane, and because each channel gates independently and stochastically, any channel might switch state at any moment in time. There are two different general approaches to this problem. Microscopic models describe individual ion channels as continuous-time Markov chains where stochastic transitions occur between discrete states. In contrast, stochastic differential equation (SDE) models describe ionic currents as the sum of a deterministic Hodgkin–Huxley-style component plus a stochastic noise component.

24.4.1 Microscopic Models of Ion Channel Gating

In microscopic descriptions each ion channel is considered a distinct object which can exist in one of several discrete states. These states typically correspond to different conformations of the ion channel protein. At least one of the states is "open" (the channel passes current), and at least one state is "closed" (the channel does not pass current). The channel is considered memoryless (the Markov property), so that the length of time the channel has spent in the present state has no bearing on its future behavior. Transitions between states are driven by thermal fluctuations, and because actual transition times are on the order of nanoseconds (Hille 2001), for modeling purposes they are usually assumed to be instantaneous. For voltage-gated ion channels, the membrane voltage influences the probability of transitions between states. For example, a Na^+ channel will be more likely to switch to an open state upon depolarization. Most microscopic simulation methods use Monte Carlo techniques. The most conceptually simple way to do this is to first, after choosing a small time step Δt, convert each transition rate, k, in a channel's kinetic scheme to a transition probability, p by assuming $p \approx k\Delta t$. Then, for each time step, a random number x between zero and one is drawn for each channel. If $x < p$, the state transition is performed. However, for realistic neuronal simulations this approach is prohibitively expensive because it requires the generation of at least one random number for each ion channel at each time step. For example, if a neuron model containing one million two-state ion channels were to be simulated with a time step of 1 μs, then one millisecond of simulation time would require generation of $10^6 \times \frac{1ms}{1\mu s} = 10^9$ random numbers, each to be compared to transition probabilities, in addition to the normal calculation of the voltage dynamics.

A solution to this problem comes from algorithms developed to accelerate simulation of stochastic chemical reactions (Gillespie 1977, 2001; Cao et al. 2006). In early algorithms, following each channel transition random numbers are drawn to determine both when the next transition will occur, and which transition will occur (Skaugen and Walloe 1979; Chow and White 1996). Then the simulation steps forward in time by the calculated interval and performs the corresponding state transition. Although this algorithm is exact, meaning that it has zero approximation errors, it does not scale well to many ion channels. More recently, we developed software called the Parallel Stochastic Ion Channel Simulator (PSICS) to enable more efficient simulation of stochastic ion channel gating in spatially distributed neuron models (Cannon et al. 2010). This simulator, which is used for the studies described in detail here, includes several approaches to speed up simulations without compromising accuracy:

1. It groups together the ion channels in each state per spatial compartment, enabling stochastic calculation of the *numbers* of channels making a certain transition instead of performing the calculation for each channel independently. This grouping is allowable because sufficiently nearby ion channels can be considered interchangeable from an electrical point of view. That is, they are subject to and influence the same local membrane potential.
2. It uses a version of the tau-leap method (Cao et al. 2006), where multiple state-transitions are allowed within a single time step, even for a single ion channel. This allows for longer time steps and hence faster simulation. However, the changes at the end of the time step must be evaluated to make sure that voltage would not have changed so much during the time step that the assumption of fixed transition probabilities was unjustifiable.
3. It ignores state transitions that are highly improbable, according to a specified threshold.
4. It precomputes lookup tables for the state-transition probabilities over the anticipated voltage range.

When these approaches are combined, simulation of detailed and realistic compartmental models of neurons becomes possible in a reasonable period of time (Cannon et al. 2010).

24.4.2 Stochastic Differential Equation Models of Ion Channel Gating

The SDE approach for simulating ion channels was pioneered with the hope of providing faster simulation algorithms than the Monte Carlo techniques (Fox and Lu 1994; Fox 1997). In early descriptions, the membrane conductance gating variable dynamics were reduced to a Langevin equation:

$$\frac{\mathrm{d}x}{\mathrm{d}t} = \alpha(1-x) - \beta x + \eta(V, t)$$

where x is a given Hodgkin–Huxley gating variable, α and β are voltage-dependent transition rates, and $\eta(V, t)$ is a zero-mean Gaussian noise term. Note that this is equivalent to the classic Hodgkin–Huxley description plus a stochastic noise term.

Although relatively fast to compute, in this form the algorithm shows poor approximation to the exact solutions, even in the limit of a large population of ion channels (Mino et al. 2002; Bruce 2009). This is because it does not accurately capture correlations in fluctuations between channel states (Bruce 2009; Goldwyn et al. 2011; Linaro et al. 2011). In general, the difficulties in the SDE approach arise in deciding where to insert the noise term (as a current, conductance, or channel subunit fluctuation), deciding how the noise intensity should scale as voltage changes, and deciding how to accurately account for temporal correlations in the noise. The primary discrepancies have only recently been resolved by modeling the noise statistics based on single channel properties rather than on channel subunit properties as in earlier studies (Goldwyn and Shea-Brown 2011). Despite these advances, further approximations in the SDE approach await resolution: for example, at present the noise statistics may lag the voltage dynamics during fast transients (Linaro et al. 2011). In addition, the relative speeds and approximation errors of modern SDE approaches versus modern Markov-chain simulation algorithms (Cannon et al. 2010) remain unquantified, making it unclear which approach is best for a given simulation problem. Indeed, SDE approaches are yet to be applied successfully to investigation of stochastic ion channel gating in dendrites.

24.5 Dendritic Morphology Determines the Influence of Stochastic Ion Channel Gating

Does stochastic gating of ion channels influence membrane potential dynamics in dendrites and what is the impact of dendrite morphology on the functional consequences of stochastic ion channel gating? Previous simulations suggest that stochastic ion channel gating causes larger fluctuations in the membrane potential of dendritic neurons compared to spherical neurons of a similar surface area (van Rossum et al. 2003), but these simulations did not evaluate different dendritic structures or distinct neuron types. By developing software for efficient simulation of stochastic gating of ion channels in spatially extended model neurons, we have been able to address these issues directly (Cannon et al. 2010). To do so, we compared membrane potential activity of 29 reconstructed neurons corresponding to six different neuron types. To isolate the influence of neuronal morphology we inserted the same density of ion channels in the membrane of each simulated neuron. Ion channel models and distributions were implemented according to a previously published neocortical pyramidal neuron model (Mainen and Sejnowski 1996). The leak conductance was modeled as a mixture of voltage-independent Na^+ and K^+ channels with open probabilities of 0.7 and density of 0.016 μm^{-2}. Active channels included fast Na^+ channels (1 μm^{-2}), non-inactivating K^+ channels (0.05 μm^{-2}), and high-voltage Ca^{2+} channels (0.15 μm^{-2}). Because ion channels distributions were identical across neurons, differences in voltage fluctuations must reflect differences in morphology.

We found that with these parameters stochastic ion channel gating caused the resting membrane potential of all simulated neurons to fluctuate with standard deviation up to 0.8 mV. This is in agreement with experimental measurement of voltage noise from cultured neurons (Diba et al. 2004) and from neurons in brain

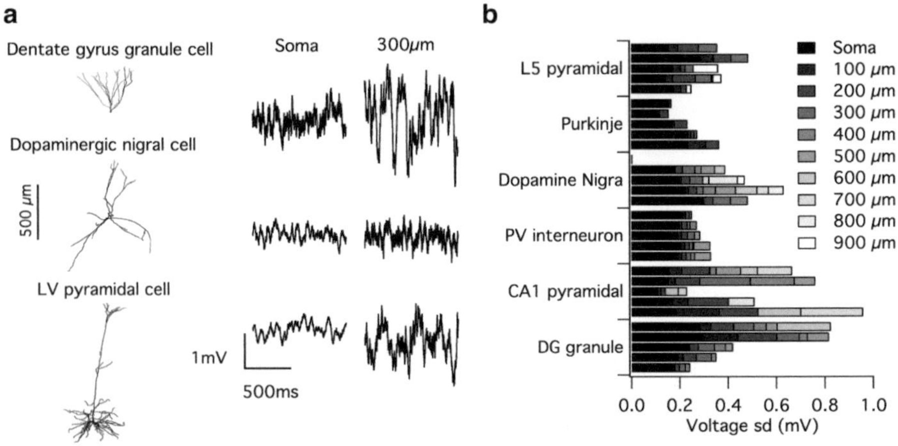

Fig. 24.2 The impact of stochastic ion channel gating depends on neuronal morphology. (**a**) Example resting membrane potential traces (*right*) from three model neurons with dendritic arbours from different cell types (*left*). Traces are from soma and at a dendritic location 300 μm from soma. (**b**) Standard deviation of the resting voltage for 29 reconstructed neurons from six different cell types. Greyscale level indicates dendritic distance from soma. Note that the amplitude of fluctuations depends on cell type, and is generally larger in the dendrites than at the soma. Adapted from Cannon et al. (2010)

slices (Jacobson et al. 2005). Comparison of voltage fluctuations at different locations along the dendrites of the simulated neurons reveals that the amplitude of the fluctuations increases with distance from the soma (Fig. 24.2). Thus, stochastic gating of ion channels can contribute to the voltage noise in neurons and this noise is greatest in more distal dendrites. When we compared voltage fluctuations between different neuron types we found striking differences. For example, noise from stochastic ion channel gating was relatively small in simulated Purkinje cells and parvalbumin positive interneurons, but was much larger in CA1 pyramidal neurons (Fig. 24.2). This suggests that the impact of stochastic gating is sensitive to differences in morphology that are found between different neuron types. These effects are likely mediated both by differences in the degree of axial charge flow and differences in local membrane impedance profiles across neural morphologies (as discussed earlier). Both of these properties are influenced by dendritic diameter, length and branching pattern (Koch 1999).

To test the effects of channel kinetics on voltage noise we reduced the time constants for the gating of leak channels tenfold. This manipulation changes only the rate of switching between open and closed states while leaving the mean fraction of open channels unaltered. This change caused the amplitude of the voltage fluctuations to increase approximately threefold, but dependence of noise on dendritic location and differences between neuron types were maintained, indicating that in realistic neuronal morphologies channel kinetics will influence the amplitude of voltage fluctuations resulting from stochastic ion channel gating (Cannon et al. 2010).

Together these simulations demonstrate that neuronal geometry critically determines the influence of stochastic gating on membrane potential dynamics. They also show that geometry and ion channel properties interact, with slow gating ion channels causing larger membrane potential fluctuations. While these results demonstrate that models which include physiologically realistic ion channel and membrane properties lead to consistent functional effects of stochastic ion channel gating, the sensitivity of these effects to differences in morphology and channel properties suggests that predictions for specific cell types should be treated with care. In particular, these results indicate the importance of obtaining accurate reconstructions of neuronal morphology and more data about the single channel properties of neuronal ion channels.

24.6 Stochastic Dendrites Integrate Probabilistically

How might stochastic gating of ion channels influence neuronal computations? To address this question, it is important to establish the effects of stochastic channel gating on transformation of synaptic input into an action potential output. We therefore adopted a detailed model of a hippocampal CA1 pyramidal neuron that is well constrained by experimental data, but has previously only been simulated using deterministic ion channels (Jarsky et al. 2005). We implemented the model in a fully stochastic configuration and further refined it by adding a realistic distribution of HCN channels, which mediate the hyperpolarization activated current (I_h) (Cannon et al. 2010). These channels are expressed in an increasing gradient from soma to distal dendrites (Magee 1998; Golding et al. 2005; Lorincz et al. 2002; Nolan et al. 2004). We then investigated the response of the model to activation of 1,503 excitatory synaptic inputs located randomly throughout the basal and apical dendrites. The activity of each input followed a Poisson process and the amplitude of the inputs was adjusted so that the firing rate of the stochastic neuron was approximately 20 Hz. We chose these parameters to mimic activity levels that might occur during active theta states in the hippocampus. We reasoned that if stochastic gating of ion channels influences neuronal computation then switching the model between stochastic and deterministic configurations should modify its transformation of synaptic input into spike output.

When we compare responses to repeated presentation of the same pattern of synaptic inputs, we find as expected that the deterministic version of the model produces identical responses on each trial. In contrast, the version of the model in which all ion channels gate stochastically generates responses that differ from trial to trial (Cannon et al. 2010). In general action potentials occur at similar times in each trial, but unlike the deterministic model where the probability of spiking at a particular time point is either one or zero, in the stochastic model the probability of spiking at any time point varies between one and zero (Fig. 24.3). At times when spikes are triggered in the deterministic model they are "dropped" on a fraction of trials by the stochastic model, whereas at other times at which spikes are absent in

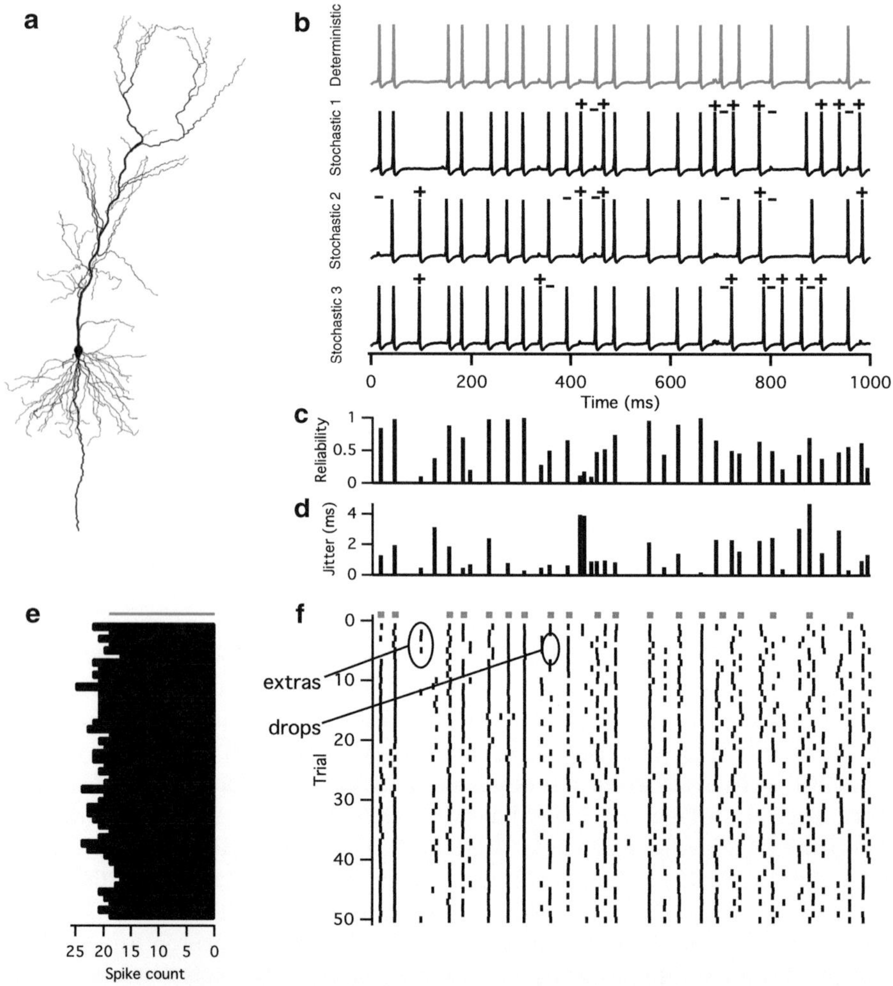

Fig. 24.3 Channel noise affects synaptic integration in a morphologically realistic neuron. (**a**) Morphology of the CA1 neuron model. (**b**) Example voltage traces from deterministic model (*grey*), and stochastic trials 1, 2, and 3 (*black*). *Plus symbols* mark "dropped" spikes and *minus symbols* mark "extra" spikes occurring on stochastic trials when compared to the deterministic trial. (**c–d**) Reliability (**c**), jitter (**d**) for each spike event. (**e**) Spike count for each trial in (**f**). (**f**) Raster plot of action potential times recorded at soma during 1 s of model time. One deterministic trial in *grey squares* and 50 stochastic trials in *black ticks*. Adapted from Cannon et al. (2010)

the deterministic model "extra" spikes occur in the stochastic model. We quantified the reliability of each spike as the fraction of trials on which it occurred (Fig. 24.3c). Stochastic ion channel gating also introduces considerable jitter into the timing of action potentials, quantified as the standard deviation of the spike timing (Fig. 24.3d). Whereas we find that the probability of spike initiation is modified when ion channels gate stochastically, the waveform of action potentials does not differ.

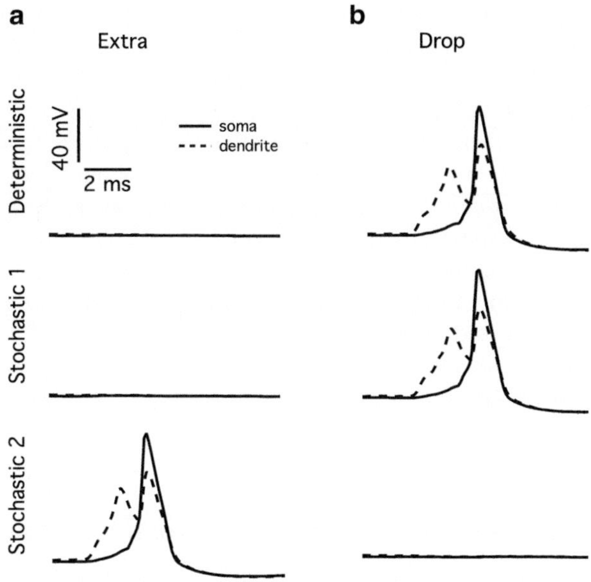

Fig. 24.4 Probabilistic dendritic events underlie probabilistic axonal firing from stochastic ion channel gating. *Rows* are example voltage traces from deterministic (*top*), stochastic trial 1 (*middle*), and stochastic trial 2 (*bottom*) simulations from Fig. 24.3. Extra (**a**) and dropped (**b**) axonal action potentials from probabilistic dendritic spikes. Adapted from Cannon et al. (2010)

This is in agreement with previous simulations of the responses of stochastic neurons which used injected current rather than distributed synaptic input to initiate action potentials (Diba et al. 2006). Together, these data predict that in real neurons, stochastic ion channel gating will cause responses to a given synaptic input to be probabilistic. This is in contrast to deterministic simulations which generate identical all-or-nothing responses to repeated presentation of the same input.

Where in the neuron does the variability in spiking originate from and do different types of ion channel, or ion channels found at different locations, preferentially contribute to the functional effects of stochastic ion channel gating? By recording the membrane potential simultaneously in each primary dendrite of the model neuron (~20 μm from the soma), the origin of the "extra" and "dropped" spikes can be addressed. This reveals that both "extra" and "dropped" somatic spikes are preceded by all-or-nothing dendritic depolarizations indicating that they originate from dendritic spikes that propagate towards the soma (Fig. 24.4).

To address the roles of particular ion channels we simulated models in which subsets of ion channels gate stochastically while the remaining ion channels gate deterministically. We found that stochastic gating of any single ion channel type causes "dropped" and "extra" spikes, but no single ion channel type alone could fully account for all of the "dropped" and "extra" spikes in the fully stochastic model. Stochastic gating of either Na^+ channels, A-type or delayed rectifier potassium channels alone caused more than 50 % of the number of "dropped" and "extra" spikes in the fully stochastic model, indicating strong redundancy in the contribution of particular ion channel types to probabilistic integration of synaptic inputs. When only axonal channels gated stochastically then the number of "dropped" and "extra" spikes was reduced by approximately 50 %, whereas when only dendritic ion channels gated stochastically the number of "dropped" and "extra" spikes was similar to in the fully stochastic model, indicating that stochastic gating of dendritic ion channels is particularly important for probabilistic integration of distributed synaptic inputs.

Together, these data suggest that stochastic gating of dendritic ion channels may have profound effects on single neuron computations. In contrast to deterministic models where the same synaptic inputs produce the same output, when ion channels gate stochastically then the response to synaptic inputs is probabilistic. The considerable redundancy revealed by substantial numbers of "extra" and "dropped" spikes when only a single channel type gates stochastically, suggests that the basic results are robust to the exact choice of parameters. Indeed, because our choice of simulation parameters is quite conservative, it is perhaps more likely that in real neurons, ion channels with a greater conductance would further increase the impact of stochastic gating.

24.7 Other Sources of Electrical Noise in Neurons

Before discussing further functional implications of these results, it is worth considering the likely influence of noise from stochastic ion channel gating relative to that of other potential neuronal sources of noise. In general these other noise sources can be divided into two categories: synaptic and non-synaptic. Non-synaptic sources of noise include thermal or "Johnson" noise, and ion channel shot noise. Johnson noise arises from thermal fluctuations in the potential difference across a conductor (e.g., the cell membrane) and is calculated to be negligible for neural membranes (Lecar and Nossal 1971; Manwani and Koch 1999). Ion channel shot noise is generated by momentary fluctuations in the flow of discrete ions, but the high rate of ion flow in physiological currents ($\sim 10^6$/s) implies that shot noise is likely negligible. Thus stochastic channel gating should be considered the dominant non-synaptic source of electrical noise. Synaptic sources of variability are numerous and include probabilistic presynaptic release of neurotransmitter vesicles, heterogeneous location of vesicle release within synapses, heterogeneity in the size of vesicles, diffusion of the neurotransmitter, stochastic binding of neurotransmitter to postsynaptic receptors, and stochastic gating of ionotropic receptors. These factors may lead to substantial variability in synaptic currents (Franks et al. 2003; Lisman et al. 2007; Faisal et al. 2008) and it will be important to establish their influence relative to that of stochastic ion channel gating.

24.8 Summary and Future Directions

Given that the functional consequences of stochastic ion channel gating are difficult to predict, but might in principle affect neuronal computation, why have they not received more attention? The success and adaptability of the original

Hodgkin–Huxley formalism for describing membrane conductances compared with the relative scarcity of studies demonstrating physiologically relevant implications of stochastic ion channel gating has justified use of deterministic approaches to understanding neuronal computations. However, the neuronal compartments where stochastic effects are likely to be greatest, such as dendritic branches, spines, and synaptic terminals, are also the least accessible to experimental analysis. Hence this confidence in deterministic frameworks may turn out not to be supported by the data. Of equal importance, the computational demands of simulating neurons with fully stochastic ion channel models have until recently made it impractical to carry out detailed simulations of morphologically realistic models of neurons in which all channels gate stochastically.

Further functional consequences of stochastic channel gating in dendrites may remain to be discovered. Questions for future research include:

- What are the implications of channel noise for neural coding? Several theoretical studies have now shown that stochastic ion channel gating can induce unreliability in neural responses, and imprecision in response timing (Schneidman et al. 1998; Dudman and Nolan 2009; Cannon et al. 2010). If information is encoded in the timing of spike sequences, then stochastic channel gating might set an upper limit on the information rate because spike timing cannot be more precise than the inherent jitter. If information were instead encoded in spike rates, then channel gating might also limit information coding by adding noise to any spike count over a short time interval (Fig. 24.3e). Alternatively, the probabilistic synaptic integration caused by stochastic channel gating may be used by the nervous system in some beneficial way. For example, one possibility is that channel noise is used to implement a "sampling" process where fluctuations in a neuron's firing pattern reflect uncertainty in the brain's inferences about the external world (Fiser et al. 2010; Buesing et al. 2011).
- How big an impact do the fluctuations from stochastic channel gating have relative to the contributions from other sources of electrical noise? For example, it is possible that probabilistic vesicle release will prove at least as significant. The answer to this question may strongly depend on cell type specific factors such as ion channel composition, dendritic morphology, and synaptic properties.
- Where in the cell does stochastic ion channel gating have the greatest impact? Our simulation results suggest that stochastic ion channel gating is of particular importance for dendritic computations. This is consistent with general biophysical considerations which suggest that voltage noise from channel fluctuations will be largest when channel numbers are small and the local membrane impedance is large. Indeed we expect that channel noise will have greatest impact in small protrusions such as dendritic spines, which to our knowledge have not yet been considered in fully stochastic simulations.
- Does stochastic channel gating introduce unreliability into intracellular biochemical signaling? Several voltage-gated ion channels, particularly those with a high permeability to calcium, serve dual roles as mediators of electrical and biochemical signaling. Because these ion channels gate stochastically, downstream sensors of these signals must also be activated stochastically, unless they

act to integrate the biochemical signals over a slower timescale to average out stochastic fluctuations. In particular, local Ca^{2+} microdomain signaling may be drastically affected by stochastic channel gating because Ca^{2+} microdomains are believed to require a few hundred microseconds to reach their steady-state concentration (Neher 1998)—a timescale comparable to the switching times of many dendritic ion channels (Magee and Johnston 1995).

In conclusion the ability to simulate stochastic ion channels distributed through neurons with arbitrary morphology has enabled direct comparison between realistic models of synaptic integration. The results indicate that in principle stochastic gating of ion channels has profound effects on neuronal computation, converting neurons from all-or-nothing to probabilistic integrators of synaptic input. Key challenges for the future are to establish the range of neuronal computations that are influenced by stochastic ion channel gating and to determine the consequences for higher level models of neuronal computation. Addressing these challenges will require more detailed and accurate experimental data and further improvements in simulation methods to enable efficient stochastic simulation of whole neuronal circuits. A final challenge is to develop experimental tests that enable simulation-based predictions of the consequences of stochastic ion channel gating to be tested in experimental models (Dorval and White 2005).

References

Bhalla US (2004a) Signaling in small subcellular volumes. I. Stochastic and diffusion effects on individual pathways. Biophys J 87(2):733–744

Bhalla US (2004b) Signaling in small subcellular volumes. II. Stochastic and diffusion effects on synaptic network properties. Biophys J 87(2):745–753

Bittner KC, Andrasfalvy BK, Magee JC (2012) Ion channel gradients in the apical tuft region of CA1 pyramidal neurons. PLoS One 7(10):e46652

Bruce IC (2009) Evaluation of stochastic differential equation approximation of ion channel gating models. Ann Biomed Eng 37(4):824–838

Buesing L, Bill J, Nessler B, Maass W (2011) Neural dynamics as sampling: a model for stochastic computation in recurrent networks of spiking neurons. PLoS Comput Biol 7(11):e1002211

Cannon RC, O'Donnell C, Nolan MF (2010) Stochastic ion channel gating in dendritic neurons: morphology dependence and probabilistic synaptic activation of dendritic spikes. PLoS Comput Biol 6(8)

Cao Y, Gillespie DT, Petzold LR (2006) Efficient step size selection for the tau-leaping simulation method. J Chem Phys 124(4):044109

Chen X, Johnston D (2004) Properties of single voltage-dependent K+ channels in dendrites of CA1 pyramidal neurones of rat hippocampus. J Physiol 559(Pt 1):187–203

Chow CC, White JA (1996) Spontaneous action potentials due to channel fluctuations. Biophys J 71(6):3013–3021

DeFelice LJ (1981) Introduction to membrane noise. Plenum, New York

Diba K, Lester HA, Koch C (2004) Intrinsic noise in cultured hippocampal neurons: experiment and modeling. J Neurosci 24(43):9723–9733

Diba K, Koch C, Segev I (2006) Spike propagation in dendrites with stochastic ion channels. J Comput Neurosci 20(1):77–84

Dorval AD Jr, White JA (2005) Channel noise is essential for perithreshold oscillations in entorhinal stellate neurons. J Neurosci 25(43):10025–10028
Dudman JT, Nolan MF (2009) Stochastically gating ion channels enable patterned spike firing through activity-dependent modulation of spike probability. PLoS Comput Biol 5(2):e1000290
Engel D, Jonas P (2005) Presynaptic action potential amplification by voltage-gated Na+ channels in hippocampal mossy fiber boutons. Neuron 45(3):405–417
Faisal AA, White JA, Laughlin SB (2005) Ion-channel noise places limits on the miniaturization of the brain's wiring. Curr Biol 15(12):1143–1149
Faisal AA, Selen LP, Wolpert DM (2008) Noise in the nervous system. Nat Rev Neurosci 9(4): 292–303
Fiser J, Berkes P, Orban G, Lengyel M (2010) Statistically optimal perception and learning: from behavior to neural representations. Trends Cogn Sci 14(3):119–130
Fox RF (1997) Stochastic versions of the Hodgkin–Huxley equations. Biophys J 72(5):2068–2074
Fox RF, Lu Y (1994) Emergent collective behavior in large numbers of globally coupled independently stochastic ion channels. Phys Rev E Stat Phys Plasmas Fluids Relat Interdiscip Topics 49(4):3421–3431
Franks KM, Stevens CF, Sejnowski TJ (2003) Independent sources of quantal variability at single glutamatergic synapses. J Neurosci 23(8):3186–3195
Gillespie DT (1977) Exact stochastic simulation of coupled chemical-reactions. Abstr Paper Am Chem Soc 173:128
Gillespie DT (2001) Approximate accelerated stochastic simulation of chemically reacting systems. J Chem Phys 115(4):1716–1733
Golding NL, Mickus TJ, Katz Y, Kath WL, Spruston N (2005) Factors mediating powerful voltage attenuation along CA1 pyramidal neuron dendrites. J Physiol 568(Pt 1):69–82
Goldwyn JH, Shea-Brown E (2011) The what and where of adding channel noise to the Hodgkin–Huxley equations. PLoS Comput Biol 7(11):e1002247
Goldwyn JH, Imennov NS, Famulare M, Shea-Brown E (2011) Stochastic differential equation models for ion channel noise in Hodgkin-Huxley neurons. Phys Rev E Stat Nonlin Soft Matter Phys 83:041908
Hille B (2001) Ion channels of excitable membranes, 3rd edn. Sinauer, Sunderland, MA
Hines ML, Carnevale NT (1997) The NEURON simulation environment. Neural Comput 9(6): 1179–1209
Hodgkin AL, Huxley AF (1952) A quantitative description of membrane current and its application to conduction and excitation in nerve. J Physiol 117(4):500–544
Hoffman DA, Magee JC, Colbert CM, Johnston D (1997) K+ channel regulation of signal propagation in dendrites of hippocampal pyramidal neurons. Nature 387(6636):869–875
Jacobson GA, Diba K, Yaron-Jakoubovitch A, Oz Y, Koch C, Segev I, Yarom Y (2005) Subthreshold voltage noise of rat neocortical pyramidal neurones. J Physiol 564:145–160
Jarsky T, Roxin A, Kath WL, Spruston N (2005) Conditional dendritic spike propagation following distal synaptic activation of hippocampal CA1 pyramidal neurons. Nat Neurosci 8:1667–1676
Johansson S, Arhem P (1994) Single-channel currents trigger action potentials in small cultured hippocampal neurons. Proc Natl Acad Sci USA 91(5):1761–1765
Johnston D, Narayanan R (2008) Active dendrites: colorful wings of the mysterious butterflies. Trends Neurosci 31(6):309–316
Johnston D, Christie BR, Frick A, Gray R, Hoffman DA, Schexnayder LK, Watanabe S, Yuan LL (2003) Active dendrites, potassium channels and synaptic plasticity. Philos Trans R Soc Lond B Biol Sci 358(1432):667–674
Koch C (1999) Biophysics of computation: information processing in single neurons. Computational neuroscience. Oxford University Press, New York
Kole MH, Hallermann S, Stuart GJ (2006) Single Ih channels in pyramidal neuron dendrites: properties, distribution, and impact on action potential output. J Neurosci 26:1677–1687
Lecar H, Nossal R (1971) Theory of threshold fluctuations in nerves. II. Analysis of various sources of membrane noise. Biophys J 11(12):1068–1084

Linaro D, Storace M, Giugliano M (2011) Accurate and fast simulation of channel noise in conductance-based model neurons by diffusion approximation. PLoS Comput Biol 7(3):e1001102

Lindner B, Garcia-Ojalvo J, Neiman A, Schimansky-Geier L (2004) Effects of noise in excitable systems. Phys Rep 392(6):321–424

Lisman JE, Raghavachari S, Tsien RW (2007) The sequence of events that underlie quantal transmission at central glutamatergic synapses. Nat Rev Neurosci 8(8):597–609

Lorincz A, Notomi T, Tamas G, Shigemoto R, Nusser Z (2002) Polarized and compartment-dependent distribution of HCN1 in pyramidal cell dendrites. Nat Neurosci 5:1185–1193

Magee JC (1998) Dendritic hyperpolarization-activated currents modify the integrative properties of hippocampal CA1 pyramidal neurons. J Neurosci 18:7613–7624

Magee JC (2000) Dendritic integration of excitatory synaptic input. Nat Rev Neurosci 1(3):181–190

Magee JC, Johnston D (1995) Characterization of single voltage-gated Na+ and Ca2+ channels in apical dendrites of rat CA1 pyramidal neurons. J Physiol 487:67–90

Magistretti J, Ragsdale DS, Alonso A (1999) Direct demonstration of persistent Na+ channel activity in dendritic processes of mammalian cortical neurones. J Physiol 521:629–636

Mainen ZF, Sejnowski TJ (1996) Influence of dendritic structure on firing pattern in model neocortical neurons. Nature 382(6589):363–366

Manwani A, Koch C (1999) Detecting and estimating signals in noisy cable structure, I: neuronal noise sources. Neural Comput 11:1797–1829

Mino H, Rubinstein JT, White JA (2002) Comparison of algorithms for the simulation of action potentials with stochastic sodium channels. Ann Biomed Eng 30:578–587

Neher E (1998) Usefulness and limitations of linear approximations to the understanding of Ca++ signals. Cell Calcium 24:345–357

Neher E, Sakmann B (1976) Single-channel currents recorded from membrane of denervated frog muscle fibres. Nature 260(5554):799–802

Nolan MF, Malleret G, Dudman JT, Buhl DL, Santoro B, Gibbs E, Vronskaya S, Buzsaki G, Siegelbaum SA, Kandel ER, Morozov A (2004) A behavioral role for dendritic integration: HCN1 channels constrain spatial memory and plasticity at inputs to distal dendrites of CA1 pyramidal neurons. Cell 119(5):719–732

Nusser Z (2009) Variability in the subcellular distribution of ion channels increases neuronal diversity. Trends Neurosci 32(5):267–274

Raj A, van Oudenaarden A (2008) Nature, nurture, or chance: stochastic gene expression and its consequences. Cell 135(2):216–226

Rall W (1959) Branching dendritic trees and motoneuron membrane resistivity. Exp Neurol 1:491–527

Rall W (1962) Theory of physiological properties of dendrites. Ann N Y Acad Sci 96:1071–1092

Reyes A (2001) Influence of dendritic conductances on the input–output properties of neurons. Annu Rev Neurosci 24:653–675

Robinson RB, Siegelbaum SA (2003) Hyperpolarization-activated cation currents: from molecules to physiological function. Annu Rev Physiol 65:453–480

Schneidman E, Freedman B, Segev I (1998) Ion channel stochasticity may be critical in determining the reliability and precision of spike timing. Neural Comput 10:1679–1703

Skaugen E, Walloe L (1979) Firing behaviour in a stochastic nerve membrane model based upon the Hodgkin–Huxley equations. Acta Physiol Scand 107(4):343–363

Spruston N (2008) Pyramidal neurons: dendritic structure and synaptic integration. Nat Rev Neurosci 9:206–221

van Rossum MC, O'Brien BJ, Smith RG (2003) Effects of noise on the spike timing precision of retinal ganglion cells. J Neurophysiol 89:2406–2419

Wahl-Schott C, Biel M (2009) HCN channels: structure, cellular regulation and physiological function. Cell Mol Life Sci 66:470–494

White JA, Budde T, Kay AR (1995) A bifurcation analysis of neuronal subthreshold oscillations. Biophys J 69:1203–1217

White JA, Klink R, Alonso A, Kay AR (1998) Noise from voltage-gated ion channels may influence neuronal dynamics in the entorhinal cortex. J Neurophysiol 80:262–269

Chapter 25
Cellular and Dendritic Memory Allocation

George Kastellakis and Panayiota Poirazi

Abstract The physical basis of memory storage in the brain, the *memory engram*, is a long-standing question in neuroscience, yet an established theory of its realization is still missing. It is only recently that advances in genetic tools and cellular imaging have allowed the visualization of cellular correlates of learning and memory, thus beginning to uncover the rules that underlie the formation of new memory engrams in the brain. As data cumulate, memory formation appears to be a complex process that spans multiple layers, from the neural network down to the subcellular level. Dendrites in particular are suggested to play a central role, not only in the localized processing of incoming signals but also in controlling, both spatially and temporally, the plasticity of their synaptic connections. In this chapter, we discuss recent evidence regarding the formation of memory traces and their properties, paying special attention to the dendritic structure that underlies cellular memory engrams as well as the implications of these findings for the development of cognitive and theoretical models of learning and memory.

G. Kastellakis
Institute of Molecular Biology and Biotechnology, Foundation for Research and Technology, N. Plastira 100, Vassilika Vouton, P.O. Box 1385, Heraklion 70013, Crete, Greece

Department of Biology, University of Crete, Heraklion, Crete, Greece
e-mail: gkastel@gmail.com

P. Poirazi (✉)
Institute of Molecular Biology and Biotechnology, Foundation for Research and Technology, N. Plastira 100, Vassilika Vouton, P.O. Box 1385, Heraklion 70013, Crete, Greece
e-mail: poirazi@imbb.forth.gr

H. Cuntz et al. (eds.), *The Computing Dendrite: From Structure to Function*, Springer Series in Computational Neuroscience 11, DOI 10.1007/978-1-4614-8094-5_25,

25.1 Introduction

Initial lesion studies that searched for the memory engram found that permanent memories are not stored in a particular location of the brain. Based on these early findings, Lashley postulated the "theory of mass action" according to which the brain acts "as a whole" during learning and memory is not localized to a certain brain compartment (Lashley 1950). Since then, however, a large body of work has shown that at least some forms of new memories are directed or allocated to identifiable neurocircuits, which comprise the cellular substrate of memory engrams (Dash et al. 1990; Han et al. 2007, 2009; Quirk et al. 1995). The key mechanism underlying the formation of such memory engrams, namely, the activity-dependent modification of synaptic connections between neurons, was first proposed by Donald Hebb in the 1940s (Hebb 1949). According to Hebb's postulate, when a presynaptic neuron persistently causes the postsynaptic cell to fire, then the connection between them is strengthened. A few decades later, the phenomena of long-term potentiation (LTP) (Bliss and Lomo 1973) and long-term depression (LTD) (Stent 1973) of synaptic efficacy were proposed as the most plausible mechanisms for implementing Hebbian learning. According to this connectionist view of memory formation, the engram is stored in the strengths of synaptic contacts between neurons, which are modified in an activity-dependent manner. McCulloch and Pitts were the first to propose a mathematical model of a neuronal network in which neurons are treated as point processes (completely void of a dendritic tree) and in which associative memories are stored in the connection strengths between neurons (McCulloch and Pitts 1943). Despite the theoretical appeal of artificial neural networks and their applications as generic function approximators (Hornik et al. 1989), their failure to account for the processing performed by dendritic trees renders them less representative models of brain function (Hinton 1989).

In single-compartment neural network models, dendritic processes are either completely ignored or treated as passive conduits whose function is to transfer incoming stimuli to the soma, which, in turn, acts as the only nonlinear unit in these point neuron models. Several studies in recent decades, however, have established that dendritic subunits can act as local processing devices which contribute to memory encoding and recall by actively and dynamically reshaping their local response function as well as the output of the cell (Branco and Häusser 2010; Gómez González et al. 2011; Lavzin et al. 2012; Losonczy and Magee 2006; Pissadaki et al. 2010; Poirazi et al. 2003a, b; Polsky et al. 2004; Sidiropoulou and Poirazi 2012; Spruston 2008). Additionally, synaptic plasticity mechanisms can interact with local dendritic nonlinearities, allowing dendritic subunits to store complex patterns of multiple incoming stimuli via the use of local cooperative plasticity rules (Govindarajan et al. 2011; Makara et al. 2009; Sjöström et al. 2008; Spruston 2008). Thus, the connectionist view of memory storage where synaptic connections are the sole reservoir for memory storage must be revised to account for the new possibilities offered by active dendritic trees. In this book chapter we discuss recent findings regarding memory formation, starting from the cellular level and stepping down towards the level of dendrites and spines, in an attempt to establish a link between activity-dependent dendritic modifications and the allocation of memories within neuronal circuits.

25.2 The Cellular Memory Engram

25.2.1 Ensembles of Neurons Can Store Memories

Several studies have recently shown that memories are stored in specific neuronal ensembles in which neurons are recruited in a competitive manner, revealing for the first time the cellular trace created during learning (Han et al. 2007, 2009; Josselyn 2010; Zhou et al. 2009). As discussed in the following section, this competitive recruitment of neurons into the populations containing the memory is subserved by the transcription factor cAMP response element-binding protein (CREB), whose overexpression interferes with neuronal excitability.

25.2.2 The Role of CREB in Memory Allocation

CREB is a transcription factor that regulates a large number of genes and has multiple functions in different organs, along with a well-studied role in neuronal plasticity. CREB was first associated with memory in Aplysia (Dash et al. 1990), but knockout mutations in mice also cause profound memory deficits and block the late phase of LTP consolidation (Bourtchuladze et al. 1994). Conversely, increasing CREB levels enhances learning in general (Josselyn et al. 2001). At the cellular level, CREB overexpression in vivo increases hippocampal NMDA currents and LTP as well as dendritic spine density (Marie et al. 2005), while at the network level, CREB plays a key role in memory allocation. Neurons that overexpress CREB before fear learning are three times more likely to be included in the ensemble of neurons that is activated during subsequent fear memory retrieval (Han et al. 2007). Importantly, despite the increased CREB levels, the size of the neuronal population that comprises the memory trace is kept constant, suggesting that fear memory allocation is a competitive process in which neurons with increased CREB levels are preferentially recruited over those with baseline levels. Moreover, fear memory is lost if CREB-overexpressing cells are reversibly or irreversibly inactivated (Han et al. 2009; Zhou et al. 2009) showing that CREB-expressing cells are the ones containing the memory engram. Finally, experiments involving local and acute expression of CREB have demonstrated that CREB is sufficient to rescue spatial memory in the hippocampus (Sekeres et al. 2010). The mechanism via which CREB influences memory allocation is postulated to be the ensuing increase in the excitability of the infected cells. This is evidenced by the lower action potential threshold and afterhyperpolarization current of CREB-overexpressing cells, which are proposed to facilitate plasticity during fear memory induction, thus increasing the probability of being included in a memory trace (Silva et al. 2009; Zhou et al. 2009).

Corroborating these memory allocation studies, a number of experiments confirm the existence of a cellular basis of memory as expressed at the behavioral level. Using a viral vector on transgenic mice, Liu et al. selectively labeled the neurons activated during contextual fear training with light-activated ChR2 ion channels. Subsequent reactivation of these labeled neurons with light alone induced a behavioral fear response (Liu et al. 2012). Garner et al. used transgenic mice to introduce an excitatory receptor exclusively activated by designer drug (DREADD) in cells that are activated in a specific context. First, they expressed the DREADD in the population of cells that were spontaneously active in one context (ctxA). Subsequently, that population was artificially activated via DREADD ligand injection, while mice were being fear conditioned in a second context (ctxB). This manipulation resulted in the formation of a hybrid fear memory that required both the chemical and context stimulus in order to be expressed (Garner et al. 2012).

These findings establish that memory engrams are stored in specific cellular populations for certain types of memory and serve as models for future studies of more complex forms of memory. However, the subcellular structure of the memory engram remains unknown. In the following sections we review evidence from numerous studies that highlight the potentially critical role of dendrites in regulating, in a spatially and temporally restricted manner, incoming signals and synaptic plasticity, both of which are prerequisites for the formation of memory engrams.

25.3 The Subcellular Structure of the Memory Engram

25.3.1 Dendrites Play a Key Role in Memory Allocation

The localization of memory traces in sparse neuronal subpopulations reveals only the higher level of its multilayered structure. Long-term potentiation and depression, which are considered the correlates of Hebbian memory storage, act at the level of individual synapses. As the majority of excitatory synapses to neocortical pyramidal neurons terminate on their dendritic tree (Larkman 1991), dendrites provide an environment rich in biophysical mechanisms that affects the subcellular structure of engrams. Thus, plasticity mechanisms acting at both the cellular and the dendritic levels need to be taken into account when investigating the formation of memory engrams. In the following paragraphs we review evidence regarding the computational capabilities of dendrites and argue that these elaborate structures are key players in the process that underlies memory formation.

25.3.2 Computations in Dendrites

The passive properties of dendrites as well as their morphology endow them with certain nonlinear capabilities that enhance single-neuron computations compared to a linearly thresholded summation unit (point neuron). For example, Wilfrid Rall was the first to propose that dendrites act as delay lines for incoming signals, and this delay can be thought to "label" distinct inputs so that the order of activation of inputs can be reflected in the time course of somatic voltage changes (Rall et al. 1967). This is because interactions between nearby inputs lead to sublinear summation of individual responses providing a saturating mechanism for depolarization. Additionally, shunting inhibition can have a divisive effect on excitatory signals (Fatt and Katz 1953), creating a computational mechanism at the dendritic level. Finally, the strong attenuation of incoming signals at dendritic branch points acts as a natural compartmentalizing factor (Branco and Häusser 2010).

On top of the nonlinearities provided by passive characteristics, dendrites can transform incoming signals through a plethora of active channels that are present in most pyramidal neuron types (Spruston 2008). These active mechanisms contribute to the generation of various types of dendritic spikes, local regenerative phenomena that may be critical to the induction of plasticity (Antic et al. 2010; Losonczy and Magee 2006; Polsky et al. 2004; Remy and Spruston 2007; Schiller et al. 2000; Stuart and Sakmann 1994). These active mechanisms also allow the integration of backpropagating action potentials and the forward propagation of dendritic spikes and thus may further contribute to plasticity induction and memory storage (Spruston 2008). As revealed in the following paragraphs, the contribution of dendritic functions in memory formation is an active line of research that is seeking to map the subcellular structure of memory engrams.

25.3.3 Spatial Arrangement of Incoming Connections

A requirement for understanding the formation of memory engrams at the subcellular level is to discover the structure and distribution of incoming synaptic inputs that terminate onto dendritic branches and how these properties are shaped by experience. By combining high-speed two-photon imaging with electrophysiological recordings, it has become possible to observe the arrangement of synaptic inputs in dendritic branches in vivo. It was found that the synaptic input to visual cortex neurons that share the same orientation preference is widely distributed throughout the dendritic tree. Synaptic inputs of different orientation preferences are interspersed in the dendrites, without showing signs of regularity in the order with which synapses are arranged in the dendritic branch (Jia et al. 2010). Similarly, in the barrel cortex of rats, whisker inputs activate a set of multiple "hot spots" in the dendritic arbor of single neurons, and each hotspot is activated stochastically with low probability (Varga et al. 2011). In the auditory cortex of mice, neighboring spines receive synaptic inputs coding for different and widely varying frequencies or ranges of frequencies without having an orderly arrangement in dendrites (Chen et al. 2011). Taken together, these results suggest that sensory afferent inputs are not orderly arranged in dendritic branches of the receiving neurons, but instead inputs with different sensory tuning appear to be randomly scattered in dendritic branches. This lack of order at the dendritic level may appear at odds with the well-known observation that sensory cortical neurons have characteristic receptive fields and form orderly cortical maps in which receptive fields shift progressively between neighboring neurons. Additionally, as we shall see in Sect. 25.3.6, a number of recent studies provide ample evidence in support of the functional clustering of synapses in dendritic branches, both in vitro and in vivo, which is also at odds with the aforementioned observations. One possible explanation for this discrepancy could be that the clustering of co-activated synapses within dendritic branches represents individual features (i.e., certain orientations or sound frequencies) that are not necessarily neighboring in the respective sensory space but appear simultaneously in sensory signals that are frequently experienced by the animal. Natural speech for example consists of repetitions of sets of frequencies that are not along a continuum. The coappearance of these distinct frequencies would result in the formation of dendritic clusters of synapses with different sensory tuning, in agreement with the findings of Chen et al. (2011).

25.3.4 Structural Plasticity in Dendritic Spines

In cortical pyramidal cells, dendritic spines are the terminal loci where excitatory synaptic connections are formed and they are thought to be the elementary units of memory storage (Yuste and Denk 1995). In contrast to neuronal cells and dendritic arbors, which are relatively stable in mature animals (Trachtenberg et al. 2002), dendritic spines are continuously and spontaneously created in the brain (Grutzendler et al. 2002; Marrs et al. 2001), forming a physical substrate for the generation and elimination of synapses that subserves the functional rewiring of neuronal circuits (Matsuzaki et al. 2004). Single spines typically contain a single synapse which is in most cases the single point of axodendritic contact (Takahashi et al. 2012), and their volume has been found to correlate with the synaptic efficacy in mature synapses (Matsuzaki et al. 2001). Spine dynamics are closely linked to the plasticity of synaptic inputs that terminate on them. Glutamate uncaging experiments have demonstrated that the size of a dendritic spine correlates with its functional efficacy both in vitro (Matsuzaki et al. 2001) and in vivo (Noguchi et al. 2011). Thus, the spine enlargement that follows LTP results in greater functional EPSP (Matsuzaki 2007). Additionally, Ca^{2+} transients are confined within the spine heads and are greater in thinner spines, allowing for independent plasticity induction in spines (Sabatini et al. 2002).

Using glutamate uncaging and calcium imaging it has become possible to observe spine dynamics during LTP consolidation. Late LTP facilitation between neighboring spines as well as the phenomenon of bidirectional synaptic tagging and capture has been observed recently (Govindarajan et al. 2011; Harvey and Svoboda 2007). This LTP facilitation takes place between spines that are less than 70 μm apart and is biased to occur within the same dendritic branch in the presence of Mg^{+}, thus providing further credence to the hypothesis that the single dendritic branch is the fundamental unit of learning and memory formation in the brain. Overall, these studies suggest that spine dynamics enable dendritic spines to rapidly encode incoming stimuli and suggest rules that govern their structural plasticity (Kasai et al. 2010). The tight relation between spine structural and functional plasticity renders spines the primary candidate sites of structural plasticity, in contrast to the structure of dendrites and axons which changes little during adulthood (Trachtenberg et al. 2002). Dendrites on the other hand contribute to the spine plasticity by constraining synaptic plasticity both in time and space and enabling cooperativity so that it occurs within specific dendritic compartments.

25.3.5 Spatially Constrained Consolidation of LTP and LTD

The changes in synaptic efficacy that accompany LTP after learning require at least two phases. The persistent change in synaptic efficacy that lasts for more than a few hours is termed late LTP, in order to distinguish it from early LTP, a form of synaptic potentiation that lasts only for a few hours before decaying back to baseline levels (Reymann and Frey 2007). Late LTP has been found to be dependent on de novo protein synthesis. The production of these plasticity-related proteins (PRPs) may require the activation of nuclear genes, or alternatively, PRPs may be translated locally by the cellular machinery that resides in dendrites (Miller et al. 2002; Ostroff et al. 2002; Steward and Schuman 2007; Sutton and Schuman 2006). For this purpose a surprisingly rich set of mRNAs resides in the dendrites, enabling individual dendritic compartments to translate PRPs at request, without depending on somatic translation (Cajigas et al. 2012). The local compartmentalization of the endoplasmic reticulum can restrict the spatial mobility of membrane cargo near spines and dendritic branch points (Cui-Wang et al. 2012). It is therefore possible to observe NMDA- and protein synthesis-dependent LTP in isolated dendrites of CA1 neurons (Vickers et al. 2005). Dendritic protein synthesis appears to be specialized in producing proteins with synaptic functions rather than general housekeeping proteins, for example key kinases (CaMKIIa, PKMz), cytoskeletal proteins (Arc, MAP2), and neurotransmitter receptors of the AMPA and NMDA families (Andreassi and Riccio 2009; Bramham 2008; Wang et al. 2010). Inhibition of both protein translation and mRNA transcription has been shown to disrupt the consolidation of LTP (Kelleher et al. 2004).

According to the synaptic tagging and capture (STC) model, synaptic stimulation during learning creates a temporary synaptic "tag" in the activated synapses, marking them as candidates for permanent potentiation (Frey and Morris 1997). The consolidation to late LTP and stabilization of the tagged synapses requires the presence of PRPs that must be available in the dendrite within a time window in order to be captured by the tagged synapses. These proteins may be produced via strong (late LTP inducing) synaptic stimulation or via pharmacological means. The PRPs can readily be captured by tagged synapses, even if their translation was not induced by the tagging stimulus, leading to the phenomenon of cross-capture, whereby an early LTP- or early LTD-inducing stimulus is converted in its late form due to the presence of protein products generated by a previous or a subsequent late LTP/LTD-inducing stimulus (Frey and Frey 2008; Redondo and Morris 2011; Sajikumar et al. 2005). It has been suggested that such spatial and temporal limits lead to stable formation of engrams that consist of synapses which are organized in clusters within individual dendritic branches (Govindarajan et al. 2006).

25.3.6 Clustering of Synaptic Inputs

The ability of dendrites to perform local computations has led to the proposal that dendritic branches should be considered as the fundamental processing unit of the nervous system (Branco and Häusser 2010). However, in order for neurons to exploit the additional processing capabilities afforded by dendritic spiking mechanisms, their synaptic inputs need to be distributed spatially and temporally in a way that facilitates dendritic spike generation (Branco and Häusser 2010; Poirazi and Mel 2001). This proposed clustering of synaptic inputs is a crucial property that allows such facilitation (Poirazi et al. 2003b) and has been shown to enhance neuronal storage capacity (Poirazi and Mel 2001). Numerous recent studies support the synaptic clustering hypothesis, according to which inputs with functional similarities are organized by plasticity in clusters within the dendrites of pyramidal neurons (Govindarajan et al. 2006; Makino and Malinow 2011; Poirazi and Mel 2001; Takahashi et al. 2012; Winnubst and Lohmann 2012).

Cooperative plasticity in dendrites provides a mechanism that favors the hypothesis of synaptic clustering. Synaptic plasticity and its consolidation require changes in AMPA receptor distribution, remodeling of actin cytoskeleton, and protein synthesis. These changes require NMDA signaling, Ca^{2+}, numerous kinases (CaMKII, PKA, PKC, MAPK), phosphatases, as well as activation of molecular pathways such as Ras and Rap (Kennedy et al. 2005). The spatial extent of these molecular pathways can be limited. Ca^{2+} concentration changes during synaptic stimulation are restricted to individual dendritic spines (Yuste and Denk 1995), and the spread of Ras activity is confined to the spines of a small dendritic region of 10 μm (Harvey et al. 2008). The spatial confinement of synaptic plasticity consolidation may thus provide a mechanism that facilitates the clustering of potentiated synapses after learning.

Optogenetic techniques have recently made it possible to observe the localization of synaptic inputs as they are formed both in vivo and in vitro and provide evidence that supports the functional clustering of synaptic inputs in the brain. In organotypic cultures of the developing hippocampal cortex, which is characterized by spontaneous bursts of activity, neighboring synaptic inputs tend to be more coactive, creating functional clusters. The observed clustering is eliminated by the NMDA antagonist AP5 or the Na^+ channel blocker TTX, suggesting that it is a result of activity-dependent plasticity (Kleindienst et al. 2011). Using fluorescently tagged AMPA subunits in order to visualize AMPA trafficking during normal experiences in vivo, Makino and Malinow 2011 found that GluR1 subunits were enriched in groups of neighboring spines in the barrel cortex of mice as a result of experience but not during sensory deprivation, suggesting that synaptic potentiation occurs in groups of neighboring synapses that form clusters. Additional in vivo studies in the mouse barrel cortex revealed that neighboring spines in pyramidal neuron dendrites were significantly co-activated and formed functional synaptic "assemblets" which were spread out in the dendritic tree and were abolished in the presence of AP5 (Takahashi et al. 2012). Anatomical studies also found that synaptic clusters occur more often than random and that enlarged, mushroom-shaped spines appear more often in these clusters (Yadav et al. 2012). Finally, during a repetitive motor learning task, it was found that a third of the newly formed spines occurs in clusters, most of which contain two spines, and that clustered spines are more stable than nonclustered ones (Fu et al. 2012). Overall, there is strong accumulating evidence that the computationally advantageous clustering of functionally relevant synaptic inputs is indeed an organizing principle in the dendrites of multiple neuron types and may arise through experience-related cooperative plasticity that takes place in dendritic branches (Magee 2011). Therefore the rules underlying the formation of synaptic clusters must be taken into account along with other types of synaptic plasticity in order to understand the processes leading to the formation of memory engrams.

25.3.7 Plasticity of Dendritic Excitability

While structural plasticity can support the rapid encoding of the characteristics of incoming signals in synapses, other biophysical mechanisms can also regulate the effect of localized groups of synapses on neuronal output by modulating the excitable properties of dendrites. Dendritic spikes in CA1 pyramidal neurons for example are strongly regulated by prior activity of the neuron via Na^+ channel inactivation which can attenuate dendritic excitability globally in the cell (Remy et al. 2009). Changes in synaptic input on the other hand can lead to permanent alterations in the physiological properties and the distributions of ionic channels locally in dendrites (Zhang and Linden 2003). This form of intrinsic neuronal plasticity can be induced by electrical stimulation in vitro or through exposure to an enriched environment (Beck and Yaari 2008). Indeed, it has been shown that the properties of voltage-gated channels in the main apical dendrites of CA1 pyramidal neurons are altered after LTP-inducing excitatory stimulation. A-type currents are persistently down-regulated leading to increased dendritic excitability (Frick et al. 2004). This type of plasticity can be induced locally via synaptic activity as is evidenced by the potentiation of branch coupling strengths: the repetitive elicitation of dendritic spikes leads to a slow but permanent increase in the coupling between local dendritic spikes and the soma in hippocampal CA1 cells through NMDA-dependent regulation of A-type potassium currents (Losonczy et al. 2008), an intrinsic plasticity mechanism which has been termed *branch strength potentiation* (BSP).

Theoretical models have indicated that this form of branch-specific plasticity can enable self-organization of dendritic inputs in branches, in a way that can facilitate binding of input features in dendrites (Legenstein and Maass 2011). Experimentally, exposure of animals to an enriched environment leads to the enhancement of dendritic spike propagation of a subset of dendritic branches in rat CA1 neurons, a mechanism that might be utilized to store recent experiences in dendrites (Makara et al. 2009).

25.3.8 Homeostatic Plasticity

Homeostatic mechanisms that are continuously active in a cell can also influence the formation of memory traces via the fine-tuning or the normalization of synaptic strengths both globally, at the cellular level, and locally, within a branch. Uncontrolled synaptic potentiation may lead to destabilization of neural network activity, a phenomenon that is compensated by mechanisms that act to revert the neural firing rate back into physiological levels. This is achieved via multiple mechanisms affecting the intrinsic membrane excitability, presynaptic transmitter release, balance between excitation and inhibition, connectivity, and synaptic strengths (Turrigiano and Nelson 2004). Synaptic scaling, a slow process that readjusts synaptic efficacies in order to keep the firing rate of the neuron within physiological limits, has been proposed to be such a homeostatic mechanism (Turrigiano 2008). Synaptic scaling has been observed at the level of network and individual neurons (Burrone et al. 2002) and more locally at the level of single synapses (Hou et al. 2011), indicating the ability of single synapses to regulate their own synaptic potency. Multiple models of homeostatic scaling of plasticity have been proposed, suggesting either a whole-cell process (Ibata et al. 2008) or a set of rules that act locally between neighboring synapses (Rabinowitch and Segev 2008). Homeostatic plasticity of inhibition occurs in concert with homeostatic enhancement of excitability, as evidenced by chronic TTX administration in the hippocampus (Echegoyen et al. 2007). Homeostatic mechanisms can also affect the probability of presynaptic vesicle fusion via enhanced calcium influx during spikes, effectively enhancing synaptic strength (Zhao et al. 2011). Finally, homeostasis of intrinsic excitability can lead to upregulation of I_h and subsequent decrease of excitability during LTP induction in dendrites of CA1 neurons (Fan et al. 2005), while the opposite occurs after LTD induction (Brager and Johnston 2007).

The aforementioned homeostatic mechanisms act in parallel with plasticity induction and protein production to shape synaptic strengths during memory formation. They act on different and varying time scales, performing a continuous restructuring of incoming connection strengths, local excitability, and dendritic coupling strengths as new memories are encoded in populations of cells. Their effect on memory formation and expression is not yet assessed experimentally due to the spatial and temporal constraints of experimental approaches. Therefore, computational models that incorporate plasticity rules of different temporal and spatial dimensions from the neural network down to the sub-dendritic level can provide insights that help understand their action.

25.4 Computational Modeling of Memory Allocation

25.4.1 Computational Models of Dendritic Function

Historically, the role of dendrites in neuronal computation was first approached computationally due to the difficulty of recording at the dendrite level. Neuronal models that take into account dendritic nonlinearities have assigned various possible functions to dendrites, branches, and spines. Early studies assessed the signal-enhancing capabilities that arise when spines are considered as excitable structures (Segev and Rall 1988; Shepherd et al. 1985). Using a compartmental neuronal model Bartlett Mel showed that, by clustering its synaptic inputs, the neuron could implement a "sum of products" computation (Mel 1993). Later, Traub et al. (1994) used a CA3 computational model to show that the timing of inhibition can control Ca^{2+} concentration, thus crucially affecting synaptic plasticity. Agmon-Snir et al. (1998) studied the role of dendrites in neurons of the auditory brainstem and showed that dendrites enhance the cell's coincidence detection properties. The receptive fields of complex cells can arise from integration of synaptic inputs from the LGN to dendrites of complex cells, as shown in a pyramidal model cell (Mel et al. 1998). Dendritic clustering, dendritic spike generation, and their role in neuronal capacity have been examined in a series of computational studies of CA1 neurons, indicating that synapse clustering can greatly enhance the memory storage capacity of single neurons and additionally that the firing rate of CA1 neurons can be reliably approximated with a simpler model where dendritic branches are represented as point nonlinearities in a two-layer model of neuronal cells (Poirazi et al. 2003a, b; Poirazi and Mel 2001; Sidiropoulou et al. 2006).

25.4.2 Computational Models of Plasticity and Memory Formation

Synaptic plasticity was traditionally modeled as an instantaneous procedure, mainly because the underlying biological processes incorporate large and complex networks of molecular interactions that are not yet fully understood (Ajay and Bhalla 2006; Bhalla and Iyengar 1999; Shouval et al. 2010; Smolen et al. 2006). Recently, more realistic yet simplified phenomenological models of memory formation that examine the influence of multiple plasticity rules in memory consolidation have begun to emerge. Barrett et al. propose a stochastic state-based model of synaptic plasticity consolidation that can exhibit synaptic capture (Barrett et al. 2009). Clopath et al. implemented a model of spike timing-dependent synaptic plasticity and synaptic capture (STC) and found that synaptic consolidation is not independent in synapses but acts as a whole-cell phenomenon (Clopath et al. 2008). These models studied plasticity at the cellular population level without taking into account dendritic plasticity. Dendritic plasticity has also recently been introduced to computational models of memory. In a rate coding network model that incorporates multiple dendritic branches, dendritic compartmentalization was shown to enable accurate memory formation that is not sensitive to the intensity of the stimulus on the condition that dendritic inhibition is global or is coupled with somatic inhibition (Morita 2008). In another study, Legenstein et al. have incorporated branch-specific homeostatic plasticity rules in a neuron model with nonlinear dendrites and dendritic-branch strength potentiation rules. Their results show that dendritic nonlinearities and dendritic-branch plasticity induce a competition which allows the neuron to bind multiple input features within specific dendritic branches (Legenstein and Maass 2011). While dendritic plasticity models are just beginning to emerge, they provide valuable tools towards understanding the mechanistic underpinnings of memory at the dendrite level.

25.5 Perspective

As we have briefly seen in this chapter, the plasticity that underlies learning and memory acts at multiple levels and time scales (Fig. 25.1). Fast spine dynamics are shaped by synaptic activity, the dendritic branch excitation properties are regulated by activity and homeostatic mechanisms while neuronal excitability is affected by previous learning, levels of proteins and CREB, and, at the network level, regulation of inhibition. These parallel plasticity processes (structural and intrinsic excitability) are interrelated, as dendritic excitability properties are affected by synaptic activity and vice versa (Sjöström et al. 2008). An emerging view is that while single synapses can express plasticity independently, they are too weak to function by themselves as memory units. Instead, the dendritic branches, which appear to be relatively independent units of function and plasticity, are putatively the fundamental processing units of the brain (Branco and Häusser 2010; Govindarajan et al. 2006; Losonczy et al. 2008; Poirazi and Mel 2001). The observation of synaptic clusters implies that synapses act in groups, formed by cooperative plasticity and local protein synthesis, which exert a nonlinear effect in the output of the cell (Losonczy and Magee 2006). This effect is in turn regulated by intrinsic plasticity mechanisms that modify the excitability properties of dendritic branches where groups of synapses reside. This model entails that memory traces consist of clusters of functionally related synapses that are transformed by the local biophysical properties of dendritic branches. According to this hypothesis, active dendrites serve as the units of plasticity governed by the temporal dynamics of the synaptic tagging and capture model, the plasticity of dendritic excitability and homeostatic mechanisms. New memories are directed to neuronal populations governed by the dynamics of CREB activation (Silva et al. 2009). CREB levels, in turn, are increased temporarily after memory induction (Stanciu et al. 2001) leading to increased neuronal excitability and a lower threshold for plasticity, thus making expressing neurons "attractors" of new memories. Therefore, memories that appear in close temporal proximity are expected to be stored within overlapping neuronal populations and

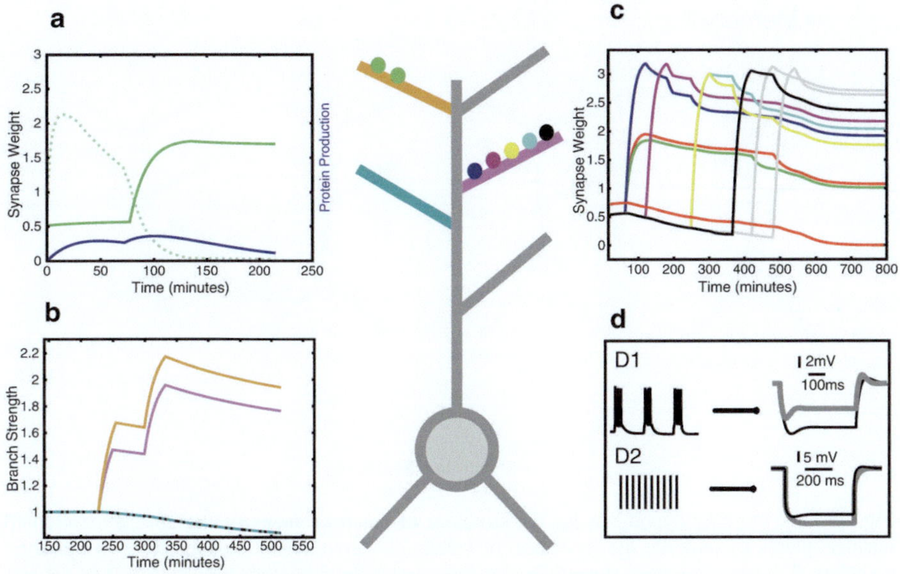

Fig. 25.1 Multiple plasticity rules shape synapse strength and excitability in dendritic branches. (**a**) Synaptic tagging and capture allow cooperative plasticity to occur across time scales of hours. *Dotted green*: early LTP, *solid green*: late LTP, *blue*: plasticity-related protein concentration. (**b**) Branch strength potentiation allows compartmentalized plasticity in dendritic branches. Branches that generate dendritic spikes increase their branch coupling strengths (*yellow*, *purple*) in contrast to idle branches (*cyan*). (**c**) Homeostatic mechanisms normalize synaptic weights as new learning occurs (traces represent the time evolution of synaptic strengths of ten different synapses). (**d**) Dendritic excitability is differentially modulated depending on the plasticity induction protocol. (d1) Maximal LTP induction homeostatically regulates the I_h. Theta burst pairing of backpropagating action potentials and synaptic stimulation (*left*) induces reduction of excitability and a decrease in the voltage response at the soma (*right*, *black*: before, *grey*: after stimulation). (d2) Moderate LTP after high-frequency stimulation (*left*) decreases dendritic I_h, reducing voltage response in the stimulated dendrite (*right*, *black*: before, *grey*: after stimulation). Adapted from Papoutsi et al. (2012), with permission

form synaptic clusters within common dendritic branches (Fig. 25.2). This model proposes a mechanistic explanation for the creation of episodic memories through dendritic plasticity.

In summary, experimental studies begin to uncover the structure of memories for basic forms of learning. There is a large body of work on the existence and possible manipulation of memory engrams at the cellular level (Josselyn 2010; Silva et al. 2009). On the other hand, dendritic activation patterns (Fu et al. 2012; Kleindienst et al. 2011; Magee 2011; Winnubst and Lohmann 2012) and localized plasticity properties (Govindarajan et al. 2011; Losonczy et al. 2008; Makara et al. 2009) make dendrites the key candidates to serve as the fundamental unit of learning and memory formation in the brain. Technical advances in dendritic visualization and manipulations as well as theoretical and computational studies are expected to elucidate further the role of dendritic plasticity in brain function and its implications.

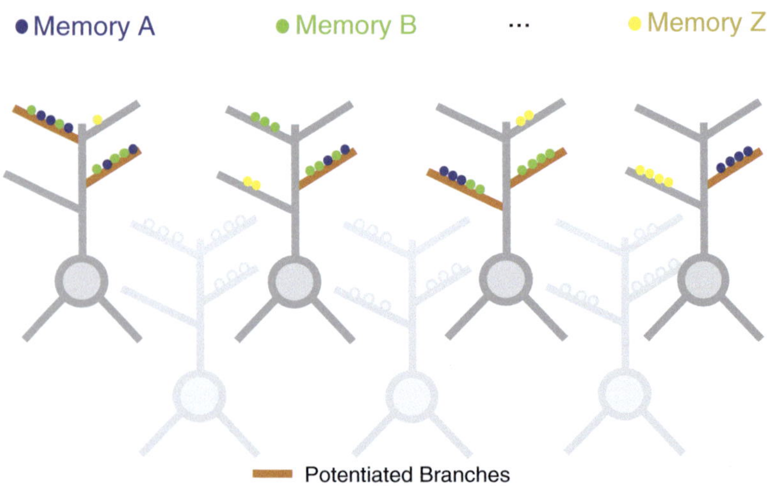

Fig. 25.2 A working hypothesis for the structure of dendritic memory engrams. Synaptic and intrinsic plasticity underlie the formation of stable, clustered synaptic connections in dendritic branches. By increasing their excitability, branches can exert a potent nonlinear effect to somatic output. Synaptic protein capture can thus aid the formation of dendritic memory traces whereby memories created close in time (e.g., Memory A and B) are expected to form synaptic clusters within common dendrites (*blue* and *green* synapses) whereas memories that are separated by long temporal intervals (e.g., Memory A and Z) are expected to be stored in non-overlapping branches (*blue* and *yellow* synapses) and probably in non-overlapping neurons

References

Agmon-Snir H, Carr CE, Rinzel J (1998) The role of dendrites in auditory coincidence detection. Nature 393:268–272

Ajay SM, Bhalla US (2006) Synaptic plasticity in vitro and in silico: insights into an intracellular signaling maze. Physiology 21:289–296

Andreassi C, Riccio A (2009) To localize or not to localize: mRNA fate is in 3'UTR ends. Trends Cell Biol 19:465–474

Antic SD, Zhou W-LW-L, Moore AR et al (2010) The decade of the dendritic NMDA spike. J Neurosci Res 88:2991–3001

Barrett AB, Billings GO, Morris RGM, Van Rossum MCW (2009) State based model of long-term potentiation and synaptic tagging and capture. PLoS Comput Biol 5:e1000259

Beck H, Yaari Y (2008) Plasticity of intrinsic neuronal properties in CNS disorders. Nat Rev Neurosci 9:357–369

Bhalla US, Iyengar R (1999) Emergent properties of networks of biological signaling pathways. Science 283:381–387

Bliss TV, Lomo T (1973) Long-lasting potentiation of synaptic transmission in the dentate area of the anaesthetized rabbit following stimulation of the perforant path. J Physiol 232:331–356

Bourtchuladze R, Frenguelli B, Blendy J et al (1994) Deficient long-term memory in mice with a targeted mutation of the cAMP-responsive element-binding protein. Cell 79:59–68

Brager DH, Johnston D (2007) Plasticity of intrinsic excitability during long-term depression is mediated through mGluR-dependent changes in I(h) in hippocampal CA1 pyramidal neurons. J Neurosci 27:13926–13937

Bramham CR (2008) Local protein synthesis, actin dynamics, and LTP consolidation. Curr Opin Neurobiol 18:524–531
Branco T, Häusser M (2010) The single dendritic branch as a fundamental functional unit in the nervous system. Curr Opin Neurobiol 20:494–502
Burrone J, O'Byrne M, Murthy VN (2002) Multiple forms of synaptic plasticity triggered by selective suppression of activity in individual neurons. Nature 420:414–418
Cajigas IJ, Tushev G, Will TJ et al (2012) The local transcriptome in the synaptic neuropil revealed by deep sequencing and high-resolution imaging. Neuron 74:453–466
Chen X, Leischner U, Rochefort NL et al (2011) Functional mapping of single spines in cortical neurons in vivo. Nature 475:501–505
Clopath C, Ziegler L, Vasilaki E et al (2008) Tag-trigger-consolidation: a model of early and late long-term-potentiation and depression. PLoS Comput Biol 4:e1000248
Cui-Wang T, Hanus C, Cui T et al (2012) Local zones of endoplasmic reticulum complexity confine cargo in neuronal dendrites. Cell 148:309–321
Dash PK, Hochner B, Kandel ER (1990) Injection of the cAMP-responsive element into the nucleus of Aplysia sensory neurons blocks long-term facilitation. Nature 345:718–721
Echegoyen J, Neu A, Graber KD, Soltesz I (2007) Homeostatic plasticity studied using in vivo hippocampal activity-blockade: synaptic scaling, intrinsic plasticity and age-dependence. PloS One 2:e700
Fan Y, Fricker D, Brager DH et al (2005) Activity-dependent decrease of excitability in rat hippocampal neurons through increases in I(h). Nat Neurosci 8:1542–1551
Fatt P, Katz B (1953) The electrical properties of crustacean muscle fibres. J Physiol 120:171–204
Frey S, Frey JU (2008) "Synaptic tagging" and "cross-tagging" and related associative reinforcement processes of functional plasticity as the cellular basis for memory formation. Prog Brain Res 169:117–143
Frey U, Morris RG (1997) Synaptic tagging and long-term potentiation. Nature 385:533–536
Frick A, Magee J, Johnston D (2004) LTP is accompanied by an enhanced local excitability of pyramidal neuron dendrites. Nat Neurosci 7:126–135
Fu M, Yu X, Lu J, Zuo Y (2012) Repetitive motor learning induces coordinated formation of clustered dendritic spines in vivo. Nature
Garner AR, Rowland DC, Hwang SY et al (2012) Generation of a synthetic memory trace. Science 335:1513–1516
Gómez González JF, Mel BW, Poirazi P (2011) Distinguishing linear vs. non-linear integration in CA1 radial oblique dendrites: it's about time. Front Comput Neurosci 5:44
Govindarajan A, Kelleher RJ, Tonegawa S (2006) A clustered plasticity model of long-term memory engrams. Nat Rev Neurosci 7:575–583
Govindarajan A, Israely I, Huang S-Y, Tonegawa S (2011) The dendritic branch is the preferred integrative unit for protein synthesis-dependent LTP. Neuron 69:132–146
Grutzendler J, Kasthuri N, Gan WB (2002) Long-term dendritic spine stability in the adult cortex. Nature 420:812–816
Han J-H, Kushner SA, Yiu AP et al (2007) Neuronal competition and selection during memory formation. Science 316:457–460
Han J-H, Kushner SA, Yiu AP et al (2009) Selective erasure of a fear memory. Science 323:1492–1496
Harvey CD, Svoboda K (2007) Locally dynamic synaptic learning rules in pyramidal neuron dendrites. Nature 450:1195–1200
Harvey CD, Yasuda R, Zhong H, Svoboda K (2008) The spread of Ras activity triggered by activation of a single dendritic spine. Science 321:136–140
Hebb DO (1949) The organization of behavior. Wiley and Sons. New York
Hinton GE (1989) Connectionist learning procedures. Artif Intell 40:185–234
Hornik K, Stinchcombe M, White H (1989) Multilayer feedforward networks are universal approximators. Neural Netw 2:359–366
Hou Q, Gilbert J, Man H-Y (2011) Homeostatic regulation of AMPA receptor trafficking and degradation by light-controlled single-synaptic activation. Neuron 72:806–818

Ibata K, Sun Q, Turrigiano GG (2008) Rapid synaptic scaling induced by changes in postsynaptic firing. Neuron 57:819–826

Jia H, Rochefort NL, Chen X, Konnerth A (2010) Dendritic organization of sensory input to cortical neurons in vivo. Nature 464:1307–1312

Josselyn S (2010) Continuing the search for the engram: examining the mechanism of fear memories. J Psychiatry Neurosci 35:221–228

Josselyn SA, Shi C, Carlezon WA et al (2001) Long-term memory is facilitated by cAMP response element-binding protein overexpression in the amygdala. J Neurosci 21:2404–2412

Kasai H, Hayama T, Ishikawa M et al (2010) Learning rules and persistence of dendritic spines. Spine 32:241–249

Kelleher RJ, Govindarajan A, Tonegawa S (2004) Translational regulatory mechanisms in persistent forms of synaptic plasticity. Neuron 44:59–73

Kennedy MB, Beale HC, Carlisle HJ, Washburn LR (2005) Integration of biochemical signalling in spines. Nat Rev Neurosci 6:423–434

Kleindienst T, Winnubst J, Roth-alpermann C et al (2011) Activity-dependent clustering of functional synaptic inputs on developing hippocampal dendrites. Neuron 72:1012–1024

Larkman AU (1991) Dendritic morphology of pyramidal neurones of the visual cortex of the rat: III. Spine distributions. J Comp Neurol 306:332–343

Lashley K (1950) In search of the engram. Symposia of the society for experimental biology. 4:454–482

Lavzin M, Rapoport S, Polsky A et al (2012) Nonlinear dendritic processing determines angular tuning of barrel cortex neurons in vivo. Nature 490:397–401

Legenstein R, Maass W (2011) Branch-specific plasticity enables self-organization of nonlinear computation in single neurons. J Neurosci 31:10787–10802

Liu X, Ramirez S, Pang PT et al (2012) Optogenetic stimulation of a hippocampal engram activates fear memory recall. Nature

Losonczy A, Magee JC (2006) Integrative properties of radial oblique dendrites in hippocampal CA1 pyramidal neurons. Neuron 50:291–307

Losonczy A, Makara JK, Magee JC (2008) Compartmentalized dendritic plasticity and input feature storage in neurons. Nature 452:436–441

Magee JC (2011) Observations on clustered synaptic plasticity and highly structured input patterns. Neuron 72:887–888

Makara JK, Losonczy A, Wen Q, Magee JC (2009) Experience-dependent compartmentalized dendritic plasticity in rat hippocampal CA1 pyramidal neurons. Nat Neurosci 12:1485–1487

Makino H, Malinow R (2011) Article compartmentalized versus global synaptic plasticity on dendrites controlled by experience. Neuron 72:1001–1011

Marie H, Morishita W, Yu X et al (2005) Generation of silent synapses by acute in vivo expression of CaMKIV and CREB. Neuron 45:741–752

Marrs GS, Green SH, Dailey ME (2001) Rapid formation and remodeling of postsynaptic densities in developing dendrites. Nat Neurosci 4:1006–1013

Matsuzaki M (2007) Factors critical for the plasticity of dendritic spines and memory storage. Small 57:1–9

Matsuzaki M, Ellis-Davies GC, Nemoto T et al (2001) Dendritic spine geometry is critical for AMPA receptor expression in hippocampal CA1 pyramidal neurons. Nat Neurosci 4:1086–1092

Matsuzaki M, Honkura N, Ellis-Davies GCR, Kasai H (2004) Structural basis of long-term potentiation in single dendritic spines. Nature 429:761–766

McCulloch WS, Pitts W (1943) A logical calculus of the ideas immanent in nervous activity. Bull Math Biophys 5:115–133

Mel BW (1993) Synaptic integration in an excitable dendritic tree. J Neurophysiol 70:1086–1101

Mel BW, Ruderman DL, Archie KA (1998) Translation-invariant orientation tuning in visual "complex" cells could derive from intradendritic computations. J Neurosci 18:4325–4334

Miller S, Yasuda M, Coats JK et al (2002) Disruption of dendritic translation of CaMKIIalpha impairs stabilization of synaptic plasticity and memory consolidation. Neuron 36:507–519

Morita K (2008) Possible role of dendritic compartmentalization in the spatial working memory circuit. J Neurosci 28:7699–7724

Noguchi J, Nagaoka A, Watanabe S et al (2011) In vivo two-photon uncaging of glutamate revealing the structure-function relationships of dendritic spines in the neocortex of adult mice. J Physiol 589:2447–2457

Ostroff LE, Fiala JC, Allwardt B, Harris KM (2002) Polyribosomes redistribute from dendritic shafts into spines with enlarged synapses during LTP in developing rat hippocampal slices. Neuron 35:535–545

Papoutsi A, Sidiropoulou K, Poirazi P (2012) Memory beyond synaptic plasticity: the role of intrinsic neuronal excitability. In: Memory mechanisms in health and disease: mechanistic basis of memory, 1st edn. World Science Publishing, Singapore, p 444

Pissadaki EK, Sidiropoulou K, Reczko M, Poirazi P (2010) Encoding of spatio-temporal input characteristics by a CA1 pyramidal neuron model. PLoS Comput Biol 6:e1001038

Poirazi P, Mel BW (2001) Impact of active dendrites and structural plasticity on the memory capacity of neural tissue. Neuron 29:779–796

Poirazi P, Brannon T, Mel BW (2003a) Arithmetic of subthreshold synaptic summation in a model CA1 pyramidal cell. Neuron 37:977–987

Poirazi P, Brannon T, Mel BW (2003b) Pyramidal neuron as two-layer neural network. Neuron 37:989–999

Polsky A, Mel BW, Schiller J (2004) Computational subunits in thin dendrites of pyramidal cells. Nat Neurosci 7:621–627

Quirk GJ, Repa C, LeDoux JE (1995) Fear conditioning enhances short-latency auditory responses of lateral amygdala neurons: parallel recordings in the freely behaving rat. Neuron 15:1029–1039

Rabinowitch I, Segev I (2008) Two opposing plasticity mechanisms pulling a single synapse. Trends Neurosci 31:377–383

Rall W, Burke RE, Smith TG et al (1967) Dendritic location of synapses and possible mechanisms for the monosynaptic EPSP in motoneurons. J Neurophysiol 30:1169–1193

Redondo RL, Morris RGM (2011) Making memories last: the synaptic tagging and capture hypothesis. Nat Rev Neurosci 12:17–30

Remy S, Spruston N (2007) Dendritic spikes induce single-burst long-term potentiation. Proc Natl Acad Sci USA 104:17192–17197

Remy S, Csicsvari J, Beck H (2009) Activity-dependent control of neuronal output by local and global dendritic spike attenuation. Neuron 61:906–916

Reymann KG, Frey JU (2007) The late maintenance of hippocampal LTP: requirements, phases, "synaptic tagging", "late-associativity" and implications. Neuropharmacology 52:24–40

Sabatini BL, Oertner TG, Svoboda K (2002) The life cycle of Ca^{2+} ions in dendritic spines. Neuron 33:439–452

Sajikumar S, Navakkode S, Sacktor TC, Frey JU (2005) Synaptic tagging and cross-tagging: the role of protein kinase Mzeta in maintaining long-term potentiation but not long-term depression. J Neurosci 25:5750–5756

Schiller J, Major G, Koester HJ, Schiller Y (2000) NMDA spikes in basal dendrites of cortical pyramidal neurons. Nature 404:285–289

Segev I, Rall W (1988) Computational study of an excitable dendritic spine. J Neurophysiol 60:499–523

Sekeres MJ, Neve RL, Frankland PW, Josselyn SA (2010) Dorsal hippocampal CREB is both necessary and sufficient for spatial memory. Learn Mem 17:280–283

Shepherd GM, Brayton RK, Miller JP et al (1985) Signal enhancement in distal cortical dendrites by means of interactions between active dendritic spines. Proc Natl Acad Sci USA 82:2192–2195

Shouval HZ, Wang SS-H, Wittenberg GM (2010) Spike timing dependent plasticity: a consequence of more fundamental learning rules. Front Comput Neurosci 4:1–13

Sidiropoulou K, Poirazi P (2012) Predictive features of persistent activity emergence in regular spiking and intrinsic bursting model neurons. PLoS Comput Biol 8:e1002489

Sidiropoulou K, Pissadaki EK, Poirazi P (2006) Inside the brain of a neuron. EMBO Rep 7:886–892
Silva AJ, Zhou Y, Rogerson T et al (2009) Molecular and cellular approaches to memory allocation in neural circuits. Science 326:391–395
Sjöström PJ, Rancz EA, Roth A, Häusser M (2008) Dendritic excitability and synaptic plasticity. Physiol Rev 88:769–840
Smolen P, Baxter DA, Byrne JH (2006) A model of the roles of essential kinases in the induction and expression of late long-term potentiation. Biophys J 90:2760–2775
Spruston N (2008) Pyramidal neurons: dendritic structure and synaptic integration. Nat Rev Neurosci 9:206–221
Stanciu M, Radulovic J, Spiess J (2001) Phosphorylated cAMP response element binding protein in the mouse brain after fear conditioning: relationship to Fos production. Brain Res Mol Brain Res 94:15–24
Stent GS (1973) A physiological mechanism for Hebb's postulate of learning. Proc Natl Acad Sci USA 70:997–1001
Steward O, Schuman EM (2007) Protein synthesis at synaptic sites on dendrites. Annu Rev Neurosci 24:299–325
Stuart G, Sakmann B (1994) Active propagation of somatic action potentials into neocortical pyramidal cell dendrites. Nature 367:69–72
Sutton MA, Schuman EM (2006) Dendritic protein synthesis, synaptic plasticity, and memory. Cell 127:49–58
Takahashi N, Kitamura K, Matsuo N et al (2012) Locally synchronized synaptic inputs. Science 335:353–356
Trachtenberg JT, Chen BE, Knott GW et al (2002) Long-term in vivo imaging of experience-dependent synaptic plasticity in adult cortex. Nature 420:788–794
Traub RD, Jefferys JG, Miles R et al (1994) A branching dendritic model of a rodent CA3 pyramidal neurone. J Physiol 481:79–95
Turrigiano GG (2008) The self-tuning neuron: synaptic scaling of excitatory synapses. Cell 135:422–435
Turrigiano GG, Nelson SB (2004) Homeostatic plasticity in the developing nervous system. Nat Rev Neurosci 5:97–107
Varga Z, Jia H, Sakmann B, Konnerth A (2011) Dendritic coding of multiple sensory inputs in single cortical neurons in vivo. Proc Natl Acad Sci USA
Vickers CA, Dickson KS, Wyllie DJA (2005) Induction and maintenance of late-phase long-term potentiation in isolated dendrites of rat hippocampal CA1 pyramidal neurones. J Physiol 568:803–813
Wang DO, Martin KC, Zukin RS (2010) Spatially restricting gene expression by local translation at synapses. Trends Neurosci 33:173–182
Winnubst J, Lohmann C (2012) Synaptic clustering during development and learning: the why, when, and how. Front Mol Neurosci 5:70
Yadav A, Gao YZ, Rodriguez A et al (2012) Morphologic evidence for spatially clustered spines in apical dendrites of monkey neocortical pyramidal cells. J Comp Neurol 520:2888–2902
Yuste R, Denk W (1995) Dendritic spines as basic functional units of neuronal integration. Nature 375:682–684
Zhang W, Linden DJ (2003) The other side of the engram: experience-driven changes in neuronal intrinsic excitability. Nat Rev Neurosci 4:885–900
Zhao C, Dreosti E, Lagnado L (2011) Homeostatic synaptic plasticity through changes in presynaptic calcium influx. J Neurosci 31:7492–7496
Zhou Y, Won J, Karlsson MG et al (2009) CREB regulates excitability and the allocation of memory to subsets of neurons in the amygdala. Nat Neurosci 12:1438–1443

Chapter 26
Synaptic Plasticity and Pattern Recognition in Cerebellar Purkinje Cells

Giseli de Sousa, Reinoud Maex, Rod Adams, Neil Davey, and Volker Steuber

Abstract Many theories of cerebellar learning assume that long-term depression (LTD) of synapses between parallel fibres (PFs) and Purkinje cells is the basis for pattern recognition in the cerebellum. Here we describe a series of computer simulations that use a morphologically realistic conductance-based model of a cerebellar Purkinje cell to study pattern recognition based on PF LTD. Our simulation results, which are supported by electrophysiological recordings in vitro and in vivo, suggest that Purkinje cells can use a novel neural code that is based on the duration of silent periods in their activity. The simulations of the biologically detailed Purkinje cell model are compared with simulations of a corresponding artificial neural network (ANN) model. We find that the predictions of the two models differ to a large extent. The Purkinje cell model is very sensitive to the amount of LTD induced, whereas the ANN is not. Moreover, the pattern recognition performance of the ANN increases as the patterns become sparser, while the Purkinje cell model is unable to

---

G. de Sousa
Connectionism and Cognitive Science Laboratory, Department of Informatics and Statistics, Federal University of Santa Catarina, Florianopolis SC, Brazil

Biocomputation Group, Science and Technology Research Institute, University of Hertfordshire, Hatfield, Herts, UK
e-mail: giseli@inf.ufsc.br

R. Maex
Group for Neural Theory, Department of Cognitive Sciences, École Normale Supérieure, Paris, France

Biocomputation Group, Science and Technology Research Institute, University of Hertfordshire, Hatfield, Herts, UK
e-mail: reinoud.maex@ens.fr

R. Adams • N. Davey • V. Steuber (✉)
Biocomputation Group, Science and Technology Research Institute, University of Hertfordshire, Hatfield, Herts, UK
e-mail: r.g.adams@herts.ac.uk; n.davey@herts.ac.uk; v.steuber@herts.ac.uk

H. Cuntz et al. (eds.), *The Computing Dendrite: From Structure to Function*, Springer Series in Computational Neuroscience 11, DOI 10.1007/978-1-4614-8094-5_26,

recognise very sparse patterns. These results highlight that it is important to choose a model at a level of biological detail that fits the research question that is being addressed.

26.1 Introduction

Cerebellar Purkinje cells are amongst the neurons with the most elaborate dendrites and an ideal preparation for the study of dendritic computation. In Purkinje cell dendrites, excitatory inputs from approximately 150,000 parallel fibres (PFs) and a single climbing fibre (CF) are integrated with inhibitory inputs from molecular layer interneurons (MLIs) and recurrent Purkinje cell collaterals, and the resulting spikes that are transmitted down the Purkinje cell axon provide the output from cerebellar cortex to the rest of the brain. Understanding synaptic integration and information processing by Purkinje cell dendrites is a prerequisite for understanding the computational role of the cerebellum as a whole.

The cerebellum contributes to many diverse brain functions, including motor control, in which it is responsible for the smoothness and precision of body movements. Although this is still under active debate (see for example Schonewille et al., 2011), it is often thought that in order for new motor skills to be accomplished learning must take place in the cerebellum. In general, it is known that learning occurs in neural systems when synapses change their strength. The classical Marr–Albus–Ito cerebellar learning theory (Marr, 1969; Albus, 1971; Ito, 1984) proposes that motor learning is based on plasticity at PF synapses onto Purkinje cells. This theory suggests that a Purkinje cell can learn to discriminate between different activity patterns presented by its thousands of afferent PFs based on long-term depression (LTD) at PF synapses. LTD decreases the strength of PF synapses by reducing the number of postsynaptic AMPA receptors; this has been shown to occur when Purkinje cell synapses receive excitatory input from both the CF and PF at roughly the same time (Ito et al., 1982; Ito, 2001). Thus, it has been suggested that the CF provides a teaching signal. As a result of PF LTD, the Purkinje cell firing rate has been assumed to be reduced in response to the presentation of a learned PF pattern (as compared to a novel pattern), and thus the Purkinje cell is expected to exert less inhibition on the cerebellar nuclei. As a consequence of this process, the cerebellar output would be increased, so implementing the motor learning. However, Purkinje cells in vivo are characterised by continuous irregular simple spike firing at varying rates (Shin et al., 2007; Armstrong and Rawson, 1979; Goossens et al., 2001), and it is by no means clear how synaptic plasticity could be used to store and retrieve input patterns in such a continuously spiking neuron. Moreover, the complex interactions between voltage and ligand gated ion channels and intracellular calcium that occur in the Purkinje cell dendrites and soma, and the distribution of PF inputs across the large dendrite, could affect the recall of stored patterns.

In this chapter we summarise and elaborate on previous work addressing the question how PF LTD can enable Purkinje cells to respond differently to stored as opposed to novel PF input patterns (Steuber et al., 2007; Sousa et al., 2009). We describe results of simulations of a conductance-based multi-compartmental Purkinje cell model with a morphologically realistic representation of a dendritic tree, and compare these with results from an artificial neural network (ANN) model that uses the same LTD-based learning rule, but that does not contain any representation of dendritic morphology or realistic neuronal physiology. In particular, we describe how the actual amount of synaptic depression can affect the performance of the Purkinje cell model, and we contrast this with how it affects the performance of the ANN.

26.2 Purkinje Cell Model

The simulations described in this chapter used the conductance-based multi-compartmental model of a cerebellar Purkinje cell by De Schutter and Bower (1994a,b). This model is based on a morphological reconstruction and a passive model of a Purkinje cell by Rapp et al. (1994) and includes ten different types of voltage-dependent channels that are modelled using Hodgkin–Huxley type equations. The soma compartment contains a fast and persistent Na^+ conductance, a delayed rectifier, a transient A-type K^+ conductance, a non-inactivating M-type K^+ conductance, an anomalous rectifier, and a low-threshold T-type Ca^{2+} conductance. The dendritic compartments contain a Purkinje cell-specific high-threshold P-type and a low-threshold T-type Ca^{2+} conductance, two different types of Ca^{2+}-activated K^+ (KCa) conductances and an M-type K^+ conductance. Like real Purkinje cells, the model can generate three types of activity: tonic spiking, bursts, and pauses. These activity patterns make up the trimodal spiking behaviour that has been observed in Purkinje cells (for review, see Engbers et al., 2013). In the model, the tonic spiking is based on the activation of somatic fast Na^+ and delayed rectifier K^+ channels that results from small current injections or weak and irregular synaptic activation, while bursts are generated by the interaction of dendritic P-type Ca^{2+}channels and KCa channels in response to strong current injections or synaptic input.

Each cell was originally modelled with 147,400 dendritic spines, which were activated randomly by a sequence of PF inputs at an average frequency of 0.28 Hz. In our pattern recognition simulations, the background excitation was balanced by tonic inhibition, which made the model fire simple spikes at an average frequency of 48 Hz. Due to the large number of dendritic spines, which made the simulations computationally expensive, a simplified version of the model was constructed by decreasing the number of spines to 1 % of the original number, resulting in a model with 4,550 compartments. Each of the 1,474 spines in this scaled down version of the Purkinje cell model received input through 100 PF synapses that were activated at the same rate as in the full model, which resulted in a total rate of PF excitation for each spine of 28 Hz. As this simplified model gave very similar results to the full model, it was used in the simulations presented here. The simulations were performed using the GENESIS neural simulator (Bower and Beeman, 1998), with additional routines implemented in C++ and MATLAB.

26.3 Modelling Learning and Pattern Recognition

In this study, pattern recognition means the process whereby a neuron, represented by a computational model, responds differentially to particular sets of patterns. The patterns here are represented initially by a number of random binary inputs, which are learned by two different models: an ANN and a morphologically realistic Purkinje cell model. These two models were used in order to compare the pattern recognition performance between them and to evaluate how dendritic morphology and neuronal dynamics can affect performance, as was done previously in studies of pattern recognition (Graham, 2001; Steuber and De Schutter, 2001; Steuber et al., 2007).

The pattern recognition simulations were performed in two steps. In the first step, a number of random binary input patterns were generated, initially 200, and half of these patterns were learned by a corresponding ANN. To reflect the dimensionality of the PF input to the Purkinje cell model, the input patterns had an arity of 147,400 with a varying sparsity or number of active inputs (confusingly, sparsity is normally measured as the fraction of "on" bits in a pattern, so that a sparsity of 0.9 is actually a highly non-sparse patterns with 90 % of the bits being "on"). Most of the simulations used input patterns with 1,000 active inputs (a sparsity of about 0.007), as these patterns resulted in the best performance in the initial simulations of the Purkinje cell model.

The ANN used was a modified version of an associative net with feed-forward connections between its inputs and output (Willshaw et al., 1969) and was trained by applying an LTD learning rule. The simulations of the ANN consisted of two phases: learning and recall.

Initially all the weights to the unit were set to 1. In the learning phase, the weights of all the active inputs were adjusted using a multiplicative learning rule; the network learned each pattern by reducing the weight to a fraction of its previous value, determined by a specific LTD factor (in Fig. 26.1, the LTD factor chosen was 0.5). More formally the learning rule was: To learn the set of patterns $\mathbf{x}^{\mu}$, for each of $\mathbf{x}^{\mu}$:

$$w_i' := \begin{cases} w_i \text{LTD\_factor} & \text{if } x_i^{\mu} = 1 \\ w_i & \text{otherwise} \end{cases} \tag{26.1}$$

In the recall phase, the ANN response was calculated by summing all the synaptic weights associated with active inputs. So the output of the ANN, y, was given by $y = \sum x_i w_i$, where x_i are the inputs and w_i are the weights, which is a linear transfer function. As shown for the recall phase in Fig. 26.1, the stored patterns resulted in lower responses than novel patterns (middle and right graphs, respectively).

In the second step of the pattern recognition simulations, the vector of synaptic weights was transferred from the ANN onto AMPA receptor conductances in the multi-compartmental Purkinje cell model. This represents learning the PF patterns by depressing the corresponding AMPA receptor conductances during LTD induction. To test the recall of learned patterns, the Purkinje cell model was then presented with a corresponding pattern of synchronous AMPA receptor activation at the PF synapses. As with the ANN this was done for both the stored and novel patterns.

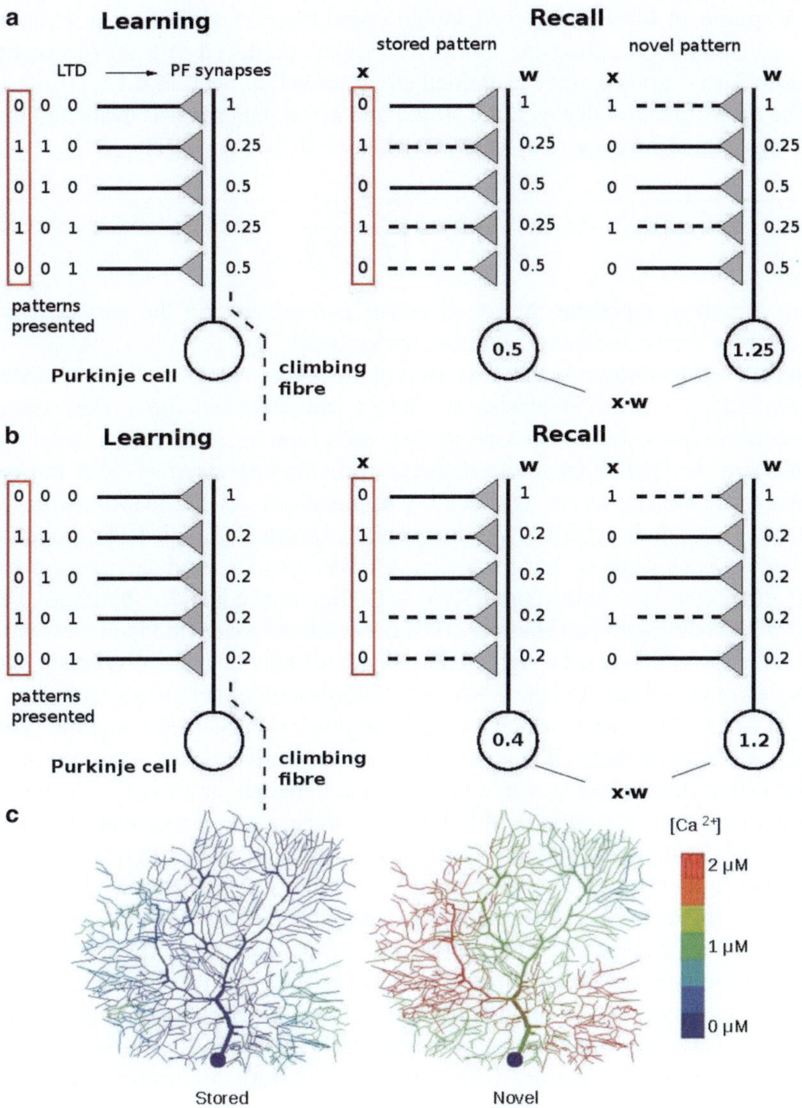

Fig. 26.1 ANN with multiplicative and fixed LTD learning rule and Purkinje cell model. (**a**) ANN with the original multiplicative learning rule (Steuber and De Schutter, 2001; Steuber et al., 2007). Learning phase: three patterns are stored by multiplying the synaptic weights for every active input in every stored pattern by an LTD factor of 0.5. Recall phase: the network response is calculated as the inner product of the input and weight vector. The resulting responses show a lower output value for the stored pattern when compared to the novel pattern. (**b**) ANN with a fixed LTD learning rule (Sousa et al., 2009). In the learning phase, the synaptic weight is set to a fixed LTD value (here 0.2) for the first active input and only for the first active input in the stored PF patterns. The patterns are recalled as in (**a**). (**c**) Purkinje cell model. Snapshot of the Ca^{2+} concentration in all compartments of the Purkinje cell model 10 ms after presentation of a stored (*left*) and novel (*right*) pattern. The stronger synaptic activation of the model by the novel pattern leads to larger Ca^{2+} influx compared to the stored pattern, more activation of KCa channels and a longer pause in firing (see text and Fig. 26.2)

The response of the Purkinje cell model could then be quantified in a variety of ways, such as by counting the number of spikes produced in a post-presentation window. The various metrics examined are described in Sect. 26.4.1.

The discrimination between the stored and novel patterns was evaluated by calculating a signal-to-noise (s/n) ratio (Dayan and Willshaw, 1991):

$$\mathrm{s/n} = 2\frac{(\mu_s - \mu_n)^2}{(\sigma_s^2 + \sigma_n^2)} \tag{26.2}$$

where μ_s and μ_n represent the mean values and σ_s^2 and σ_n^2 the variances of the responses to stored and novel patterns, respectively.

In the LTD learning rule that was used in the original set of simulations (Steuber and De Schutter, 2001; Steuber et al., 2007), and described above, the weights of activated synapses were decreased by half each time a pattern was learned, which represented the AMPA receptor conductance being depressed by 50 % during the learning process. However, consecutive applications of this learning rule could result in very small synaptic weights, while experiments with LTD induction in cerebellar slices suggest that the mean AMPA receptor conductances are hardly ever depressed to less than 50 % of the pre-induction baseline (Wang et al., 2000; Mittmann and Häusser, 2007). Therefore, we decided to investigate the effect of a lower bound on weight changes and different degrees of synaptic plasticity on learning and recall in the ANN and Purkinje cell model. In this study, we used a set of fixed LTD values, which varied from 0 to 0.8. This meant that the synaptic weight was set to a fixed LTD value when the first active input was stored in a particular synapse; subsequent active inputs did not change the weight. Consequently, in this case a given weight could change at most once during learning. So, the weights of the network initially were set to one, $w_i := 1$, and they were adjusted according to the fixed LTD learning rule $w_i' = \text{LTD\_value}$ whenever $x_i^\mu = 1$ for any pattern μ. One consequence of using this learning rule (which will be important in our subsequent argument) is that the output of the ANN is exactly the same for all *stored* patterns, giving zero variance. However this will not be the case for the Purkinje cell model, as it is inherently noisy, due to the background input.

In Fig. 26.1, an example of the new fixed LTD learning rule is presented, where an LTD value of 0.5 was used. The results of this study are presented in the next section (Sect. 26.4).

26.4 Pattern Recognition in the Purkinje Cell Model

26.4.1 Pattern Recognition Based on Simple Spike Pauses

The simulations of LTD-based learning and recall of PF input patterns in the Purkinje cell model led to surprising results (Steuber et al., 2007). Presentations of PF input patterns with large enough numbers of active inputs (at least 750 active PFs

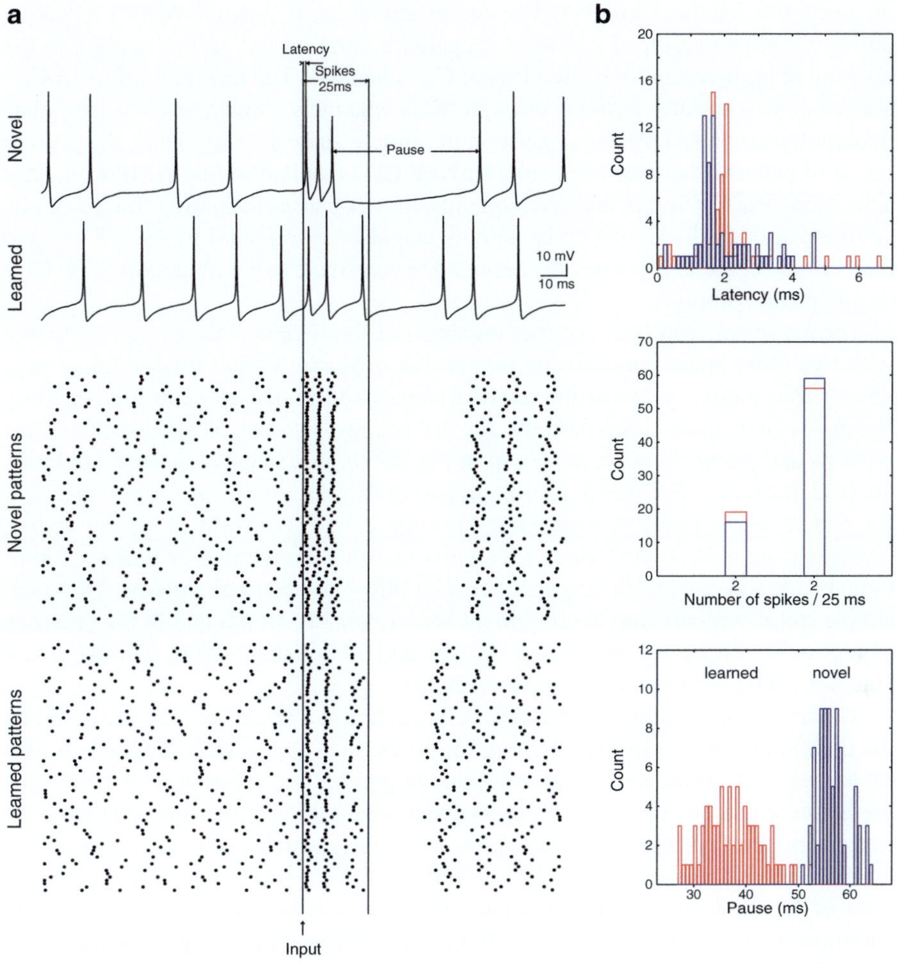

Fig. 26.2 Responses of the Purkinje cell model to novel and learned PF input patterns. (**a**) *Upper panel*: The presentation of a learned and a novel PF pattern results in a burst-pause sequence in the Purkinje cell model, with a pause in response to the novel pattern that is longer than that for the learned pattern. *Lower panel*: Raster plot of spike responses to presentations of 75 learned and 75 novel patterns (100 PF patterns with 1,000 active PFs were stored using the original multiplicative learning rule from Steuber et al., 2007). (**b**) Response distribution for three different spike features. *Upper*: Latency of first spike after pattern presentation. *Middle*: Number of spikes in the first 25 ms. *Lower*: Length of pause. Modified from Steuber et al. (2007)

out of a total of 147,400 PFs, that is a sparsity of at least 0.005) so that they could be detected in the presence of the ongoing simple spike activity in the Purkinje cell model led to stereotypical responses composed of a short burst of spikes followed by a pause. Moreover, the model responded with shorter pauses to learned patterns than to novel patterns, as can be seen in Fig. 26.2a for stored and novel PF input patterns with 1,000 active inputs. The generation of pauses in the model was based on a negative feedback process. The strong activation of synaptic AMPA receptors during the presentation of a pattern triggered a burst of Na^+ spikes and led to the opening of high-voltage activated P-type Ca^{2+} channels. The resulting influx of Ca^{2+} activated an outward current through KCa channels, which caused an after-hyperpolarisation (AHP) and a pause in the simple spike activity. Thus, the presentation of patterns that had been stored by PF LTD resulted in less AMPA receptor activation, less Ca^{2+}influx and consequently shorter pauses in spiking. Interestingly, similar pauses in the simple spike activity can also be triggered by the CF input to the Purkinje cell that is commonly assumed to contribute to the induction of PF LTD during pattern storage.

The pause duration as a carrier of information about input patterns was compared with two other spike features: the number of spikes in a fixed time-window (e.g. 25 ms) after pattern presentation and the latency of the first spike in the response. Examples of response distributions for the latency, number of spikes in a 25 ms window and pause duration are given in Fig. 26.2b, where the respective s/n ratios are 0.33, 0.21 and 15.6. These results suggested that the pause was the best criterion for pattern recognition by cerebellar Purkinje cells. This form of neural coding diverges from the classical view, which assumes that the number or timing of individual spikes is used to distinguish between novel and learned patterns. However, simple spike numbers may be important for weaker PF stimuli and in the presence of less or no PF background input (Walter and Khodakhah, 2006; Mittmann and Häusser, 2007; Walter and Khodakhah, 2009).

The model prediction that Purkinje cells could use a novel neural code based on the duration of pauses in their activity raised the question of the robustness of this finding against parameter variations, and, importantly, the question whether this prediction could be confirmed by experimental data. Extensive parameter explorations using the Purkinje cell model showed that the pause-based coding was indeed a robust result and not affected by variability in the synaptic responses, temporal jitter in the input spikes, physiologically realistic on-beam inhibition by MLIs and variations in the PF background rate. In fact, an increased PF background rate led to faster Purkinje cell spiking and therefore less variation in the pauses and an improved pattern recognition performance, in contrast to pattern recognition by the ANN where noise in the inputs deteriorated the performance.

More support for a pause-based recall of PF patterns that had been stored by LTD of their synapses onto Purkinje cells came from electrophysiological recordings of Purkinje cells in cerebellar slices and in awake behaving mice. In Purkinje cell recordings in cerebellar slices, stimulating PF inputs with increasing stimulus intensities resulted in increased simple spike pause durations. This result supported the model prediction that weakening the PF input by LTD should lead to pauses with shorter lengths. Moreover, a standard LTD induction protocol (coincident stimulation of PF and CF input for 5 min at 1 Hz) led to a shortening of pauses in the Purkinje cell responses, both with on-beam inhibition blocked and on-beam inhibition intact. Finally, recordings from awake behaving mice revealed that simple spike trains from Purkinje cells in LTD-deficient mice contained longer pauses than the ones from mice where LTD was intact, providing additional indirect evidence for the reduction of pause durations by PF LTD and a potential pause-based code (Steuber et al., 2007).

In a follow-up study, we used the Purkinje cell model with a modified learning rule to analyse the effect of different levels of fixed synaptic plasticity on pattern recognition by Purkinje cells; these results are presented in the next section.

26.4.2 *Effect of Fixed LTD at Different LTD Values*

The multiplicative LTD learning rule that had been used in the simulations described in the previous section could result in very small AMPA receptor conductances in the Purkinje cell model. This is in apparent contrast to data from LTD induction experiments in cerebellar slices, where the mean Purkinje cell response to PF stimulation is usually not depressed to less than 50 % of the baseline response before LTD induction (Wang et al., 2000; Mittmann and Häusser, 2007). However, this is contentious as other work has shown that many PF synapses are silent (Isope and Barbour, 2002). In a follow-up study, we therefore explored the effect of different amounts of LTD on pattern recognition. We changed the learning rule so that for every active PF in a stored pattern the corresponding synaptic weight was set to a fixed LTD value that served as a lower bound. We initially investigated the effect of varying two parameters on pattern recognition: the fixed LTD value and the number of active PFs, the sparsity of the input patterns. Firstly, to study the effect of the extent of LTD at PF synapses, we varied the LTD value over a range of values from 0 to 0.8, where 0 represented that the synaptic weight are set to 0 and 0.8 represented that the synaptic weight was decreased from 1 to 0.8. For this experiment, we used the same numbers of active PFs (1,000, which represents 0.7 % of the total number of PF inputs) and PF patterns (100 novel and 100 stored) as in most of the previous work (Steuber et al., 2007). Figure 26.3 shows the results for the ANN (a) and Purkinje cell model (b) where the performance of each model was calculated by averaging their s/n ratio over ten runs with different sets of patterns.

From these results, we can see that the ANN was insensitive to the LTD value (Fig. 26.3a). In contrast, the pattern recognition capacity, based on the duration of pauses, in the Purkinje cell model improved when the LTD value decreased, finding an optimal performance when the synaptic weights of active PFs were set to zero (Fig. 26.3b). The reason for the difference in sensitivity of the ANN and the Purkinje model to varying amounts of LTD became apparent when the squared difference of the mean responses to stored and novel patterns ($(\mu_s - \mu_n)^2$, the numerator of 2) and the summed variance of the responses ($\sigma_s^2 + \sigma_n^2$, the denominator of 2) of the two models were plotted against the LTD saturation value (Fig. 26.3c, d).

In the Purkinje cell model, increasing LTD values reduced the difference in pause duration between stored and novel patterns, with variances that were affected to a lesser extent (Fig. 26.3d). Given that the responses of the Purkinje cell model were noisy, and the variance of the responses was always much higher than in the ANN, this led to the drastic reduction in s/n ratio for weak LTD shown in Fig. 26.3.

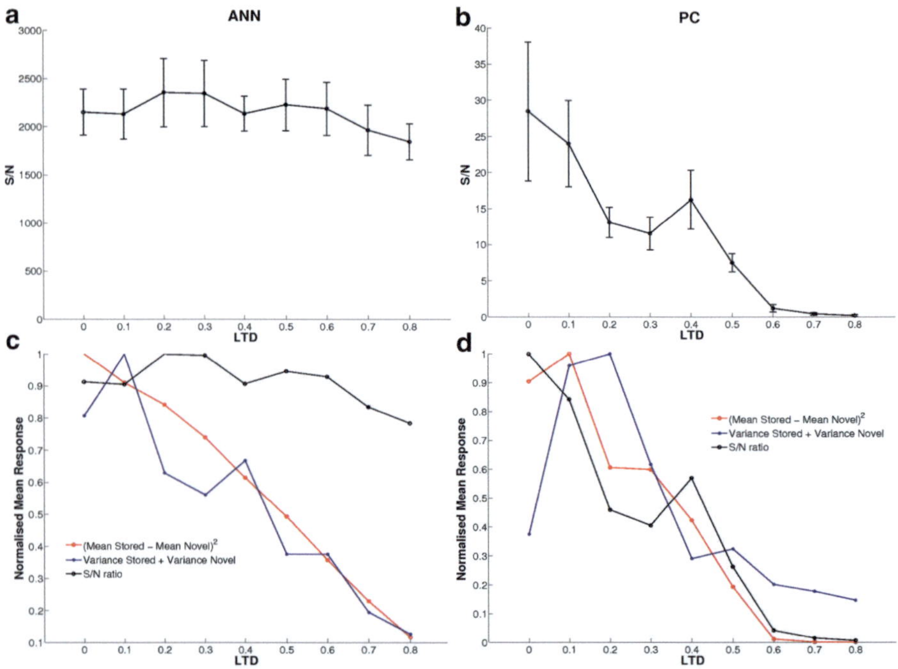

Fig. 26.3 Pattern recognition performance of the ANN and Purkinje cell model for a range of fixed LTD values. The performance was evaluated by calculating the mean s/n ratio for the ANN (**a**) and for pause duration in the PC model (**b**), averaging over ten different sets of 100 stored and 100 novel PF patterns with 1,000 active PFs. Error bars indicate standard deviation. (**c,d**): Normalised s/n ratio (*black*), squared difference between the mean responses (numerator of (26.2), *red*) and sum of the variance of the responses (denominator of (26.2), *blue*) to stored and novel patterns for the ANN (**c**) and for pause duration in the Purkinje cell model (**d**). (**c**) In the ANN, an increase in LTD value leads to decreases in both numerator and denominator that cancel out, resulting in an approximately constant s/n ratio. (**d**) In the Purkinje cell model, an increase in LTD value affects the denominator more than the numerator and thus leads to a reduced s/n ratio

In the ANN, the fixed LTD learning rule resulted in invariant responses to stored patterns ($\sigma_s^2 = 0$), and in a variance of responses to novel patterns σ_n^2 that exhibited a linear relationship to the squared difference of the mean responses to stored and novel patterns $(\mu_s - \mu_n)^2$. Thus, in the ANN, the numerator and denominator of 26.2 were affected equally by the fixed LTD value, which meant that the s/n ratio was not affected by the amount of LTD induced (Fig. 26.3c). In fact, it can be shown analytically that the s/n ratio of the ANN can be approximated as

$$\text{s/n} = 2\frac{A(S-D)}{D} \tag{26.3}$$

where S is the total number of synapses, A is the number of active synapses per pattern and D is the number of depressed synapses after learning, which can be approximated as

$$D = S\left(1 - \left(1 - \frac{A}{S}\right)^{L}\right) \tag{26.4}$$

where L is the number of learned patterns (for an analytical demonstration, see Appendix). So, for the configuration of the ANN used here, $S = 147{,}400$, $A = 1{,}000$, $L = 100$, D will be approximately 72,779, and the s/n ratio will be 2,050, which corresponds to the number shown in Fig. 26.3a. It is important to note that the s/n ratio is independent of the actual amount of synaptic depression. By contrast, in the Purkinje cell model the s/n ratio is dependent on the actual extent of LTD as the variance of the responses does not change as much as the mean difference between the responses to learned and novel patterns.

In a second experiment, we measured the effect of variation in the number of active PFs. The range of active PFs was varied from 500 to 5,000, corresponding to 0.3–3.4 % of all PF inputs (147,400), where the same synaptic conductance was used regardless of the number of active PFs. The LTD saturation values were varied over the same range as in previous experiments (from 0 to 0.8). Comparing once more the effects on pattern recognition by the ANN and the Purkinje cell model based on pause duration, we can see from Fig. 26.4a, c that the performance of the ANN decreased with an increasing number of activated PFs, and that its performance was independent of the LTD factor across the whole range of sparsities tested. In contrast, in the Purkinje cell model, lower LTD factors produced better s/n values for pattern recognition, especially in an optimal range of active PF numbers between 1,000 and 2,000 (Fig. 26.4b, d). Figure 26.4c also shows that the analytical prediction (26.3) provides a good fit with the results of the numerical simulation.

The rapid deterioration in performance of the Purkinje cell model for PF patterns with less than 1,000 active synapses can be explained by the fact that the synaptic conductances were not scaled when varying the number of active PFs, so that the resulting PF input patterns were too weak to activate the Purkinje cell sufficiently. A further experiment has shown that in the case of 500 active PFs, the best performance was achieved (s/n~ 38) when the synaptic conductance was doubled from the original conductance value (0.7 nS), which was set initially for 1,000 active PFs (results not shown here).

26.5 Conclusion

Ever since the theoretical work by Marr (1969) and Albus (1971) more than 40 years ago, much of cerebellar research has been influenced by the assumption that PF LTD forms the basis of cerebellar learning, enabling Purkinje cells to store and recall PF input patterns. Several theoretical and computational studies have built on

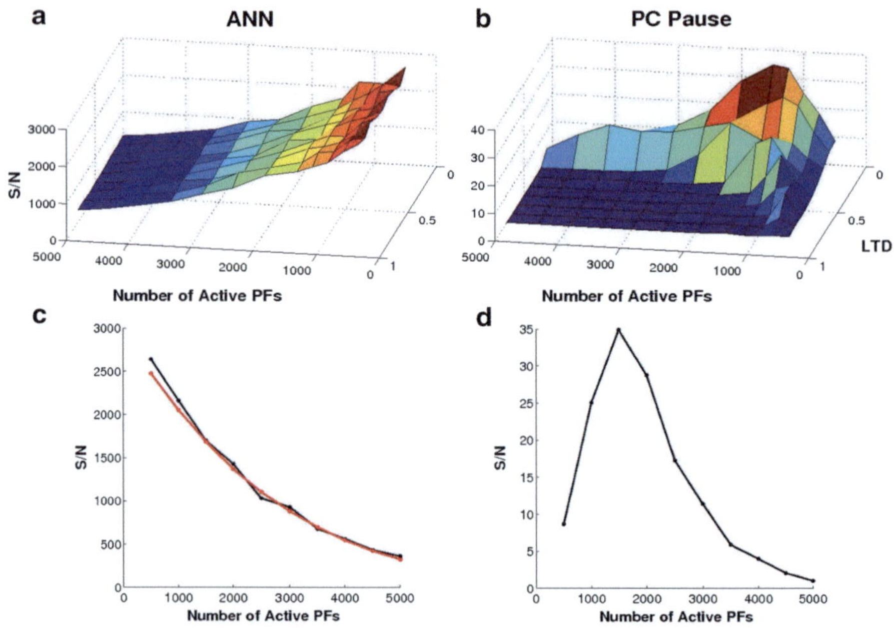

Fig. 26.4 Pattern recognition performance of the ANN (**a,c**) and the Purkinje cell model (**b,d**) using pause duration as response criterion. The performance of the ANN (**a**) and the Purkinje cell model (**b**) over ten trials is shown for a range of LTD values (0–0.8) and varying numbers of active PFs (500–5,000). While the performance of the ANN is independent of the LTD value and decreases monotonically with the number of active PFs (**a**), the Purkinje cell model performs best for an LTD value of 0 and between 1,000 and 2,000 active PFs (**b**). (**c,d**) Pattern recognition performance of the ANN (**c**) and the Purkinje cell model (**d**) for an LTD value of 0 and varying numbers of active PFs. (**c**) Compares simulation data from the ANN (*black*) and the predicted values based on (Equ. 26.3) (*red*), showing that the analytical prediction is in good agreement with the numerical simulations

the work by Marr and Albus, typically using simplified models such as ANNs or integrate-and-fire networks (Gilbert, 1974; Medina et al., 2000; Schweighofer and Ferriol, 2000). However, the integration of synaptic inputs and generation of spike responses by Purkinje cells are governed by complex interactions of ion channels and intracellular Ca^{2+} in their dendritic tree. Moreover, PF input patterns are likely to activate synapses that are distributed across the Purkinje cell dendrite, and Purkinje cells in vivo fire simple spikes at varying rates (Shin et al., 2007; Armstrong and Rawson, 1979; Goossens et al., 2001). All of these factors are expected to affect the response of Purkinje cells to PF patterns that have been stored by LTD.

Although experiments in cerebellar slices have indicated that a change in the Purkinje cell firing rate can carry some information about the strength of PF inputs and the PF synaptic weights (Walter and Khodakhah, 2006; Mittmann and Häusser, 2007; Walter and Khodakhah, 2009), the large number of PF inputs that each Purkinje cell receives (approximately 150,000) means that a complete understanding of pattern recognition by Purkinje cells requires simulations of a conductance-based Purkinje cell model with a morphologically realistic representation of a dendritic tree. In this chapter, we have described a series of simulations of a biophysically realistic Purkinje cell model (De Schutter and Bower, 1994a,b) that were used to study the LTD-based storage and recall of PF input patterns. These simulations predict that the feature of the Purkinje cell spike response that gives the strongest differentiation between stored and novel PF patterns is the length of a pause in spiking that follows the presentation of PF patterns, with shorter pauses in response to PF patterns that have been stored by LTD (Steuber et al., 2007). The predictions of these simulations were robust against parameter variations and different kinds of noise in the simulations, and they were confirmed using experiments in cerebellar slices and in awake behaving mice (Steuber et al., 2007).

Although the initial set of simulations used a biophysically detailed Purkinje cell model that had been tuned to replicate a wide range of behaviours in vitro and in vivo (De Schutter and Bower, 1994a,b), they applied a simplified LTD learning rule, which involved dividing the synaptic weights of active PF inputs by two every time a PF pattern was learned. This could result in very small synaptic weights and did not fit experimental data on LTD induction in cerebellar slices, where the mean AMPA receptor conductances saturate and are hardly ever depressed to less than 50 % of their pre-depression baseline values (Mittmann and Häusser, 2007; Wang et al., 2000). This suggested that it would be interesting to study the impact of the actual level of synaptic depression that was allowed.

We found that the ability of the Purkinje cell model to discriminate between learned and novel PF input patterns was very sensitive to the value at which LTD saturated. The best performance resulted from LTD saturation values of zero, which corresponds to silencing the PF synapses completely. Interestingly, large numbers of silent PF synapses have been observed by monitoring microscopically identified PF-Purkinje cell connections in cerebellar slices (Isope and Barbour, 2002). Our simulation results indicate that the discrepancy between the existence of these silent synapses and the apparent saturation of LTD in induction experiments needs to be resolved to understand the connection between LTD and cerebellar learning.

In contrast, a corresponding ANN was unaffected by the amount of LTD induced. When we looked in detail at the difference between the response of the Purkinje cell and ANN, it was apparent that for the ANN both the numerator and denominator of the s/n ratio decreased at the same rate as the LTD depression was decreased. However, for the Purkinje cell the numerator dropped more rapidly than the denominator, so that the mean response to stored and novel patterns became very similar.

Our final experiment added the sparseness of the patterns as one of the parameters. As is well known, with the type of learning rule employed here, sparse patterns will give better results: sparse patterns are more likely to be orthogonal (not share any "on" bits) and this is confirmed for the ANN in Fig. 26.4a, where performance gets worse as patterns become less sparse, independently of the value of synaptic depression. For the Purkinje cell the response to sparseness is initially similar—increasing sparseness leads to better performance. However, unlike the ANN this effect is not present with very sparse patterns, where performance begins to fall due to the inability of the patterns to activate the Purkinje cell sufficiently.

The discrepancy of the results from the simple ANN and biophysically detailed Purkinje cell model highlights that it is important to choose a model at a level of detail that fits the research question that is being addressed. When the results can be affected by the specific physiology and morphology of a neuron, it is clearly better to use a model where the physiology and morphology are represented explicitly. This approach has been followed in the work that has been summarised in this chapter, leading to the prediction that the cerebellum can use pauses in Purkinje cell activity to encode information. However, a complete understanding of the functional role of pause- versus spike-number-based coding for cerebellar computation will require computer simulations and experiments that investigate how these different features of the Purkinje cell spike trains can be read out by their target neurons in the cerebellar nuclei (Luthman et al., 2009; Schutter and Steuber, 2009; Steuber and Jaeger, 2013).

Appendix: Calculation of the Signal-to-Noise Ratio for the ANN

We here derive a simple approximation for the signal-to-noise ratio of the ANN responses to stored and novel patterns. We assume, as in Fig. 26.4, that the synapses take binary values: they are either depressed or non-depressed, and without loss of generality (see below) we assign them weights zero and one. In this case, the responses to stored patterns are always zero, as is their variance. Hence we need only to calculate the mean and variance of the responses to novel patterns. For this we proceed in two steps: (a) find the number D of depressed synapses; (b) draw A synapses (A being the number of active synapses per pattern) from a pool of $n = S$ synapses of which D are depressed. Let us start with the second problem, so D is supposed to be known.

Calculating the response to a novel pattern is now equivalent to drawing, without replacement, A balls from an urn containing S balls of which D are white (depressed) and $S - D$ balls are red (non-depressed). The distribution of the number of red balls drawn on each trial is known to be hypergeometric. However, if S be very large compared to A, the numbers approximate a binomial distribution with the probability p of drawing a non-depressed synapse equal to $(1 - D/S)$. A binomial distribution has mean $n\,p$ and variance $n\,p\,(1-p)$, hence the average output of the ANN for novel patterns will be $A(1 - D/S)$ with variance $A(1 - D/S)(D/S)$. The signal-to-noise ratio becomes:

$$\begin{aligned} \mathrm{s/n} &= \frac{(\mu_s - \mu_n)^2}{0.5(\sigma_s^2 + \sigma_n^2)} \\ &= 2\frac{(A(1 - D/S))^2}{A(1 - (D/S)(D/S)} \\ &= 2\frac{A(S - D)}{D} \end{aligned}$$

Note that the signal-to-noise ratio does not depend on the exact values of the weights assigned to the binary synapses. If the depressed and non-depressed synapses were to be assigned weights a and b, respectively, instead of 0 and 1, we would obtain from the binomial distributions $\mu_s = an$, $\mu_n = an(1-p)+bnp$, $\sigma_s^2 = 0$, and $\sigma_n^2 = (a-b)^2 np(1-p)$, which after substitution yields the same value for s/n.

We now turn to the problem of assessing the number D of depressed synapses. If the number of learned patterns L is very small and/or the patterns are very sparse (A small), then the degree of overlap between learned patterns will also be small, so that D becomes approximately equal to LA (each depressed synapse is unique).

Then the signal-to-noise ratio reduces to:

$$\begin{aligned} \mathrm{s/n} &= 2\frac{A(S-LA)}{LA} \\ &= 2\frac{(S-LA)}{L} \\ &\approx 2\frac{S}{L} \end{aligned}$$

where it is assumed that $L * A \ll S$. Hence the "s/n" signal-to-noise ratio is proportional to the number S of synapses and inversely proportional to the number L of learned patterns.

If LA is large, we cannot assume LA to be a good approximation of D. But let us assume all synapses are independently depressed with probability A/S for each pattern, then each synapse has probability $(1-A/S)^L$ of not having been depressed by any of the L patterns, and consequently the mean number D of depressed synapses will approximate $S-S(1-A/S)^L$.

References

Albus JS (1971) A theory of cerebellar function. Math Biosci 10:25–61

Armstrong DM, Rawson JA (1979) Activity patterns of cerebellar cortical neurones and climbing fibre afferents in the awake cat. J Physiol 289(1):425–448

Bower JM, Beeman D (1998) The book of GENESIS: exploring realistic neural models with the general neural simulation system, 2nd edn. Springer, New York

Dayan P, Willshaw DJ (1991) Optimising synaptic learning rules in linear associative memories. Biol Cybern 65(4):253–265

De Schutter E, Bower J (1994a) An active membrane model of the cerebellar Purkinje cell. I. Simulation of current clamps in slice. J Neurophysiol 71(1):375–400

De Schutter E, Bower J (1994b) An active membrane model of the cerebellar Purkinje cell: II. Simulation of synaptic responses. J Neurophysiol 71(1):401–419

Engbers JD, Fernandez FR, Turner RW (2013) Bistability in Purkinje neurons: ups and downs in cerebellar research. Neural Netw 47:18–31

Gilbert PF (1974) A theory of memory that explains the function and structure of the cerebellum. Brain Res 70(1):1–18

Goossens J, Daniel H, Rancillac A, van der Steen J, Oberdick J, Crépel F, De Zeeuw CI, Frens MA (2001) Expression of protein kinase c inhibitor blocks cerebellar long-term depression without affecting Purkinje cell excitability in alert mice. J Neurosci 21(15):5813–5823

Graham BP (2001) Pattern recognition in a compartmental model of a CA1 pyramidal neuron. Network 12(4):473–492

Isope P, Barbour B (2002) Properties of unitary granule cell-Purkinje cell synapses in adult rat cerebellar slices. J Neurosci 22(22):9668–9678

Ito M (1984) The cerebellum and neural control. Raven Press, New York

Ito M (2001) Cerebellar long-term depression: characterization, signal transduction, and functional roles. Physiol Rev 81(3):1143–1195

Ito M, Sakurai M, Tongroach P (1982) Climbing fibre induced depression of both mossy fibre responsiveness and glutamate sensitivity of cerebellar Purkinje cells. J Physiol 324(1): 113–134

Luthman J, Adams R, Davey N, Maex R, Steuber V (2009) Decoding of Purkinje cell pauses by deep cerebellar nucleus neurons. BMC Neurosci 10(Suppl 1):P105

Marr D (1969) A theory of cerebellar cortex. J Physiol 202:437–470

Medina JF, Garcia KS, Nores WL, Taylor NM, Mauk MD (2000) Timing mechanisms in the cerebellum: testing predictions of a large-scale computer simulation. J Neurosci 20(14): 5516–5525

Mittmann W, Häusser M (2007) Linking synaptic plasticity and spike output at excitatory and inhibitory synapses onto cerebellar Purkinje cells. J Neurosci 27(21):5559–5570

Rapp M, Segev I, Yarom Y (1994) Physiology, morphology and detailed passive models of guinea-pig cerebellar Purkinje cells. J Physiol 474(1):101–118

Schonewille M, Gao Z, Boele HJ, Veloz MFV, Amerika WE, Simek AA, Jeu MTD, Steinberg JP, Takamiya K, Hoebeek FE, Linden DJ, Huganir RL, Zeeuw CID (2011) Reevaluating the role of LTD in cerebellar motor learning. Neuron 70(1):43–50

Schutter ED, Steuber V (2009) Patterns and pauses in Purkinje cell simple spike trains: experiments, modeling and theory. Neuroscience 162(3):816–826

Schweighofer N, Ferriol G (2000) Diffusion of nitric oxide can facilitate cerebellar learning: a simulation study. Proc Natl Acad Sci USA 97(19):10661–10665

Shin SL, Hoebeek FE, Schonewille M, De Zeeuw CI, Aertsen A, De Schutter E (2007) Regular patterns in cerebellar Purkinje cell simple spike trains. PLoS ONE 2(5):e485

Sousa G, Adams R, Davey N, Maex R, Steuber V (2009) The effect of different forms of synaptic plasticity on pattern recognition in the cerebellar cortex. In: Kolehmainen M, Toivanen P, Beliczynski B (eds) Adaptive and natural computing algorithms, Lecture notes in computer science, vol 5495. Springer Berlin, pp 413–422

Steuber V, De Schutter E (2001) Long-term depression and recognition of parallel fibre patterns in a multi-compartmental model of a cerebellar Purkinje cell. Neurocomputing 38:383–388

Steuber V, Jaeger D (2012) Modeling the generation of output by the cerebellar nuclei. Neural Netw 47:112–119

Steuber V, Mittmann W, Hoebeek FE, Silver RA, De Zeeuw CI, Häusser M, De Schutter E (2007) Cerebellar LTD and pattern recognition by Purkinje cells. Neuron 54(1):121–136

Walter JT, Khodakhah K (2006) The linear computational algorithm of cerebellar Purkinje cells. J Neurosci 26(50):12861–12872

Walter JT, Khodakhah K (2009) The advantages of linear information processing for cerebellar computation. Proc Natl Acad Sci USA 106(11):4471–4476

Wang SSH, Denk W, Häusser M (2000) Coincidence detection in single dendritic spines mediated by calcium release. Nat Neurosci 3(12):1266–1273

Willshaw D, Buneman O, Longuet-Higgins H (1969) Non-holographic associative memory. Nature 222:960–962

Chapter 27
Response of Gap Junction-Coupled Dendrites: A Sum-Over-Trips Approach

Yulia Timofeeva and Stephen Coombes

Abstract Dendrites form the major components of neurons. They are complex branching structures that receive and process thousands of synaptic inputs from other neurons. The impulse response function for branched dendritic trees can be calculated using a so-called sum-over-trips approach. In this chapter we extend this formalism to treat networks of dendritic trees connected via dendro-dendritic gap junctions. To illustrate the usefulness of this extended formalism for understanding how gap junctions can contribute to signal integration in neural networks, we consider how they affect somatic voltages in a simple two neuron network with gap junction coupling between distal dendrites. We find that proximal input on one cell can strongly innervate the soma of that cell though the spread of charge to the gap junction-coupled cell is weak. In contrast distal inputs on one cell weakly innervate the soma of that cell though charge can spread effectively to the gap junction-coupled cell.

27.1 Introduction

The nervous system is an extremely complex structure that includes neurons specialised for generating electrical signals in response to chemical and electrical inputs and transmitting them to other cells. Despite the fact that the experimental

Y. Timofeeva
Department of Computer Science and Centre for Complexity Science,
University of Warwick, Coventry, CV4 7AL, UK
e-mail: y.timofeeva@warwick.ac.uk

S. Coombes (✉)
School of Mathematical Sciences, University of Nottingham,
Nottingham, NG7 2RD, UK
e-mail: stephen.coombes@nottingham.ac.uk

H. Cuntz et al. (eds.), *The Computing Dendrite: From Structure to Function*,
Springer Series in Computational Neuroscience 11, DOI 10.1007/978-1-4614-8094-5_27,

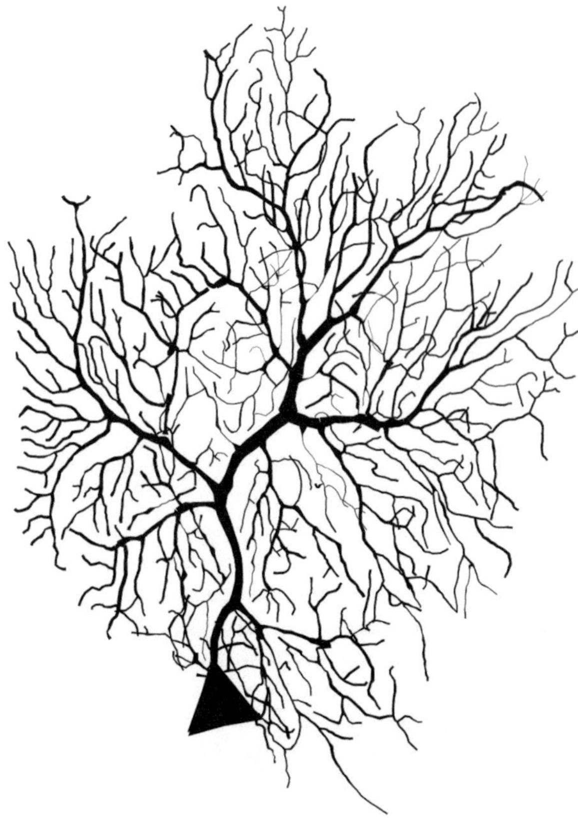

Fig. 27.1 Dendrites form the major components of neurons. They are complex branching structures that receive and process thousands of synaptic inputs from other neurons. It is well known that dendritic morphology plays an important role in the function of dendrites. Visit NeuroMorpho.Org to see a repository of digitally reconstructed neurons (Ascoli et al., 2007). Here we show an example of a digitally reconstructed Purkinje cell (Martone et al., 2003), NeuroMorpho.Org ID: NMO_00864

and theoretical work for the understanding of signal processing in neurons was begun more than a century ago, the role of an individual neuron in neural computation has long been debated. Traditionally, relatively simple computational properties have been attributed to the individual cell and complex computations to networks of these simple elements. However, this assumption is oversimplified in view of the properties of real neurons and the computations they perform (London and Häusser, 2005). In particular neurons are spatially extended and can have elaborate dendritic structures as illustrated in Fig. 27.1.

Dendrites are involved in receiving and integrating thousands of inputs (via both *chemical* and *electrical* synapses) from other cells as well as in determining the extent to which action potentials are produced by neurons. The theoretical work of Rall during the 1950s and 1960s that introduced a new framework (cable theory and compartmentalisation) for modelling these complex structures was revolutionary. He demonstrated that the structural and electrical properties of dendrites play a critical role in the way a neuron processes its synaptic inputs (Segev et al., 1995). Indeed there is now much activity within the computational neuroscience community in studying networks of compartmental neurons with synaptic coupling, such as the

Blue Brain project (Markram, 2006). This is a direct attempt to reverse-engineer the mammalian brain, in order to understand brain function and dysfunction through detailed computer simulations. Such studies emphasise that the need to give biological interpretation to model output requires biological realism to be taken seriously in all model components, and has generated much activity in the mathematical neuroscience community in an attempt to attain similar levels of understanding without recourse to brute-force simulations (Rabinovich et al., 2006). However, it is fair to say that both approaches have paid correspondingly little attention to the *electrical synapse* or *gap junction*. An electrical synapse is a mechanically and electrically conductive link between two adjacent nerve cells that is formed at a fine gap between the pre- and post-synaptic cells known as a gap junction and permits a direct electrical connection between them. Each gap junction contains numerous connexin hemichannels which cross the membranes of both cells. With a lumen diameter of about 1.2–2.0 nm, the pore of a gap junction channel is wide enough to allow ions and even medium-sized molecules like signalling molecules to flow from one cell to the next thereby connecting the two cells' cytoplasm. Their discovery was first demonstrated between escape-related giant nerve cells in *crayfish* in the late 1950s. They are now known to be abundant in the retina and cerebellar cortex of vertebrates and have been directly demonstrated between inhibitory neurons in the neocortex (Galarreta and Hestrin, 1999) (particularly between fast-spiking cells and low-threshold spiking cells). In fact it would appear that they are now ubiquitous throughout the human brain, being found in the hippocampus (Fukuda and Kosaka, 2000), the inferior olivary nucleus in the brain stem (Sotelo et al., 1974), the spinal cord (Rash et al., 1996), the thalamus (Hughes and Crunelli, 2007) and have recently been shown to form axo-axonic connections between *excitatory* cells in the hippocampus (on mossy fibers) (Hamzei-Sichani et al., 2007). Without the need for receptors to recognise chemical messengers gap junctions are much faster than chemical synapses at relaying signals. The synaptic delay for a chemical synapse is typically in the range 1–100 ms, while the synaptic delay for an electrical synapse may be only about 0.2 ms. Not very much is known about the functional aspects of gap junctions, but they are thought to be involved in the *synchronisation* of neurons (Alvarez et al., 2002; Bennet and Zukin, 2004) and are believed to contribute to both normal (Hormuzdi et al., 2004) and abnormal physiological brain rhythms, including epilepsy (Velazquez and Carlen, 2000). However, it is known that network dynamics can be tuned by the location of the gap junction on the dendritic tree (Traub et al., 2001; Saraga et al., 2006).

Seminal work by Abbott et al. (1991, 1992) has shown how to calculate response functions for passive branched dendritic trees with arbitrary geometry. In Sect. 27.2 we briefly review the cable equation and the "sum-over-trips" formalism. Next in Sect. 27.3 we describe how this can be extended to cover networks of dendritic trees connected via dendro-dendritic gap junctions. In Sect. 27.4 we consider the application to two cells coupled by gap junctions between their distal dendrites, and explore how this network integrates different patterns of input activity. We find that proximal input on one cell can strongly innervate the soma of that cell though the spread of charge to the gap junction-coupled cell is weak. In contrast distal inputs on one cell weakly innervate the soma of that cell though charge can spread effectively to the gap junction-coupled cell. Finally in Sect. 27.5 we consider possible future directions for the work in this chapter.

27.2 Modelling a Single Branched Dendritic Tree

A nerve fibre consists of a long thin, electrically conducting core surrounded by a thin membrane whose resistance to transmembrane current flow is much greater than that of either the internal core or the surrounding medium. Injected current can travel long distances along the dendritic core before a significant fraction leaks out across the highly resistive cell membrane. Conservation of electric current in an infinitesimal cylindrical element of nerve fibre yields a second-order linear partial differential equation known as the *cable equation* (Rall, 1962). Let $V(x,t)$ denote the membrane potential at position x along a uniform passive cable at time t measured relative to the resting potential of the membrane. Let τ be the cell membrane time constant (measured in ms), D the diffusion constant (in μm²/ms), then the basic uniform (infinite) cable equation is

$$\frac{\partial V(x,t)}{\partial t} = -\frac{V(x,t)}{\tau} + D\frac{\partial^2 V(x,t)}{\partial x^2} + I(x,t), \quad x \in (-\infty,\infty). \tag{27.1}$$

The constants τ and $D = \lambda^2/\tau$, where λ (in μm) is the electrotonic space constant, can be found in terms of the electrical parameters of the cell membrane: the specific membrane capacitance C (μF/cm²), the resistance across a unit area of passive membrane R (Ω cm²) and the specific cytoplasmic resistivity R_a (Ω cm). We have that $\tau = RC$ and $D = a/(4R_aC)$, where a (in μm) denotes the diameter of the branch. The source term $I(x,t)$ describes an external input. Diffusion along the dendritic tree generates an effective spatio-temporal distribution of delays as expressed by the associated Green's function of the cable equation.

In response to a delta-Dirac pulse at x' at $t = 0$ and taking $V(x,0) = 0$ the dendritic potential behaves as $V(x,t) = G(x - x',t)$, where

$$G(x,t) = \frac{1}{\sqrt{4\pi Dt}} \mathrm{e}^{-t/\tau} \mathrm{e}^{-x^2/(4Dt)}, \qquad t \geq 0. \tag{27.2}$$

The Green's function $G(x,t)$ determines the linear response to an instantaneous injection of unit input at a given point on the tree. Using the linearity of the cable equation one may write the general solution as

$$V(x,t) = \int_0^t \mathrm{d}t' \int_{-\infty}^{\infty} \mathrm{d}x' G(x-x',t-t') I(x',t') + \int_{-\infty}^{\infty} \mathrm{d}x' G(x-x',t) V(x',0). \tag{27.3}$$

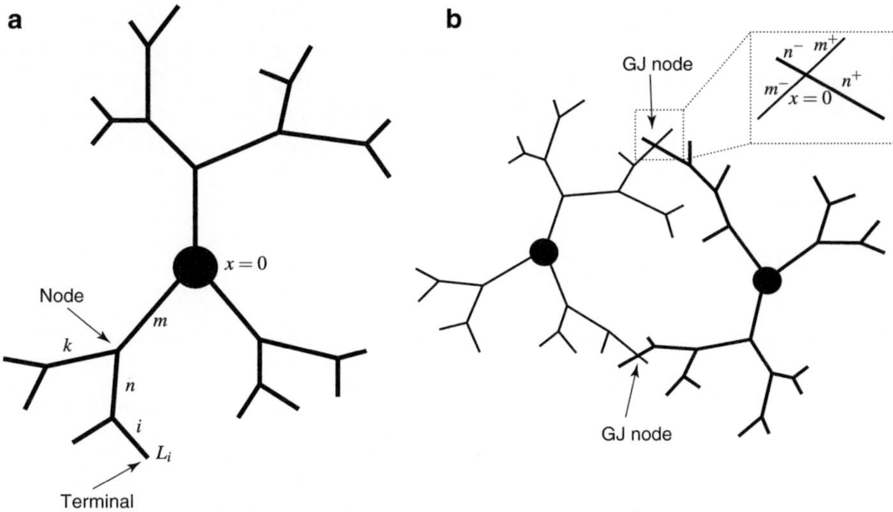

Fig. 27.2 (**a**) A single neuron consisting of a branched dendritic structure with the lumped soma. (**b**) A network of two cells connected by gap junctions (GJs)

This equation describes a convolution of the input $I(x,t)$ and some initial conditions $V(x,0)$ with the response function $G(x,t)$.

Abbott et al. (1991) have shown how to calculate response functions for passive branched dendritic trees, extending the approach above for a single unbranched dendrite. The machinery to do this borrows heavily from the path-integral formalism for describing Brownian motion, and has been used to give simple diagrammatic rules for obtaining the Green's function for a passive tree with a dendritic structure such as that shown in Fig. 27.2a. Formally speaking the response on a tree at position x on branch i to (delta-Dirac) input at position y on branch j is

$$G_{ij}(x,y,t) = \sum_{\text{trips}} A_{\text{trip}} G(L_{\text{trip}}(i,j,x,y),t), \tag{27.4}$$

where L_{trip} (in μm) is the length of a *path* along the tree. A path (or a trip) must start at point x on branch i and end at point y on branch j. Starting out from point x a trip may travel in either direction along branch i, but it may subsequently change direction only at a branching point or a terminal (and may pass through the points x and y). When a trip arrives at a branching node, it may pass through the node to any other segment radiating from the node or it may reflect from the node back along the same segment on which it entered. Any trip which approaches a terminal simply reverses its direction and reflects back. The trip coefficients are chosen in such a way to enforce all appropriate boundary conditions (at branching points and terminal nodes). There are an infinite number of paths and trip coefficients A_{trip} that can be generated and although the longer paths will typically contribute less and less to the solution the convergence of the "sum-over-trips" method is not monotonic if the terms in (27.4) are added in monotonic order of increasing length L_{trip} (Caudron et al., 2012). In Sect. 27.3 we summarise the rules for generating the trip coefficients as well as provide the new ones for handling gap junction coupling between dendritic trees. The linearity of the model means that voltage anywhere in the tree can be obtained by spatial integration of an injected current with the Green's function of the tree.

27.3 Dendro-Dendritic Gap Junction Coupling

We start by considering a single neuron consisting of a soma connected to a branched dendritic structure (see Fig. 27.2a). We assume that the dendrites are passive and so the dynamics of the membrane voltage along a finite branch i of length L_i is described by the linear cable equation:

$$\frac{\partial V_i}{\partial t} = -\frac{V_i}{\tau_i} + D_i \frac{\partial^2 V_i}{\partial x^2} + I_i(x,t), \qquad x \in [0, L_i]. \tag{27.5}$$

Note that we allow for the electrical properties and source term on each branch to be different (distinguished by the subscript i). Equation (27.5) for each branch must be accompanied with additional equations describing the dynamics of voltage at two ends of each branch, i.e. at $x = 0$ and at $x = L_i$. If the proximal or distal end of a branch is a branching node point the continuity of the potential across a node and Kirchoff's law of conservation of current are imposed. For example, for a node in Fig. 27.2a this means the following set of equations:

$$V_m(L_m, t) = V_n(0,t) = V_k(0,t), \tag{27.6}$$

enforcing continuity of voltage, and

$$\left.\frac{1}{r_m}\frac{\partial V_m}{\partial x}\right|_{x=L_m} = \left.\frac{1}{r_n}\frac{\partial V_n}{\partial x}\right|_{x=0} + \left.\frac{1}{r_k}\frac{\partial V_k}{\partial x}\right|_{x=0}, \tag{27.7}$$

enforcing conservation of current, where $r_m = 4R_a/(\pi a_m^2)$ (in $\Omega\,\mathrm{cm}^{-1}$) is the axial resistance on branch m. If a branch terminates at $x = L_i$ we either have a no-flux boundary condition known as a sealed (or closed) end ($\partial V_i(x,t)/\partial x\,|_{x=L_i} = 0$) or a zero value boundary condition known as a killed (or open) end ($V_i(L_i,t) = 0$). We model the soma as an equipotential surface with passive properties described by a capacitance $c_s = \pi a_s^2 C$ and conductance $g_s = (\pi a_s^2)/R$, where a_s denotes the diameter of the soma. Such a lumped soma model can be treated as a special node point with the somatic membrane voltage V_s satisfying $V_s(t) = V_j(0,t)$ (continuity of voltage) and

$$c_s \frac{\partial V_s}{\partial t} = -g_s V_s + \sum_j \left.\frac{1}{r_j}\frac{\partial V_j}{\partial x}\right|_{x=0}, \tag{27.8}$$

where the sum is over all branches connected to the soma (conservation of current). We refer the reader to (Tuckwell, 1988; Ermentrout and Terman, 2010; Sterratt et al., 2011) for an overview of boundary conditions at node points, terminals and a lumped soma.

The single neuron model above can be naturally extended to a network of cells coupled by gap junctions (see Fig. 27.2b). Gap-junctional coupling can be considered as a special node point on an extended branching structure. This special node point, the GJ node, connects branches from two separate dendritic trees and it requires the following set of boundary conditions (assuming that it is placed at $x = 0$, see the zoomed insert of Fig. 27.2b):

$$V_{m^-}(0,t) = V_{m^+}(0,t), \qquad V_{n^-}(0,t) = V_{n^+}(0,t), \tag{27.9}$$

and

$$\frac{1}{r_m}\left(\left.\frac{\partial V_{m^-}}{\partial x}\right|_{x=0} + \left.\frac{\partial V_{m^+}}{\partial x}\right|_{x=0} \right) = g_{\mathrm{GJ}}(V_{m^-}(0,t) - V_{n^-}(0,t)), \tag{27.10}$$

$$\frac{1}{r_n}\left(\left.\frac{\partial V_{n^-}}{\partial x}\right|_{x=0} + \left.\frac{\partial V_{n^+}}{\partial x}\right|_{x=0} \right) = g_{\mathrm{GJ}}(V_{n^-}(0,t) - V_{m^-}(0,t)), \tag{27.11}$$

where g_{GJ} is the conductance of the gap junction. The expressions in (27.9) reflect continuity of the potential across individual branches m and n, and (27.10)–(27.11) enforce conservation of current. These types of model, either at a single neuron level or at the level of a whole network, can be solved by brute-force numerical simulations. However, since the equations that describe the system are linear, insight into the model's behaviour can be obtained by analytically constructing the network Green's function (generalising the approach for a single dendritic tree). Denoting this function by $G_{ij}(x,y,t)$ then the voltage at any location along the cell or cells for multiple stimuli is given as

$$V_i(x,t) = \sum_j \int_0^t G_{ij}(x, y_j, t-s) I_j(s) \mathrm{d}s, \tag{27.12}$$

where y_j is a location of a spatially localised stimulus $I_j(t)$ on branch j (and we have assumed that initial data is identically zero).

The "sum-over-trips" method described above in Sect. 27.2 shows how the response function on a branching structure can be found in terms of the response function of an unbranched dendrite. In this case the Green's function can be calculated directly in the time domain. However, this method was originally developed for passive dendrites of a single cell without the inclusion of a soma. The incorporation of a lumped soma and the inclusion of gap junctions lead to a modified branching structure in which some special node points (the somatic node and the GJ node) appear. This requires additional boundary conditions which makes it hard to find a closed form solution in the time domain. Fortunately, a generalisation of the original "sum-over-trips" method to the frequency domain allows one to make explicit progress in obtaining the response function (Coombes et al., 2007; Timofeeva et al., 2013).

For simplicity we assume here that the membrane time constant τ, the diffusion coefficient D, the diameter of each branch a and the axial resistance r are the same along the whole tree and refer the reader to (Timofeeva et al., 2013) for the case of nonidentical membrane parameters. We aim to construct the Green's function $\hat{G}_{ij}(x,y,\omega)$ which is the Laplace transform of the original Green's function $G_{ij}(x,y,t)$ defined in terms of the spectral parameter ω as

$$\hat{G}_{ij}(x,y,\omega) = \int_0^\infty \mathrm{e}^{-\omega t} G_{ij}(x,y,t)\mathrm{d}t. \tag{27.13}$$

The response function $\hat{G}_{ij}(x,y,\omega)$ can be found using the "sum-over-trips" formalism as

$$\hat{G}_{ij}(x,y,\omega) = \sum_{\text{trips}} A_{\text{trip}}(\omega)\hat{G}(L_{\text{trip}},\omega), \qquad \hat{G}(x,\omega) = \frac{\mathrm{e}^{-\gamma(\omega)|x|}}{2D\gamma(\omega)}, \tag{27.14}$$

where $\gamma(\omega) = \sqrt{(\omega + 1/\tau)/D}$. We recognise $\hat{G}(x,\omega)$ as the Laplace transform of $G(x,t)$ defined by (27.2). There are an infinite number of trips that can be generated and allowed to pass through any type of node (branching, somatic or GJ) and reflect from any type of node (branching, somatic or GJ) and terminals. The trip coefficients $A_{\text{trip}}(\omega)$ are chosen according to the following set of rules (see Timofeeva et al. 2013 for more details):

- Initiate $A_{\text{trip}}(\omega) = 1$.
  Branching node
- For any branching node connecting N branches, at which the trip passes from one branch to another branch, A_{trip} is multiplied by a factor $2/N$.
- For any branching node connecting N branches, at which the trip approaches a node and reflects off this node back along the same branch, A_{trip} is multiplied by a factor $2/N - 1$.
  Terminal
- For every terminal which always reflects any trip, A_{trip} is multiplied by $+ 1$ for the sealed-end boundary condition or by $- 1$ for the killed-end boundary condition.
  Somatic node
- For the somatic node with N connected branches at which the trip passes through the soma from one branch to another branch, A_{trip} is multiplied by a factor $2p_s(\omega,N)$.
- For the somatic node with N connected branches at which the trip approaches the soma and reflects off the soma back along the same branch, A_{trip} is multiplied by a factor $2p_s(\omega,N) - 1$.

GJ node

- For the GJ node at which the trip passes through the gap junction from one branch to another branch, A_{trip} is multiplied by a factor $p_{GJ}(\omega)$.
- For the GJ node at which the trip approaches the gap junction, passes it and then continues along the same branch, A_{trip} is multiplied by a factor $1 - p_{GJ}(\omega)$.
- For the GJ node at which the trip approaches the gap junction and reflects off the gap junction back along the same branch, A_{trip} is multiplied by a factor $-p_{GJ}(\omega)$.

Here the frequency dependent parameters $p_s(\omega,N)$ and $p_{GJ}(\omega)$ are given as

$$p_s(\omega,N) = \frac{\gamma(\omega)/r}{N\gamma(\omega)/r+\gamma_s(\omega)}, \qquad \gamma_s(\omega)=c_s\omega+g_s, \tag{27.15}$$

$$p_{GJ}(\omega) = \frac{1}{2(\gamma(\omega)/(g_{GJ}r)+1)}. \tag{27.16}$$

The network Green's function in the time domain can be obtained by (numerically) inverse Laplace transforming $\hat{G}_{ij}(x,y,\omega)$. Analysis on mathematical convergence of the "sum-over-trips" method for a single cell can be found in Abbott (1992) and Caudron et al. (2012) and can be readily generalised to the network level.

27.3.1 A Semi-infinite Branch

For the case of a semi-infinite dendritic branch with the terminal point at one end the "sum-over-trips" approach generates only two trips for any pair of points (x,y) along this branch (Fig. 27.3a). Considering the sealed-end boundary condition at the terminal we find

$$\hat{G}_{11}(x,y,\omega)=\hat{G}(x-y,\omega)+\hat{G}(x+y,\omega). \tag{27.17}$$

Taking the inverse Laplace transform of this solution we obtain the response function in the time domain as

$$G_{11}(x,y,t)=G(x-y,t)+G(x+y,t), \tag{27.18}$$

where G is given by (27.2). For the case of a lumped soma attached to a semi-infinite branch (Fig. 27.3b), the solution in the frequency domain needs to be modified as

$$\hat{G}_{11}(x,y,\omega)=\hat{G}(x-y,\omega)+(2p_s(\omega,1)-1)\hat{G}(x+y,\omega). \tag{27.19}$$

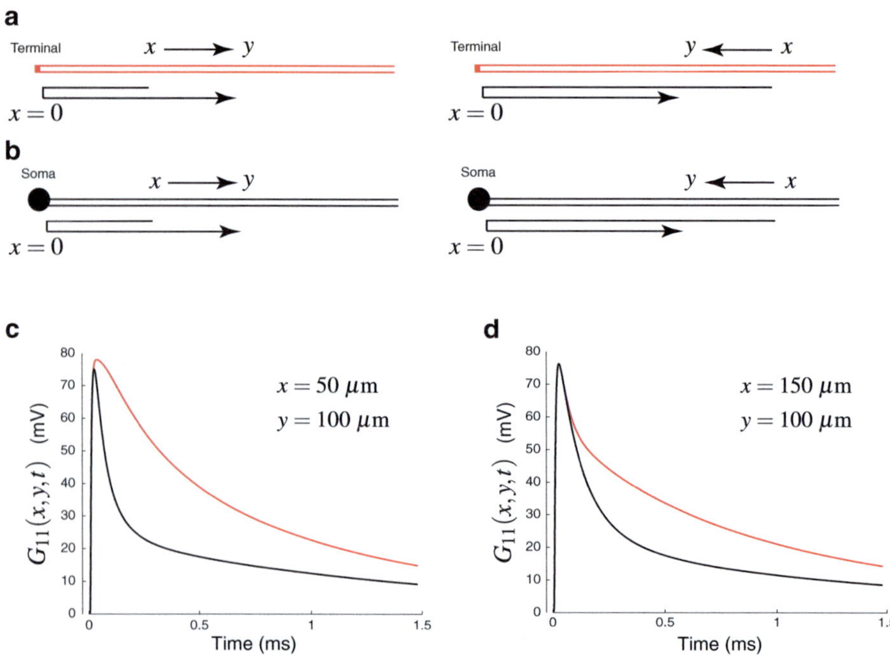

Fig. 27.3 (**a**) A semi-infinite dendritic branch with a sealed end boundary condition at $x = 0$. (**b**) A semi-infinite dendritic branch with the lumped soma at $x = 0$. (**c–d**) The response functions for the model with the terminal (*red curves*) and for the model with the lumped soma (*black curves*). Parameters: $C = 1\,\mu\text{F/cm}^2$, $R = 2{,}000\,\Omega\,\text{cm}^2$, $R_a = 100\,\Omega\,\text{cm}$, $a = 2\,\mu\text{m}$ (giving $D = 50{,}000\,\mu\text{m}^2/\text{ms}$ and $\tau = 2\,\text{ms}$), $a_s = 25\,\mu\text{m}$, location of the stimulus $y = 100\,\mu\text{m}$, $x = 50\,\mu\text{m}$ (**c**) and $x = 150\,\mu\text{m}$ (**d**)

In this simplified case we can perform the inverse Laplace transform analytically to obtain:

$$G_{11}(x,y,t) = G(x-y,t) + \frac{e^{-t/\tau}}{Drc_s}\left[Q_1 e^{\lambda_+(\lambda_+ t - |x+y|/\sqrt{D})}\text{erfc}\left(\frac{|x+y|}{\sqrt{4Dt}} - \lambda_+\sqrt{t}\right)\right.$$
$$\left. + Q_2 e^{\lambda_-(v_- t - |x+y|/\sqrt{D})}\text{erfc}\left(\frac{|x+y|}{\sqrt{4Dt}} - \lambda_-\sqrt{t}\right)\right] - G(x+y,t), \tag{27.20}$$

where $\lambda_\pm = [-1 \pm \sqrt{1 - 4c_s D r^2 (g_s - c_s/\tau)}]/(2c_s r\sqrt{D})$, $Q_1 = \lambda_+/(\lambda_+ - \lambda_-)$ and $Q_2 = \lambda_-/(\lambda_- - \lambda_+)$. Figure 27.3c, d demonstrates how the inclusion of a lumped soma affects the voltage response in a model with a semi-infinite dendritic branch. We note that the inclusion of a realistic soma model can significantly sculpt the shape of the dendritic response to impulsive (delta-Dirac) input. In particular, after an initial transient, the membrane voltage decay can be much faster for the model that includes a lumped soma.

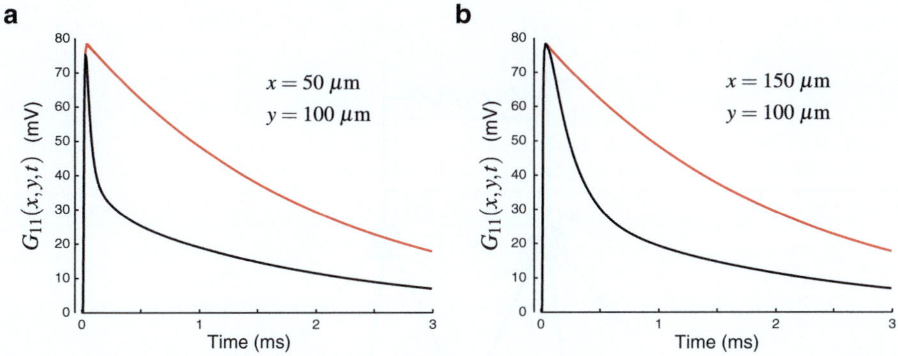

Fig. 27.4 The response functions for the model with a finite dendritic branch. *Red curves*: a finite branch with two terminals. *Black curves*: a finite branch with lumped soma at $x = 0$. Parameters as in Fig. 27.3 and $L = 200\,\mu$m. Location of the stimulus $y = 100\,\mu$m, $x = 50\,\mu$m (**a**) and $x = 150\,\mu$m (**b**)

27.3.2 A Finite Branch

Here we consider the case that a dendritic branch in the models in Fig. 27.3a, b has a finite length L and is terminated with a sealed-end boundary condition. There are an infinite number of trips from x to y that can be constructed due to reflections at both ends of the branch. In the case of a finite dendrite with two terminals the solution is simply a sum involving the function $\hat{G}\,(L_{\text{trip},}\omega)$ as the trip coefficients are + 1. However, if the soma is included each reflection at $x = 0$ requires us to add a coefficient $(2p_s(\omega,1) - 1)$ so that the (Laplace transformed) Green's function is

$$
\begin{aligned}
\hat{G}_{11}(x,y,\omega) = \sum_{n=0}^{\infty} (2p_s(\omega,1)-1)^n [&\hat{G}(y-x+2nL,\omega) \\
&+(2p_s(\omega,1)-1)\hat{G}(y+x+2nL,\omega) \\
&+(2p_s(\omega,1)-1)\hat{G}(-(y-x)+2L(n+1),\omega) \\
&+\hat{G}(-(y+x)+2L(n+1),\omega)].
\end{aligned}
\tag{27.21}
$$

In Fig. 27.4 we plot the Green's function at two different locations along a dendritic branch for the model with and without a soma. We see a similar behaviour to that of Fig. 27.3, namely that the inclusion of a lumped soma can lead to a faster temporal decay of dendritic voltage.

27.4 Two Neuron Network with Gap Junction Coupling

Here we demonstrate how the "sum-over-trips" formalism can be applied to a network of cells coupled by gap junctions to provide some insight into the network dynamics. As an example we consider a model of two identical cells, each of which consists of a passive soma and N attached passive semi-infinite dendrites as shown in Fig. 27.5a.

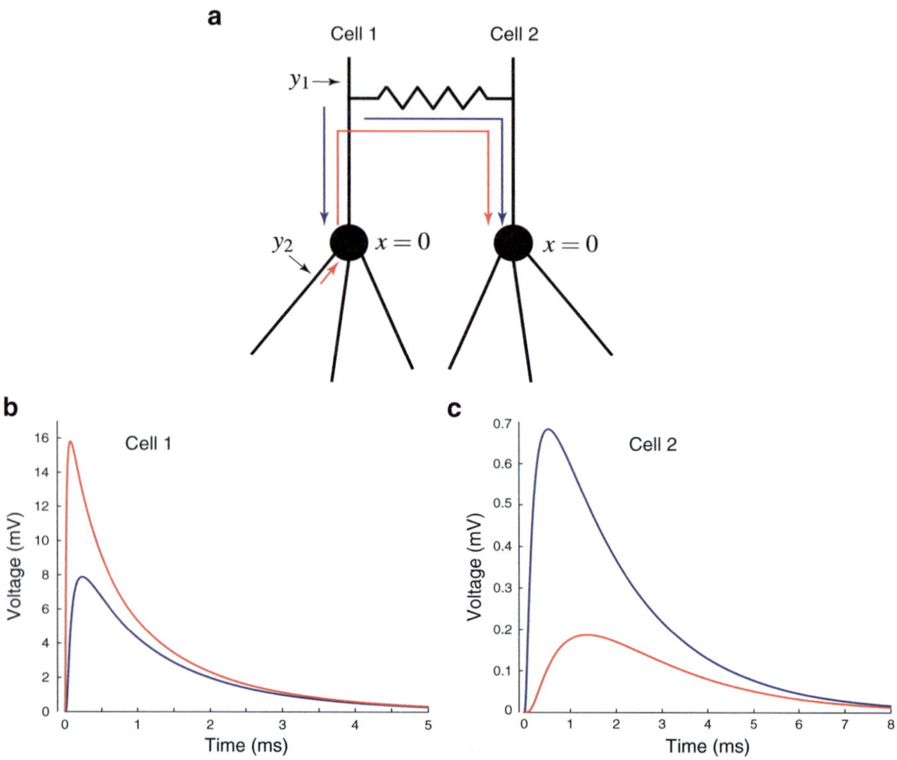

Fig. 27.5 (**a**) Two identical cells coupled by a gap junction. (**b–c**) Voltage profiles in the soma of Cell 1 (**b**) and in the soma of Cell 2 (**c**) in response to a delta-pulse at the location y_1 (*blue curve*) and at the location y_2 (*red curve*). Biophysical parameters of each cell as in Fig. 27.3. $L_{GJ} = 100\,\mu$m, $R_{GJ} = 1/g_{GJ} = 200$ MΩ, $y_1 = 125\,\mu$m, $y_2 = 50\,\mu$m, number of branches connected to each soma $N = 4$

The cells are coupled by a dendro-dendritic gap junction located at some distance L_{GJ} away from their cell bodies. We assume that this network receives either a distal input at the location y_1 or a proximal input at the location y_2, which are both applied to Cell 1. Knowing the Green's function for this network it is easy to compute the voltage response along the whole structure for any form of these two inputs by performing the integration in (27.12). To study how distal and proximal inputs affect the somatic responses of both cells, we find the Green's functions in the soma of Cell 1, namely $G_{11}(0,y_1,t)$ and $G_{11}(0,y_2,t)$, and in the soma of Cell 2, namely $G_{21}(0,y_1,t)$ and $G_{21}(0,y_2,t)$, for a delta-Dirac pulse either at the location y_1 or y_2. Applying the "sum-over-trips" method it is possible to construct compact solutions for these Green's functions in the frequency domain. The somatic response functions for Cell 1 take the following forms:

$$\hat{G}_{11}(0,y_1,\omega) = 2p_s(1-p_{GJ})\hat{G}(y_1,\omega) + \sum_{n=0}^{\infty} 2^n(-p_{GJ}(2p_s-1))^{n+1}(1-2p_{GJ})(2p_s) \times \hat{G}(y_1+2(n+1)L_{GJ},\omega) \tag{27.22}$$

and

$$\hat{G}_{11}(0,y_2,\omega) = 2p_s\hat{G}(y_2,\omega) + \sum_{n=0}^{\infty} 2^n(-p_{GJ})^{n+1}(2p_s-1)^n(2p_s)^2 \times \hat{G}(y_2+2(n+1)L_{GJ},\omega). \tag{27.23}$$

For Cell 2, compact somatic solutions for the individual inputs can be found as

$$\hat{G}_{21}(0,y_1,\omega) = 2p_s p_{GJ}\hat{G}(y_1,\omega) + \sum_{n=0}^{\infty} 2^n(-p_{GJ}(2p_s-1))^{n+1}(2p_{GJ}-1)(2p_s) \times \hat{G}(y_1+2(n+1)L_{GJ},\omega) \tag{27.24}$$

and

$$\hat{G}_{21}(0,y_2,\omega) = \sum_{n=0}^{\infty} 2^n(-p_{GJ}(2p_s-1))^n p_{GJ}(2p_s)^2 \times \hat{G}(y_2+2(n+1)L_{GJ},\omega). \tag{27.25}$$

In these formulae $\hat{G}(x,\omega)$ is as described in (27.14), and $p_s = p_s(\omega,N)$ and $p_{GJ} = p_{GJ}(\omega)$ are given by (27.15) and (27.16). More details about how such compact solutions can be constructed making use of combinatorics can be found in Timofeeva et al. (2013). Taking the inverse Laplace transform of (27.22)–(27.25) we obtain the Green's functions for each soma in the time domain. In Fig. 27.5b, c we plot these somatic response functions for distal input at the location y_1 (blue profiles) and for proximal input at the location y_2 (red profiles). Proximal input strongly depolarises Cell 1, but spread of charge to Cell 2 is weak. On the other hand, distal input localised closely to the gap junction depolarises Cell 2 significantly better and at the same time the spread of charge to Cell 1 is weakened. It is worth mentioning here that the strength of synaptic conductances on single neurons can increase with distance from the soma (Magee and Cook, 2000), though we take both inputs to be identical in this example. The model dynamics are consistent with recent results, reported in Vervaeke et al. (2012) which suggest that gap junctions enable distal inputs to drive the activity of a neural network more effectively. It is important to recognise that the passive properties of the cell membrane provide the fundamental basis for neural integration. In our example the somas are not equipped with any active conductances for generating proper action potentials. However, the passive somatic responses can be linked with the level of excitability in each cell by introducing a notion of threshold for spike generation.

The "sum-over-trips" approach for constructing the Green's function on the entire network provides an alternative to the brute-force numerical simulations based on the compartmental method. The Green's function in the time domain captures the essential characteristics of synaptic efficacy such as the peak depolarisation and the time at which this peak occurs. Moreover, the Laplace representation of the response function allows us to use the method of moments (Agmon-Snir, 1995) for studying spatio-temporal integration properties of a network using the relation

$$M_n = \int_{-\infty}^{\infty} t^n G_{ij}(x,y,t)\mathrm{d}t = \left. \frac{\mathrm{d}^n G_{ij}(x,y,\omega)}{\mathrm{d}\omega^n} \right|_{\omega=0}, \tag{27.26}$$

where M_n is the nth moment of the response function $G_{ij}(x,y,t)$.

27.5 Future Directions

In this chapter we have extended the "sum-over-trips" approach to describe networks of branched dendritic trees connected via gap junctions. As an illustration of the usefulness of this analytical approach we have shown how to calculate the response of a simple two neuron network and have found results consistent with recent data reported in Vervaeke et al. (2012). There are a number of natural extensions of the work in this chapter to incorporate further important aspects of biological realism. The one that requires the least mathematical changes to the framework developed so far is the inclusion of quasi-active membrane, along the lines in Coombes et al. (2007). Quasi-active membrane models can more accurately describe neurons that exhibit resonances whereby subthreshold oscillatory behaviour is amplified for inputs at preferential frequencies. This extends the more usual "RC" (resistor-capacitor) circuit description of passive membrane to the "LRC" (inductance-resistor-capacitor) case. A more substantial mathematical challenge is the treatment of active dendrites and tractable models of action potential generating membrane at the soma and dendritic "hot-spots". Recent progress in developing minimal models of excitable tissue capable of generating accurate action potential shapes has been made in Coombes (2008), using piece-wise linear planar models. This piece-wise linearity fits nicely with the "sum-over-trips" approach (which is premised on linearity of the underlying dendritic voltage model) and will allow for the analytical treatment of a minimal model of an active soma coupled to a branched dendritic tree along the lines described in Svensson and Coombes (2009). An alternative to using models of excitable membrane is to use integrate-and-fire type models, which has inspired the development of the "spike-diffuse-spike" formalism (Coombes and Bressloff, 2003; Timofeeva et al., 2006; Timofeeva, 2010) for active "hot-spots". Extending this to somatic models is more challenging when one wishes to incorporate the back coupling to the tree (Bressloff, 1995), though recent progress on this problem has been made by Schwemmer and Lewis (2012).

A programme of work based on the above observations will allow for a more thorough investigation of how network dynamics of spatially extended neurons coupled by gap junctions are affected by geometric complexity and intrinsic properties of dendrites as well as by gap junction modulation. In particular it will help us to understand the role of location and strength of gap junctions on generating different patterns of neural network activity.

References

Abbott LF (1992) Simple diagrammatic rules for solving dendritic cable problems. Physica A 185:343–356

Abbott LF, Fahri E, Gutmann S (1991) The path integral for dendritic trees. Biol Cybern 66:49–60

Agmon-Snir H (1995) A novel theoretical approach to the analysis of dendritic transients. Biophys J 69:1633–1656

Alvarez AV, Chow CC, Bockstaele EJV, Williams JT (2002) Frequency-dependent synchrony in locus ceruleus: role of electrotonic coupling. Proc Natl Acad Sci USA 99(6):4032–4036

Ascoli GA, Donohue DE, Halavi M (2007) NeuroMorpho.Org: a central resource for neuronal morphologies. J Neurosci 27(35):9247–9251

Bennet MVL, Zukin RS (2004) Electrical coupling and neuronal synchronization in the mammalian brain. Neuron 41:495–511

Bressloff PC (1995) Dynamics of a compartmental integrate-and-fire neuron without dendritic potential reset. Physica D 80:399–412

Caudron Q, Donnelly SR, Brand SPC, Timofeeva Y (2012) Computational convergence of the path integral for real dendritic morphologies. J Math Neurosci 2(11)

Coombes S (2008) Neuronal networks with gap junctions: a study of piece-wise linear planar neuron models. SIAM J Appl Dyn Syst 7:1101–1129

Coombes S, Bressloff PC (2003) Saltatory waves in the spike-diffuse-spike model of active dendritic spines. Phys Rev Lett 91:028102(1–4)

Coombes S, Timofeeva Y, Svensson CM, Lord GJ, Josic K, Cox SJ, Colbert CM (2007) Branching dendrites with resonant membrane: a "sum-over-trips" approach. Biol Cybern 97:137–149

Ermentrout BG, Terman DH (2010) Mathematical foundations of neuroscience. Springer, Berlin

Fukuda T, Kosaka T (2000) Gap junctions linking the dendritic network of GABAergic interneurons in the hippocampus. J Neurosci 20:1519–1528

Galarreta M, Hestrin S (1999) A network of fast-spiking cells in the neocortex connected by electrical synapses. Nature 402:72–75

Hamzei-Sichani F, Kamasawa N, Janssen WGM, Yasumura T, Davidson KGV, Hof PR, wearne SL, Stewart MG, Young SR, Whittington MA, Rash JE, Traub RD (2007) Gap junctions on hippocampal mossy fiber axons demonstrated by thin-section electron microscopy and freeze-fracture replica immunogold labeling. Proc Natl Acad Sci USA 104:12548–12553

Hormuzdi SG, Filippov MA, Mitropoulou G, Monyer H, Bruzzone R (2004) Electrical synapses: a dynamic signaling system that shapes the activity of neuronal networks. Biochim Biophys Acta 1662:113–137

Hughes SW, Crunelli V (2007) Just a phase they're going through: the complex interaction of intrinsic high-threshold bursting and gap junctions in the generation of thalamic α and θ rhythms. Int J Psychophysiol 74:3–17

London M, Häusser M (2005) Dendritic computation. Annu Rev Neurosci 28:503–532

Magee JC, Cook EP (2000) Somatic EPSP amplitude is independent of synapse location in hippocampal pyramidal neurons. Nat Neurosci 3:895–903

Markram H (2006) The blue brain project. Nat Rev Neurosci 7:153–160

Martone ME, Zhang S, Gupta A, Qian X, He H, Price DL, Wong M, Santini S, Ellisman MH (2003) The cell-centered database: a database for multiscale structural and protein localization data from light and electron microscopy. Neuroinformatics 1(4):379–395

Rabinovich MI, Varona P, Selverston AI, Abarbanel HDI (2006) Dynamical principles in neuroscience. Rev Mod Phys 78:1213–1265

Rall W (1962) Theory of physiological properties of dendrites. Ann N Y Acad Sci 96:1071–1092

Rash JE, Dillman RK, Bilhartz BL, Duffy HS, Whalen LR, Yasumura T (1996) Mixed synapses discovered and mapped throughout mammalian spinal cord. Proc Natl Acad Sci USA 93:4235–4239

Saraga F, Ng L, Skinner FK (2006) Distal gap junctions and active dendrites can tune network dynamics. J Neurophysiol 95:1669–1682

Schwemmer MA, Lewis TJ (2012) Bistability in a leaky integrate-and-fire neuron with a passive dendrite. SIAM J Appl Dyn Syst 11:507–539

Segev I, Rinzel J, Shepherd GM (eds) (1995) The theoretical foundations of dendritic function: selected papers of Wilfrid Rall with commentaries. MIT, Cambridge

Sotelo C, Llinas R, Baker R (1974) Structural study of inferior olivary nucleus of the cat: morphological correlates of electrotonic coupling. J Neurophysiol 37:541–559

Sterratt D, Graham B, Gillies A, Willshaw D (2011) Principles of computational modelling in neuroscience. Cambridge University Press, Cambridge

Svensson CM, Coombes S (2009) Mode locking in a spatially extended neuron model: active soma and compartmental tree. Int J Bifurcat Chaos 19:2597–2607

Timofeeva Y (2010) Travelling waves in a model of quasi-active dendrites with active spines. Physica D 239(9):494–503

Timofeeva Y, Lord GJ, Coombes S (2006) Spatio-temporal filtering properties of a dendritic cable with active spines. J Comput Neurosci 21:293–306

Timofeeva Y, Coombes S, Michieletto D (2013) Gap junctions, dendrites and resonances: a recipe for tuning network dynamics. J Math Neurosci 3:15

Traub RD, Kopell N, Bibbig A, Buhl EH, LeBeau FEN, Whittington MA (2001) Gap junctions between interneuron dendrites can enhance synchrony of gamma oscillations in distributed networks. J Neurosci 21:9478–9486

Tuckwell HC (1988) Introduction to theoretical neurobiology volume I. Cambridge University Press, Cambridge

Velazquez JLP, Carlen PL (2000) Gap junctions, synchrony and seizures. Trends Neurosci 23:68–74

Vervaeke K, Lőrincz A, Nusser Z, Silver RA (2012) Gap junctions compensate for sublinear dendritic integration in an inhibitory network. Science 335:1624–1628

Chapter 28
Automated Parameter Constraining of Single-Neuron Models

Shaul Druckmann

Abstract Single-neuron models and especially detailed single-neuron models often have a large number of free parameters that must be constrained. Despite the seemingly straightforward nature of this problem, the main challenge is properly defining and quantifying the match between a given model and a target data set. In this chapter we present the problem of constraining the parameters of single-neuron models, focusing on conductance-based models. We introduce and review the steps of the procedure and present multiple objective optimization, a technique that allows a more general and powerful approach to expressing the comparison between models and experiments. Using this approach one is able to map out the trade-offs achievable by the model between the different goals of the optimization and decide accordingly whether to be satisfied with the current model or consider different models and model parameters. We discuss how to interpret the results of a parameter constraining attempt and open questions in neural model parameter constraining.

28.1 Introduction

At the risk of preaching to the choir, we shall begin by contending that unraveling the inner workings of the single neuron is one of the most fundamental questions in theoretical neuroscience. How do these computational elements perform their different functions? How much of the computational burden does the single element take upon itself and how much is solely due to network interactions? Lord Kelvin is famously known to have said, "I am never content until I have constructed a mechanical model of the subject I am studying. If I succeed in making one, I understand; otherwise I do not." In this chapter we shall consider a small, but crucial, step along this path—constraining the parameters of single-neuron models, the basic biophysical

S. Druckmann (✉)
Janelia Farm Research Campus, HHMI, Ashburn, USA
e-mail: druckmanns@janelia.hhmi.org

H. Cuntz et al. (eds.), *The Computing Dendrite: From Structure to Function*, Springer Series in Computational Neuroscience 11, DOI 10.1007/978-1-4614-8094-5_28,

models of neural computation. The goal of this chapter is not to provide a simple answer to this complicated and general question (since we believe none such exists) but rather to present the salient issues in order for the reader to be able to recognize, understand, and wisely employ parameter constraining approaches.

28.1.1 Brief Introduction to Conductance-Based Models and Their Parameters

Conductance-based models are biophysical models that, as their name implies, directly describe neuronal dynamics in terms of the different conductances assumed to be present in the neuron. This class of models differs from phenomenological biophysical models, such as the integrate-and-fire model introduced by Lapicque in 1907 [a translation of the original 1907 paper can be found in Lapicque (2007)], as in the latter certain behaviors (e.g., action potential generation) are introduced by direct intervention rather than being accounted for by a description of the underlying ion channels. The main advantage of conductance-based models is that they offer a direct and crucial link between the macroscopic electrical behavior of the neuron under study and the microscopic level of the ion channels responsible for this behavior.

The bare bones of a conductance-based model consist first and foremost of the assumed morphology, typically obtained via reconstruction of stained neurons filled by injection of a cell tracer such as biocytin. Occasionally the full morphology is replaced with a more schematic and simplified version of the true morphology to simplify analysis and decrease the required computations to simulate the model (Bahl et al. 2012). Second, some assumptions must be posed regarding the neuron's ion channel composition. This can range from simple assumptions such as a fully passive model to highly detailed ones including over a dozen different channels differentially expressed in the soma, axon, and dendrites (Poirazi et al. 2003).

The dynamics of such models are governed by well-known equations [see for instance Koch and Segev (1998) and references therein and introduction chapter]. To introduce the mathematical nomenclature consider for example the dynamics of a simple passive single compartment model:

$$C_m \dot{V}_m = -\frac{1}{R_m}\left(V_m - V_r\right) + I_{ext}$$

where V_m is the membrane voltage, and the dot notation, $\dot{V}_m$, represents its temporal derivative dV/dt (Volt/second), C_m is the membrane capacitance in units of Farad (F), R_m is the membrane resistance (Ω), V_r is the resting membrane potential (Volt), and I_{ext} (Amp) is an externally applied current. This simple equation can be readily solved for a known current.

The equation above is one specific example of the more general equation of the change in voltage reflecting the sum of currents (the sign, negative/positive, is a convention):

$$C_m^i \dot{V}^i = -\sum_{type} I_{type}^i$$

where the sum over I_{type}^i represents the sum of different types of currents in the ith compartment arriving from various sources: passive ion channels, voltage-dependent ion channels, intracellular current flow, externally injected currents, etc. The current contributed by each channel is governed by dynamical equations based on the classical Hodgkin and Huxley papers (Hodgkin and Huxley 1952) but using more convenient notation:

$$I_a = \bar{g}_a m^\alpha h^\beta \left(V_m - E_a\right)$$

$$\dot{m} = \frac{-\left(m - m_{inf}\right)}{m_{tau}}; \quad \dot{h} = \frac{-\left(h - h_{inf}\right)}{h_{tau}}$$

where $\bar{g}_a$ represents the maximal conductance of channel a and E_a its reversal potential. Both of these are constant (i.e., non-time dependent) quantities. m^α represents the fraction of activation (within the range of 0–1) taken to the power α. h^β represents the fraction of inactivation taken to the power β. In the lower equation m_{inf} indicates the steady-state value of m and m_{tau} the time constant of m, both typically voltage dependent. The form of the voltage dependency is different for each channel; m_{inf} is typically of sigmoidal form, whereas the form of m_{tau} varies considerably but often is approximately Gaussian.

The conductance of each channel is a parameter in the current equations above. Unfortunately, it is extremely difficult to experimentally characterize the density of the ion channels along the surface of the membrane. Hence, it is typically left as a free parameter, making the parameter constraining procedure a central part of the generation of conductance-based models. A single-conductance value may be sufficient to describe the conductance value across the entire neuron, or more realistically it can be allowed to change between different neurite regions to reflect the changing density of ionic channels. In addition, other aspects of the model that may not be well constrained, for instance those related to morphology (e.g., spine density) or to single-channel dynamics (e.g., uncharacterized voltage dependency), can be left as free parameters as well (and potentially also variable between different neurite regions).

Traditionally, this was performed by a process of manual parameter tuning, whereby the modeler changed the values of the different parameters until a suitable match with the model's goal was achieved (Mainen et al. 1995). There are numerous drawbacks to this approach. First, it requires a significant amount of expertise and a large investment of time. Second, due to the subjective nature of the approach, biases of different nature might be implicitly introduced. Last, but not least, the effort spent in arriving at a given model makes it unlikely that alternative possibilities for modeling the phenomenon, such as different ion channel composition and increased/reduced details, will be explored.

It is therefore not surprising that the alternative option of performing automated parameter constraining by harnessing the increasingly formidable available computational resources was raised already in the mid-1990s (Vanier and Bower 1999; Bhalla and Bower 1993; Foster et al. 1993; Tawfik and Durand 1994). Recently, there has been a revived interest in applying automated parameter constraining (Achard and De Schutter 2006; Weaver and Wearne 2006; Keren et al. 2005; Huys et al. 2006; Huys and Paninski 2009) and especially to directly fit experimental data (Keren et al. 2009; Bahl et al. 2012; Druckmann et al. 2007; Ambros-Ingerson et al. 2008). Notably, the "single-neuron modeling competition," launched in 2007, aims to advance quantitative neuronal modeling by offering public challenges and the opportunity to compare results from different labs dealing with models of differing levels of complexity (Jolivet et al. 2008a, b; Gerstner and Naud 2009).

28.2 Constraining Parameters of Conductance-Based Models

Broadly speaking, an automated model parameter constraining procedure consists largely of four parts: the target data to be modeled (Fig. 28.1a), the measure by which we find the model to be successful (Fig. 28.1b), the model including its associated parameters (Fig. 28.1c), and finally the approach to adjust the parameters in order to improve the model (Fig. 28.1d). Additionally, once the parameter constraining process has been performed, some model selection criteria must be applied to determine whether the optimization was successful or further attempts are required (Fig. 28.1e).

Two of the aspects, the target data and model assumptions, though clearly crucially important, are problem specific, and we will therefore touch upon them only briefly. Of the two remaining issues at least in our hands the distance measure has been ultimately the more important factor. Intuitively, the reason behind its importance is that the main benefit of automated parameter constraining is that tens of thousands of parameter combinations can be tested without human intervention. Accordingly, it is imperative that the ones that are considered the most interesting by the automated measure (i.e., have the lowest distance value to the data) will be those that indeed offer the best approximation (we will discuss later what is meant by the best approximation, but for now let us just say that these should be at least considered good approximations if a human were examining the same parameter values). Therefore, it is arguably better to spend time carefully considering the distance function and adopting a crude parameter choice approach (such as random sampling) than vice versa. Though both of these aspects are discussed in this chapter, when attempting a parameter constraining procedure we would recommend that the user at least initially focus on the choice of distance functions used in the procedure as opposed to the optimization algorithm.

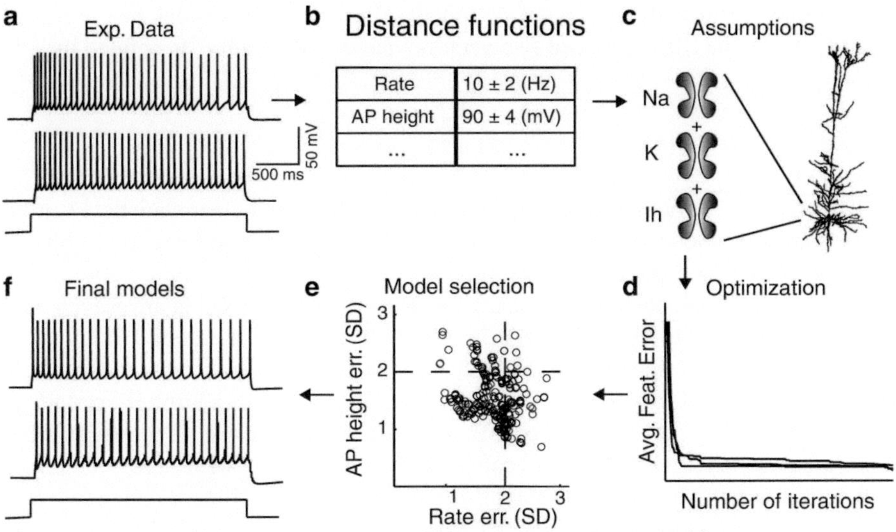

Fig. 28.1 Overview of model parameter constraining procedure. (**a**) Example of target data on which the model is based upon. Two repetitions of a step current injection are shown along with the current injection that generated them (*bottom*). Two traces with fairly large differences were chosen to highlight the variability. (**b**) The difference between any prospective model and target data is measured by the distance functions chosen for the comparison. In this example feature-based distance functions are shown. (**c**) The main assumptions behind the model consist of a reconstructed morphology and an assumed set of membrane ion channels (including their kinetics but not their densities). (**d**) An optimization approach is used to test multiple tentative models (i.e., sets of parameter values) and progressively find solutions with lower distance values to the target data. (**e**) At the end of the optimization, a criterion is used to select the tested tentative models that are deemed successful solutions. In this example a value of less than two experimental standard deviations in each feature is used as a criterion. (**f**) An example of the response of two different successful models to a step current input as in (**a**). Two models with fairly large differences were chosen to highlight the variability. Figure adapted from Druckmann et al. (2011)

28.3 Distance Functions

What is the right distance function to use for neural modeling? This deceptively simple question is actually quite difficult to answer. A crucial point to consider from the onset is that due to the highly complex nature of neurons and the often-simplified nature of models, any model will be merely an approximation of the true neuron's dynamics. Unfortunately, answering the question of what makes a good or bad approximation of complex data such as neural recordings is extremely difficult without very strong assumptions. Nonetheless, that is the question that must be addressed in order to define what is a good distance function. Accordingly, it is not surprising that the choice of distance function is highly nontrivial and has been the subject of much research and debate (Achard and De Schutter 2006; Ambros-Ingerson et al. 2008; Druckmann et al. 2007; Huys et al. 2006; Keren et al. 2005; LeMasson and Maex 2001; Weaver and Wearne 2006).

The most commonly used distance function in most realms is the mean squared error (MSE). The MSE between two vectors, *a* and *b* of length *n*, is defined by

$$MSE = \frac{1}{n}\sum_{i=1}^{n}\left(a_i - b_i\right)^2$$

This form of error is popular since it can be placed in a clear statistical setting (Gaussian noise model), is easy to optimize analytically, and can serve as an approximation to the true error around fixed points (Boyd and Vandenberghe 2004).

The problems associated with using MSE to fit neuron models are clearly laid out in (LeMasson and Maex 2001). Briefly stated, the issue is that small jitters in the time of the spike that can occur even when comparing two repetitions of the response of the *same* neuron to the *same* stimulus (Fig. 28.2a) get very strongly penalized since there is a spike at the wrong time and a missing spike at the right time. Therefore, a model with a flat line can actually have a lower MSE and is therefore considered a better approximation by this particular distance function than two repetitions from the same neuron (Fig. 28.2b).

To overcome this pathological sensitivity to timing, the authors suggested to replace the explicit dependence on time by measuring the distance between the two vectors not in the straightforward way of comparing each time point but rather of looking at the voltage trace in the V (voltage), dV/dt state space (Fig. 28.2c). This removes time as an explicit parameter and solves the issue of jittered traces having very large error values.

However, this approach is not perfect either. A slightly more subtle but also important point is that at low firing rates the spikes represent very little of the voltage trace. Indeed, assuming a firing rate of 5 Hz, with an average spike width of 1 millisecond, spikes take up only 5 % of the voltage trace and most of the weight will be put on the subthreshold behavior of the neuron (Fig. 28.2c). If the spikes are isolated both of these problems can be overcome. One approach is to do so directly, by "cutting out" the data and model spikes and comparing the voltage excursions directly (Weaver and Wearne 2006) or combining the two approaches: cutting out the spikes and comparing them alone in the V, dV/dt state space (Bahl et al. 2012). Unfortunately, completely removing time from a trace may be problematic since it smears the difference between, for instance, an adapting and non-adapting neuron's spike train. Moreover, cutting spikes and comparing them one to one leaves unanswered the question of what to do with discrepancies in spike number (of course, one can think of many straightforward ways to penalize these kinds of differences between traces by separate distance functions).

A different approach is that of "feature-based" distance functions (Druckmann et al. 2007). Feature-based distance functions are distance functions that take into account the variability of the responses of a neuron to a single (frozen) stimulus. For any feature (firing rate, AP height, AP width, time to first AP, etc.) a distance measure is obtained by first extracting the mean and standard deviation of the feature value for all experimental repetitions. Following that, the distance between the model-generated response and the experimental data is measured by the distance of

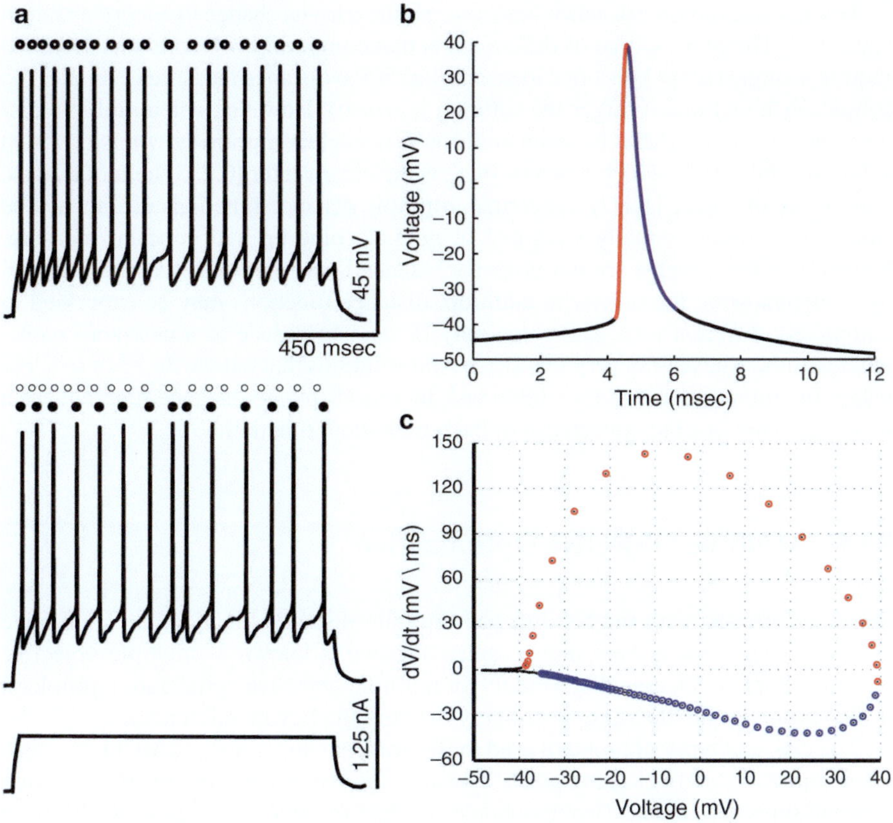

Fig. 28.2 Difficulties with simple error functions. (**a**) Two voltage responses (*top* and *mid*) from the same neuron to an identical step current (*bottom*). Two traces with fairly large differences were chosen to highlight the variability. Spike times for each response are shown as *black-filled circles* above trace. The spike times of the top trace have been transposed to the bottom trace but marked with *empty circles*, in order to simplify comparison. Note temporal jitter of spike times. (**b**) Zoom in on a single-action potential. Up-stroke marked in *red* and down-stroke in *blue*. (**c**) Same action potential as in (b) but shown in the *V*, d*V*/d*T* space. Same color scheme has been used. Grid exemplifies one possible binning of the space. Note how up-stroke is represented by only a few points, down-stroke by about three times more and most in sub-threshold range

the value from the experimental mean in units of experimental standard deviation. The advantage of these distance functions is that they can capture many different aspects of neuronal dynamics and take into account their variability. The disadvantage is that it is unclear which and how many features would be effective to constrain a particular model.

In summary, one way or another, we find ourselves with the realization that we will require multiple distance functions.

Would using multiple distance functions require a trivial change to our optimization approach? The main source of difference is that comparing two solutions seems to require a single (scalar) distance measure (i.e., if the distance measure is smaller the solution is better and if larger the solution is worse). However, when multiple distance values are available for each solution it is unclear a priori how to say which solution is better and which is worse (e.g., is the error vector [0.2, 5] better or worse than the error vector [1,1]?). Of course, multiple distance functions can always be summed up (either equally weighted or not) to come up with a scalar measure. However, if the weights are not properly balanced, one distance measure may end up overshadowing the others. In addition, distance functions may be expressed in completely different units and some may be easy to reduce to a near-zero value, whereas others have only very few successful solutions that reduce the error to a low value. In summary, it is not a simple task to sum disparate distance functions in a way that maintains the importance of each individual function.

28.4 Multiple Objective Optimization

When confronted with the need to use multiple distance functions an alternative approach to summing them into a scalar distance is known as multiple objective optimization (MOO; note that MOOP, for multiple objective optimization problem, is also found in the literature most likely to avoid the bovine reference).

The classical types of problem used to illustrate the importance of MOO are problems with two clearly clashing goals. Consider for instance trying to build a column that will support a ceiling. Two reasonable criteria to evaluate a design would be those of column strength and cost. Namely, we would like to make sure that the ceiling is well supported, i.e., minimize the fragility of the column. Second, we would like to construct the column as cheaply as possible, i.e., minimize its cost. It is clear that these two criteria are potentially in conflict. A massive column will be non-fragile but will require a lot of material and therefore have high cost. Since these two criteria are completely disparate, i.e., expressed in different units and on different scales, weighting them one against the other in an a priori fashion is clearly very difficult. Instead, we would like to know what are all the possible solutions and how they trade-off these two criteria against each other, allowing us to make an informed decision (Deb 2001).

It is not a coincidence that we find ourselves immediately considering trade-offs. Indeed, finding the best trade-offs between conflicting (or potentially conflicting) criteria is at the heart of the MOO approach. Conceptually this is a very appealing notion for neural modeling since it avoids trying to arbitrarily weight different criteria against each other. In other words, it avoids questions such as: which is more important for a model correct spike timing or correct shape of AP? Indeed, assigning a specific weight to the importance of a distance function that quantifies the accuracy of the shape of APs in comparison to a distance function that quantifies the temporal fidelity of AP timing seems impossible. Yet this is precisely what we must do in order to sum the

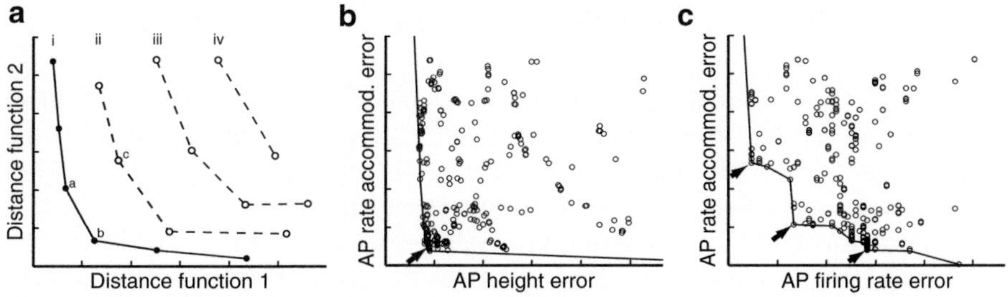

Fig. 28.3 Pareto fronts in multiple objective optimization. (**a**) A schematic representing a hypothetical set of solution distance values and their division into Pareto-optimal fronts. Each of the fronts is marked by a roman numeral (*i–iv*). (**b**) 2-dimensional projection of a set of distance values from a six-objective optimization (AP rate accommodation is a measure of the change in firing from beginning to end of stimulus, see Druckmann et al. (2007)). Note that the Pareto front has a "corner" to it, indicating that no real trade-off exists between these objectives and single optimum exists (marked by *arrow*). (**c**) Different 2-dimensional projection of a set of distance values from same six-objective optimization as in (**b**). Note that the Pareto front has a fractured shape, indicating that trade-offs indeed exist. Panels **b** and **c** adapted from Druckmann et al. (2007)

two distance functions if we seek to arrive at a scalar distance function. MOO allows us a way out of this conundrum.

The key difference between single and multiple objective optimization is the replacement of the greater-than (>) or lesser-than (<) relationship that is well defined when considering two scalars (but not when considering two vectors) by that of *domination*. Consider two solutions, A and B. Assuming that there are M objectives (or distance functions), $f_i, i=1 \ldots M$, solution A is said to dominate solution B if the following two conditions hold:

1. $\exists j \text{ such that } f_j(A) < f_j(B)$
2. $\forall k \in \{1, 2, \ldots. M\}\ f_k(A) \leq f_k(B)$

Or stating the above in words, solution A is said to dominate solution B if solution A is strictly better than solution B in at least one objective and no worse than solution B in all other objectives. Note that three possible relations between A and B now exist. Either A dominates B or B dominates A, and most importantly neither A dominates B nor B dominates A. When does the latter relation happen? Mutual non-domination occurs when solution A is better than B in one objective, but solution B is better than A in another one, in other words when A and B represent different trade-offs amongst the objectives. For instance, in Fig. 28.3a solutions a and b do not dominate each other, whereas they both dominate solution c. Indeed, the purpose of multiple objective optimization is not to find the single best solution as in scalar optimization but rather to map the best possible trade-offs between the different objectives. This best set of trade-offs is referred to as the "Pareto-optimal front" (filled dots in Fig. 28.3).

28.5 Optimization Approaches: Genetic Algorithms

Considering that most optimization approaches can be applied quite widely there are numerous options to choose from. However, since this is not the focus of the current chapter we shall restrict ourselves to discussing genetic algorithms (GAs) that we have empirically found very useful. These algorithms seek to optimize a given error function by considering in parallel a set of tentative solutions while combining, propagating, and perturbing them in a manner reminiscent of (and inspired by) evolution of organisms.

The concept of GAs was introduced by J. Holland in the 1970s, and they are by now algorithms of widespread use. Briefly stated, these algorithms operate by going through multiple iterations (generations) during each of which the fitness of each solution (an organism) in a large group of solutions (a population) is evaluated and the next iteration of solutions is generated according to the fitness of the current population in an effort to discover more successful solutions. Since many solutions are considered in parallel the risk of falling into local minima is smaller than, for instance, straightforward gradient descent. Indeed, these algorithms are well suited for problems that have nonlinear dependencies and rich, complicated error surfaces, both salient properties of neuronal model parameter constraining (Achard and De Schutter 2006; Druckmann et al. 2007).

The basic logic of GA optimization is quite simple. Starting with an initial set of solutions, one performs multiple iterations of evaluating the current solutions and using the best solutions to attempt to find better solutions (see pseudo code in Table 28.1). Below we briefly describe each step:

Evaluation (*fitness selection*)—In the case of neuronal parameter constraining, this step typically comprises for every solution simulating the response of a model with fixed parameters to a series of stimuli and comparing the results to the

Table 28.1 Pseudocode for basic genetic algorithm

Basic genetic algorithm Pseudocode
For generation in generations
For solution in solutions
evaluate solution
propagate successful solutions
evolve propagated solutions

Genetic algorithms are a class of stochastic optimization algorithms that repeatedly test a set of solutions (a population) through multiple iterations (generations). In each iteration all solutions are evaluated. Then the successful solutions are used to generate the basis for the next iteration of solutions. These tentative solutions are then perturbed stochastically to yield the next iteration of solutions, and the cycle continues until it reaches a predetermined number of generations experimental response of the target neuron. This step can be performed very effectively using parallel computing clusters since the evaluation of each solution is completely independent of the others.

Propagation (*reproduction*)—The goal of this operation is to make multiple duplicates of successful solutions (which will later be perturbed) and eliminate poorer solutions in an effort to move towards more valuable solutions in the next generation. To explain the concept let us consider a single-objective optimization attempt and an attempt to minimize a penalty function (as opposed to maximizing a value function). In such a case, more successful solutions are simply solutions with a lower error value in this single objective. One straightforward approach for propagating successful solution is to stochastically select which solutions will propagate according to a probability proportional to the solution's success. In order to do so, one can normalize the population by assigning a score to each solution that is simply a fixed maximal penalty value minus each solution's penalty value. Summing these scores across the population and dividing by the sum assigns a probability of duplication to each solution by which more successful solutions are more likely to propagate.

Evolution (*mutation*)—The goal is to semi-randomly change parameter values in an effort to stochastically strike upon a more successful solution. This operation represents the main search aspect of the algorithm. Often the evolution is performed through one or both of the operators: mutation and crossover. Mutation in principle consists simply of jittering the parameter values. However, there are many methods to do so with respective important parameters such as the degree of change (i.e., the standard deviation of Gaussian noise added to each parameter) or whether parameters are changed independently or with some (possibly learned) covariation. Regarding crossover operators, similarly to genetic crossover the basic idea is taking two solutions (i.e., two lists of parameter values) and creating a combined solution by choosing a point on the list and taking the parameter values up until that point from one solution and beyond that point from the other. We note that using crossover assigns meaning to the particular order of the parameter list, a possible complication.

We stress that GA optimization is a mature and rich field with many suggested algorithms and variants. We refer the interested reader to several possible textbooks (Deb 2001; Goldberg 1989; Holland 1992).

The main difference between a single-objective GA and a multi-objective GA is the process of solution selection and propagation. Instead of simply copying solutions proportionally to their scalar error value one needs a more sophisticated criterion for the multiple objectives. In MOO the most natural property of a favorable solution is one that dominates many other solutions. This property can be transformed into a selection process in many ways (Deb 2001) for instance having the number of solutions that a given solution dominates as a fitness value and the population size minus that number as an error value. Below we describe in more detail a specific approach that has the advantage of not allowing the most favorable solutions to drop out of the population (which is typically a possibility due to the stochastic nature of the algorithm); this property is known in the technical literature as "elitism."

The intuition behind this approach to the selection and propagation process is straightforward. We consider the current set of solutions and the one from the previous generation as a whole. Our goal is to select only half of the solutions (since we are considering solutions from two generations), while favoring the more successful solutions. Accordingly, we divide the population into a series of tentative Pareto-optimal fronts, i.e., sets of solutions that are mutually non-dominating but dominate the other solutions in the set (Fig. 28.3a, there are four Pareto-optimal fronts, marked i–iv). We then add entire Pareto-optimal fronts to our set of retained solutions until we find that we have no more room for the next front. At that point we must select a subset of the front. The criterion of selection is that of maximal spread, i.e., we want to avoid selecting solutions that are very similar to each other. Similarity can be defined in the objective space (i.e., the values of the distance functions), the solution space (i.e., the values of the parameters), or both. Such measures of similarity (or dissimilarity rather) are termed "crowding distances" (Deb 2001) since they aim to avoid a "crowd" of similar solutions overtaking the population.

The particular algorithm we have used is a custom implementation in NEURON of an algorithm named NSGA-II (Deb et al. 2002). It uses the selection approach described above and more sophisticated measures of stochastically perturbing the solutions which are meant to emulate the behavior of binary GAs that have some favorable properties. Details can be found in (Deb et al. 2002).

28.6 Running Parameter Constraining

Currently the most popular program for simulating conductance-based models is NEURON (Carnevale and Hines 2005). Besides NEURON's general expressive power that allows nearly any biophysical model to be simulated, two features of NEURON are particularly useful in the context of parameter constraining. First, NEURON allows for parallel computation, and in particular it is straightforward to implement "embarrassingly parallel" tasks such as the shipping of different parameter values to different processors for model evaluation. Second, NEURON allows for variable-time-step integration, a numerical method approach which can yield major speedup in parameter constraining. In brief, instead of using a fixed and very small integration time step, the size of the time step is chosen adaptively according to the rate of change in the integration. This option is highly useful when performing parameter constraining since many parameter combinations may yield only weakly nonlinear behavior (e.g., low sodium and high potassium channel parameter combinations) that can be integrated with longer integration steps for faster model evaluation. NEURON's hoc language is rich enough to allow direct programming of GAs in hoc. Alternatively, the ability to combine NEURON and Python (Hines et al.

2009) can be used to perform the more generic GA programming in Python (for an elegant implementation see http://projects.g-node.org/emoo/).

In terms of computation time, a standard GA uses a population size on the order of 100 and should be allowed to run for several hundred iterations. Thus, the amount of time it takes to simulate the response of a single model to the tested data set should be multiplied by 10^4 for the run time of the parameter constraining attempt. If a model requires 1 min of simulation time (a reasonable time for a simple model), this corresponds to roughly a week on a single-core machine. On a cluster, if there are as many processors as organisms in the population the entire population can be evaluated simultaneously and the run time divided by that factor. Therefore on a 100-core cluster, given a simulation time of 1 min for a single model, the parameter constraining should take a few hours. We note that it is advisable to repeat the parameter constraining attempt several times due to its stochastic nature.

28.7 Interpreting the Results of a MOO Parameter Constraining Attempt

The first question at the end of a parameter constraining attempt is how successful are the final models. At face value, this should be indicated by the final error scores, i.e., how well the model matches the target data. However, while important, this factor alone can be misleading. Since conductance-based models can contain dozens of parameters, overfitting is a serious potential issue. The classical approach to avoid overfitting is to constrain the model using only a part of the data (the training set) and then to simulate the response of the selected models (unmodified) to a second set of previously unseen data (generalization set). One simple approach to do this in neuroscience is to probe the neuron with different types of current stimuli (e.g., steps and ramps) and use only one type of stimulus for training and the other for generalization (Druckmann et al. 2011). Indeed, this is the approach we recommend as the final test for the validity of the models.

If the parameter constraining attempt failed to sufficiently reduce the error function between data and model during training a number of issues should be checked. First, since the algorithm is stochastic a repeated attempt at optimization may yield better results. Second, two obvious parameters that control the strength of the optimization are the number of solutions checked in parallel (number of organisms) and the number of iterations performed by the algorithm (number of generations); these may need to be increased. Third, as the number of distance functions used increases so does the risk of the optimization getting stuck since there are multiple trade-off combinations and not enough solutions to properly express them. If the number of solutions cannot be further increased it is worthwhile to attempt and sum together some of the distance functions in order to have a smaller number of parallel distance functions. Intuitively, distance functions that test the same measure over different stimuli (and especially different intensities of the same stimulus) are good candidates to sum together without too much adverse an effect (Druckmann et al. 2007).

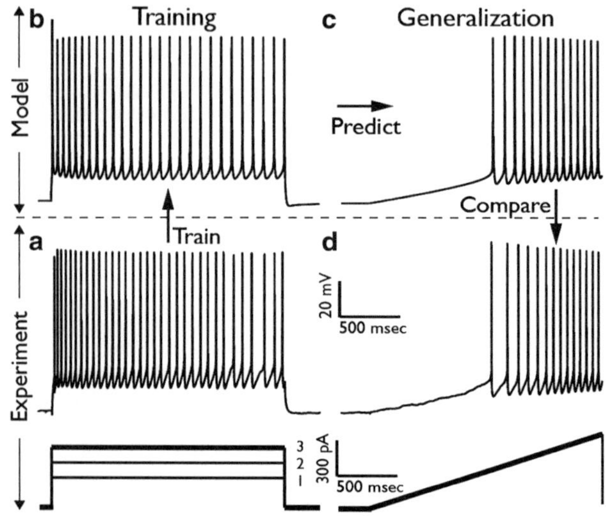

Fig. 28.4 Training and generalization paradigm. (**a**, **b**) Training phase involves using a defined set of experimental data to generate models. (**a**) In this case, experimental voltage responses recorded from rat layer 5 pyramidal cell to three depolarizing current steps (*bottom*) are used as the training set; the experimental response to the largest step current, #3, is shown. (**b**) Model response to step current #3 following training on the three current steps. (**c**, **d**) Generalization phase involves testing the success of the model on predicting the response of the target neuron to a stimulus not used during parameter constraining. (**c**) Model response to a new stimulus, in this case a ramp current (*bottom trace* in **d**). (**d**) Experimental response to the same ramp current. Figure adapted from Druckmann et al. (2011)

Alternatively, one can examine the shape of the Pareto-optimal fronts at the end of the optimization to find whether actual trade-offs between the different distance functions occur (Fig. 28.3b, c). If the Pareto front projected down to a subset of distance functions has a "corner" to it (Fig. 28.3b) then any weighting of these two distance functions will yield the same optimum (marked by arrow in Fig. 28.3b) and hence keeping these distance functions separate is not crucial. In contrast, if the Pareto-optimal front has a curve or a fractured shape (Fig. 28.3c) then different relative weightings of the distance functions will yield different optima (marked by arrows in Fig. 28.3c). Such distance functions should be kept separate and should not be summed together. Finally, it is always possible that the model may not be appropriate for the data. Before that conclusion is reached it is important to check whether all the appropriate parameters were indeed set to be free parameters along with an appropriate range.

We feel and would like to stress that MOO is not just a more sophisticated form of searching through parameter space. Rather, it allows one to flexibly test, explore, find, and develop the right combination of model assumptions, data-model distance functions, and free parameters for a given neural model. These are the real problems at the heart of neural parameter constraining. Once they have been solved and finalized, any parameter search approach used upon them is likely to yield good solutions (Fig. 28.4).

28.8 Alternative Optimization Approaches: Direct Parameter Constraining

Considering fitting the current, not voltage, response of a target neuron provides an especially appealing alternative approach to parameter constraining. First note that since the Hodgkin–Huxley equations depend only on the history of the neuron voltage, if that is known (i.e., recorded) then the gating variables as a function of time can be computed directly (let us ignore for the moment channels that are dependent on additional attributes, e.g., SK channels and their dependency on calcium). Accordingly, the recorded voltage at each time point is used to calculate the gating variables, activation, and driving force of each channel. The total current experienced by the neuron is given on one hand by the contribution of all channels (and possibly also an injected current) and on the other hand by the empirical derivative of the recorded voltage. Equating these two expressions for voltage, all that remains as unknown is the linear scaling of each channel's contribution by its conductance. Therefore, finding these conductance values reduces to a simple linear regression problem. This point was noticed at least as early as Toth and Crunelli (2001) and was later generalized to include more complicated models (Huys et al. 2006; Lepora et al. 2012) and nonlinear parameters (Huys and Paninski 2009). These techniques are very promising, yet it remains unclear how effective they are since on a practical level they have yet to be tested in a robust way against real experimental data (i.e., not surrogate data obtained from a model) and on a conceptual level it is unclear how realistic are the simplified assumptions of the assumed noise model.

28.9 Open Questions

As hinted above, many aspects of parameter constraining could be rewarding for future study. A few important ones are the following:

Optimal distance functions. Each of the distance functions mentioned above has advantages and disadvantages. Though MOO allows one to combine different distance functions, it does not solve the question of which distance functions are the best to use. One possible approach is to try to understand what makes each neuron class distinct from other neuron classes and directly incorporate these features into the parameter optimization. Of course, such an approach depends on being able to define a neuronal class, an issue question of much debate (Ascoli et al. 2008; Druckmann et al. 2012; Sugino et al. 2006). One result of such analysis (and many other analyses) is that both the pattern of AP firing and the shape of individual APs carry useful information. Therefore, any approach is likely to require distance functions related to both aspects.

Relative importance of the different ion channels underlying neuronal dynamics. The approach described in this chapter allows one to generate a large amount of different models according to a given set of target experimental data. The spread of successful models in parameter space, i.e., the densities of different ion channels present in successful models, can potentially be analyzed in order to determine the relative importance of different ion channels in generating the electrical behavior under consideration (Prinz et al. 2004; Weaver and Wearne 2006). This question is of great conceptual importance since one of the main goals of conductance-based models is to relate neuron dynamics/function to the underlying biophysics.

Model uniqueness. A typical parameter constraining attempt using GAs will lead to multiple solutions. How should the multiplicity of solutions be addressed? Should the best solution be chosen and the rest ignored? Or should all solutions that fall within the experimental variability be considered valid models? A closely related question is how does the range of successful solutions change as we add more distance functions and/or target stimuli. Will the number of successful models go down all the way to a single model? Or to zero? In other words, is the parameter constraining underdetermined or overdetermined. In some systems it is quite clear that many different channel combinations yield similar behavior (Prinz et al. 2004; Schulz et al. 2006). However, one needs to keep in mind that the question of how many successful models are found is intimately related to the question of how one formalizes the comparison between model and experiment as well as where do we set our threshold, both aspects that are, at least a priori, extrinsic to the biological question.

28.10 Summary

In this chapter we presented an approach to constraining the parameters of conductance-based models. This approach can be applied to other biophysical model types as well. We have stressed that the main difficulty with properly constraining such models is formalizing the way in which we quantify to what degree is a given model a good or a bad approximation of the data. We presented MOO, an approach that allows a richer vocabulary of expressing the comparison between models and experiments. We briefly described one optimization tool that is well suited to MOO, genetic algorithm optimization. This chapter did not provide a single, "one-size-fits-all" solution to the problem of constraining biophysical models. This is intentional, since we strongly believe that, at least at the moment, the field is not at a state where one can use an off-the-shelf automated parameter constraining algorithm as a black box which inputs data and outputs a model. Rather, the advantage of the automated parameter constraining algorithm is that it shifts the weight from the tedious task of guessing and checking hundreds of parameter combinations to the more fruitful and challenging task of trying to define what type of an approximation is a good biophysical model, which will allow the computer to properly search through the parameter combinations.

The results above were presented, for the sake of simplicity, from a soma-centric point of view. The approach described in this chapter can be extended in a straightforward manner to deal with dendritic phenomena and parameters (Hay et al. 2011). Looking forward, as we seek to integrate more detailed, biophysically realistic models into larger and larger neural circuit models, e.g., Markram (2006) and Bibbig et al. (2001), it is crucial to have powerful and flexible tools to constrain single-neuron models if we are to hope for network-level accuracy. We hope this chapter will inspire and instruct parameter constraining of conductance-based models in accordance with rich experimental data.

Appendix

All simulations for this chapter were performed in the NEURON simulation environment (Carnevale and Hines 2005). The basic equations and pseudo code have been given in the main text. More details regarding the models, data, and fitting procedures are given in the following references, Druckmann et al. (2007); Druckmann et al. (2008); and Druckmann et al. (2011).

References

Achard P, De Schutter E (2006) Complex parameter landscape for a complex neuron model. PLoS Comput Biol 2(7):e94

Ambros-Ingerson J, Grover LM, Holmes WR (2008) A classification method to distinguish cell-specific responses elicited by current pulses in hippocampal CA1 pyramidal cells. Neural Comput 20(6):1512–1536

Ascoli GA, Alonso-Nanclares L, Anderson SA, Barrionuevo G, Benavides-Piccione R, Burkhalter A, Buzsaki G, Cauli B, Defelipe J, Fairen A, Feldmeyer D, Fishell G, Fregnac Y, Freund TF, Gardner D, Gardner EP, Goldberg JH, Helmstaedter M, Hestrin S, Karube F, Kisvarday ZF, Lambolez B, Lewis DA, Marin O, Markram H, Munoz A, Packer A, Petersen CC, Rockland KS, Rossier J, Rudy B, Somogyi P, Staiger JF, Tamas G, Thomson AM, Toledo-Rodriguez M, Wang Y, West DC, Yuste R (2008) Petilla terminology: nomenclature of features of GABAergic interneurons of the cerebral cortex. Nat Rev Neurosci 9(7):557–568

Bahl A, Stemmler MB, Herz AVM, Roth A (2012) Automated optimization of a reduced layer 5 pyramidal cell model based on experimental data, J Neurosci Methods 210;(1):22–34

Bhalla US, Bower JM (1993) Exploring parameter space in detailed single neuron models: simulations of the mitral and granule cells of the olfactory bulb. J Neurophysiol 69(6):1948–1965

Bibbig A, Faulkner HJ, Whittington MA, Traub RD (2001) Self-organized synaptic plasticity contributes to the shaping of gamma and beta oscillations in vitro. J Neurosci 21(22):9053–9067

Boyd S, Vandenberghe L (2004) Convex optimization. Cambridge University Press, Cambridge, New York

Carnevale NT, Hines ML (2005) The NEURON book. Cambridge University Press, Cambridge, NY

Deb K (2001) Multi-objective optimization using evolutionary algorithms. Wiley-Interscience series in systems and optimization, 1st edn. Wiley, Chichester, NY

Deb K, Pratap A, Agarwal S, Meyarivan T (2002) A fast and elitist multiobjective genetic algorithm: NSGA-II. IEEE Trans Evol Comput 6(2):182–197

Druckmann S, Banitt Y, Gidon A, Schurmann F, Markram H, Segev I (2007) A novel multiple objective optimization framework for constraining conductance-based neuron models by experimental data. Front Neurosci 1(1):7–18

Druckmann S, Berger TK, Hill S, Schurmann F, Markram H, Segev I (2008) Evaluating automated parameter constraining procedures of neuron models by experimental and surrogate data. Biol Cybern 99(4–5):371–379

Druckmann S, Berger TK, Schurmann F, Hill S, Markram H, Segev I (2011) Effective stimuli for constructing reliable neuron models. PLoS Comput Biol 7(8):e1002133

Druckmann S, Hill S, Schürmann F, Markram H, Segev I (2012) A hierarchical structure of cortical interneuron electrical diversity revealed by automated statistical analysis. Cereb Cortex

Foster WR, Ungar LH, Schwaber JS (1993) Significance of conductances in Hodgkin-Huxley models. J Neurophysiol 70(6):2502–2518

Gerstner W, Naud R (2009) How good are neuron models? Science 326(5951):379–380

Goldberg DE (1989) Genetic algorithms in search, optimization, and machine learning. Addison-Wesley Publishing Company, Reading, Massachusetts

Hay E, Hill S, Schurmann F, Markram H, Segev I (2011) Models of neocortical layer 5b pyramidal cells capturing a wide range of dendritic and perisomatic active properties. PLoS Comput Biol 7(7):e1002107

Hines M, Davison AP, Muller E (2009) NEURON and python. Front Neuroinform 3:1

Hodgkin AL, Huxley AF (1952) A quantitative description of membrane current and its application to conduction and excitation in nerve. J Physiol 117(4):500–544

Holland JH (1992) Adaptation in natural and artificial systems: an introductory analysis with applications to biology, control, and artificial intelligence. MIT Press, Cambridge, MA

Huys QJ, Ahrens MB, Paninski L (2006) Efficient estimation of detailed single-neuron models. J Neurophysiol 96(2):872–890

Huys QJ, Paninski L (2009) Smoothing of, and parameter estimation from, noisy biophysical recordings. PLoS Comput Biol 5(5):e1000379

Jolivet R, Roth A, Schurmann F, Gerstner W, Senn W (2008a) Special issue on quantitative neuron modeling. Biol Cybern 99(4–5):237–239

Jolivet R, Schurmann F, Berger TK, Naud R, Gerstner W, Roth A (2008b) The quantitative single-neuron modeling competition. Biol Cybern 99(4–5):417–426

Keren N, Bar-Yehuda D, Korngreen A (2009) Experimentally guided modelling of dendritic excitability in rat neocortical pyramidal neurones. J Physiol 587(Pt 7):1413–1437

Keren N, Peled N, Korngreen A (2005) Constraining compartmental models using multiple voltage recordings and genetic algorithms. J Neurophysiol 94(6):3730–3742

Koch C, Segev I (1998) Methods in neuronal modeling: from ions to networks. MIT Press, Cambridge, MA

Lapicque L (2007) Quantitative investigations of electrical nerve excitation treated as polarization. 1907. Biol Cybern 97(5–6):341–349

LeMasson G, Maex R (2001) Introduction to equation solving and parameter fitting. In: De Schutter E (ed) Computational neuroscience. CRC Press, Boca Raton, FL, pp 1–24

Lepora NF, Overton PG, Gurney K (2012) Efficient fitting of conductance-based model neurons from somatic current clamp. J Comput Neurosci 32(1):1–24

Mainen ZF, Joerges J, Huguenard JR, Sejnowski TJ (1995) A model of spike initiation in neocortical pyramidal neurons. Neuron 15(6):1427–1439

Markram H (2006) The blue brain project. Nat Rev Neurosci 7(2):153–160

Poirazi P, Brannon T, Mel BW (2003) Arithmetic of subthreshold synaptic summation in a model CA1 pyramidal cell. Neuron 37(6):977–987

Prinz AA, Bucher D, Marder E (2004) Similar network activity from disparate circuit parameters. Nat Neurosci 7(12):1345–1352

Schulz DJ, Goaillard JM, Marder E (2006) Variable channel expression in identified single and electrically coupled neurons in different animals. Nat Neurosci 9(3):356–362

Sugino K, Hempel CM, Miller MN, Hattox AM, Shapiro P, Wu C, Huang ZJ, Nelson SB (2006) Molecular taxonomy of major neuronal classes in the adult mouse forebrain. Nat Neurosci 9(1):99–107

Tawfik B, Durand DM (1994) Nonlinear parameter estimation by linear association: application to a five-parameter passive neuron model. IEEE Trans Biomed Eng 41(5):461–469

Toth T, Crunelli V (2001) Estimation of the activation and kinetic properties of INa and IK from the time course of the action potential. J Neurosci Methods 111(2):111–126

Vanier MC, Bower JM (1999) A comparative survey of automated parameter-search methods for compartmental neural models. J Comput Neurosci 7(2):149–171

Weaver CM, Wearne SL (2006) The role of action potential shape and parameter constraints in optimization of compartment models. Neurocomputing 69(10–12):1053–1057

Chapter 29
Morphological Reduction of Dendritic Neurons

Kathryn R. Hedrick and Steven J. Cox

Abstract Computational models are important tools for determining dendritic properties and for understanding their functional roles. However, these models are limited by simulation time and storage requirements, particularly when modeling neuronal networks. We review reduced models of the neuron that accurately report the transmembrane potential at a few specified locations while retaining dendritic properties, including the spatial distribution of synaptic inputs throughout the dendritic tree. These models are rooted in two classes of methods from linear algebra: methods based on the singular value decomposition and moment-matching methods. The reduced models can be used to further elucidate dendritic function as they greatly reduce the computational cost associated with simulating networks of morphologically accurate neurons. We demonstrate this capability by simulating a network of hippocampal pyramidal cells and interneurons coupled through chemical synapses and electrical gap junctions.

K.R. Hedrick
Johns Hopkins University, 720 Rutland Avenue, Baltimore, MD 21205, USA
e-mail: kathryn.r.hedrick@gmail.com

S.J. Cox (✉)
Rice University, 6100 Main Street, Houston, TX 77005, USA
e-mail: cox@rice.edu

H. Cuntz et al. (eds.), *The Computing Dendrite: From Structure to Function*, Springer Series in Computational Neuroscience 11, DOI 10.1007/978-1-4614-8094-5_29,

29.1 Introduction

Neuronal activity largely depends on the properties of dendrites, including their electrical properties, their morphology, and the spatial distribution of synapses. Electrically, dendrites have passive properties, such as membrane capacitance and axial resistivity, and active properties, such as voltage-dependence of ionic currents. Morphology affects how a local excitatory postsynaptic potential (EPSP) is transmitted to the soma, and the efficacy of neuronal input depends on the location of the synapse in the dendritic tree. For example, an EPSP near the soma is more likely to evoke an action potential (AP) than is an EPSP of equal strength in distal dendrites, and EPSPs from two synapses may combine nonlinearly depending on their proximity to each other as well as the ionic currents present near the synapses.

Elucidating dendritic properties and functionality often requires a partnership between experimental and computational studies, but simulation time and storage requirements constrain the detail captured by computational models. To retain morphology, one must discretize in space to form compartments, computing the synaptic currents, ionic currents, and transmembrane potential within each compartment. If discretizing to the level of a spine (about 1 μm in length), 1,000 compartments are needed to model just 1 mm of dendrite, and tracking ionic currents can easily lead to ten variables per compartment. The number of variables can become intractable when modeling the entire dendritic tree of each neuron within a network.

This high computational cost has inspired a vast collection of reduced models. A successful reduced model approximates the output of a neuron, such as membrane potential or timing of action potentials, using a system with a much smaller dimension and simulation time by sacrificing or approximating certain dendritic properties. Reduced models are as old as computational neuroscience itself. The quasi-active model, which linearizes ionic currents about rest, already appears in the seminal paper in which Hodgkin and Huxley first characterize active ionic currents (Hodgkin and Huxley, 1952). This linear approximation is accurate for subthreshold stimulation, and it is still widely used today. The reduction of dendritic morphology also dates back to the 1950s, when Wilfrid Rall derived conditions for which the dendritic tree can be mapped to an equivalent cable (Rall, 1959). However, the assumptions prescribed by Rall were too restrictive, and further attempts to relax the assumptions (Schierwagen, 1989; Poznanski, 1991) were insufficient to apply the reduction to a general dendritic tree.

Since Rall, the primary focus of reduced models has been reducing the dendritic morphology. The most common technique used is merging compartments. In its simplest form, this means coarsening the spatial discretization of the neuron, effectively lumping together the synaptic currents, ionic currents, and transmembrane potential over larger regions in the dendrites. The more sophisticated reduced models use the characteristics particular to the given neuron. For example, Pinsky and Rinzel reduced the 19-compartment model of a CA3 pyramidal cell constructed by Traub et al. (1991) and Traub and Miles (1991) to two compartments. Their reduction was based on the observation that most fast-spiking ionic currents are located in or near the soma while most calcium channels are relegated to the distal dendrites (Pinsky and Rinzel, 1994). Similarly, while Traub and Miles also derived a detailed 51-compartment model of a CA3 interneuron (Traub and Miles, 1995), the most commonly used model is a single-compartment model from Wang and Buzsáki that reproduces the high spike rate of interneurons (Wang and Buzsáki, 1996; Skinner and Saraga, 2010).

Several reduced models have recently emerged that retain the spatial distribution of synaptic inputs within the dendrites. These models, which are the focus of this chapter, are constructed using methods from numerical linear algebra. The methods are divided into two classes: methods based on the singular value decomposition (SVD) and moment-matching methods. Models from both classes accurately report the transmembrane potential at a few specified locations of the neuron.

The models are illustrated in Fig. 29.1. In reality, the transmembrane potential, $v(x,t)$, and the synaptic conductance, $g(x,t)$, vary continuously in space and time. In a discrete model, the potential, $v_i(t)$, and synaptic conductance, $g_i(t)$, vary in time within each compartment i. In a reduced model constructed by merging compartments, each compartment covers more dendrite, leading to fewer compartments at the expense of the spatial specificity of synapses. By contrast, in a morphologically reduced model, the number of compartments is reduced while retaining the spatial specificity by projecting the system to a similar system of much smaller dimension. SVD-based methods can be applied to a neuron with any set of ionic currents; however, the reduced model lacks a biological interpretation. While moment-matching methods apply only to passive (linear) or quasi-active (linearized) systems, the electrophysiology of the reduced model is similar to that of full model. For a detailed review and comparison of the two classes of reduction techniques, see Antoulas et al. (2001) and Antoulas (2005).

In Sect. 29.2, we present the general neuronal models that are to be reduced. In Sects. 29.3 and 29.4, we review the reduced models of the neuron using SVD-based methods and moment-matching methods, respectively, providing a brief derivation of each reduced model. We close in Sect. 29.5 by demonstrating the generality and capability of moment-matching methods by reducing a network of quasi-active hippocampal cells interconnected through chemical synapses and electrical gap junctions.

29.2 Discrete Models of the Neuron

We begin by describing the discrete active model and its linear counterpart, the discrete quasi-active model. These models are standard, and a detailed derivation of each can be found in several texts, including (Gabbiani and Cox, 2010). We present the models using the classic Hodgkin and Huxley description of the squid giant axon (Hodgkin and Huxley, 1952), but the models generalize for any ionic currents.

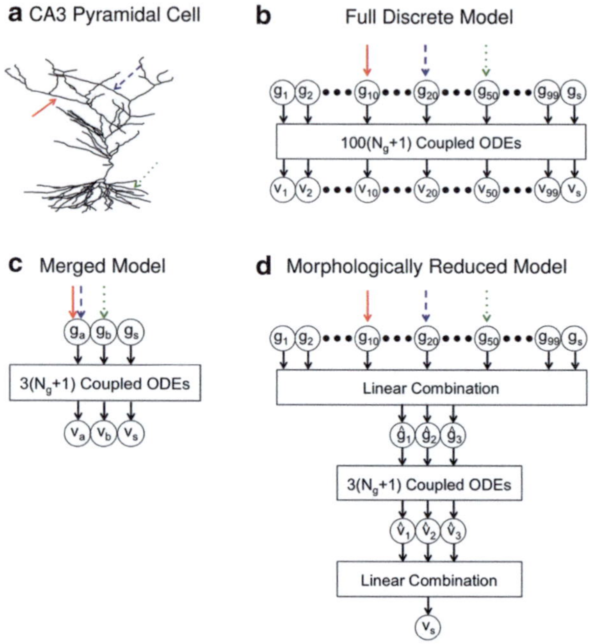

Fig. 29.1 Schematics of neuronal models. (**a**) Two-dimensional projection of a rat CA3 pyramidal cell. Neuronal input drives the conductance at each synapse, denoted $g(x,t)$. The transmembrane potential, $v(x,t)$, depends on the synaptic conductance as well as the neuron's intrinsic properties. A neuron typically receives thousands of inputs throughout the dendritic tree. We highlight three such synapses with *arrows*. (**b**) The full discrete model is formed by discretizing in space into compartments. For this example, there are 100 compartments, and the soma is compartment 100. The synaptic conductance in each compartment i is denoted $g_i(t)$. The potential at each compartment, $v_i(t)$, is computed through solving $100(N_g + 1)$ coupled ordinary differential equations (ODEs), where N_g denotes the number of gating variables for all active ionic currents (see Sect. 29.2.1). Note that each synapse indicated by an *arrow* drives a different compartment. (**c**) Reduced model constructed by merging compartments. For this example, the compartments represent apical dendrites (g_a,v_a), basal dendrites (g_b,v_b), and the soma (g_s,v_s). There are now only $3(N_g + 1)$ coupled ODEs to solve, but spatial specificity is lost, as illustrated by the *solid red* and *dashed blue arrows* driving the same compartment. (**d**) Reduced model constructed through an SVD-based or moment-matching method. The conductances at each compartment are linearly combined to form the conductance at three *reduced compartments*. The potentials at each reduced compartment are computed by solving $3(N_g + 1)$ coupled ODEs, and the somatic potential is computed through linearly combining these potentials. In morphologically reduced models, computational complexity is reduced without sacrificing spatial specificity

29.2.1 Discrete Active Model

By Kirchoff's Current Law (KCL), the transmembrane potential, v_j, in compartment $j \le n$ is governed by the differential equation,

$$\begin{aligned} A_j\big(C_m v_j'(t) + g_L(v_j(t) - V_L)\big) + \sum_{k=1}^{n} G_{\text{ax}}(j,k) v_k(t) + I_j^{\text{act}}(t) \\ = -g_j(t)(v_j(t) - V_{\text{syn},j}). \end{aligned} \tag{29.1}$$

The first two terms describe the capacitive and leakage current leaving compartment j through the cell membrane, where A_j (cm^2) is the surface area, C_m (μF/cm^2) is the membrane capacitance, g_L (mS/cm^2) is the membrane leakage conductance, and V_L (mV) is the membrane reversal potential. The third term is the axial current flowing from compartment j to neighboring compartments, where G_{ax} is called the Hines matrix and encodes the branching structure of any general dendritic tree (Hines, 1984; Gabbiani and Cox, 2010). The fourth term is the current flowing across the cell membrane through active (voltage-dependent) ionic channels (29.2). The final term is the current flowing into the cell through the synapse at compartment j, which is proportional to the synaptic conductance, g_j (mS), and biased by the synaptic reversal potential, $V_{\text{syn,j}}$ (mV).

The squid giant axon has a sodium and a delayed rectifier potassium channel, leading to the active current at compartment j,

$$\begin{aligned} I_j^{\text{act}}(v_j, m_j, h_j, q_j) &= \overline{g}_{\text{Na}} m_j^3 h_j (v_j - V_{\text{Na}}) + \overline{g}_K q_j^4 (v_j - V_K), \\ \tau_w(v_j) w_j'(t) &= w_\infty(v_j) - w_j(t) \text{ for } w \in \{m, h, q\}. \end{aligned} \tag{29.2}$$

The sodium and potassium currents are proportional to their peak conductances (mS), $\overline{g}_{\text{Na}}$ and $\overline{g}_{\text{K}}$, respectively, and are biased by their reversal potentials, V_{Na} and V_{K}. The gating variables (m,h,q), describe the activation or inactivation of the ionic currents. Given a fixed potential, any gating variable, $w \in \{m,h,q\}$, approaches its steady-state value, w_∞, at the rate τ_w. All parameters were determined empirically. We assume for simplicity of notation that the peak conductances are uniform throughout the tree, although the system can be generalized for a nonuniform distribution.

Equations (29.1) and (29.2) combine to form the discrete active model, which has the state vectors, $v = [v_1,\ldots,v_n]^T \in \mathbb{R}^n$ and $\eta = [m_1\ h_1\ q_1,\ldots,m_n\ h_n\ q_n]^T \in \mathbb{R}^{3n}$. This means that for a cell with $n = 100$ compartments, one must track 100 variables for the transmembrane potential and 300 gating variables, leading to 400 coupled ODEs.

The state vectors are governed by the full nonlinear system,

$$\begin{aligned} Cv'(t) + Gv(t) + I^{\text{act}}(v, \eta) = -\text{diag}(g(t))(v(t) - V_{\text{syn}}),\ y(t) = e_{\text{SIZ}}^T v(t), \\ \text{diag}(\tau_\eta(v))\eta'(t) = \eta_\infty(v) - \eta(t). \end{aligned} \tag{29.3}$$

This equation is a succinct way of writing (29.1) and (29.2) for all compartments. Let $\overline{A}$ be the average surface area across all dendritic compartments, and set $D = \text{diag}(A/\overline{A})$, or a diagonal matrix with $A/\overline{A}$ along the diagonal. The capacitance and conductance matrices are given by

$$C = \overline{A} C_m D \text{ and } G = \overline{A} g_L D + G_{\text{ax}}.$$

Active currents are contained in the nonlinear function, $I^{\text{act}} : \mathbb{R}^{n \times 3n} \rightarrow \mathbb{R}^n$, given by (29.2) with the additional term, $-\ A g_L V_L$. Although (29.3) tracks the potential throughout the tree, we define the output of the system, $y(t)$, to be the transmembrane potential at the spike initiation zone (SIZ), usually assumed to be the soma. The output is obtained through multiplication by e_{SIZ}, the canonical unit vector with the single nonzero element corresponding to the SIZ.

29.2.2 The Discrete Quasi-Active Model

The active model can be as accurate as is one's knowledge of the neuron's morphology and electrical properties. However, the model is nonlinear and thus one cannot take advantage of the efficient methods for solving linear systems. A common approximation is the quasi-active model, which linearizes the active model about its resting state. A detailed derivation of the quasi-active model can be found in Gabbiani and Cox (2010), and the electrophysiology of the model is described in Koch (1999).

Suppose the synaptic conductance at compartment j is given by $\varepsilon\, g_j(t)$ for some small $\varepsilon > 0$. The potential and gating variables ($w \in \{m,h,q\}$) can then be written as $v_j(t) = v_{\mathrm{R},j} + \varepsilon\tilde{v}_j(t) + O(\varepsilon^2)$ and $w_j(t) = w_{\mathrm{R},j} + \varepsilon\tilde{w}_j(t) + O(\varepsilon^2)$, where $\tilde{v}$ and $\tilde{w}$ are the linear perturbations from the resting states, v_R and $w_\mathrm{R} = w_\infty(v_\mathrm{R})$, respectively.

Let $u_j(t) \equiv -g_j(t)(v_{\mathrm{R},j} - V_{\mathrm{syn,j}})$ for all j. By matching terms of order ε in (29.1) and (29.2), the linear perturbation from rest at compartment j is governed by

$$\begin{aligned} &A_j\left(C_m\tilde{v}_j'(t) + (g_L + g_{m_1} + g_{q_1})\tilde{v}_j + i_{m,j} + i_{h,j} + i_{q,j}\right) + \sum_{k=1}^{n} G_{\mathrm{ax}}(j,k)\tilde{v}_k = u_j, \\ &L_w i_{w,j}'(t) - \tilde{v}_j + i_{w,j}/g_{w_2} = 0 \quad \text{for} \quad w \in \{m,h,q\}. \end{aligned} \tag{29.4}$$

A current flows from each compartment to ground, or extracellular fluid, according to the membrane capacitance, C_m, and constant conductances, g_L, g_{m_1}, and g_{q_1}. For each gating variable, the current, $A_j i_{w,j} = \left(A_j g_{w_2} / w_\infty'(v_{\mathrm{R},j})\right)\tilde{w}_j$, flows to ground through the inductor, L_w, and resistor with conductance, g_{w_2}, placed in series. The remaining two terms describe the axial current through the neuron and the synaptic input. For more details as well as a detailed circuit diagram, see Koch (1999) and Hedrick and Cox (2013). (Note that while the quasi-active model contains an inductor, there is no known biological element corresponding to an inductor.)

Finally, (29.4) can be written as the linear system,

$$\mathcal{C}z'(t) + \mathcal{G}z(t) = \mathcal{B}u(t), \ \hat{y}(t) = L^T z(t), \tag{29.5}$$

with the state vector $z = [i_m; i_h; i_q; \tilde{v}] \in \mathbb{R}^{4n}$. The system's matrices are given by

$$\mathcal{C} = \begin{bmatrix} \bar{A}L_m D & & & \\ & \bar{A}L_h D & & \\ & & \bar{A}L_q D & \\ & & & \bar{A}C_m D \end{bmatrix} \text{ and } \mathcal{G} = \begin{bmatrix} \bar{A}/g_{m_2} D & & & -\bar{A}D \\ & \bar{A}/g_{h_2} D & & -\bar{A}D \\ & & \bar{A}/g_{q_2} D & -\bar{A}D \\ \bar{A}D & \bar{A}D & \bar{A}D & G_{44} \end{bmatrix} \tag{29.6}$$

where $G_{44} = \bar{A}(g_L + g_{m_1} + g_{q_1})D + G_{\mathrm{ax}}$. The input matrix, $\mathcal{B} \in \mathbb{R}^{4n\times n}$, has nonzero elements only in the last $n \times n$ submatrix, which contains the identity matrix. The output vector, $L \in \mathbb{R}^{4n\times 1}$, has nonzero elements only in the last $n \times 1$ submatrix, which contains the unit vector, e_{SIZ}. The output, y, is the linear perturbation from rest of the transmembrane potential at the SIZ.

29.3 SVD-Based Model Reduction

To model a neuron with n compartments and N_g gating variables, solving either the active or quasi-active model requires solving $n(N_g + 1)$ coupled differential equations, a task that can be computationally expensive as $n(N_g + 1)$ is often at least on the order of 10^3. The goal of a reduced model is to track the transmembrane potential at the SIZ given a system of much smaller dimension. Kellems et al. presented the first morphological reduction of the neuron that retains the spatial specificity of inputs (Kellems et al., 2009, 2010). We briefly review the SVD-based methods they used to reduced the quasi-active neuron in Sect. 29.3.1 and the active neuron in Sect. 29.3.2. For a detailed description of the SVD, Cholesky factorization, or balanced truncation, see Gabbiani and Cox (2010).

29.3.1 *Reduction of the Quasi-Active Model*

Balanced truncation (BT) is a method from control theory for reducing linear systems (Rahn, 2001; Dullerud and Paganini, 2000). BT centers around two notions: controllability and observability. Given the initial state, $z(0) = 0$, $\bar{z}$ is a *controllable state* if there exists an input, $u(t)$, such that $z(\tau) = \bar{z}$ for some $\tau > 0$. Given $u(t) = 0$ for all t and $z(0) = \bar{z}$, $\bar{z}$ is an *observable state* if the output, $y(\tau)$, is nonzero for some $\tau > 0$. Only states that are both controllable and observable are relevant, for an uncontrollable state is not affected by input, and an unobservable state does not affect the output. For example, due to the interplay of ionic currents, dendrites have intrinsic bounds on the transmembrane potential. A state is uncontrollable if it contains potentials outside these bounds since no synaptic input can drive the neuron to such a state. Assuming the SIZ is the soma and the cell receives no synaptic input, a small depolarization in distal dendrites that is not transmitted to the soma is unobservable since one cannot determine this initial depolarization from voltage recordings at the soma.

BT reduces the dimension of a system by quantifying controllability and observability and keeping only the more relevant states. Consider the quasi-active system of (29.5) written in the standard form,

$$z'(t) = Az(t) + Bu(t),\ y(t) = L^T z(t), \tag{29.7}$$

where $A = -\mathcal{C}^{-1}\mathcal{G}$ and $B = \mathcal{C}^{-1}\mathcal{B}$. Define the controllability and observability Grammians,

$$P = \int_0^\infty \mathrm{e}^{At} BB^* \mathrm{e}^{A^*t}\, dt \text{ and } Q = \int_0^\infty \mathrm{e}^{A^*t} LL^* \mathrm{e}^{At}\, dt,$$

respectively, where $*$ denotes the conjugate transpose. The range of P and Q is the subspace consisting of all controllable and observable states, respectively.

The first step of BT is to transform the system into a *balanced system*, or a system for which the Grammians are equal and diagonal. For such a system, there exists a single measure for both controllability and observability. This transformation is done by computing the Cholesky factors, $P = U U^*$ and $Q = V V^*$, and computing the SVD of the mixed product, $U^*V = Z \Sigma Y^*$. The transformation matrix is given by $T = \Sigma^{-1/2} Y^* V^*$. Substituting $\hat{z} = Tz$ into (29.7) leads to the balanced counterpart to the quasi-active system,

$$\hat{z}'(t) = \hat{A}z(t) + \hat{B}u(t),\ y(t) = \hat{L}^T \hat{z}(t), \tag{29.8}$$

where $\hat{A} = TAT^{-1}$, $\hat{B} = TB$, and $\hat{L}^T = L^T T^{-1}$.

The Grammians of this balanced system, $\hat{P} = TPT^*$ and $\hat{Q} = T^{-*}QT^{-1}$, are equal and diagonal, given by $\hat{P} = \hat{Q} = \Sigma$. The diagonal elements of Σ, called the *Hankel singular values*, provide a measure for controllability and observability, where larger values correspond to more relevant states. Since they are nonnegative and in descending order, one truncates the last $n - k$ singular values to construct the reduced system of dimension k,

$$\xi'(t) = \hat{A}_{kk}\ \xi(t) + \hat{B}_k u(t),\ \ \hat{y}(t) = \hat{L}_k^T\ \xi(t),$$

where $\hat{A}_{kk}$ is the initial $k \times k$ submatrix of $\hat{A}$, and $\hat{B}_k$ and $\hat{L}_k$ are the initial k rows of $\hat{B}$ and $\hat{L}$, respectively. The reduced state vector, $\xi \in R^k$, has no apparent biological interpretation, but the output of the system approximates the SIZ potential.

The reduced model is both accurate and efficient. Kellems et al. showed that it can accurately track the subthreshold potential, perform parameter sweeps for synaptic scaling, and determine the neuron's resonant frequency. As an example, their reduced model of a projection neuron in the rat entorhinal cortex has up to five digits of accuracy while reducing the dimension by a factor of 269 and the simulation time by a factor of 4 (Kellems et al., 2009).

29.3.2 Reduction of the Active Model

The quasi-active model is accurate given small input currents, and it can be used effectively for network simulations by coupling it with a spiking threshold scheme such as the integrate and fire (IAF) model. However, as the model is linear, it cannot capture nonlinearities in the neuronal dynamics, such as the action potential. For this, one must use the active model, given by (29.3). Kellems also reduced the active model using a combination of two methods: the principal orthogonal decomposition (POD) (Liang et al., 2000; Kunisch and Volkwein, 2002) and the direct empirical interpolation method (DEIM) (Chaturantabut and Sorensen, 2010).

Principal Orthogonal Decomposition

Equation (29.3) can be written in the standard form,

$$v'(t) = Hv(t) + N(v,\eta) - G_{\text{syn}}(t)(v(t) - V_{\text{syn}}),\ y(t) = e_{\text{SIZ}}^T v(t)$$
$$\text{diag}(\tau_\eta(v))\eta'(t) = \eta_\infty(v) - \eta(t), \tag{29.9}$$

where $H = -C^{-1}G$ and $G_{\text{syn}} = C^{-1}\text{diag}(g)$. The nonlinear terms are contained in the function, $N : \mathbb{R}^{n \times nN_g} \to \mathbb{R}^n$, given by $N = -C^{-1}I^{\text{act}}$.

POD is the standard method for reducing nonlinear systems, such as (29.9). Given a reducer $X \in \mathbb{R}^{n \times r}$ such that $r \ll n$, substituting $v = X\xi$ into (29.9) leads to the reduced system of dimension r,

$$\xi'(t) = \hat{H}\xi(t) + X^T N(X\xi,\eta) - \hat{G}_{\text{syn}}(t)\xi(t) + X^T G_{\text{syn}}(t)V_{\text{syn}},\ \hat{y}(t) = e_{\text{SIZ}}^T X\xi(t)$$
$$\text{diag}(\tau_\eta(X\xi))\eta'(t) = \eta_\infty(X\xi) - \eta(t), \tag{29.10}$$

where $\hat{H} = X^T HX$ and $\hat{G}_{\text{syn}} = X^T G_{\text{syn}} X$.

The accuracy of the reduced system, or the extent to which $\hat{y} \approx y$, depends on the choice of the reducer, X. According to POD, the reducer is constructed by first taking N_s snapshots of the potential to form the matrix, $S = [v(t_1)\dots v(t_{N_s})] \in \mathbb{R}^{n \times N_s}$. The columns of the reducer are the best orthonormal basis for S, meaning they solve the optimization problem,

$$\min_{\{X_i\}_{i=1}^r} \sum_{j=1}^{N_s} \left\| v(t_j) - \sum_{i=1}^{N_s} (v(t_j)^T X_i) X_i \right\|_2^2.$$

The columns are computed from the SVD of the snapshot matrix, $S = U\Sigma V^*$. The singular values, which are found on the diagonal of Σ, are nonnegative and in descending order, and the best rank r approximation of S is given by $S \approx U_r \Sigma_{r,r} V_r^*$, where U_r and V_r are the first r columns of U and V, respectively, and $\Sigma_{r,r}$ is the first $r \times r$ submatrix of Σ. The reducer is given by $X = U_r$, and the accuracy of the reduced system depends on the rate of decay of the system's singular values.

While the state vector, ξ, has dimension $r \ll n$, the nonlinear function, N, depends on $X\xi \in \mathbb{R}^n$ and, even worse, on $\eta \in \mathbb{R}^{nN_g}$. Hence, the reduced system only appears to be reduced, but in reality it has the same complexity as the original system. DEIM is a method recently developed to overcome this problem.

Direct Empirical Interpolation Method

DEIM is a discrete variant of the empirical interpolation method, a reduction method for finite element applications (Barrault et al., 2004). DEIM considers discrete points in space (dendritic compartments for our application) and can be applied to any general set of ODEs.

Given a basis matrix, $W \in \mathbb{R}^{n \times k}$ for $k \ll n$, the nonlinear term can be approximated by

$$N(X\xi(t), \eta(t)) \approx Wc(t) \tag{29.11}$$

for some set of coefficients, $c(t)$. This overdetermined system can be solved uniquely for c by including only k rows of W. Explicitly, given k spatial interpolation points, $p_1,\dots,p_k$, let $P = [e_{p_1},\dots,e_{p_k}]$, where each column is the canonical unit vector with the single nonzero element, $e_{p_i}(p_i) = 1$. After multiplying both sides of (29.11) by P^T, c has the unique solution, $c = (P^T W)^{-1} P^T N$. Equation (29.11) then becomes

$$N(X\xi, \eta) \approx W(P^T W)^{-1} P^T N(X\xi, \eta) \in \mathbb{R}^n.$$

This may not appear like a simplification, but a closer look reveals the dramatic decrease in the computations required. Each row j of N encodes the active ionic currents for compartment j and depends only on v_j and w_j for all gating variables, w. Therefore, $P^T N$ depends on $P^T X\xi \in \mathbb{R}^k$ rather than on $X\xi \in \mathbb{R}^n$. Similarly, instead of tracking $\eta \in \mathbb{R}^{nN_g}$, the gating variables in all compartments, one need only to track $\hat{\eta} \in \mathbb{R}^{kN_g}$, where $\hat{\eta}$ contains the gating variables in compartments $\{p_i\}_{i=1}^k$.

Connecting all the pieces, the combined approach of POD and DEIM leads to the reduced system,

$$\begin{gathered} \xi'(t) = \hat{H}\xi(t) + \hat{N}(\hat{\xi}, \hat{\eta}) - \hat{G}_{\text{syn}}(t)\xi(t) + X^T G_{\text{syn}}(t) V_{\text{syn}},\ \hat{y}(t) = e_{\text{SIZ}}^T X\xi(t) \\ \text{diag}(\tau_{\hat{\eta}}(\hat{\xi}))\hat{\eta}'(t) = \hat{\eta}_\infty(\hat{\xi}) - \hat{\eta}, \end{gathered} \tag{29.12}$$

where $\hat{\xi} = P^T X\xi \in \mathbb{R}^k$, and $\hat{N} = X^T W(P^T W)^{-1} P^T N \in \mathbb{R}^r$. The success of this spatial interpolation scheme depends on the appropriate choice of the basis matrix, W, and interpolation points, $\{p_i\}$. DEIM specifies that the basis be constructed using POD on snapshots of N since POD provides the optimal basis. DEIM then provides an algorithm for iteratively constructing interpolation points that lead to an accurate reduced system with a global error bound. For more details, see Chaturantabut and Sorensen (2010).

Kellems et al. showed that the reduced model can accurately capture the subthreshold potentials and the nonlinear action potentials. They report up to two orders of dimension reduction with an order of magnitude speed-up (Kellems et al., 2010).

29.4 Moment-Matching Model Reduction

As described in Sect. 29.3.1, BT can be used to reduce the quasi-active model while retaining the rich spatial structure in the input signal. However, BT requires one to transform the system before truncating, and the cost and storage requirements associated with computing the system's Grammians become prohibitive for large systems. In contrast, moment-matching methods iteratively project the system onto a subspace of reduced dimension, a reduction process that is far less complex in terms of the computations performed and the storage requirements.

Moment-matching methods have recently been used to reduce the quasi-active model (Kellems et al., 2009; Yan and Li, 2011; Hedrick and Cox, 2013). Like SVD-based reductions, the reduced models retain the spatial distribution of inputs while approximating the potential at a few specified locations of the neuron. The reduction procedure is straightforward, and under certain conditions, the reduced model has an electrophysiological interpretation.

29.4.1 Reduced Systems

Consider the quasi-active system of (29.5). Given a reduced dimension, r, and a reducer, $\mathcal{X} \in \mathbb{R}^{n(N_g+1)\times r}$, substituting $z = \mathcal{X}\hat{z}$ into (29.5) leads to the reduced system,

$$\hat{\mathcal{C}}\hat{z}'(t) + \hat{\mathcal{G}}\hat{z}(t) = \hat{\mathcal{B}}u(t),\ \hat{y}(t) = \hat{L}^T\hat{z}(t), \tag{29.13}$$

where $\hat{\mathcal{C}} = \mathcal{X}^T\mathcal{C}\mathcal{X}, \hat{\mathcal{G}} = \mathcal{X}^T\mathcal{G}\mathcal{X}$, $\hat{\mathcal{B}} = \mathcal{X}^T\mathcal{B}$, and $\hat{L} = \mathcal{X}^T L$.

The system's accuracy, or the extent to which $\hat{y} \approx y$, depends on the choice of reducer. Kellems et al. used the iterative rational Krylov algorithm (IRKA) to construct the reducer (Kellems et al., 2009; Gugercin et al., 2008). While this reducer is effective, Yan and Li demonstrated that the simple reducer based on the PRIMA algorithm for reducing circuits with resistors, inductors, and capacitors (RLC circuits) is similarly effective (Yan and Li, 2011; Odabasioglu et al., 1998). A numerically stable way to construct this reducer is to use the Arnoldi iteration (Trefethen and Bau, 1997) to orthogonalize the basis vectors,

$$\mathcal{X}_{\text{basis}} = [\mathcal{G}^{-T}L,\ (\mathcal{G}^{-T}\mathcal{C})\mathcal{G}^{-T}L,\ \ldots,\ (\mathcal{G}^{-T}\mathcal{C})^{r-1}\mathcal{G}^{-T}L],\ \mathcal{X} = \text{orth}(\mathcal{X}_{\text{basis}}). \tag{29.14}$$

These basis vectors apply to any general quasi-active system.

One shortcoming of all reduced models presented thus far is that the reduced state vector has no apparent interpretation. In the case of the quasi-active model, the reduction procedure transforms a well-defined RLC circuit to a mysterious black box that has the same input–output relationship. We recently presented an alternative reducer that endows the reduced system with an RLC circuit structure similar to that of the original system (Hedrick and Cox, 2013). The reduction procedure is modified from a structure-preserving reduction technique (Li and Bai, 2005), and the reduced system is applicable when the density of active channels is uniform throughout the dendrites.

Its reducer is constructed in two steps. First, the matrix, $Y \in \mathbb{R}^{n \times r}$, is constructed by using the Arnoldi procedure to orthogonalize the basis vectors,

$$Y_{\text{basis}} = \left[G_{\text{eff}}^{-1}e_{\text{SIZ}},\ (G_{\text{eff}}^{-1}D)G_{\text{eff}}^{-1}e_{\text{SIZ}},\ \dots,\ (G_{\text{eff}}^{-1}D)^{r-1}G_{\text{eff}}^{-1}e_{\text{SIZ}}\right],$$
$$Y = \text{orth}(Y_{\text{basis}}),\ G_{\text{eff}} = \bar{A}\left(g_L + \sum_w \left(g_{w_1} + g_{w_2}\right)\right)D + G_{\text{ax}}, \tag{29.15}$$

the summation being taken over all gating variables. Second, the reducer, $\mathcal{Y} \in \mathbb{R}^{n(N_g+1)\times r(N_g+1)}$, is a block diagonal matrix with Y on each $n \times r$ diagonal submatrix.

All three reduced systems have similar reductions in dimension and simulation time, accurately capturing the subthreshold potential (Kellems et al., 2009; Yan and Li, 2011; Hedrick and Cox, 2013). As an example, our reduced model of a rat CA1 pyramidal cell has a relative error on the order of 10^{-4} while reducing both the dimension and simulation time by a factor of 50 (Hedrick and Cox, 2013).

We briefly describe the theory underlying moment-matching methods in Sect. 29.4.2, and we describe the electrophysiology of the latter reduced system in Sect. 29.4.3.

29.4.2 Theory

Since u is the input vector into both the original system (29.5) and the reduced system (29.13), the outputs of each system will agree if the systems' transfer functions agree. The transfer function of a system is a linear mapping from its inputs to its outputs in the frequency domain. Let $\mathscr{L} f$ denote the Laplace transform of a given function, f. The transfer functions of the full and reduced systems, H and $\hat{H}$, satisfy $\mathscr{L}\, y(s) = H(s)\mathscr{L}\, u(s)$ and $\mathcal{L}\hat{y}(s) = \hat{H}(s)\mathcal{L}u(s)$. The error in the outputs is related to the error in the transfer functions via

$$\max_t |\, y(t) - y(t)\,|^2 \leq \left(\int_{-\infty}^{\infty} \|\, H(i\omega) - H(i\omega)\,\|_2^2 d\omega\right)\left(\int_{-\infty}^{\infty} \|\, u(i\omega)\,\|_2^2\, d\omega\right). \tag{29.16}$$

Assuming u has bounded energy, the reduced system is accurate if the transfer functions agree along the imaginary axis.

Moment-matching methods minimize the error bound given by (29.16) by interpolating H at specified interpolation points. According to IRKA, the interpolation points are set to be the negative of the poles of the transfer function, which are found iteratively. For full details of IRKA, as well as a review of moment-matching methods, see Gugercin et al. (2008).

As shown in Yan and Li (2011) and Hedrick and Cox (2013), an accurate reduced system can be constructed by setting a single interpolation point of zero. The Taylor expansions of H and $\hat{H}$ about zero reveal the moments, M_i and $\hat{M}_i$, where

$$H(s)=\sum_{j=0}^{\infty}(-s)^j M_j,\ M_j=L^T(\mathcal{G}^{-1}\mathcal{C})^j\mathcal{G}^{-1};$$

$$\hat{H}(s)=\sum_{j=0}^{\infty}(-s)^j\hat{M}_j,\ \hat{M}_j=\hat{L}^T(\hat{\mathcal{G}}^{-1}\hat{\mathcal{C}})^j\hat{\mathcal{G}}^{-1}.$$

The reducer is constructed to match as many of the leading moments as possible, for if $\hat{M}_j=M_j$ for $j=0,\ldots,r-1$, then $\hat{H}(s)=H(s)+O(s^r)$.

For a given matrix A, let sp(A) denote the range of A, or the span of its columns, and let $\mathcal{X}_{\text{basis}}$ denote the r basis vectors given by (29.14). If $\mathcal{X}\in\mathbb{R}^{n\times r}$ such that $\text{sp}(\mathcal{X}_{\text{basis}})\subseteq\text{sp}(\mathcal{X})$, then $\hat{M}_j=M_j$ for $j=0,\ldots,r-1$. This is a classic result of moment-matching methods (Villemagne and Skelton, 1987; Grimme, 1997; Li and Bai, 2005), and a proof simplified for the quasi-active neuron can be found in Hedrick and Cox (2013). Both reducers presented in Sect. 29.4.1, $\mathcal{X}$ and $\mathcal{Y}$, satisfy this property and thus lead to accurate reduced systems. These reduction methods are often referred to as Krylov subspace projection methods since the reducer projects the original system to the Krylov subspace, $K_r(\mathcal{G}^{-T}\mathcal{C},\mathcal{G}^{-T}L)=\text{sp}(\mathcal{X}_{\text{basis}})$.

29.4.3 Electrophysiology

Consider again the classic Hodgkin and Huxley system, and assume the reduced system was constructed using the reducer, $\mathcal{Y}$, presented in Sect. 29.4.1. The matrices in the reduced system have the same structure as do the matrices in (29.6), where

$$\hat{\mathcal{C}}=\begin{bmatrix}\bar{A}L_m I_r & & & \\ & \bar{A}L_h I_r & & \\ & & \bar{A}L_q I_r & \\ & & & \bar{A}C_m I_r\end{bmatrix}\text{ and }\hat{\mathcal{G}}=\begin{bmatrix}\bar{A}/g_{m_2}I_r & & & -\bar{A}I_r\\ & \bar{A}/g_{h_2}I_r & & -\bar{A}I_r\\ & & \bar{A}/g_{q_2}I_r & -\bar{A}I_r\\ \bar{A}I_r & \bar{A}I_r & \bar{A}I_r & G_{44}\end{bmatrix}.$$

I_r is the $r\times r$ identity matrix, and $\hat{G}_{44}=\bar{A}(g_L+\hat{g}_L+g_{m_1}+g_{q_1})I_r+\hat{G}_{\text{ax}}$. Explicit expressions for $\hat{g}_L$ and $\hat{G}_{\text{ax}}$ can be found in Hedrick and Cox (2013).

Let the reduced state vector be given by $\hat{z}=[\hat{i}_m;\hat{i}_h;\hat{i}_q;\hat{v}]^T\in\mathbb{R}^{4r}$. The reduced system tracks the transmembrane potential, $\hat{v}$, and active ionic currents, $\hat{i}_w$, in each of the r reduced compartments. Similar to (29.4), the membrane potential and active currents for each compartment j are governed by

$$\bar{A}_j\left(C_m\hat{v}_j'(t)+(g_L+\hat{g}_L+g_{m_1}+g_{q_1})\hat{v}_j+\hat{i}_{m,j}+\hat{i}_{h,j}+\hat{i}_{q,j}\right)+\sum_{k=1}^{n}\hat{G}_{\text{ax}}(j,k)\hat{v}_k=Y_j^T u,$$

$$L_w\hat{i}\,'_{w,j}(t)-\hat{v}_j+\hat{i}_{w,j}/g_{w_2}=0\text{ for }w\in\{m,h,q\}.$$

For both the full and reduced quasi-active systems, the membrane has the capacitance C_m, and currents flow to ground across resistors with conductances g_L, g_{m_1}, and g_{q_1}. The three ionic currents to ground flow through an inductor with inductance L_w and a resistor with conductance g_{w_2} for $w \in \{m,h,q\}$. The reduced system has an additional leakage current proportional to $\hat{g}_L$. All currents to ground are proportional to the compartmental surface area, A_j for the full system and $\bar{A} = \text{mean}(A)$ for the reduced system. The primary difference between the two systems is that $\hat{G}_{\text{ax}}$ is dense, implying that each pair of reduced compartments are connected through a resistor with conductance, $\hat{G}_{\text{ax}}(i,j)$. Since accuracy is typically achieved with $r < 10$, this increased density has little effect on the simulation time.

Each compartment of the full system is driven by the input current, $u_j(t) = -g_j(t)(v_{\text{R},j} - V_{\text{syn},j})$. The columns of Y filter the inputs to provide the driving current for reduced compartment j, given by $Y_j^T u(t)$. In particular, the first column of the reducer acts as the dendritic filter, attenuating inputs according to the distance between the synapse and SIZ. Finally, all reduced compartments contribute to the SIZ potential, $y \approx \sum_{j=1}^{r} Y(\text{SIZ}, j)\hat{v}_j$. See Hedrick and Cox (2013) for detailed circuit diagrams comparing the full and reduced models.

29.5 Reduction of an Interconnected Hippocampal Network

We close by demonstrating the generality and potential of moment-matching methods by reducing a hippocampal network with both chemical synapses and electrical gap junctions. We combine the reduction techniques described in Yan and Li (2011) and Hedrick and Cox (2013) to construct a reduced model that accurately predicts the network activity while retaining the underlying electrophysiology.

Two questions arise when simulating such a network. First, how well do both the full and reduced quasi-active models approximate such network dynamics as synchrony and current across a gap junction? Second, how do the accuracy and gain of the reduced model scale with the network size and number of gap junctions? In Sect. 29.5.1, we describe how the models can be extended to incorporate gap junctions. We then address both questions through concrete examples in Sect. 29.5.2.

29.5.1 *Extension to Incorporate Gap Junctions*

Several modeling studies indicate that the strength and location of gap junctions affect synchrony among interneurons, which in turn can drive population rhythms in the hippocampus (Saraga and Skinner, 2004; Saraga et al., 2006; Zahid and Skinner, 2009; Skinner and Saraga, 2010). The quasi-active models described in Sects. 29.2.2 and 29.4.1 can incorporate gap junctions given a few modifications.

The Discrete Quasi-Active Model

We use the basket cell model derived by Wang and Buzsáki, one of the most commonly used models of a hippocampal interneuron (Wang and Buzsáki, 1996; Skinner and Saraga, 2010). The model includes the traditional Hodgkin and Huxley sodium and potassium currents used in Sect. 29.2, but the parameters are modified to reproduce the high spike rate characteristic of an interneuron. As the distribution of active channels is unknown in CA3 interneurons, we assume the channel density is uniform throughout the neuron. Using this model, an isolated interneuron has an intrinsic firing rate of 16.85 Hz; consequently, the cell has no resting potential about which to linearize.

We first adapt the active model. For simplicity of notation, assume two homogeneous interneurons are connected by a single gap junction at compartment p^1 of cell 1 and p^2 of cell 2, and let g_{gap} denote the associated conductance ($m\,S$). The intrinsic current, or the current crossing the cell membrane given no inputs, is given by $I_0 = Ag_L(v_{\text{R}} - V_L) + I^{\text{act}}(v_{\text{R}}, m_{\text{R}}, h_{\text{R}}, q_{\text{R}})$, where I^{act} is given by (29.2), $w_{\text{R}} = w_\infty(v_{\text{R}})$ for all gating variables, and $v_{\text{R}} = -78$ mV (roughly midway between the minimal potential and the potential at which the cell evokes an AP). For cell 1, the potential and active current at compartment j are governed by (29.1) and (29.2) with the following two changes: I_0 is subtracted from I_j^{act}, and the input in (29.1) is replaced by

$$I_j^{\text{stim},1}(t) = -g_j^1(t)(v_j^1(t) - V_{\text{syn,j}}) - I_0 + g_{\text{gap}}(v_{p^2}^2(t) - v_{p^1}^1(t))\delta_{jp^1}, \tag{29.17}$$

where $\delta_{jp^1} = 1$ if $j = p^1$ and zero otherwise.

If $I^{\text{stim}} = 0$, the cell has the resting state, $(v_{\text{R}}, m_{\text{R}}, h_{\text{R}}, q_{\text{R}})$. As described in Sect. 29.2.2, we linearize about this resting state to arrive at the quasi-active model given by (29.5). For each cell k, the state vector, z^k, and input matrix, B^k, are unchanged. $\mathcal{C}^k$ and $\mathcal{G}^k$ are given by (29.6), but $g_{\text{gap}} e_{p^k} e_{p^k}^T$ is added to G_{44}^k We set the output of the neuron to be the SIZ and gap junction potentials. Accordingly, $L^k \in R^{4n \times 2}$ with nonzero elements in the last $n \times 2$ submatrix, given by $[e_{\text{SIZ}}\ \ e_{p^k}]$. Finally, the input into compartment j of cell 1 is given by

$$u_j^1(t) = -g_j^1(t)(v_{\text{R}} - V_{\text{syn,j}}) - I_0 + g_{\text{gap}} v_{p^2}^2(t)\delta_{jp^1}. \tag{29.18}$$

The governing equation for cell 2 is analogous. When this model is coupled with the IAF scheme, as described in Sect. 29.5.2, the isolated quasi-active neuron has an intrinsic spike rate of 16.852 Hz.

The Reduced Quasi-Active Model

According to the last term in (29.17), the current across a gap junction depends on the gap junction potentials. Hence, the reduced model must approximate the gap junction potentials as well as the SIZ potential.

This problem is similar to the case of the passive model (no active ionic channels), for which the synaptic current depends on the potential at the synapse. Yan and Li suggest that one set the system's outputs to be the synaptic potentials as well as the

SIZ potential, leading to a reduced dimension proportional to the number of synapses (Yan and Li, 2011). The reduced model is accurate, but the gain is quickly lost as the number of synapses grows. We offered an alternative reduced system with the single output, $y = v_{\mathrm{SIZ}}$. This reduced system sacrifices some of the nonlinearities in the synaptic currents, but its dimension is independent of the number of synapses (Hedrick and Cox, 2013).

Here, our approach is a hybrid between the two reduction techniques. Assuming the number of gap junctions for each cell is small, we take the former approach in setting all gap junction potentials as outputs of the system. To retain the electrophysiology, we use a reducer adapted from (29.15). Given a single gap junction at compartment p^k, block i of the basis vectors for cell k is given by

$$\begin{aligned} Y^k_{\mathrm{basis}}(i) &= ((G^k_{\mathrm{eff}})^{-1}D)^{i-1}(G^k_{\mathrm{eff}})^{-1}[e_{\mathrm{SIZ}} \;\; e_{p^k}] \in \mathbb{R}^{n\times 2}, \text{ where} \\ G^k_{\mathrm{eff}} &= \bar{A}(g_L + \sum_w (g_{w_1} + g_{w_2}))D + G_{\mathrm{ax}} + g_{\mathrm{gap}} e_{p^k} e^T_{p^k}. \end{aligned} \tag{29.19}$$

As described in Sect. 29.4.1, the reducer, $\mathcal{Y}$, is a block diagonal matrix with Y^k in each diagonal submatrix, where Y^k is constructed by orthogonalizing Y^k_{basis} via the block Arnoldi iteration (Saad, 2003). The reduced system is then given by (29.13). There are now two outputs for each cell k, where $\hat{y}^k_1$ and $\hat{y}^k_2$ approximate the potential at the SIZ and gap junction, respectively.

Figure 29.2 illustrates the full and reduced quasi-active models. The electrophysiology is largely unchanged. The input mapping, currents to ground, and axial currents are identical to those described in Sect. 29.4.3. For each cell k, the rows of Y^k corresponding to the SIZ and gap junction map the reduced compartmental potentials, $\hat{v}^k$, to the output potentials, $\hat{y}^k_1$ and $\hat{y}^k_2$, respectively. The current flowing across the gap junction into cell 1 is then given by

$$I^1_{\mathrm{gap}} = g_{\mathrm{gap}}(y^2_2 - y^1_2) \approx g_{\mathrm{gap}}(\hat{y}^2_2 - \hat{y}^1_2).$$

The gap junction current into cell 2 is analogous.

29.5.2 Numerical Simulations

We ran a series of numerical simulations to demonstrate the accuracy and gain of the reduced system when modeling neuronal networks. We first compared the synchrony between two interneurons given the active, quasi-active, and reduced quasi-active models. We then reduced neuronal networks of interconnected pyramidal cells and basket cells in the CA3.

For all simulations, the quasi-active model was coupled with the IAF scheme. When the SIZ potential reaches a set threshold value, the potential throughout the quasi-active neuron emits an AP, as recorded from the active model. Reproducing the AP shape in the quasi-active model is essential for capturing the current across the gap junction.

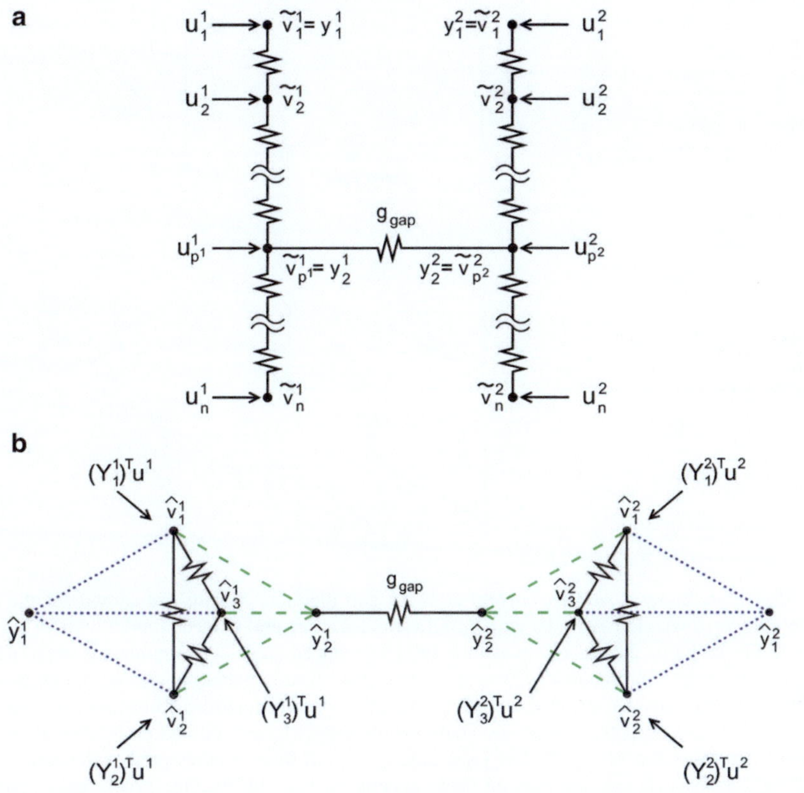

Fig. 29.2 Schematic of two quasi-active neurons coupled through a gap junction. (**a**) Full model. Each cell k is shown as a cable of n compartments with associated potentials, $\tilde{v}^k$, representing the linear perturbations from the resting potential. The input current, u_j^k, drives each compartment j according to (29.18). The SIZ and gap junction are in compartments 1 and p^k, respectively, and their potentials are the outputs of the system, y_1^k and y_2^k. (**b**) Reduced model. Each reduced cell k has three interconnected compartments with associated potentials, $\hat{v}^k$. The columns of Y^k map the inputs, u^k, to their reduced representation. The *dashed green* and *dotted blue lines* indicate the summation of all compartmental potentials weighted by the rows of Y^k corresponding to the gap junction and SIZ, respectively, such that $\hat{y} \approx y$. The currents to ground (not shown) are preserved

Synchrony

Two homogeneous interneurons were modeled according to Sect. 29.5.1. The morphology data for each cell was obtained from neuromorpho.org, cell 98-SR-HiDe (a rat interneuron in the CA3 (Ascoli et al., 2009)). We first compared the phase response curve (PRC) generated using each model. The PRC is generated by injecting a square pulse current into a distal dendritic compartment at time $t_1 + \rho_p T$, where t_1 is the time of the first AP, ρ_p is the perturbation phase, and T is the reciprocal of the cell's intrinsic frequency. The proportion phase shift is given by $\rho_s = (t_1 + T - t_2)/T$, where t_2 is the time of the second AP.

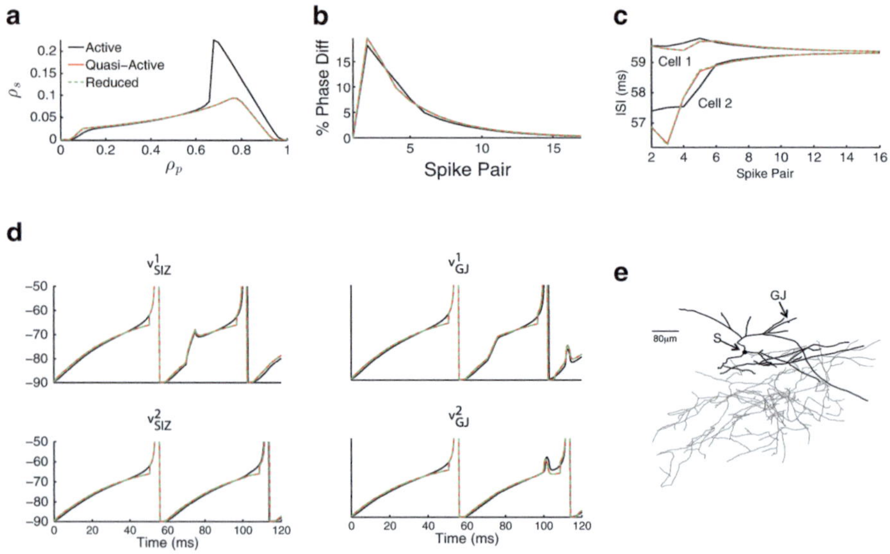

Fig. 29.3 Synchrony driven by current across a gap junction. Results are shown for the active model (*black*), quasi-active model with 1,312 variables (*red*), and reduced model with 24 variables (*green*). For all plots, the error between the full and reduced quasi-active models is negligible. (**a**) PRC for an isolated interneuron. The quasi-active linearization performs well for small perturbation phases, but cannot capture the large phase shifts due to dendritic spikes seen in the active model. (**b**) Percentage phase difference between two electrically coupled cells after a current is injected into the soma of cell 1. The total time is 1 s. All three models predict the same rate of synchronization. (**c**) Interspike interval for both cells. As the cells become synchronous, their frequencies converge to the intrinsic frequency, 16.85 Hz. (**d**) Voltage traces at the SIZ and gap junction with truncated APs. The perturbing current was applied between 70 and 75 ms. The gap junction current causes large potential changes when the APs are asynchronous. (**e**) Two-dimensional projection of the interneuron, indicating the gap junction (GJ) and soma (S), set to be the SIZ. Dendrites are shown in *black*, and the axon is shown in *gray*

As shown in Fig. 29.3a, the PRC agrees well for small perturbation phases. For larger phases, the active model evokes an AP at the gap junction, which then propagates to the soma, causing a large phase advance. Since a threshold is only applied to the quasi-active model at the SIZ, dendritic APs are not detected.

We next ran simulations similar to those presented in Saraga et al. (2006). Two interneurons are connected through a single gap junction in the distal dendrites. The cells are driven out of phase due to a 5 ms current injected into cell 1. Current flowing through the gap junction then synchronizes the cells.

As illustrated in Fig. 29.3, both the full and reduced models accurately predict the rate at which the two interneurons become synchronous. This indicates that the reduced quasi-active model may be used to study synchrony between interneurons in a fraction of the time required for the active model. For the example shown in Fig. 29.3b–d, the reduced model requires 15 % of the simulation time and 2 % of the number of state variables for the active model.

Network Simulations

The neuronal network consists of N_{pyr} pyramidal cells and N_{bc} basket cells in the CA3. Cell morphologies were obtained from neuromorpho.org, cell l48b (a rat pyramidal cell in the CA3 (Turner et al., 1995)) and cell 960217c (a rat basket cell in the CA3 (Chitwood et al., 1999)). Each compartment in the discrete models is 2 μm in length. Each pyramidal cell is driven by an independent set of random inputs, assumed to originate in the entorhinal cortex (EC). Regardless of the network size, each pyramidal cell projects to 25 pyramidal cells and 5 basket cells, and each basket cell projects to 5 pyramidal cells and 2 basket cells. Synapses on each pyramidal cell are set randomly within the proper dendritic region, where inputs from the EC project to the distal apical tuft (*stratum lacunosum-moleculare*), inputs from other pyramidal cells project to the basal dendrites (*stratum oriens*) or apical dendrites (*stratum radiatum*), and inputs from basket cells project to the cell body or proximal apical dendrites (*stratum pyramidale/lucidum*) (Johnston and Amaral, 1998; Witter and Moser, 2006; Skinner and Saraga, 2010). The locations of synapses and gap junctions are set randomly for each basket cell.

The Wang and Buzsáki model is used for basket cells, as described in Sect. 29.5.1. The ionic currents in the pyramidal cell are adapted from a model derived for CA3 pyramidal cells (Migliore et al., 1995). The quasi-active model consists of only those ionic currents that are nonzero in the subthreshold regime, giving the model delayed-rectifier, A-type, and noninactivating potassium currents with a total of five gating variables. All channels are uniformly distributed throughout the dendrites. For more details, see Hedrick (2012).

We compared the network activity given the full and reduced quasi-active models. Figure 29.4 shows examples of voltage traces. The number of cells and number of gap junctions, N_{gap}, varied between simulations. Table 29.1 provides a sampling of the results. For all simulations, a dramatic reduction was seen in the number of state variables and the simulation time.

Each reduced system accurately captures the overall network activity, both the average spike rate and the individual spike times. The spike rate was averaged over all active pyramidal cells or all basket cells, where an active pyramidal cell has a nonzero spike rate. In Table 29.1, the number of active pyramidal cells is denoted N_{act}.

The agreement between individual spike trains was determined through the coincidence factor (Kistler et al., 1997; Kellems et al., 2010), given by

$$\Gamma = \frac{N_{match} - N_{full} N_{red} (\tau / T)}{(N_{full} + N_{red})(1 - N_{full} (\tau / T)) / 2}.$$

N_{match} denotes the number of matched spikes, or the number of spikes in the reduced model that occur within $\tau = 3$ ms of the associated spike in the full model. N_{full} and N_{red} denote the total number of spikes for the full and reduced models, respectively, and $T = 1{,}000$ ms is the length of the simulation. The spike trains are equal if $\Gamma = 1$ and have no correlation in $\Gamma = 0$. Table 29.1 shows the coincidence factors averaged over all pyramidal cells or over all basket cells.

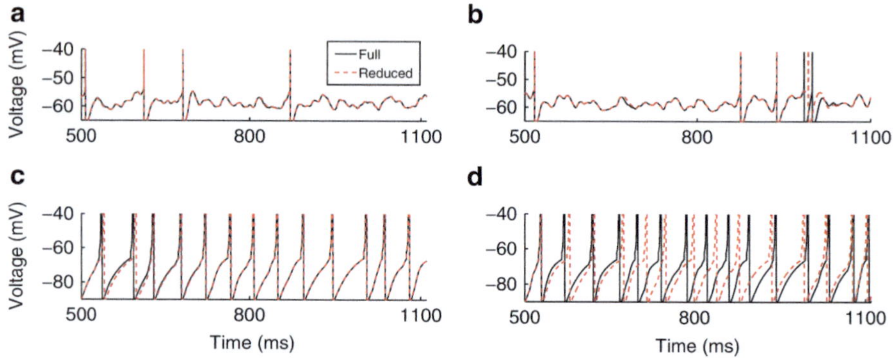

Fig. 29.4 Examples from network simulations with 40 pyramidal cells and 4 basket cells, each with one gap junction. The full quasi-active system has 3.14×10^5 variables, while the reduced system has 1,296 variables. (**a–b**) Voltage traces of two pyramidal cells. The reduced system matches (within 3 ms) all spikes of the full system except the last spike shown in (**b**). Over 1.5 s, the reduced system matches 97 % of all pyramidal cell spikes, and $\Gamma_{\text{pyr}} = 0.99$. (**c–d**) Traces of two basket cells. In (**c**), the reduced system misses the first two spikes but matches the remaining spikes. In (**d**), the reduced system becomes asynchronous with the full system, but it eventually realigns due to common input from the network. Overall, the system matches 83.9 % of all basket cell spikes, and $\Gamma_{\text{bc}} = 0.83$

Network simulations are a true test of the system's accuracy as error in one cell leads to error in its postsynaptic cells. In the case of basket cells, any error in presynaptic cells leads to a phase shift, as illustrated in Fig. 29.4d. Even so, given a large enough reduced system, the reduced model accurately predicts the network dynamics given a reduction in the number of state variables of two orders of magnitude and a reduction in the simulation time by a factor of about ten. This dramatic reduction in the storage requirements and simulation time could allow one to simulate large neuronal networks of morphologically accurate neurons.

29.6 Discussion

We have surveyed and illustrated methods that dramatically reduce the mathematical complexity of the input–output map of highly branched cells. A cell is driven by the spatio-temporal distribution of synaptic inputs, and its output is the transmembrane potential at the SIZ. The standard input–output model requires micron-scale compartmentalization of cells with thousands of microns of dendrites, leading to thousands of differential equations. This complexity is reduced via an algebraic mapping of the exact inputs onto a significantly smaller system of differential equations, followed by an algebraic mapping onto an output that is demonstrably close to the SIZ response of the full cell. In the context of synaptic integration, the algebraic

Table 29.1 Network simulations

Model	N_{pyr}	N_{bc}	N_{gap}	Variables	Time (min)	N_{act}: pyr	Rate (Hz): pyr	Rate (Hz): bc	Γ_{pyr}	Γ_{bc}
Full	100	10	0	7.84×10^5	6.99	17	10.04	24.00	1	1
Reduced				4.92×10^3	0.41	17	9.98	24.00		
Full	1,000	100	0	7.84×10^6	70.95	194	8.69	23.03	1	0.98
Reduced				4.92×10^4	6.72	194	8.69	23.03		
Full	100	10	1	7.84×10^5	7.01	20	7.57	23.43	1	0.99
Reduced				5.04×10^3	0.79	20	7.57	23.42		
Full	1,000	100	1	7.84×10^6	70.58	192	9.43	24.06	1	0.98
Reduced				7.44×10^4	7.27	192	9.43	24.07		
Full	1,000	100	3	7.84×10^6	71.08	163	9.47	22.13	1	0.97
Reduced				7.52×10^4	6.89	163	9.47	22.14		
Full	1,000	100	5	7.84×10^6	70.75	192	8.17	23.60	1	0.95
Reduced				7.68×10^4	7.07	191	8.21	23.61		

mappings encode the cell's spatial integration while the small system of differential equations encodes its time integration.

The most common reduction procedure is merging compartments. Such reduced models are usually derived for a particular neuron type and thus cannot be applied in general. Since the reduction is intuitive and based on intrinsic neuronal properties, the reduced model has in some cases helped to elucidate dendritic function. For example, Pinsky and Rinzel used their two-compartment model to theorize that bursting in CA3 pyramidal cells is due to the electrotonic separation between fast, low-threshold currents near the soma and slow, high-threshold currents in distal dendrites (Pinsky and Rinzel, 1994). However, merging compartments sacrifices the spatial component in the input signal, which is often highly structured. Furthermore, the processing capabilities of a neuron depend on the branching structure of its dendrites, and certain intrinsic properties of the neuron are lost in models that simplify the morphology (Hendrickson et al., 2011).

SVD-based reduced models apply to any neuron, regardless of the dendritic morphology and distribution of ionic channels. BT applies to linear models, such as the quasi-active model, and POD-DEIM applies to nonlinear models, such as the active model. These methods are powerful computational tools for constructing efficient reduced models that preserve the input–output map, but the conceptual link between the model and the underlying biology is severed in the reduction process.

Moment-matching reduction techniques have a simpler reduction process, requiring computation of a single reducer explicitly defined and dependent on dendritic properties. They apply to any quasi-active neuron, again regardless of dendritic morphology and distribution of ionic channels. If ionic channels have a uniform distribution, one can construct a reduced system with a similar RLC circuit structure as the original quasi-active system. The simple reduction procedure and the link between the electrophysiology of the full and reduced models could be an important step for developing efficient reduced models with the potential to further the understanding of dendritic function.

The challenge of extending the rigor, power, and structure preservation of moment-matching methods to fully active cells remains, though extensions of these methods to nonlinear systems are beginning to emerge (Gu, 2011). Beyond this, one should be able to construct efficient hybrid or adaptive schemes that couple the linear reduction of the weakly excitable dendrites to the nonlinear reduction of the strongly excitable SIZ in a fashion that captures back-propagating action potentials. Such a reduced model would permit, for example, the incorporation of plasticity models specific to the synapse. Exploiting this in networks of NMDA-dependent plastic neurons calls for model reduction of the associated slow neuronal calcium system and coupling to the fast reduced electrical system. Such a reduced system would significantly lessen the computational expense associated with simulating large networks of morphologically accurate plastic neurons.

Acknowledgements This work is supported by NSF grant DMS-1122455 and by a training fellowship from the Keck Center for Interdisciplinary Bioscience Training of the Gulf Coast Consortia (NIBIB Grant No. 1T32EB006350-01A1).

References

Antoulas A (2005) An overview of approximation methods for large-scale dynamical systems. Annu Rev Contr 29:181–190

Antoulas A, Sorensen D, Gugercin S (2001) A survey of model reduction methods for large-scale systems, vol 280. American Mathematical Society, Providence, pp 193–219

Ascoli G, Brown K, Calixto E, Card J, Galván E, Perez-Rosello T, Barrionuevo G (2009) Quantitative morphometry of electrophysiologically identified CA3b interneurons reveals robust local geometry and distinct cell classes. J Comp Neurol 515:677–695

Barrault M, Maday Y, Nguyen N, Patera A (2004) An "empirical interpolation" method: application to efficient reduced-basis discretization of partial differential equations. C R Math Acad Sci Paris 339:667–672

Chaturantabut S, Sorensen D (2010) Nonlinear model reduction via discrete empirical interpolation. SIAM M J Sci Comput 32:2737–2764

Chitwood R, Hubbard A, Jaffe D (1999) Passive electrotonic properties of rat hippocampal CA3 interneurons. J Physiol 15:743–756

Dullerud G, Paganini F (2000) A course in robust control theory. Springer, New York

Gabbiani F, Cox S (2010) Mathematics for neuroscientists. Elsevier, Boston

Grimme E (1997) Krylov projection methods for model reduction. Ph.D. thesis, University of Illinois at Urbana-Champaign, Urbana

Gu C (2011) QLMOR: a projection-based nonlinear model order reduction approach using quadratic-linear representation of nonlinear systems. IEEE Trans Comput Aided Des Integr Circ Syst 30:1307–1320

Gugercin S, Antoulas A, Beattie C (2008) H_2 model reduction for large-scale linear dynamical systems. SIAM J Matrix Anal Appl 30:609–638

Hedrick K (2012) The neural computations in spatial memory from single cells to networks. Ph.D. thesis, Rice University, Houston

Hedrick K, Cox S (2013) Structure-preserving model reduction of passive and quasi-active neurons. J Comput Neurosci 34:1–26

Hendrickson E, Edgerton J, Jaeger D (2011) The capabilities and limitations of conductance-based compartmental neuron models with reduced branched or unbranched morphologies and active dendrites. J Comput Neurosci 30:301–321

Hines M (1984) Efficient computation of branched nerve equations. Int J Biomed Comput 15:69–76

Hodgkin A, Huxley A (1952) A quantitative description of membrane current and its application to conduction and excitation in nerve. J Physiol 117:500–544

Johnston D, Amaral D (1998) Hippocampus. In: Shepherd G (ed) The synaptic organization of the brain, chap 10. Oxford University Press, New York, pp 417–458

Kellems A, Roos D, Xiao N, Cox S (2009) Low-dimensional, morphologically accurate models of subthreshold membrane potential. J Comput Neurosci 27:161–176

Kellems A, Chaturantabut S, Sorensen D, Cox S (2010) Morphologically accurate reduced order modeling of spiking neurons. J Comput Neurosci 28:477–494

Kistler W, Gerstner W, van Hemmen J (1997) Reduction of the Hodgkin-Huxley equations to a single-variable threshold model. Neural Comput 9:1015–1045

Koch C (1999) Biophysics of computation: information processing in single neurons. Oxford University Press, New York

Kunisch K, Volkwein S (2002) Galerkin proper orthogonal decomposition methods for a general equation in fluid dynamics. SIAM J Numer Anal 40:492–515

Li R, Bai Z (2005) Structure-preserving model reduction using a Krylov subspace projection formulation. Comm Math Sci 3:179–199

Liang Y, Lee H, Lim S, Lin W, Lee K, Wu C (2000) Proper orthogonal decomposition and its applications–part 1: theory. J Sound Vib 252:527–544

Migliore M, Cook E, Jaffe D, Turner D, Johnston D (1995) Computer simulations of morphologically reconstructed CA3 hippocampal neurons. J Neurophysiol 73:1157–1168

Odabasioglu A, Celik M, Pileggi L (1998) PRIMA: Passive reduced-order interconnect macromodeling algorithm. IEEE Trans Comput Aided Des Integr Circ Syst 17:645–654
Pinsky P, Rinzel J (1994) Intrinsic and network rhythmogenesis in a reduced Traub model for CA3 neurons. J Comput Neurosci 1:39–60
Poznanski R (1991) A generalized tapering equivalent cable model for dendritic neurons. Bull Math Biol 53:457–467
Rahn B (2001) A balanced truncation primer. arXiv quant-ph/0112066
Rall W (1959) Branching dendritic trees and motoneuron membrane resistivity. Exp Neurol 1:491–527
Saad Y (2003) Iterative methods for sparse linear systems. Society for Industrial and Applied Mathematics, Philadelphia
Saraga F, Skinner F (2004) Location, location, location (and density) of gap junctions in multi-compartment models. Neurocomputing 58:713–719
Saraga F, Ng L, Skinner F (2006) Distal gap junctions and active dendrites can tune network dynamics. J Neurophysiol 95:1669–1682
Schierwagen A (1989) A non-uniform equivalent cable model of membrane voltage changes in a passive dendritic tree. J Theor Biol 141(2):159–179
Skinner F, Saraga F (2010) Single neuron models: interneurons. In: Cutsuridis V, Graham B, Cobb S, Vida I (eds) Hippocampal microcircuits, Chap 16. Springer, New York, pp 399–422
Traub R, Miles R (1991) Neuronal networks of the hippocampus. Cambridge University Press, New York
Traub R, Miles R (1995) Pyramidal cell-to-inhibitory cell spike transduction explicable by active dendritic conductances in inhibitory cell. J Comput Neurosci 2:291–298
Traub R, Wong K, Miles R, Michelson H (1991) A model of a CA3 hippocampal pyramidal neuron incorporating voltage-clamp data on intrinsic conductances. J Neurophysiol 66:635–650
Trefethen L, Bau D (1997) Numerical linear algebra. Society for Industrial and Applied Mathematics, Philadelphia
Turner D, Li X, Pyapali G, Ylinen A, Buzsáki G (1995) Morphometric and electrical properties of reconstructed hippocampal CA3 neurons recorded in vivo. J Comput Neurosci 356:580–594
Villemagne C, Skelton R (1987) Model reduction using a projection formulation. Int J Contr 46:2141–2169
Wang X, Buzsáki G (1996) Gamma oscillation by synaptic inhibition in a hippocampal interneuronal network model. J Neurosci 16:6402–6413
Witter M, Moser E (2006) Spatial representation and the architecture of the entorhinal cortex. Trends Neurosci 29:671–678
Yan B, Li P (2011) Reduced order modeling of passive and quasi-active dendrites for nervous system simulation. J Comput Neurosci 31:247–271
Zahid T, Skinner F (2009) Predicting synchronous and asynchronous network groupings of hippocampal interneurons coupled with dendritic gap junctions. Brain Res 1262:115–129

Index

H. Cuntz et al. (eds.), *The Computing Dendrite: From Structure to Function*, Springer Series in Computational Neuroscience 11, DOI 10.1007/978-1-4614-8094-5,

MIX
Papier aus verantwortungsvollen Quellen
Paper from responsible sources
FSC® C105338

If you have any concerns about our products,
you can contact us on
ProductSafety@springernature.com

In case Publisher is established outside the EU,
the EU authorized representative is:
Springer Nature Customer Service Center GmbH
Europaplatz 3, 69115 Heidelberg, Germany

Printed by Libri Plureos GmbH
in Hamburg, Germany